Handbook of Environmental Engineering

Volume 16

Series Editors
Lawrence K. Wang
PhD, Rutgers University, New Brunswick, New Jersey, USA
MS, University of Rhode Island, Kingston, Rhode Island, USA
MSCE, Missouri University of Science and Technology, Rolla, Missouri, USA
BSCE, National Cheng Kung University, Tainan, Taiwan

Mu-Hao S. Wang
PhD, Rutgers University, New Brunswick, New Jersey, USA
MS, University of Rhode Island, Kingston, Rhode Island, USA
BSCE, National Cheng Kung University, Tainan, Taiwan

More information about this series at http://www.springer.com/series/7645

Lawrence K. Wang • Chih Ted Yang
Mu-Hao S. Wang
Editors

Advances in Water Resources Management

Springer

Editors
Lawrence K. Wang
Engineering Consultant and Professor
Lenox Institute of Water Technology
Newtonville, NY, USA

Chih Ted Yang
Colorado State University
Fort Collins, CO, USA

Mu-Hao S. Wang
Engineering Consultant and Professor
Lenox Institute of Water Technology
Newtonville, NY, USA

Handbook of Environmental Engineering
ISBN 978-3-319-22923-2 ISBN 978-3-319-22924-9 (eBook)
DOI 10.1007/978-3-319-22924-9

Library of Congress Control Number: 2015955826

Springer Cham Heidelberg New York Dordrecht London

Printed on acid-free paper

Springer International Publishing AG Switzerland is part of Springer Science+Business Media (www.springer.com)

Preface

The past 36+ years have seen the emergence of a growing desire worldwide that positive actions be taken to restore and protect the environment from the degrading effects of all forms of pollution—air, water, soil, thermal, radioactive, and noise. Since pollution is a direct or indirect consequence of waste, the seemingly idealistic demand for "zero discharge" can be construed as an unrealistic demand for zero waste. However, as long as waste continues to exist, we can only attempt to abate the subsequent pollution by converting it to a less noxious form. Three major questions usually arise when a particular type of pollution has been identified: (1) How serious are the environmental pollution and water resources crisis? (2) Is the technology to abate them available? and (3) Do the costs of abatement justify the degree of abatement achieved for environmental protection and water resources conservation? This book is one of the volumes of the Handbook of Environmental Engineering series. The principal intention of this series is to help readers formulate answers to the above three questions.

The traditional approach of applying tried-and-true solutions to specific environmental and water resources problems has been a major contributing factor to the success of environmental engineering, and has accounted in large measure for the establishment of a "methodology of pollution control." However, the realization of the ever-increasing complexity and interrelated nature of current environmental problems renders it imperative that intelligent planning of pollution abatement systems be undertaken. Prerequisite to such planning is an understanding of the performance, potential, and limitations of the various methods of environmental protection available for environmental scientists and engineers. In this series of handbooks, we will review at a tutorial level a broad spectrum of engineering systems (natural environment, processes, operations, and methods) currently being utilized, or of potential utility, for pollution abatement and environmental protection. We believe that the unified interdisciplinary approach presented in these handbooks is a logical step in the evolution of environmental engineering.

Treatment of the various engineering systems presented will show how an engineering formulation of the subject flows naturally from the fundamental principles and theories of chemistry, microbiology, physics, and mathematics. This emphasis on fundamental science recognizes that engineering practice has in recent years become more firmly based on scientific principles rather than on its earlier dependency on empirical accumulation of facts. It is not intended, though, to neglect empiricism where such data lead quickly to the most economic design; certain engineering systems are not readily amenable to fundamental scientific analysis, and in these instances we have resorted to less science in favor of more art and empiricism.

Since an environmental water resources engineer must understand science within the context of applications, we first present the development of the scientific basis of a particular subject, followed by exposition of the pertinent design concepts and operations, and detailed explanations of their applications to environmental conservation or protection. Throughout the series, methods of mathematical modeling, system analysis, practical design, and calculation are illustrated by numerical examples. These examples clearly demonstrate how organized, analytical reasoning leads to the most direct and clear solutions. Wherever possible, pertinent cost data have been provided.

Our treatment of environmental water resources engineering is offered in the belief that the trained engineer should more firmly understand fundamental principles, be more aware of the similarities and/or differences among many of the engineering systems, and exhibit greater flexibility and originality in the definition and innovative solution of environmental system problems. In short, the environmental and water resources engineers should by conviction and practice be more readily adaptable to change and progress.

Coverage of the unusually broad field of environmental water resources engineering has demanded an expertise that could only be provided through multiple authorships. Each author (or group of authors) was permitted to employ, within reasonable limits, the customary personal style in organizing and presenting a particular subject area; consequently, it has been difficult to treat all subject materials in a homogeneous manner. Moreover, owing to limitations of space, some of the authors' favored topics could not be treated in great detail, and many less important topics had to be merely mentioned or commented on briefly. All authors have provided an excellent list of references at the end of each chapter for the benefit of the interested readers. As each chapter is meant to be self-contained, some mild repetitions among the various texts have been unavoidable. In each case, all omissions or repetitions are the responsibility of the editors and not the individual authors. With the current trend toward metrication, the question of using a consistent system of units has been a problem. Wherever possible, the authors have used the British system (fps) along with the metric equivalent (mks, cgs, or SIU) or vice versa. The editors sincerely hope that this redundancy of units' usage will prove to be useful rather than being disruptive to the readers.

The goals of the *Handbook of Environmental Engineering* series are: (1) to cover entire environmental fields, including air and noise pollution control, solid waste processing and resource recovery, physicochemical treatment processes, biological treatment processes, biotechnology, biosolids management, flotation technology, membrane technology, desalination technology, water resources, natural control processes, radioactive waste disposal, hazardous waste management, and thermal pollution control; and (2) to employ a multimedia approach to environmental conservation and protection since air, water, soil, and energy are all interrelated.

This book (Volume 16) and its two sister books (Volumes 14–15) of the *Handbook of Environmental Engineering* series have been designed to serve as a water resources engineering reference books as well as a supplemental textbooks. We hope and expect they will prove of equal high value to advanced undergraduate and graduate students, to designers of water resources systems, and to scientists and researchers. The editors welcome comments from readers in all of these categories. It is our hope that the three water resources engineering books will not only provide information on water resources engineering, but will also serve as a basis for advanced study or specialized investigation of the theory and analysis of various water resources systems.

This book, *Advances in Water Resources Management, Volume 16*, covers the topics on multi-reservoir system operation theory and practice, management of aquifer systems connected to streams using semi-analytical models, one-dimensional model of water quality and aquatic ecosystem-ecotoxicology in river systems, environmental and health impacts of hydraulic fracturing and shale gas, bioaugmentation for water resources protection, wastewater renovation by flotation for water pollution control, determination of receiving water's reaeration coefficient in the presence of salinity for water quality management, sensitivity analysis for stream water quality management, river ice process, and mathematical modeling of water properties.

This book's first sister book, *Advances in Water Resources Engineering, Volume 14*, covers the topics on watershed sediment dynamics and modeling, integrated simulation of interactive surface water and groundwater systems, river channel stabilization with submerged vanes, non-equilibrium sediment transport, reservoir sedimentation, and fluvial processes, minimum energy dissipation rate theory and applications, hydraulic modeling development and application, geophysical methods for assessment of earthen dams, soil erosion on upland areas by rainfall and overland flow, geofluvial modeling methodologies and applications, and environmental water engineering glossary.

This book's second sister book, *Modern Water Resources Engineering, Volume 15,* covers the topics on principles and applications of hydrology, open channel hydraulics, river ecology, river restoration, sedimentation and sustainable use of reservoirs, sediment transport, river morphology, hydraulic engineering, GIS, remote sensing, decision-making process under uncertainty, upland erosion modeling, machine-learning method, climate change and its impact on water resources, land application, crop management, watershed protection, wetland for waste disposal and water conservation, living machines, bioremediation, wastewater treatment, aquaculture system management and environmental protection, and glossary and conversion factors for water resources engineers.

The editors are pleased to acknowledge the encouragement and support received from Mr. Patrick Marton, Executive Editor of the Springer Science + Business

Media, and his colleagues, during the conceptual stages of this endeavor. We wish to thank the contributing authors for their time and effort, and for having patiently borne our reviews and numerous queries and comments. We are very grateful to our respective families for their patience and understanding during some rather trying times.

Newtonville, NY, USA Lawrence K. Wang
Fort Collins, CO, USA Chih Ted Yang
Newtonville, NY, USA Mu-Hao S. Wang

Contents

Contributors

Domenico Baù, Ph.D. Department of Civil and Structural Engineering, University of Sheffield, Sheffield, United Kingdom

Erick Butler, Dr. Eng. School of Engineering and Computer Science, West Texas A&M University, Canyon, TX, USA

Xuning Guo, Ph.D. General Institute of Water Resources and Hydropower Planning and Design, Xicheng District, Beijing, People's Republic of China

Azzah Salah El-Din Hassan, M.S. Department of Geology and Geophysics, Texas A&M University, Texas, USA

Yung-Tse Hung, Ph.D., P.E., D.E.E., F.-A.S.C.E. Department of Civil and Environmental Engineering, Cleveland State University, Cleveland, OH, USA

Podjanee Inthasaro Orlando, FL, USA

Yunzhong Jiang, Ph.D. State Key Laboratory of Simulation and Regulation of Water Cycle in River Basin, China Institute of Water Resources and Hydropower Research, Beijing, China

Jao-Fuan Kao, Ph.D. Department of Environmental Engineering, College of Engineering, National Cheng Kung University, Tainan, Taiwan

Xiaohui Lei, Ph.D. State Key Laboratory of Simulation and Regulation of Water Cycle in River Basin, China Institute of Water Resources and Hydropower Research, Beijing, China

Weihong Liao, Ph.D. State Key Laboratory of Simulation and Regulation of Water Cycle in River Basin, China Institute of Water Resources and Hydropower Research, Beijing, China

Chii Cherng Liaw, B.E., M.S. Department of Environmental Engineering, National Cheng Kung University, Tainan, Taiwan

Hsue-Peng Loh, M.L.S., Ph.D. Wenko Systems Analysis, Pittsburgh, PA, USA

Nancy Loh, M.A. Wenko Systems Analysis, Pittsburgh, PA, USA

Nazih K. Shammas, Ph.D. Lenox Institute of Water Technology and Krofta Engineering Corporation, Lenox, MA, USA

Hung Tao Shen, Ph.D. Department of Civil and Environmental Engineering, Wallace H. Coulter School of Engineering, Clarkson University, Potsdam, NY, USA

David Terranova Jr, M.E. Department of Mechanical Engineering, Stevens Institute of Technology, Hoboken, NJ, USA

Hao Wang, Ph.D. State Key Laboratory of Simulation and Regulation of Water Cycle in River Basin, China Institute of Water Resources and Hydropower Research, Beijing, China

Lawrence K. Wang, Ph.D., P.E., Department of Environmental Engineering, College of Engineering, National Cheng Kung University, Tainan, Taiwan

Mu-Hao S. Wang, Ph.D., P.E., Department of Environmental Engineering, College of Engineering, National Cheng Kung University, Tainan, Taiwan

Xu Wang, Ph.D. State Key Laboratory of Simulation and Regulation of Water Cycle in River Basin, China Institute of Water Resources and Hydropower Research, Beijing, China

Ching-Gung Wen, Ph.D. Department of Environmental Engineering, College of Engineering, National Cheng Kung University, Tainan, Taiwan

Weiming Wu, Ph.D. Department of Civil and Environmental Engineering, Wallace H. Coulter School of Engineering, Clarkson University, Potsdam, NY, USA

Chih Ted Yang, Ph.D., P.E., D.W.R.E. Department of Civil and Environmental Engineering, Colorado State University, Fort Collins, CO, USA

Tongtiegang Zhao, Ph.D. State Key Laboratory of Hydro-science and Engineering, Department of Hydraulic Engineering, Tsinghua University, Haidian District, Beijing, People's Republic of China

Chapter 1
Multi-Reservoir System Operation Theory and Practice

Hao Wang, Xiaohui Lei, Xuning Guo, Yunzhong Jiang, Tongtiegang Zhao, Xu Wang, and Weihong Liao

Contents

H. Wang, Ph.D. (✉) • X. Lei, Ph.D. • Y. Jiang, Ph.D. • X. Wang, Ph.D. • W. Liao, Ph.D.
State Key Laboratory of Simulation and Regulation of Water Cycle in River Basin, China Institute of Water Resources and Hydropower Research, No. 1 Yuyuantan South Road, Haidian District, 100038 Beijing, People's Republic of China
e-mail: Wanghao@iwhr.com; lxh@iwhr.com; leixiaohui@gmail.com; lark@iwhr.com; wangxu-04@126.com; behellen@163.com

X. Guo, Ph.D.
General Institute of Water Resources and Hydropower Planning and Design, No. 2-1 north street of Liu Pu Kang, Xicheng District, 100120 Beijing, People's Republic of China
e-mail: guoxuning@whu.edu.cn

T. Zhao, Ph.D.
State Key Laboratory of Hydro-science and Engineering, Department of Hydraulic Engineering, Tsinghua University, Haidian District, 100084 Beijing, People's Republic of China
e-mail: steel.tsinghua@gmail.com

L.K. Wang, C.T. Yang, and M.-H.S. Wang (eds.), *Advances in Water Resources Management*, Handbook of Environmental Engineering, Volume 16,
DOI 10.1007/978-3-319-22924-9_1

Abstract The state-of-the-art on operation of multi-reservoir system is reviewed and multi-reservoir construction and management practice in China are introduced at the beginning. Considering the impact of human activity on the reservoir inflow, multi-reservoir operation is studied within theory framework of dualistic water cycle. The reservoir operation rule form and derivation method are the most important elements for deriving optimal multi-reservoir operation policy. Different rule curves and multi-objective optimization algorithms are discussed in this chapter. Inter-basin water transfer project becomes one of effective measures to mitigate imbalance between water supply and water demand. The multi-reservoir operation problem in inter-basin water transfer project is illustrated mainly on deriving the water transfer rule and water supply rule using bi-level model. Reservoir inflow is important information for multi-reservoir operation. The effect of inflow forecast uncertainty on real-time reservoir operation, effective forecast horizon identification and generalized marginal model of the uncertainty evolution of inflow forecast are discussed in details.

Keywords Reservoir operation • Multi-reservoir system • Reservoir operation policy • Dualistic water cycle • 2D rule curves • Equivalent reservoir • Multi-objective optimization • Water transfer rule curves • Bi-level model • Inflow forecast • Uncertainty analysis • Generalized marginal model

List of Symbols

S_t^{T}	Beginning-of-period storage of equivalent reservoir at the stage t
I_t^{T}	Stream inflows into equivalent reservoir at the stage t
R_t^{T}	Reservoir release for all water demand at the stage t
SU_t^{T}	Water spills of equivalent reservoir at the stage t
L_t^{T}	Water losses of reservoir because of evaporation and seepage
$S_{\mathrm{max}}^{\mathrm{i}}$	Maximum reservoir storage capacity
REL	Water supply reliability for water demand
RES	Water supply resiliency coefficient for water demand
ω_1, ω_2	Weighting factors
Q_t	Reservoir downstream flow at the location of protect objective

$Q_{std,\ flood}$	Reservoir standard downstream flow for the flood protect objective
N_t	Hydropower generated output at unit time
$EPow_1$	Total hydropower generation amount at the total operation period
$Q_{pro,\ navi}$	River flow required for the navigation purpose at the stage t
$Q_{pro,\ eco}$	River flow to satisfy the suitable ecology flow requirement at the stage t
Sed_{in}	Sediment amount into the reservoir at the stage t
Sed_{out}	Sediment amount out of the reservoir at the stage t
Sed_1	Sediment discharge rate
$WQ_{std,\ wq}$	Water quality standard for some indexes
WQ_t	Water quality index at the stage t
W_{avg}	Annual average amount of water supply
W_{min}	Annual minimum amount of water supply
NDS_i	Annual average transferred water amount of reservoir *i*
GSI	Generalized shortage index to reflect water shortage severity
PSF	Probabilistic streamflow forecasts
NSE	Nash–Sutcliffe efficiency coefficient
RMSE	Root Mean Square Error
H	Length of forecast lead time or forecast horizon
σ	The forecast error standard deviation
ρ_{error}	The forecast error correlation
μ	The mean of the streamflow
C_v	The coefficient of variation of the streamflow
ρ_{flow}	The correlation coefficient of the streamflow
$\underline{r}$	Minimum reservoir release
$\overline{r}$	Maximum reservoir release
d	Discount ratio of reservoir utility
s_0	Initial reservoir storage
s_T	Target storage at the end of reservoir operation horizon (N)
s'_T	Target storage at the end of reservoir inflow forecast horizon (H)

1 Introduction

1.1 State-of-the-Art Review on Operation of Multi-Reservoir System

Water resources engineers and hydrologists have long recognized that the benefits derived from the joint operation of a system of reservoirs may exceed the sum of the benefits from the independent operation of each of the reservoirs [1–148]. Independent operation implies that decisions about releases from one reservoir are not based on the state of any other reservoir. Joint operation implies that decisions about releases from one reservoir depend not only on the state of that reservoir but also on the states of the other reservoirs in the system, according to Robert et al. [1].

The major task of reservoir operation is to decide how much water should be released now and how much should be retained for future use given some available and/or forecasted information at the beginning of the current time period. In practice, reservoir operators usually follow rule curves, which stipulate the actions that should be taken conditioned on the current state of the system.

1.1.1 Analytical Analysis of Multi-Reservoir Optimal Operation

Analytical analysis is one of the most important measures for multi-reservoir joint operation, which usually provides universal and beneficial conclusion for practical application. Up to now, a large and long-existing literature employs analytical optimization methods to derive reservoir operating rules for multi-reservoir systems [2]. These can date back to rules for minimizing spill from parallel reservoirs in New York rule. During recent years, the study in this area has achieved obviously significant advantage. For example, Lund and Guzman [3] summarized such analytically derived optimal operating rules for some simple multi-reservoir systems under specific conditions and criteria. Lund [4] derived theoretical hydropower operation rules for reservoirs in parallel, in series, and single reservoirs, which offers a simplified economic basis for allocating storage and energy in multi-reservoir hydropower systems. The approach is demonstrated for an illustrative example subject to the limited conditions under which these rules hold. Draper and Lund [5] developed and discussed the properties of optimal hedging for water supply releases from reservoirs. The fundamental decision of how much water to release for beneficial use and retain for potential future use is examined analytically. Explicit correspondence is established between optimal hedging and the value of carryover storage. This more analytical view of hedging rules is useful for better understanding optimal hedging and simplifying numerical optimization of hedging operating rules. You and Cai [6] expanded a theoretical analysis and developed a conceptual two-period model for reservoir operation with hedging that includes uncertain future reservoir inflow explicitly. Some intuitive knowledge on reservoir operation is proved or reconfirmed analytically; and new knowledge is derived. This theoretical analysis provides an updated basis for further theoretical study, and the theoretical findings can be used to improve numerical modeling for reservoir operation. After that, they presented a method that derived a hedging rule from theoretical analysis with an explicit two-period Markov hydrology model, a particular form of nonlinear utility function, and a given inflow probability distribution [7]. Zhao and Cai [8] discussed the optimality conditions for standard operation policy and hedging rule for a two-stage reservoir operation problem using a consistent theoretical framework. The effects of three typical constraints, i.e., mass balance, nonnegative release, and storage constraints under both certain and uncertain conditions were analyzed. Using the derived optimality conditions, an algorithm for solving a numerical model was developed and tested with the Miyun Reservoir in China. Shiau [9] analytically derived optimal hedging for a water supply reservoir considering balance between beneficial release and carryover storage value. The analytical optimal hedging is generalized to represent two-point as well as one-point hedging. Since reservoir release was also a linear function of reservoir inflow, analytical assessment of hedging uncertainty induced by inflow is made possible. The proposed methodology was applied to the Shihmen Reservoir in northern Taiwan to illustrate effects of derived optimal hedging on reservoir performance in terms of shortage-related indices and hedging uncertainty.

1.1.2 Numerical Simulation and Optimization of Multi-Reservoir System Operation

Deterministic Optimization Operation

The application of optimization to solve reservoir operation problems has been a topic extensively studied during the last few decades. Several of these studies deal with deterministic optimization models, which do not consider the uncertainties of some variables such as future reservoir inflows [10]. Most optimization models take some type of mathematical programming technique as the basis. The basic classification of optimization techniques consists of: (1) Linear programming (LP); (2) dynamic programming (DP); and (3) nonlinear programming (NLP). Each of these techniques can be applied in a deterministic and stochastic environment. Reservoir optimization models have been applied for planning purposes as well as real-time operation. All optimization models require an objective function, decision variables, and constraints. The objective function represents a way to measure the level of performance obtained by specific changes in the decision variables [11]. The set of decision variables defines how the system is to be operated. It may define how much water is to be released and when or how much water will be allowed to flow through the outlet structures, or how much water will be kept in storage. The decision variable set is the desired output of the optimization model. The constraints on the reservoir system force the model to obey the physical laws, economic requirements, and social as well as other restrictions. Typical reservoir constraints include conservation equations; maximum and minimum releases; penstock and equipment limitations; and contractual, legal, and institutional obligations [11].

Determining optimum reservoir storage capacities and operating policies using a systems approach has generated a large number of references. On the research status of multi-reservoir optimization operation, we can refer to the review job of Yeh [12], Wurbs [13], Labadie [14] and Rani and Moreira [15]. Yeh [12] provides a state-of-the-art review of theories and applications of systems analysis techniques to the reservoir problems. Algorithms and methods surveyed in this research include linear programming, dynamic programming, nonlincar programming, and simulation. Both deterministic models were included in the review. Wurbs [13] extended the work of Yeh [12] by producing a state-of-the-art review together with an annotated bibliography of systems analysis techniques applied to reservoir operation. Their work is organized in accordance with the general practice of dividing systems analysis into the following categories: simulation, optimization, and stochastic methods. Labadie [14] assessed the state-of-the-art in optimization of reservoir system management and operations and considered future directions for additional research and application. Rani and Moreira [15] presented a survey of simulation and optimization modeling approaches used in reservoir systems operation problems. They discussed simulation, optimization and combined simulation–optimization modeling approach and to provide an overview of their applications reported in literature.

Stochastic Optimization Operation

The stochastic characteristics of multi-reservoir optimization operation are mainly due to the reservoir inflow uncertainty under such conditions that the expected values of inflows cannot appropriately represent highly variable hydrologic characteristics or when the inflows cannot be reliably forecasted for a relatively long period [10]. The methodology of stochastic optimization operation can be summarized into two categories: explicit stochastic optimization (ESO) and implicit stochastic optimization (ISO).

The ESO approach incorporates probabilistic inflow methods directly into the optimization problem, which is typically addressed by stochastic dynamic programming (SDP). SDP is an effective technique for a single reservoir with serially correlated inflows [16]. It provides the advantage of explicitly considering streamflow uncertainty in its recursive function. The main issue of applying SDP to reservoir operation optimization is how to represent uncertainty in future stream flow. Thus, many SDP studies have focused attention on this issue. For example, Kelman et al. [17] proposed a sampling SDP (SSDP) which directly incorporates inflow scenarios in DP recursive equation to reflect various characteristics of stream flows at all sites within the basin. Faber and Stedinger [18] used SSDP for a multi-reservoir system integrating Ensemble Streamflow Prediction (ESP) forecasts into a SSDP framework. The model has advantage of updating its optimal release each time a new set of ESP forecasts is available. Recently, Kim and Heo [19] presented state-of-the-art optimization models using SSDP with ESP. Zhao et al. [20] proposed an algorithm to improve the computational efficiency of both deterministic dynamic programming (DP) and stochastic dynamic programming (SDP) for reservoir operation with concave objective functions. Application of SDP methods to multi-reservoir cases bears higher computational cost than deterministic DP, due to curse of dimensionality. To overcome this use of heuristic procedures like aggregation–disaggregation of reservoirs and one-at-a-time successive decomposition is very common. Arunkumar and Yeh [21] proposed one-at-a-time decomposition SDP (similar to DPSA) approach for a multi-reservoir system. A combined decomposition iteration and simulation analysis methodology along with a constraint technique has been presented by Wang et al. [22] to solve multi-objective SDP optimization problems. Rani and Moreira [15] presented an overall review on the SDP literature.

Different from ESO, ISO uses deterministic optimization to operate the reservoir under several equally likely inflow scenarios and then examines the resulting set of optimal operating data to develop the rule curves [10]. The utilization of ISO for finding reservoir operating policies was first exploited by Young [23] in a study that utilized dynamic programming applied to annual operations. The optimal releases found by the dynamic programming model were regressed on the current reservoir storage and the projected inflow for the year. The regression equation could be thus used to obtain the reservoir release at any time given the present storage and inflow conditions. Karamouz and Houck [24] extended Young's procedure by adding one extra constraint to the optimization model specifying that the release must be within a given percentage of the release defined by the previously found operating policy. Kim and Heo [19] used ISO combined with two types of linear equations for the regression analysis to define monthly operating rules for a multipurpose reservoir. Willis et al. [25] devised a different approach that utilized the probability mass function of the optimal releases, conditioned on reservoir storage and inflow. Modern alternatives to the classical regression analysis are the application of artificial neural networks [26–29] and fuzzy rule-based modeling [30–32] to infer the operating rules. An additional advantage of fuzzy logic is that it is more flexible and allows incorporation of expert opinions, which could make it more acceptable to operators [32]. Most of the published studies show that these two techniques outperform regression-based ISO and SDP [10].

Numerical Simulation Combined with Optimization Models

With the rapid development of modern evolutionary algorithms, numerical simulation combined with optimization models becomes one of dominant and useful methods. According to the opinion of Celeste and Billib [10], this method should belong to ISO method and be called as the Parameterization–Simulation–Optimization methodology (PSO). Because of its usefulness and importance, this section illustrates the PSO method individually. The PSO technique first predefines a shape for the rule curve based on some parameters and then applies heuristic strategies to look for the combination of parameters that provides the best reservoir operating performance under possible inflow scenarios. A number of authors successfully applied the simulation–optimization principle of PSO to derive reservoir rule curves. For example, Cancelliere et al. [33] derived monthly operating rules for an irrigation reservoir using DP and ANN, which were further validated by simulating the behavior of the reservoir over a shorter period, not included in the period used for training the networks. A combined neural network simulation–optimization model with multiple hedging rules was used for screening the operation policies by Neelakantan and Pundarikanthan [34]. Koutsoyiannis and Economou [35] proposed a low dimensional Parameterization simulation–optimization approach using the methodology of parametric rule introduced by Nalbantis and Koutsoyiannis [36] Simulation was used to obtain values of the performance measure, which was optimized by a nonlinear optimization procedure. Tung et al. [37] proposed a procedure to apply genetic algorithm to optimize operation rules and applied it to the LiYuTan Reservoir in Taiwan. Momtahen and Dariane [38] proposed a direct search approach to determine optimal reservoir operating policies with a real coded genetic algorithm GA, in which the parameters of the policies were optimized using the objective values obtained from system simulations. Kangrang et al. [39] also proposed a heuristic algorithm to connect with simulation model for searching the optimal reservoir rule curves.

1.2 *Multi-Reservoir Construction and Management Practice in China*

China has long history of dam construction. Since the first reservoir Anfeng pond was built in Shou County of Anhui Province, China already has nearly 2600 years history of reservoir construction. However, the development process of dam building was rather slow before the establishment of People's Republic of China (PRC). There were only 22 dams higher than 15 m at that time. After the foundation of PRC, especially recent 30 years, the dam construction technology in China has made a great achievement. From Fig. 1.1, we can find out that the dam number of China takes a large portion of the ones of the world and a rapid building rate has being kept. These reservoirs has played fundamental role in water resources beneficial utilization and flood control.

For satisfying the energy demand and environment protection requirement, the government of China proposed hydropower development plan before 2050, which includes 13 main hydropower energy bases as shown in Fig. 1.2. Due to the topography and water resources distribution factors, the most part of hydropower energy concentrates in Southwest China.

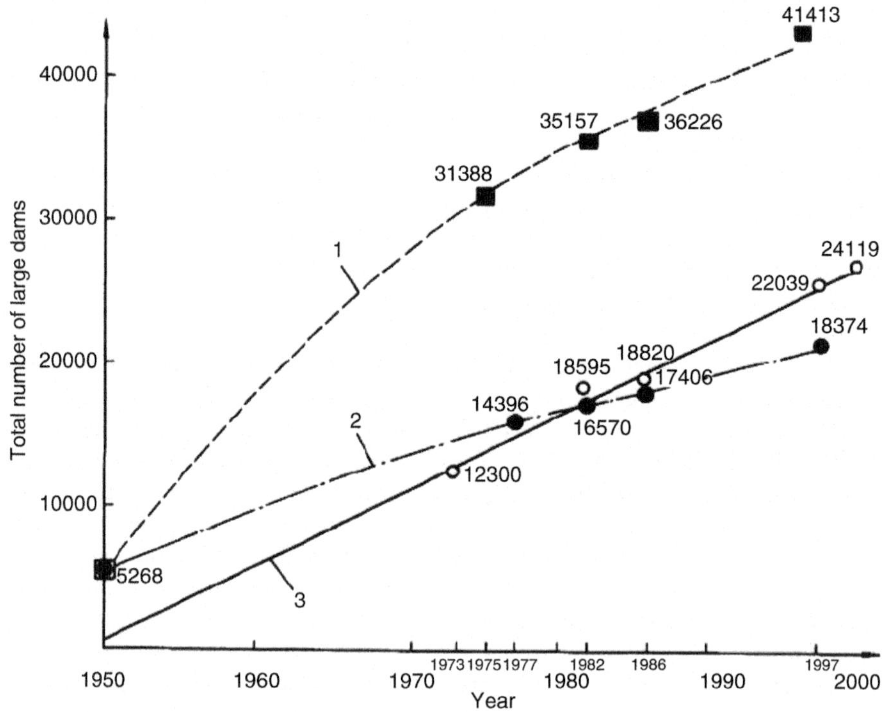

Fig. 1.1 The construction process of large dams in China and in the world (1: the number of large dams in the world, 2: in China, 3: in other countries)

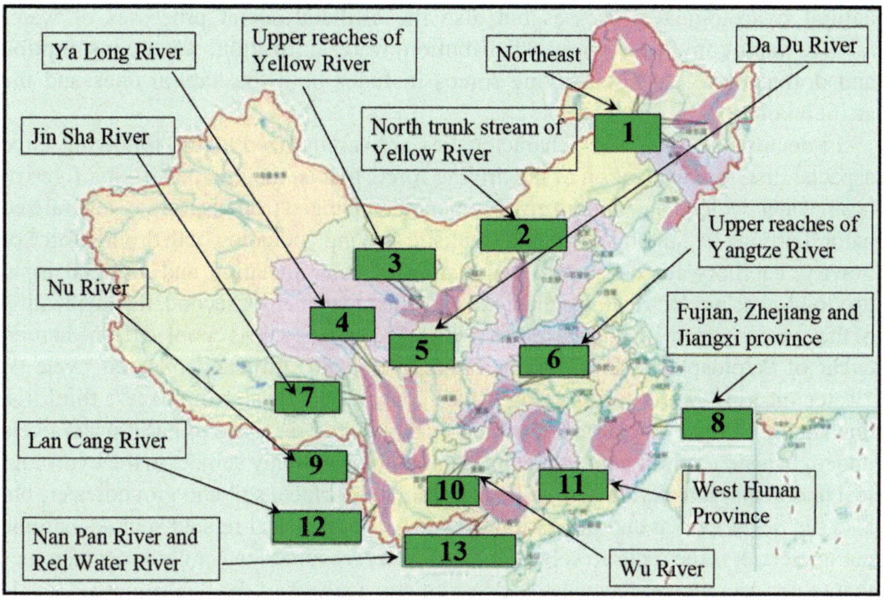

Fig. 1.2 The hydropower energy base in construction or in plan before 2050

After such a great number of reservoirs construction, the reservoir management problem, especially multi-reservoir joint operation problem, emerges as an important scientific and technological issue for reservoir managers and researchers. For example, the multi-objective optimization operation of reservoirs, reservoir operation rule forms and derivation methods, multi-reservoir joint operation problem in inter-basin water-transfer project and inflow forecast method for reservoir operation are of great significance theoretically and practically. In the following sections, those issues will be illustrated in details.

2 Multi-Reservoir Operation Within Theory Framework of Dualistic Water Cycle

2.1 Dualistic Water Cycle Theory

With the economy development and the population increase, the water cycle has been changed from the natural model to the "natural–artificial" dualistic model. The natural water cycle is consisted of precipitation, canopy interception, evapotranspiration, infiltration, surface runoff, overland flow, river flow and groundwater flow etc., and its driving forces are natural ones including radiation, gravity and wind etc. The "natural–artificial" dualistic water cycle includes not only the above natural hydrological processes but also the artificial social processes of water taking, water conveyance, water distribution, water utilization, water consumption and drainage etc., and its driving forces includes both the natural ones and the artificial ones [40].

In details, the "dualistic" characteristics are summarized as the following three aspects: first, the dualization of the driving force, that is, the internal driving force of basin water cycle in the modern environment has changed from the former centralized natural driving to "natural–artificial" dualistic-driving, including both driving force of gravity, capillary force and the evaporation of solar radiation and artificial input driving forces as electrical, mechanical, and chemical energy; second, the dualization of the cycle structure, that is the modern complete water cycle is coupled by the natural cycle of "atmosphere–slope–underground–river" and artificial collateral cycle of "water intaking–water transporting–water consumption–water drainage"; third, the dualization of the cycle parameters, that is, the overall response of basin water cycle under changed environment to precipitation input is not only subject to the hydrological and geological parameters of the natural land surface, soil and groundwater, but also the development and utilization of water resources and related socio-economic parameters. It is the focus to solve the basin water resources and environmental issues that to conduct a comprehensive and systematic analysis of the dualistic water cycle and the rules of its associated process of evolution.

In addition, the world can be also understood to be made up of society–economy system and ecology–environment system, which have mutual interaction role and feedback mechanisms between them. Within the two large systems, there exists materials and energy exchange partly through the carrier of water, which make water have five big attributes of "resources, ecology, environment, economy and society". Among them, "resources" attributes is the basic attribute of water, other attributes are due to the interaction between water and the two systems as illustrated in Fig. 1.3. These attributes of water has strong relationship with the objectives of dualistic water cycle simulation and regulation.

For the influence of intense human activity and climate variation, the water cycle process presents more and more obvious "natural and artificial" dualistic driving forces, which brings many water problems such as water scarcity, flood and water-logging, worsening water environment and degradation of water ecology system. In order to mitigate water crisis and enhance the society and economy healthy development, it is necessary to identify the evolution disciplines of water cycle and the driving mechanism. Relying on the reasonable application of complex water resources system operation theory, we can exert fully the economic, social, environmental and ecological benefits of water resources to achieve economy and society sustainable development and the harmony between human and nature. Based on these requirements, we propose the theoretical framework of dualistic water cycle simulation and regulation as in Fig. 1.4.

As shown in Fig. 1.5, the watershed water cycle is composed of "natural water cycle" and "artificial water cycle", whose intense interaction is mainly achieved by the operation of hydraulic projects. The natural water cycle includes three segments: meteorology $\rightarrow$ hydrology, hydrology $\rightarrow$ water quality and hydrology $\rightarrow$ water ecology. The artificial water cycle can be divided into two parts: flood control and

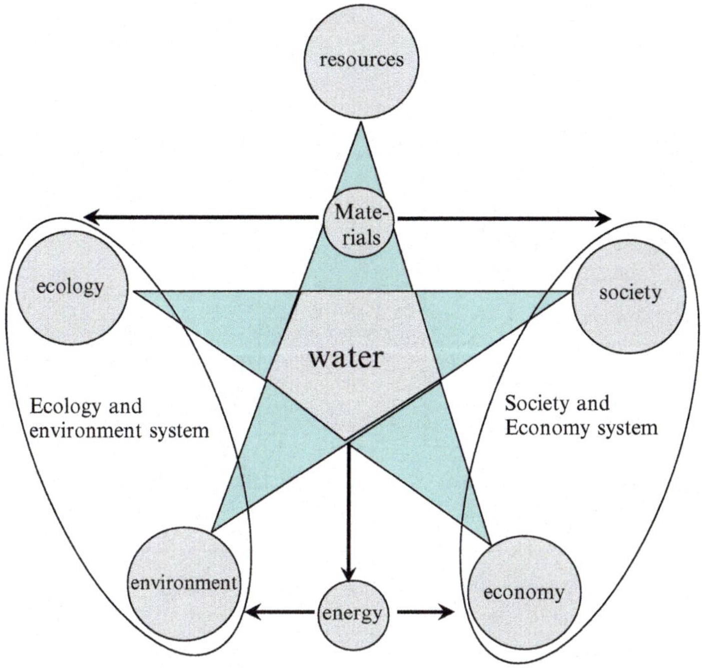

Fig. 1.3 The relationship between water and society, economy, ecology, environment system

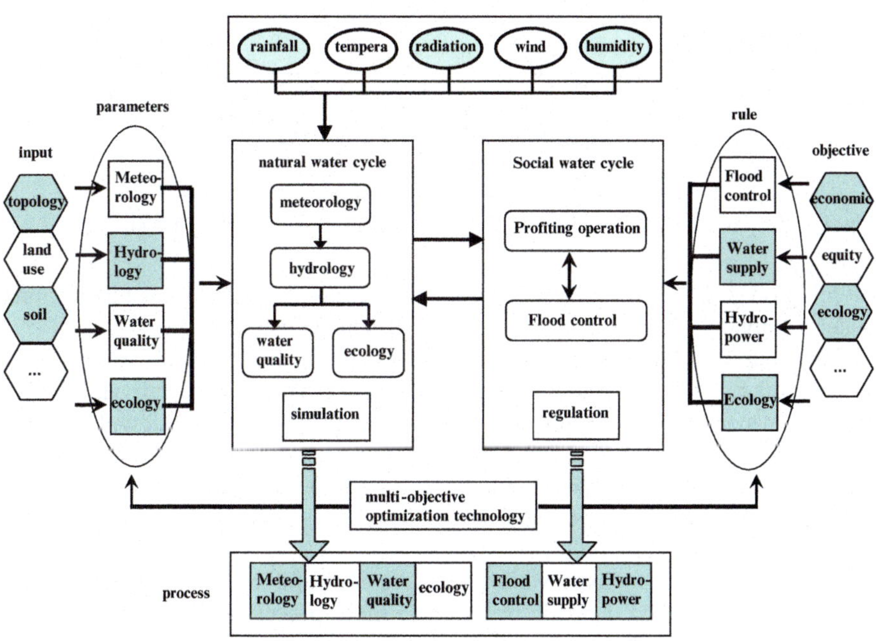

Fig. 1.4 Theoretical framework of dualistic water cycle simulation and regulation

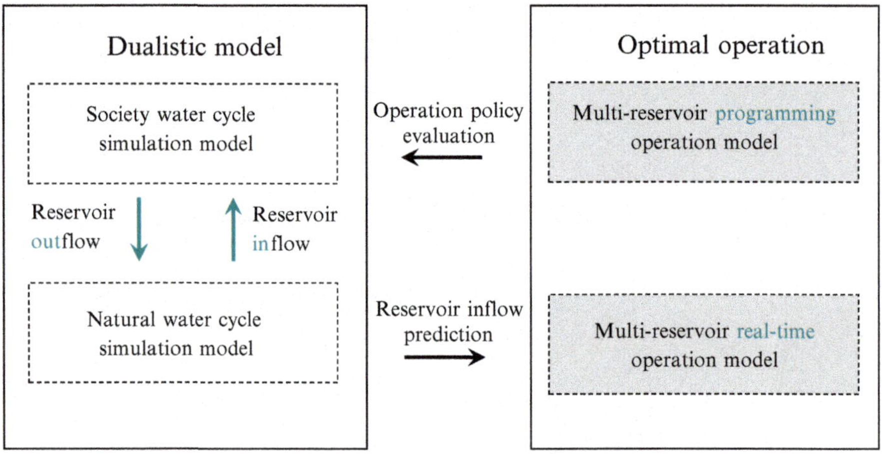

Fig. 1.5 The model system of dualistic water cycle simulation and regulation

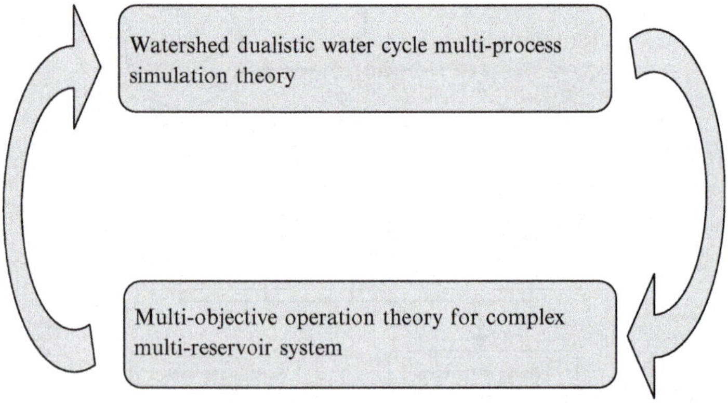

Fig. 1.6 The core theory of dualistic water cycle simulation and regulation model

profiting operation. For reservoir operation, profiting operation takes into account water supply, hydropower generation, ecology and navigation.

The coupling simulation foundation of "natural and artificial" water cycle system is the physical mechanism of dualistic water cycle and the derivative effect theory of water resources. The model system of dualistic water cycle simulation and regulation is shown in Fig. 1.5. For multi-reservoir system, the connection of dualistic model and optimal operation model is that the dualistic model can provide reservoir inflow prediction for optimal operation model and evaluate the effectiveness of system operating policy.

The core theory of dualistic water cycle simulation and regulation model includes two aspects: watershed dualistic water cycle multi-process simulation theory and multi-objective operation theory for complex multi-reservoir system as shown in Fig. 1.6. The simulation part gives the description of dualistic water cycle system from the perspective of model and the operation part can achieves the consideration of human interruption for the social water cycle process. The optimal operation of hydraulic projects can make water resources serve fully for the economy and social development and mitigate their impact on natural water cycle system.

2.2 *Main Technologies*

In this section, three kinds of main technologies to achieve the dualistic water cycle simulation and regulation are introduced, which consist of coupling technology for dualistic model, distributed hydrological modeling for inflow prediction and the technology to drive multi-reservoir operating policy.

2.2.1 Coupling Technology for Dualistic Model

The dualistic model system can taking into comprehensive consideration the natural evolution factors, high-intensity human activities and urbanization, regulation and control of hydraulic projects, etc., and can be used to describe the water cycle and water ecosystems evolution, reveals the different transformation processes of mountainous and plain areas, surface and underground, urban and rural. Because on the core model platform, by making detailed simulation of the water cycle under different historical and planning conditions, master the key and the possible effects and corresponding countermeasures from the all aspects of evolution and the process of water cycle and regulation process, so that can guide scientists in solving the problems of water resources and water ecosystems, and provide supporting tools for achieving comprehensive management objectives of the basin water resources.

Dualistic Model System Outline

The dualistic model system is developed independently by China Institute of Water resources and Hydropower Research (IWHR), referred to as Dualistic Model. The model is formed by the coupling of Water and Energy transfer Processes model (WEP), Rules-based Objected-oriented Water Allocation Simulation Model (ROWAS) and Decision Analysis for Multi-Objective System (DAMOS), the overall structure is shown in Fig. 1.7.

Dualistic model system is the software system developed specifically for the dualistic model, including the system platform of dualistic model data management functions and model calculation function. The data management function includes various types of attribute and spatial data, hydrological data, water environment data and socio-economic data, etc.; model calculation function includes the pre-processing, multi-model coupling, post-processing functions required by the model calculation.

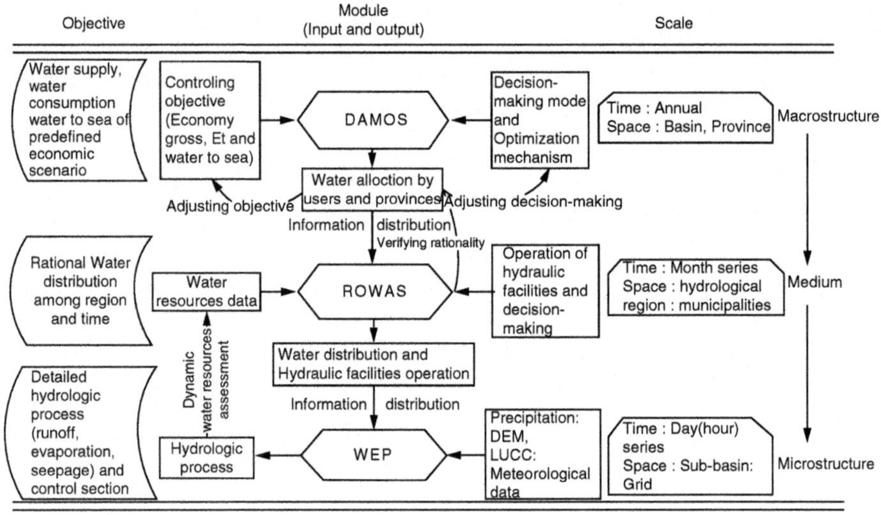

Fig. 1.7 Dualistic model structure

Characteristics of the Dualistic Model

Dualistic system model is a huge software project. The system has the following characteristics:

(1) There are many models and the complex structure, so it is difficult to develop the system. Every individual model of the dualistic model is realized by different programming languages and programming methods, such as: DAMOS model adopts common optimization software GAMS to achieve the description and solution to the optimal allocation of multi-objective water resources. ROWAS model adopts C++ to achieve a long series of simulation to the water resources supply and demand balance, and WEP model adopts Fortran language to achieve the simulation to the "natural–artificial" coupling water cycle process, water environment process and underground water process. It is necessary to couple the three models into an organic whole in order to develop the dualistic model system. It requires an appropriate transformation to every model so that every model can be integrated into the final dualistic model system. For example: develop general-purpose optimization modeling and solving framework, the DAMOS model developed by using GAMS is realized by using Java language, and the perfect integration with the application system is achieved. At the same time, data is managed in different modes for different models (DAMOS model and WEP model adopt text mode to conduct data management, and ROWAS model adopts the mutual management of text and database), to solve this practical problem, in order to couple the models into an organic whole, the system conduct unified management to the required input

and output data of every model, and build the unified data management module of multiple models on the unified database platform.

(2) Integrate a variety of software technology, and enjoy a high degree of innovation. In order to adapt to the requirement of the highly complex dualistic model calculation and data management, the dualistic model system adopts the rich client/server model to conduct system development. The development mode integrates the merits of both the fat client/server (C/S) and thin client/server (B/S), and can guarantee all functions of the dualistic model system, the user can call a variety of complex models to do calculations on the system interface, without calling the other interfaces and platform. At the same time, it can support a richer user interaction and achieve a better user response. The client end adopts the open source Eclipse RCP framework, using pure Java language for development, the database server adopts SQL Server2000, and Hibernate data access is adopted between client-servers. Java is adopted for develop so as to integrate better with practical application system, and lay a certain foundation for the future development of WEB-version based dualistic model system. At the same time, the dualistic model system integrates a wide range of software technologies, including optimization software GAMS, database software MS SQL Server, database connection components Hibernate, space display components Supermap, as well as spatial data management components ArcGIS SDE and a number of open sources GIS components MapWindow, etc.

Function of Dualistic Model System

(1) Data management function: to facilitate system development and simplify the user's familiarization to the system interface, we have adopted a general-purpose management interface for data input and output data management. The system data management is interactively reflected in the graphs, charts and other forms.
(2) Model calculation: The dualistic model system will support calculation function of DAMOS, ROWAS, WEP model, and packaging and transformation is made according to characteristics of each model respectively. Taking DAMOS model as an example, since DAMOS model adopts GAMS optimization software package in the development, but GAMS is not suitable for application system development, so the system has developed a general-purpose water resources optimization model constructing and solving package—Lp_Solve, and then rewrite the DAMOS model using the software package.
(3) In addition, the dualistic model system not only supports the calculation of the three models, but also supports the data coupling between the three models, so as to achieve automatic data exchange between the models and achieve the fully automated dualistic model. The time scale of DAMOS model is the annual value of many years, the spatial scale is province, while the time scale of ROWAS model is a long series of months, the spatial scale is the calculation unit of three-stage district and city, the time scale of WEP model is day, and the

spatial scale is the contour band within sub-basins. To a new calculation program, in the time scale of the next several decades, DAMOS model first makes optimization to the industrial structure, planting structure, water utilization, sewage pollution control, and engineering measures of every planning level year. These optimization results are provided to ROWAS model and WEP model for them to do simulation on different levels. Besides, ROWAS model will also send feedback to DAMOS model, mainly water supply and water supply guarantee rate. Information of water utilization process, drainage process and project scheduling is received after ROWAS makes water supply and demand balance calculation. This information is further passed to the WEP model for it to do simulation at even smaller time scale and spatial scale. Of course, WEP model will also send feedback to ROWAS model, mainly the resources volume information, such as: the surface inflow, ground water status, etc. Because the time scale and spatial scale of the three models are different, so data distribution needs to be made on time scale and spatial scale. To this end, we developed the data distribution procedures of the coupling between the various models.

2.2.2 Developing Distributed Hydrological Model for Inflow Prediction

The physically-based distributed hydrological model, WEP-L [40, 41], which couples simulations of natural hydrological processes and water use processes, was developed to characterize water resource variations in basins seriously affected by human impacts [40, 41]. To be applicable to a large river basin, and to overcome the implausible number of calculations caused by small grids and anamorphic simulations caused by overly rough grids, the WEP-L modeling scheme adopts calculation units of contour bands within sub-basins, in which terrain, river network, vegetation, soil, and land use data are based on spatial information data on a 1 km grid [40, 41].

After the simulation is undertaken, many problems can still be found related to the application of the distributed hydrological models [42, 43]. Some models are too complicated to operate easily or too difficult to be modified, others are limited to small basins because of the heavy burden of computation or data preparation. Three disadvantages: (1) low modularization, (2) low generalization of pre-processing programs, and (3) low automation, are possibly the key reasons for the limitations described above for WEP-L. The AutoWEP modeling scheme was therefore developed with strong generalization and expandability, pre-processing modules were improved, and an automatic parameter identification module was developed. This section describes the main improvements and modeling approach developed for AutoWEP, which can be used for inflow prediction.

To convert the WEP-L modeling method to one that can great simplify the modeling and calibration processes, enable users to reduce repetitive steps in building distributed hydrological models, upgrade the efficiency of modeling, and reach an ideal simulation precision, a completely new modeling algorithm called

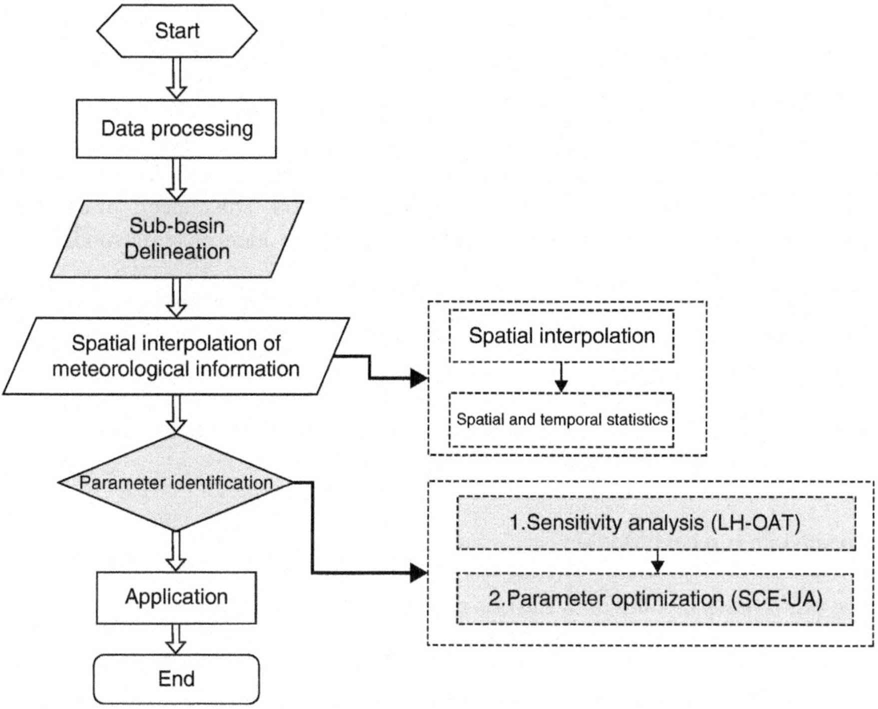

Fig. 1.8 Modeling process of AutoWEP model

AutoWEP was developed. This involved re-establishing coding structure, revising input/output parameters, and pre-processing programs. New functions were added including parameter sensitivity analysis and automatic calibration of parameters. The main improvement in the Auto WEP algorithm is the addition of the "AUTO" modules, which improves the modeling and calibration of the WEP modeling method, making it more efficient. The Auto-WEP modeling process is shown in Fig. 1.8.

For the AutoWEP algorithm, the modeling codes were rebuilt to be generalized and expandable. Furthermore, several modeling modules were updated. The main improvements of AutoWEP modeling method are summarized below.

Fortran 90 Is Used to Rebuild Modeling Codes

The development language of the AutoWEP algorithms is FORTRAN, generally a computationally efficient language for scientific research. Fortran 77, with its fixed coding form was used in the original WEP algorithm. However, such a coding style is not necessarily the most suitable or consistent with long subroutines or functions, and static arrays, which reduces the readability of the original code. The AutoWEP algorithm uses Fortran 90 coding and a modularization concept is used, which lays the foundations for algorithm expansion and the addition of open source codes.

The WEP modeling array is defined in the approach of the static array; when it is applied to different river basins, its codes must be recompiled. By introducing the concept of a dynamic array, AutoWEP modeling becomes more generalized by dynamically defining all arrays with external parameters. The concept of dynamic arrays is also applied to pre-processing and parameter identification modules of AutoWEP.

Input parameters to the WEP-L modeling routine have some redundancy, resulting in the preparation of complicated parameter files when a new model is built. By redesigning the data structure of parameter files for AutoWEP, data redundancy in input parameter files is decreased to the greatest extent. In this way, many AutoWEP input parameters can be automatically calculated.

Updated Pre-process Module

The pre-process module of WEP-L is essentially a manual exercise that makes the modeling pre-process tedious and time- consuming. Auto-WEP aims to improve the pre-processing module by making it easier to use, more practical, and functional.

To consider multiple watershed conditions, a generalized sub-basin delineation method, the Pfafstetter coding system based General Sub-basin Delineation Method (PGSDM) was developed [44, 45]. By introducing recursion algorithms to PGSDM, a river network with any grade can be delineated, and the generalization of the method is greatly improved [46].

Many manual tasks in the original modules concerning the spatial delineation of the basin can be achieved with software. These include the analysis of adjacent relationships and length scales between different sub-basins. These can be accomplished in plan view by a series of operations using common commercial programs such as ArcGIS and Excel. These operations include conversion of raster data to vectors, establishment of topological relationships, and analysis of adjacent relationships and polygon edge lengths through rasterized polygons. However, in the pre- process module of AutoWEP, the analysis can be directly achieved by a newly developed algorithm that greatly improves automation and decreases model building difficulties.

Development of a Module for Automatic Calibration of Parameters

Parameter identification in WEP-L is accomplished by a manual trial and error method that is usually time-consuming and dependent on the experience of modeler [41]. To decrease these problems, AutoWEP proposes a complete set of automatic parameter optimization methods, including: (1) automatic parameter sensitivity analysis on the basis of the LH-OAT algorithm, and (2) automatic parameter calibration on the basis of the SCE-UA algorithm.

In addition, to enhance the efficiency of automatic parameter calibration, the concept of automatically delineated parameter partitioning is proposed. The relative relationship between various types of underlying surfaces in one partition will not change during the calibration process; e.g., if the conductivity coefficient of saturated soil needs to be calibrated for a parameter partition, the original default parameter group (the value of each computational unit) will be multiplied by a correction coefficient, and a new group of conductivity coefficients of saturated soil can be obtained. Relative relationships between different sub-basin computational units in the parameter partition will remain, while the absolute value will change. However, the assumption of the calibration is that the default parameter ascertained from an underlying surface data could favorably reflect the difference between model parameters on different types of underlying surfaces, whereas the absolute value can be adjusted according to the specific situations of different river basins.

AutoWEP is composed of a pre-processing module, a model calculation module, and a parameter identification module. Pre-processing and parameter identification modules include the following steps: (1) sub-basin spatial delineation, (2) model parameter calculation, (3) spatial interpolation of meteorological data, (4) parameter sensitivity analysis, and (5) automatic parameter optimization. The basic theory and methods involved with AutoWEP are detailed below.

2.2.3 Technology to Derive Multi-Reservoir Operating Policy

At the present, the main technology to derive multi-reservoir operating rule can be divided into three categories: (1) manual correction, (2) implicit stochastic optimization algorithm, (3) simulation model embedded into optimization model. The first two kinds of methods have something in common that the reservoir operation rule is obtained by trial calculation, mathematical statistics with a complex process, low degree of automation and unsatisfactory optimization effectiveness. An efficient approach to define an operating rule is by using optimization models in combination with simulation models. A number of drawbacks of traditional methods such as linear or non-linear programming, and dynamic programming are found out [47], which arise from the complexity of modern simulation models in the form of, e.g. nonlinearity, discontinuity, and discreteness. Such limitations have recently been overcome by combining simulations with heuristic search procedures (e.g. genetic and evolutionary algorithms) and advances in computational technology. Several studies have demonstrated the efficiency of these methods in optimizing reservoir operation. Oliveira and Loucks [48] presented an approach to optimize operating rules for multi-reservoir systems using genetic algorithms (GA). Chen [47] successfully applied a real-coded GA in combination with a simulation model to optimize rule curves of a major reservoir in Taiwan. Ngo et al. [49] proposed to optimize control strategies for reservoir operation by applying a combination

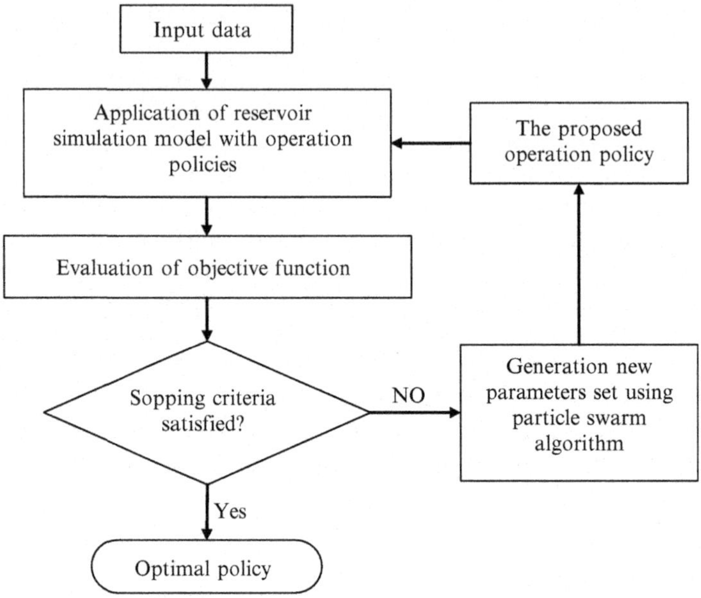

Fig. 1.9 Framework to derive the proposed optimal operating rule

of simulation and optimization models approach adopting the shuffled complex evolution (SCE) algorithm. Reddy and Kumar [50] presented a particle swarm optimization based solution to a detailed operational model for short-term reservoir operation for irrigation of multiple crops. The general framework of the simulation–optimization model used in this paper to search for the optimal operating policy is presented in Fig. 1.9.

2.3 *Dualistic Hydrology Simulation and Regulation System for Upper Reaches of Yangtze River*

In the upper reaches of Yangtze River, many large reservoirs have been built or are being constructed or planned. Three Gorges reservoir is one of the most important reservoirs in upper reaches of Yangtze River, not only because of its huge capacity but also its special location. Three Gorges reservoir locates at the boundary of upper reaches of Yangtze River. The natural water cycle process after the regulation of the multi-reservoir system presents obvious dualistic characteristics. For satisfying the strategy requirement of sustainable utilization of water resources in Yangtze River watershed, we develop a dualistic hydrology simulation and regulation system for upper reaches of Yangtze River, which takes the dualistic hydrology simulation and multi-reservoir system operation theory as the theoretical basis. As shown in

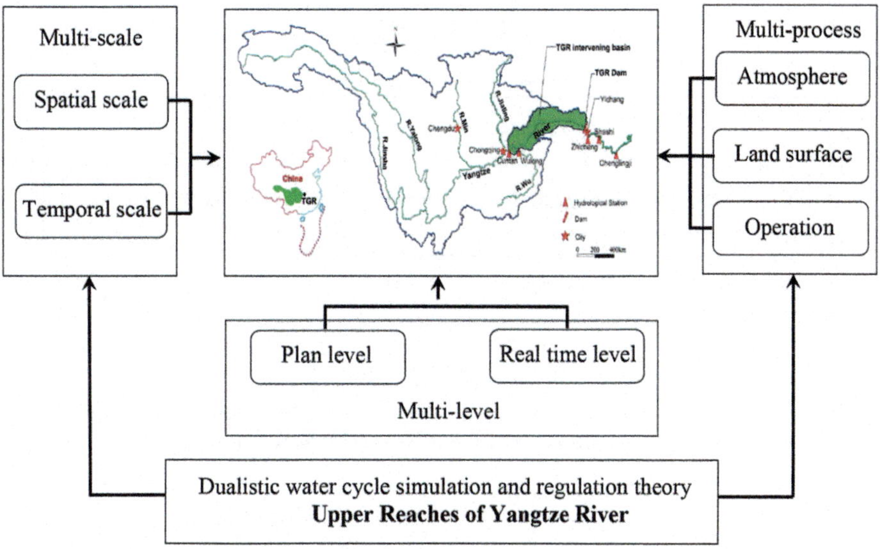

Fig. 1.10 Dualistic simulation and regulation system for upper reaches of Yangtze River

Fig. 1.10, the system is devised from the perspective of multi-scale, multi-process and multi-level for simulating and regulating water resources system of Yangtze River upper reaches.

The dualistic hydrology simulation and water resources regulation problem of upper reaches in Yangtze River need to carry out the research work from temporal and spatial multi-scales. Not only the whole watershed but also some important and specific study areas need to be studied respectively to analyze the hydrological variation characteristics with scale change. As described in Fig. 1.11, the hydrological time series of month scale, day scale and hour scale need to be modeled and generated for the real time operation and plan operation of the multi-reservoir system in upper reaches of Yangtze River.

For achieving the whole process and all element simulation of the water cycle, the system needs to be able to model the water cycle and its accompanying process. As shown in Fig. 1.12, the atmosphere process, the land surface process and operation process all can be taken into consideration in the system. In details, the atmosphere numerical simulation and forecast model includes global climate model (GCM), weather research and forecasting model (GCM) and Mesa-scale model 5 (MM5). The distributed water cycle and its accompanying process simulation model consists of distributed hydrological model EasyDHM, water pollution simulation model EasyWQ and hydrodynamic model EasyRiv. The multi-reservoir multi-objective joint operation model is constitutive of joint simulation model, multi-objective optimization model and real time operation model. The coupling technology between the different process simulations is important for the multi-process simulation of upper reaches of Yangtze River.

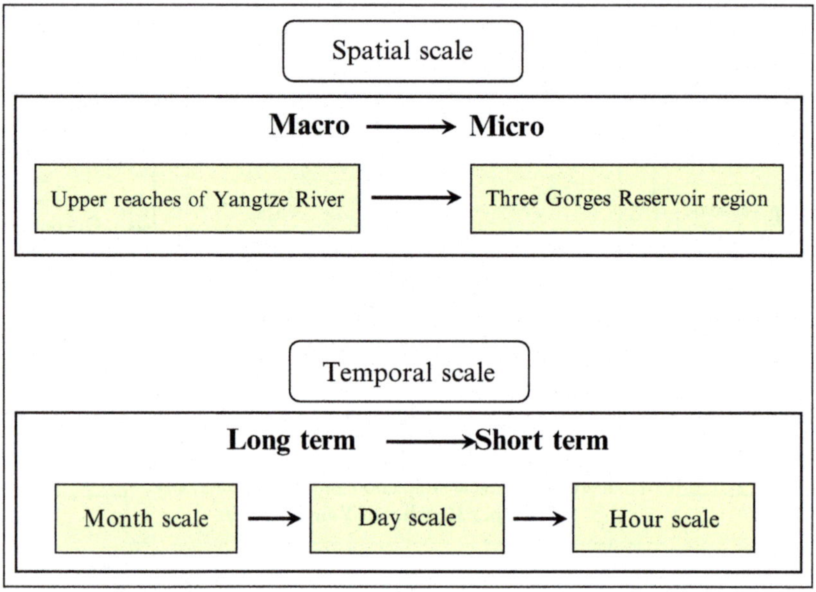

Fig. 1.11 Multi-scale modeling technology for upper reaches of Yangtze River

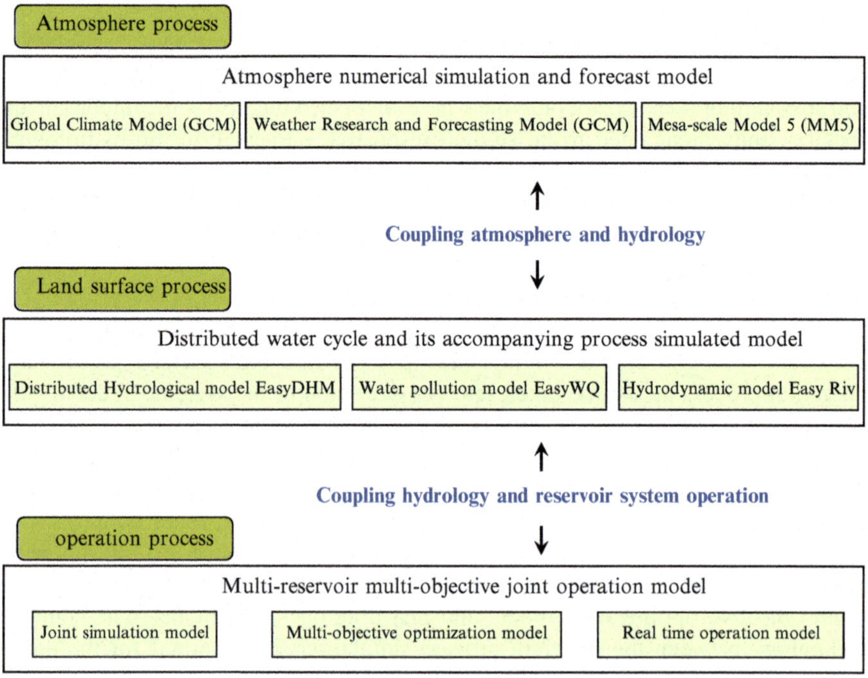

Fig. 1.12 Multi-process simulation of upper reaches of Yangtze River

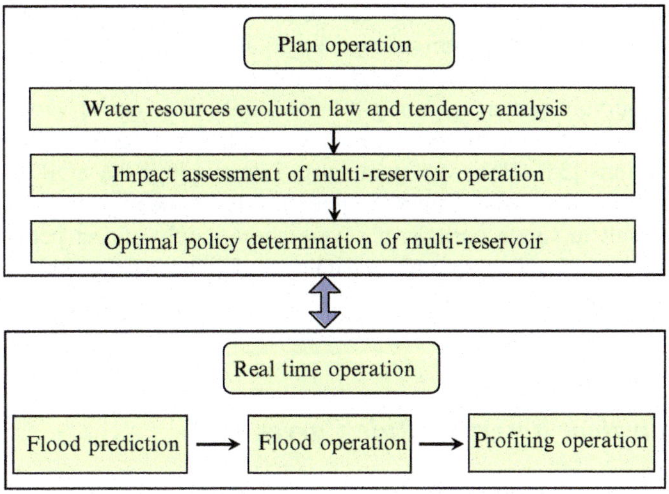

Fig. 1.13 Multi-level operation of multi-reservoir system in upper reaches of Yangtze River

The joint operation model can be divided into two levels, real time operation and plan operation, to satisfy the different operation requirement of multi-reservoir system in upper reaches of Yangtze River. As described in Fig. 1.13, the plan operation model is mainly used to analyze water resources evolution law, assess the impact of multi-reservoir operation and determine the optimal policy of the multi-reservoir system. The real time operation mainly serves for flood operation and short time scale profiting operation, which needs the inflow prediction information. The profiting operation refers to reservoir operation for the beneficial purpose such as hydropower generation, water supply, navigation or some other purposes.

3 Operation Rule Curves for Multi-Reservoir Operation

In the past, reservoirs were constructed and managed individually. However, the practical and environmental constraints require that individual reservoir operation must transit to multi-reservoir, water resources systems approaches [51]. The related research on the approaches to operate multi-reservoir has been carried out widely, reviews of which are given by Yeh [12], Wurbs [13], Labadie [14], and others. Oliveira and Loucks [48] used genetic search algorithms to derive multi-reservoir operating policies, which defined both system release and individual reservoir storage volume targets as functions of total storage in each of multiple within-year periods. Similar to the work of Oliveira and Loucks [48], Nalbantis and Koutsoyiannis [36] proposed parametric rule for planning and management of multiple-reservoir system accounting for various system operating goals.

Parametric rule uses a handful of control variables, valid through the entire control period, to distribute the system storage target into each member reservoir and calculate their water release. In contrast to most common methods used in optimal control of reservoir system requiring a large number of control variables, it can make a radical reduction of the number of control variables without yielding inferior solutions [35]. However, parametric rule employs the SOP to determine releases as much as demand if there is enough water in the multi-reservoir system. That may result in single periods of severe short supply during periods such as prolonged droughts, other extreme weather conditions, and sudden changes in water demand patterns.

3.1 Equivalent Reservoir Rule Curves

In this section, an operating policy is proposed for water-supply multi-reservoir by combining parametric rule with the hedging rule to avoid severe short supply. This operating policy includes two steps. First, the amount of water from the multi-reservoir system released for each water user is specified according to the hedging rule. In this study, the hedging rule uses the beginning-of-period storage of equivalent reservoir as a trigger to start hedging for the system common water demand. For the local water demand that can be only satisfied by some specific member reservoir, it employs the beginning-of-period storage of the corresponding member reservoir as a trigger. Second, parametric rule is employed to determine the release from each member reservoir, which should satisfy the requirement of water supply by the hedging rule. Particle swarm optimization algorithm (PSO) in combination with a simulation model is used to optimize the parameters in the proposed operating rule, which include the hedging rule curves and parameters in parametric rule. Finally, the Guanyinge, Shenwo and Tanghe (G-S-T) multi-reservoir system located at Taize River basin in China is taken as a case study to present the change of water shortage characteristics reproduced by the proposed operating rule. The operation results show that decision variables to be optimized in the proposed operating policy not only make a significant reduction compared to traditional operating rules, but severe short supply during droughts can be also controlled effectively.

Operating policies for multi-reservoir systems must specify not only the total release from the system but also the amounts, if any, to be released from each reservoir [48]. For the proposed operating rule, the hedging rule can specify the total release from the system according to the existing storage volume in the multi-reservoir system during each period and parametric rule is used to determine the amounts to be released from each member reservoir, which sums to the total release specified by the hedging rule.

3.1.1 The Hedging Rule

Droughts are considered as a normal part of climate and their occurrences are inevitable [52]. Hence, drought-induced water shortages should be paid more attention to. The hedging rule for reservoir management is just one measure to mitigate drought-related impacts.

The hedging rule for reservoir operations has been studied in different ways. For example, Srinivasan and Philipose [53] used hedging parameters to construct the hedging rule and evaluated the effects on the reservoir performance indicators. Shih and ReVelle [54, 55] determined the trigger value for a continuous hedging rule and then for a discrete hedging rule. Neelakantan and Pundarikanthan [34] presented a simulation–optimization methodology using neural network and multiple hedging rules to improve reservoir operation performance. And Tu et al. [56] considered a set of rule curves that are a function of the current storage level to trigger hedging for a multi-purpose multi-reservoir system.

The hedging rule used in this section consists of hedging rule curves and rationing factors for each water demand. Details of the hedging rule curves and its corresponding water-supply operating rule are illustrated in Fig. 1.14 and Table 1.1. In previous works [56, 57] on hedging rule curves, all planned water demands are

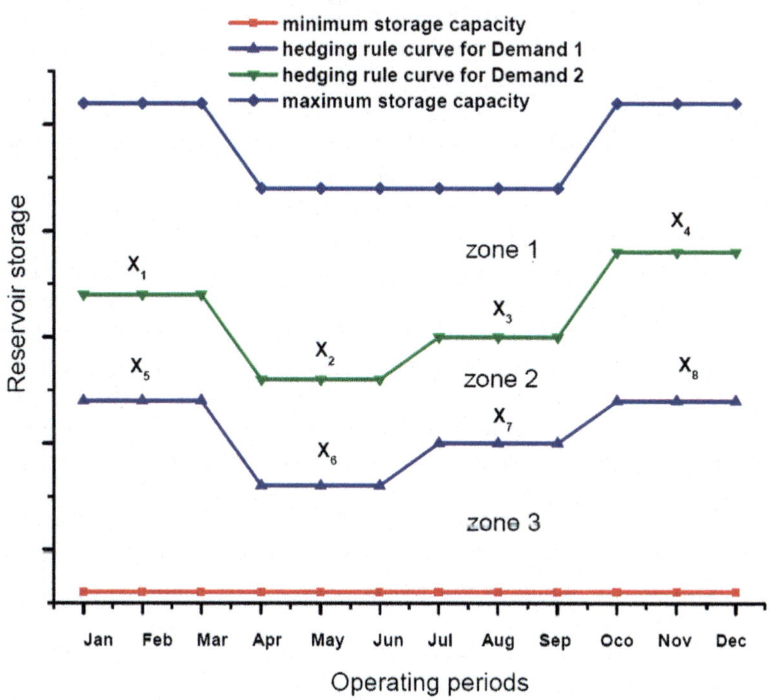

Fig. 1.14 Hedging rule curves based on reservoir storage

Table 1.1 Water-supply operating rule implied by hedging rule curves

Reservoir storage	Water supply for each demand	
	For demand 1 (D1)	For demand 2 (D2)
Zone 1	D1	D2
Zone 2	D1	α2*D2
Zone 3	α1*D1	α2*D2
Rationing factor	α1	α2

met at the same level and are rationed at the same time when drought occurs. For single purpose of water-supply operation, the water demand can be divided into various categories, such as irrigation, industry and municipal life. It should be noted that, in practice, different kind of water demand requires different water supply reliability and different degree of hedging. In this study, different hedging rule curves and rationing factors are assigned to different kinds of water demand. When drought occurs, different type of water demand owns different priority without rationing. In general, the water demand with lower priority should be rationed first and the degree of its hedging is also ought to be larger than the one of water demand with higher priority. It can be achieved through the proposed hedging rule curves in this section. Rationing factors are important parameters in the hedging rule to control the degree of hedging when drought happens. The value of rationing factors can be either obtained by optimization or determined empirically according to the experts' knowledge.

For the multi-reservoir operation, the water demand can be divided into local water demand, which can be only satisfied by some specific member reservoir, and common water demand, which can be satisfied by any reservoir in the system. The water supply for common water demand is related directly to the total water storage in the whole multi-reservoir system and the water supply for local water demand is only related to the water storage of the specific reservoir. In this study, an equivalent reservoir is developed to represent the multi-reservoir system and the beginning-of-period storage of equivalent reservoir is used as a trigger to start hedging for water supply of the common demand. A simplified equivalent reservoir representation of a multi-reservoir hydroelectric system is first proposed by Arvanitidis and Rosing [58] and the validity of this method for modeling multi-reservoir hydroelectric system optimal operation is investigated by Brandão [59]. For water-supply multi-reservoir, Robert et al. [1] proposed the concept of equivalent reservoir in the name of a fictitious system. Different from common water demand, the hedging rule curve devised for local water demand is based on the water storage of specific member reservoir. In summary, the desired releases for common or local water demand is a function of existing storage volumes in equivalent reservoir or specific member reservoir, the time of year, and water demand. And this function relationship is expressed in the form of hedging rule curves.

3.1.2 Parametric Rule to Determine the Reservoir Release

The hedging rule is aimed at determining how much water to provide for each demand. However, what amount of water release from each reservoir to satisfy them, especially for common water demand, has been still unknown. In this section, parametric rule proposed by Nalbantis and Koutsoyiannis [36] is modified to determine the amount of water release from each member reservoir at any period.

Parametric rule roughly consists of three computational stages: (1) distributing the system target storage into each reservoir of the system; (2) correcting the target storage of each reservoir so that physical constraints are not violated; (3) calculating the real storage and the release of each reservoir.

The system target storage S_{t+1}^{T} for a certain period is obtained through the continuity equation of equivalent reservoir by

$$S_{t+1}^{\mathrm{T}} = S_t^{\mathrm{T}} + I_t^{\mathrm{T}} - R_t^{\mathrm{T}} - SU_t^{\mathrm{T}} - L_t^{\mathrm{T}} \tag{1.1}$$

where S_t^{T} is beginning-of-period storage of equivalent reservoir at the stage t; I_t^{T} is stream inflows into equivalent reservoir at the stage t; R_t^{T} is equivalent reservoir release for all water demand at the stage t; SU_t^{T} is water spills of equivalent reservoir, but not include water spills from the upstream reservoir of the physical multi-reservoir system in series to the downstream reservoir; L_t^{T} is water losses of equivalent reservoir because of evaporation and seepage. Just like physical individual reservoir, equivalent reservoir has its own storage capacity ranging from full storage to dead storage. The full or dead storage of equivalent reservoir equals to sum of the ones of each individual reservoir. In Eq. (1.1), the water balance terms of equivalent reservoir including S_t^{T}, I_t^{T}, and L_t^{T} are also sum of the ones of each individual reservoir in the system, which all can be calculated or estimated in some manner. Due to the adoption of the SOP by parametric rule, R_t^{T} equals to the sum of all demand only if there is water in the equivalent reservoir. In the modified parametric rule, R_t^{T} is obtained according to the proposed hedging rule curves. When there is not enough water in equivalent reservoir or specific member reservoir, water demand will be rationed and R_t^{T} will be less than demand to avoid future severe water shortage. In this step, SU_t^{T} is first set to zero and may be adjusted in later steps.

In Eq. (1.1), the water balance terms of equivalent reservoir, except for S_{t+1}^{T}, all can be obtained in some methods. So the system target storage S_{t+1}^{T} can be got by Eq. (1.1). After that, the target storage S_{t+1}^{i} of individual reservoir *i* at the stage $t+1$ can be yielded by Eq. (1.2).

$$S_{t+1}^{i} = a_{ij} \cdot S_{t+1}^{T} + b_{ij} \quad i = 1,2\ldots m, \; j = 1,2\ldots n \tag{1.2}$$

where i, j denote, respectively, the ordinal numbers of the individual reservoir in the system and divided stages within a year; m and n correspond to the total number of individual reservoirs in the system and divided stages within a water year, in Eq. (1.2) or other equations. Both a_{ij} and b_{ij} are parameters in parametric rule to be optimized. For each individual reservoir, there is a similar water balance equation (1.3) to equation (1.1). The role of Eq. (1.3) is to determine each specific release from individual reservoir, whose sum equals to R_t^{T} in Eq. (1.1).

$$S_{t+1}^i = S_t^i + I_t^i - R_t^i - L_t^i \quad i = 1, 2 \ldots m \tag{1.3}$$

Because the sum of target storages of all the individual reservoirs equals to the target storage of equivalent reservoir, i.e. (1.4), the parameters of a_{ij}, b_{ij} in Eq. (1.2) are subject to Eqs. (1.5) and (1.6).

$$S_{t+1}^{\mathrm{T}} = \sum_{i=1}^{n} S_{t+1}^i \quad i = 1, 2 \ldots m \tag{1.4}$$

$$1 = \sum_{i=1}^{m} a_{ij} \qquad i = 1, 2 \ldots m, j = 1, 2 \ldots n \tag{1.5}$$

$$0 = \sum_{i=1}^{m} b_{ij} \qquad i = 1, 2 \ldots m, j = 1, 2 \ldots n \tag{1.6}$$

For the individual reservoir with water-supply task for local water demand, its release is not allowed to be less than $R_{t-\min}^i$, which is reservoir release for local demand according to the hedging rule. This requires the target storage distributed to this kind of reservoir by Eq. (1.2) not to be larger than $S_{\max(t+1)}^i$, which is given by Eq. (1.7).

$$S_{\max(t+1)}^i = S_t^i + I_t^i - \mathrm{R}_{t-\min}^i - L_t^i \quad i = 1, 2 \ldots m \tag{1.7}$$

As is noted, the target storage distributed to each individual reservoir must satisfy physical constraints, which can not fall down below dead storage nor exceed individual reservoir storage capacity, i.e. Eq. (1.8). When calculating water spills from equivalent reservoir, SU_t^{T} is first set to zero. If obtained S_{t+1}^{T} in Eq. (1.1) is larger than sum of the minimum between $S_{\max}^i$ and $S_{\max(t+1)}^i$, SU_t^{T} is given by Eq. (1.9). Otherwise, SU_t^{T} equals to zero.

$$S_{\min}^i \le S_{t+1}^i \le \min\left(S_{\max}^i, S_{\max(t+1)}^i\right) \quad i = 1, 2 \ldots m \tag{1.8}$$

$$SU_t^{\mathrm{T}} = S_{t+1}^T - \sum_{i=1}^{m} \min\left(S_{\max}^i, S_{\max(t+1)}^i\right) \quad i = 1, 2 \ldots m \tag{1.9}$$

Once the parameters of a_{ij}, b_{ij} have been determined by optimization, they get the same value for the fixed stage in each year. This makes distributed target storage to each individual reservoir will not always satisfy the physical constraints during a long operating horizon. To correct this inconsistency, an iteration equation is introduced by Nalbantis and Koutsoyiannis [36]. If the water amount stored in the equivalent reservoir stays between its minimum and maximum storage capacity, the target storage distributed to each individual reservoir is ought to satisfy their physical constraints after the iteration equation is supplemented.

3.2 Two-Dimension (2D) Rule Curves for Dual-Reservoir System

3.2.1 The Function of 2D Rule Curves

For multi-reservoir water-supply joint operation, the water supply decision should be made according to the storage state of the multi-reservoir system rather the member reservoir. For full consideration of the system storage, we devise on type of rule form for the two reservoir system of water supply with one piece of water demand, which is named as two dimension reservoir rule curves. As shown in Fig. 1.15, two dimension reservoir rule curves have two coordinate axes, which stand for each reservoir water storage respectively. As is noted, the symbols max_1 and max_2 represent the storage capacity of reservoir 1 and reservoir 2. The lines x_1 and y_1 are the hedging rule curves based on the storage of reservoir 1 and reservoir 2. For one dimension reservoir rule curve, the water supply will be hedged if the reservoir storage falls down below the hedging rule curve. The lines x_2 and y_2 are the rule curves for increasing water supply based on the storage of reservoir 1 and reservoir 2. For one dimension reservoir rule curve, the water supply will be increased more than water demand if the reservoir storage is higher the hedging rule curve.

Different from one dimension reservoir rule curves, the water supply decision is made for two dimension reservoir rule curves based on the relationship between the two reservoir storage combination and reservoir operation zones. In Fig. 1.15, the hedging rule curves x_1, y_1 and the rule curves for increasing water supply y_2, y_2 divide the big square of reservoir capacity into nine zones, which include 3 hedging zones, 3 normal zones and 3 increment zones. The water supply rules of hedging zones and increment zones are the same as the ones mentioned above in one dimension rule curves. The water supply is equals to water demand if the combination point of two reservoir storage locates in the normal zone. Figure 1.15 just presents the two dimension rule curves for one operation period in each year. For different operation periods, the hedging rule curves may have different positions. In summary, the two dimension reservoir rule curves give one rule form of water supply for two reservoir system with the advantage of considering the system storage together. For practical application, the two dimension reservoir rule curves should be used combined with one type of allocation rule for the system water supply task distribution between member reservoirs.

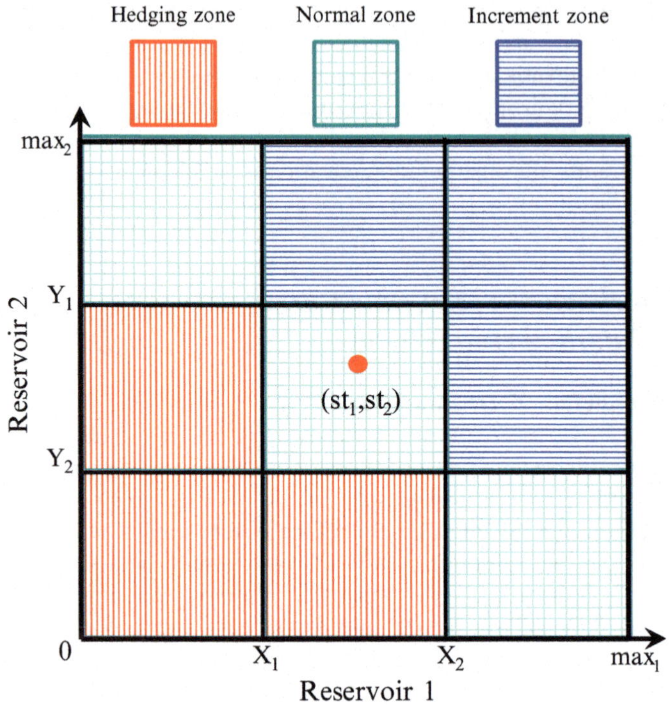

Fig. 1.15 Two dimension reservoir rule curves for dual-reservoir system

3.2.2 2D Rule Curves with Variable Allocation Ratios

The 2D rule curves shown in Fig. 1.15 are designed for the dualistic reservoir system with one piece of water demand, which has hedging zone, normal zone and increment zone. Based on the previous work, we propose another type of 2D rule curves for the dualistic reservoir system with two pieces of water demand, as described in Fig. 1.16, which also has the hedging zone and the normal zone but without the increment zone. For this type of 2D rule curves, the water demand 1 has lower water supply priority than the water demand 2. Therefore, the water demand 1 will be first hedged if there is not enough water in the reservoir system. According to the storage of the reservoir system, the water supply policy for the water demand 1 and 2 can be divided into three categories: no hedging, only the demand 1 hedging, both the demand 1 and 2 hedging. The three kinds of water supply decision are corresponding to the first part of Fig. 1.16.

For serving practical operation of dualistic reservoir system, the concept of variable allocation ratios is devised for optimal allocation of system common water supply task between each reservoir. Just like the second and the third part in Fig. 1.16, the allocation ratio in each square is constant at each operation period. The allocation ratios y and z are respectively for reservoir one and reservoir two.

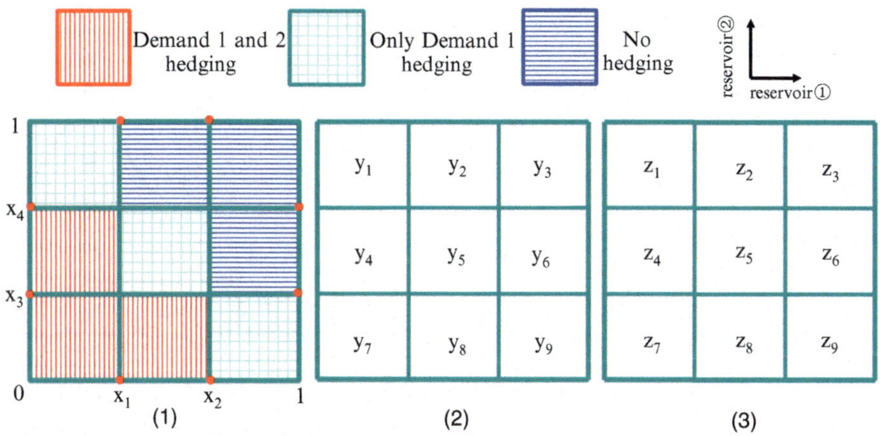

Fig. 1.16 Two dimension reservoir rule curves with variable allocation ratios

The sum of ratio y and ratio z at the same square equals to 1, which can be taken as the constraints as the optimization model. And the allocation ratios are often determined through the optimization algorithms.

The method of employing the 2D reservoir rule curves with variable allocation ratios for reservoir system operation can be summarized as two steps: First, the system manager decides how much water to provide for the system common water supply task according to 2D rule curves. Second, the common water supply task can be distributed between the specific member reservoirs, referring to the variable allocation ratios, to make sure the amount of each reservoir water release. The detailed application of 2D reservoir rule curves with variable allocation ratios will be analyzed through the following case study.

3.2.3 The Optimization Model and Result Analysis for 2D Rule Curves

In order to verify the effectiveness of 2D rule curves with variable allocation ratios, we take the dualistic water supply reservoir in Northeast China as a case study to analyze the operation results. The dualistic reservoir system consists of reservoir A and reservoir B in parallel, as illustrated in Fig. 1.17, which need to provide water supply together for downstream agriculture and industry. In addition to the common water supply task, each reservoir needs to release water to guarantee the environmental flow downstream. The flood season lasts from July to August in the study area. The capacities of reservoir A and B are 13218.9 ten thousands m^3 and 7345.5 ten thousands m^3. Reservoir inflow data and water demand data are shown in Figs. 1.18 and 1.19.

For optimal operation of dualistic reservoir system, we develop the optimization model to determine the optimal 2D rule curves and allocation ratios. The risk indexes of water supply reliability *REL* and resiliency coefficient *RES* are employed

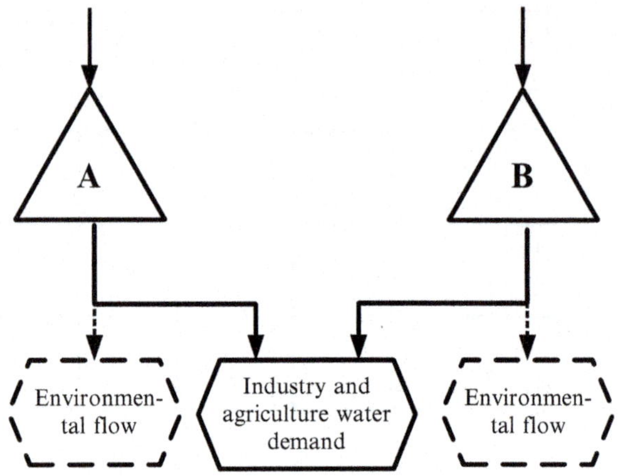

Fig. 1.17 One dualistic reservoir system in Northeast China

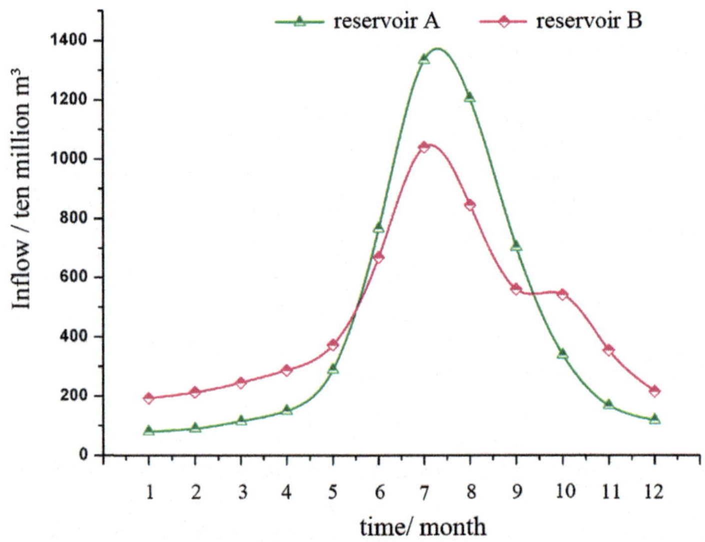

Fig. 1.18 Reservoir monthly average inflow

to evaluate the water supply disk. The analytical hierarchy process (AHP) is used to transform many indexes of different water users into one comprehensive index *R*. The optimization objective is to maximize the water supply reliability and resiliency as described in Eq. (1.10).

$$\max R = w_{\text{indu}} \ (\omega_1 REL_{\text{indu}} + \omega_2 \ RES_{\text{indu}}) + w_{\text{agri}} \ (\omega_1 \ REL_{\text{agri}} + \omega_2 RES_{\text{agri}}) \tag{1.10}$$

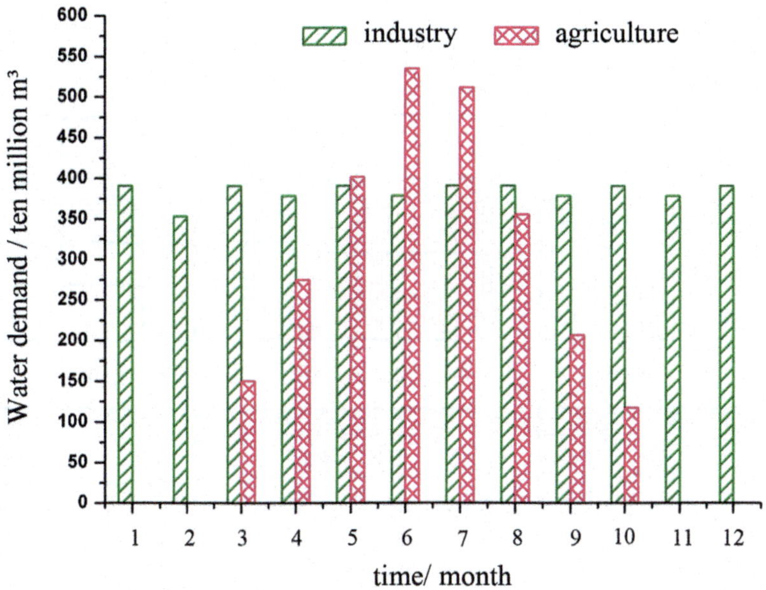

Fig. 1.19 Monthly average water demand

where REL_{indu} and REL_{agri} are the water supply reliability for industry and agriculture water demand, RES_{indu} and RES_{agri} are the water supply resiliency coefficient for industry and agriculture water demand. ω_1 and ω_2 are the weighting factors between different water supply risk indexes, w_{indu} and w_{agri} are the weighting factors between industry and agriculture.

The constraints of the optimization model include the water balance equation, reservoir storage capacity, hedging factor requirement and hedging rule curve position requirement. The optimization model is solved using the SCE algorithm based on the simulation model combined with optimization algorithm.

The optimal solution of the model consists of 2D reservoir rule curves with variable allocation ratios at each operation period in Fig. 1.20 and the allocation ratios for reservoir A and reservoir B in Table 1.2. The simulation model takes 1 month as a time step and one piece of 2D reservoir rule curves is used for 2 months. In Fig. 1.20, demand 1 and demand 2 are corresponding to agriculture and industry water demand. From January to December, the rule zones of 2D reservoir rule curves present some interesting variation. Shortly, the hedging zones in drought season are larger than the ones in flood season, which can be illustrated from the perspective of reservoir inflow changes. During flood season, the reservoir storage approaches full state and there is not necessary to restrict water supply. During drought season, there is not enough water stored in reservoir system and increasing hedging chance can avoid the catastrophic water shortage in the future.

The allocation ratios for reservoir A and reservoir B are listed in Table 1.2. Due to the larger capacity and more inflows, the allocation ratios of reservoir A are

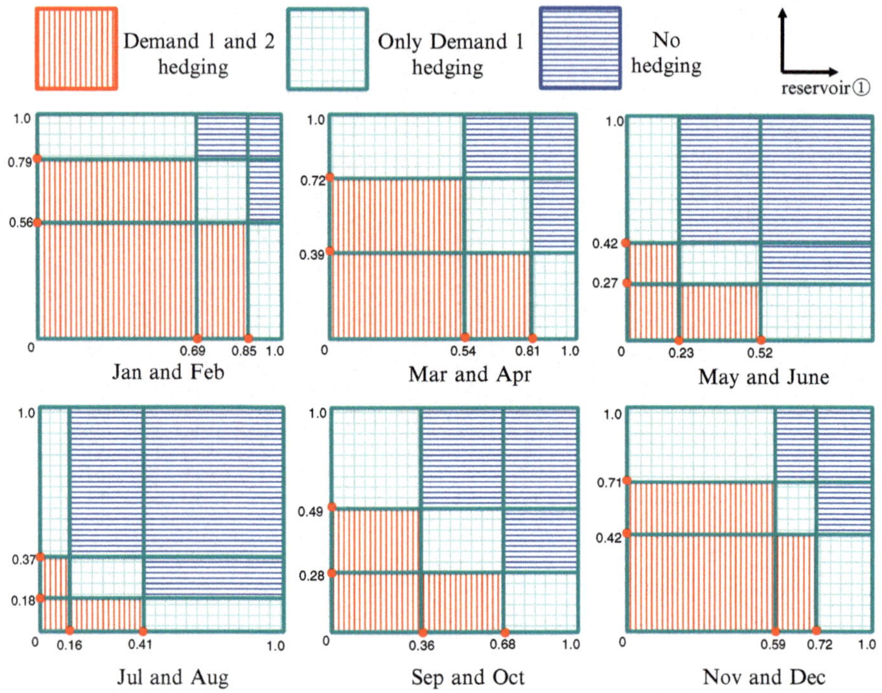

Fig. 1.20 2D reservoir rule curves with variable allocation ratios at each operation period

Table 1.2 Allocation ratios for reservoir A and reservoir B

Reservoir A	y_1	y_2	y_3	y_4	y_5	y_6	y_7	y_8	y_9
	0.52	0.61	0.63	0.54	0.64	0.67	0.57	0.69	0.71
Reservoir B	z_1	z_2	z_3	z_4	z_5	z_6	z_7	z_8	z_9
	0.48	0.39	0.37	0.46	0.36	0.33	0.43	0.31	0.29

larger than the ones of reservoir B. That means reservoir A afford more water supply task than reservoir B.

For verifying the effectiveness of 2D rule curves with variable allocation variables, we also adopt dynamic programming, 2D rule curves with fixed allocation variables and 2D rule curves with compensation regulation rule to simulate the water supply process and compute the risk indexes. The water supply results derived by different allocation rule are shown in Table 1.3. The water supply results derived by 2D rule curves with variable allocation variables are closer to the one derived by dynamic programming and better than the ones by 2D rule curves with fixed allocation variables and 2D rule curves with compensation regulation rule.

Table 1.3 Risk indexes induced by different allocation methods for common water task

Allocation methods for common water task	Industry and agriculture	Industry		Agriculture	
	R	*REL*	*RES*	*REL*	*RES*
Dynamic programming	0.92	0.99	1.00	0.84	0.68
Variable allocation methods	0.88	0.97	0.88	0.81	0.59
Fixed allocation methods	0.85	0.95	0.83	0.76	0.56
Compensation regulation	0.87	0.96	0.87	0.79	0.58

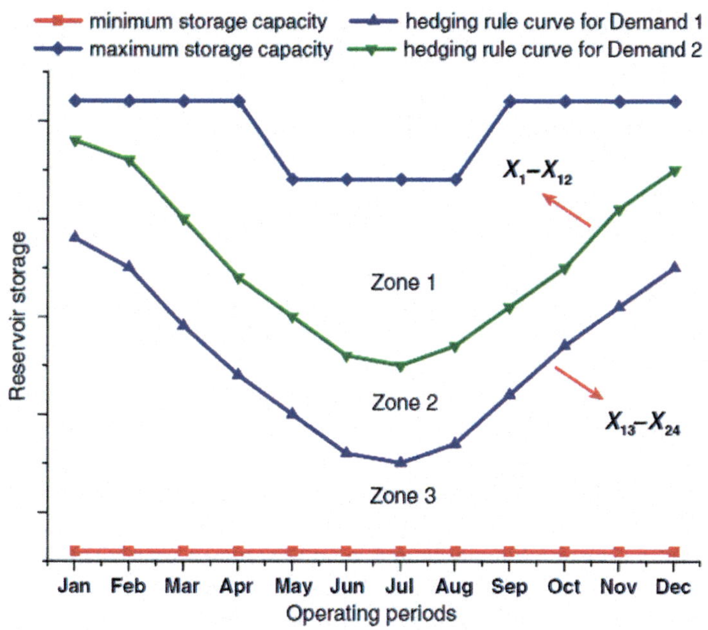

Fig. 1.21 Reservoir rule curves with one-dimension variable

3.3 Rule Curve Decision Variable Settings and Expression

The rule curve decision variable settings and expression methods can be summarized into two categories: (1) reservoir rule curves with one-dimension variables and (2) reservoir rule curves with two-dimension variables. For the first method, the rule curves is expressed using the reservoir water level or water storage at different operation periods in each year as shown in Fig. 1.21. In other words, the points at each rule curve cannot be moved on the horizon and can be only moved vertically. So the decision variables of the first method are one-dimension.

The first method has its own advantages and disadvantages. For example, the optimization model is simple and easy to build and solve. However, this method needs superabundant decision variables and the rule curves produced by this

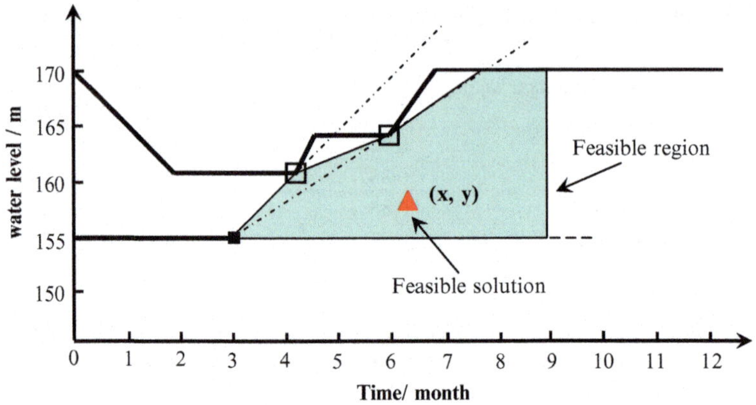

Fig. 1.22 Reservoir rule curves with two-dimension variable

method fluctuate significantly, which make the rule curves cannot be put into practices directly and restrict the effectiveness of the reservoir operating policy. For reducing the decision variable numbers, the reservoir water level at different operation periods are expressed using one variable just like Fig. 1.20. In this way, we need to make sure what periods can be aggregated, because the aggregation pattern may influence the operation results.

Based on the disadvantage of the rule curves with one dimension variables, the rule curves with two dimension variables are proposed as illustrated in Fig. 1.22. The two dimension variables means that the control point at the rule curves is determined by the reservoir water level and the time. For this method, the control points can be moved along in the horizon and vertical direction, whose location can be expressed with a combination of time and water level variables.

The second approach is able to greatly reduce the number of decision variables and decreases the optimization scale, but it increases the dimension of decision variables because the one-dimensional variables of water level is converted to two-dimensional variables in the form of a combination of time and water level. The two dimension variable settings makes feasible region of the optimization problem narrower and improve the search difficulty of optimal solution.

The advantage of this method is also obvious, e.g., fewer variables and the produced rule curves without the necessity of correction, which provides a basis for improving optimization model efficiency and finding global optimal solution. Therefore, as long as the search difficulty of optimal solution is solved, efficiency and effectiveness of operation chart optimization may be improved. Another problem worthy of noting is avoiding the crossover between the different rule curves. For dealing with the crossover problem of rule curves, the method with two dimension variables is much more difficulty than the method with one dimension variables.

4 Multi-Objective Optimization Operation of Multi-Reservoir System

The operation of multi-reservoir system presents a large variety of multi-objective optimization problems (MOOPs) that require powerful optimization tools to fully characterize the existing trade-offs. Because the traditional optimization methods use a point-by-point search approach and their outcome is a single optimal solution [60], they are not appropriate to solve MOOPs. Recently, evolutionary algorithms (EAs) and swarm intelligence techniques are becoming increasingly popular for solving MOOPs. Their population-based stochastic search techniques are more appropriate to solve MOOPs [61].

4.1 Mathematic Expression of Multi-Objective Function

Multi-reservoir joint operation can alter the temporal and spatial distribution of water resources, which drives the economy-society system development and influences the evolution of ecology and environment system. Corresponding to the four attributes "economic, social, ecological, environmental" of water resources, multi-reservoir joint operation has seven big objectives "flood control, water supply, hydropower generation, navigation, ecology, sediment and environment".

4.1.1 The Operation Objective for Flood Control

The operation objective for flood control includes minimizing the over stander rate of reservoir downstream flow, minimizing the over stander rate of reservoir downstream water level, and minimizing the reservoir highest flood level. The operation objective for flood control can be expressed mathematically as the following:

(1) Minimizing the over stander rate of reservoir downstream flow

$$\min\ (\mathrm{flood}_1) = \frac{\sum_{t=1}^{m} 1\$\left(Q_t > Q_{std,flood}\right)}{m} \tag{1.11}$$

where $Q_{std,\ flood}$ is the reservoir downstream standard flow for the downstream protect objective; Q_t is the reservoir downstream flow at the location of protect objective; $1\$(Q_t > Q_{std,\ flood})$ means if $Q_t > Q_{std,\ flood}$, the count of number is equals to 1; else the count of number is equals to 0. $\sum_{t=1}^{m} 1\$\left(Q_t > Q_{std,flood}\right)$ is the total periods of the reservoir downstream flow at the location of protect objective over the reservoir downstream standard flow. Flood_1 is the over stander rate of reservoir downstream flow.

(2) Minimizing the over stander rate of reservoir downstream water level

$$\min\left(\text{flood}_2\right) = \frac{\sum_{t=1}^{m} 1\$\left(Z_t > Z_{std,flood}\right)}{m} \tag{1.12}$$

where $Z_{\text{std, flood}}$ is the reservoir downstream standard water level for the downstream protect objective; Z_t is the reservoir downstream water level at the location of protect objective; $1\$(Z_t > Z_{\text{std, flood}})$ means if $Z_t > Z_{\text{std, flood}}$, the count of number is equals to 1; else the count of number is equals to 0. $\sum_{t=1}^{m} 1\$\left(Z_t > Z_{\text{std,flood}}\right)$ is the total periods of the reservoir downstream water level at the location of protect objective over the reservoir downstream standard water level. Flood_2 is the over stander rate of reservoir downstream water level.

(3) Minimizing the reservoir highest flood level

$$\min\left(\text{flood}_3\right) = \max\left(Z_t\right) \tag{1.13}$$

where Z_t is the reservoir water level at the flood season, Flood_3 is the reservoir highest flood level.

4.1.2 The Operation Objective for Water Supply

The operation objective for water supply consists of maximizing the water supply amount at the drought season and minimizing the reduction of reservoir outflow into the river after the flood season.

(1) Maximizing the water supply amount at the drought season

$$\max\left(\text{WSup}_1\right) = \sum_{t \in \text{drought season}} Q_t * t \tag{1.14}$$

where Q_t is the water supply amount at the period t in the drought season, WSup_1 is the total water supply amount at the drought season.

(2) Minimizing the reduction of reservoir outflow into the river after the flood season

$$\min\left(\text{WSup}_2\right) = \max \overline{Q_n - Q_t} \tag{1.15}$$

where Q_t is the reservoir outflow into the river after the flood season, Q_n is the river natural flow after the flood season, WSup_2 is the reduction of reservoir outflow into the river after the flood season.

4.1.3 The Operation Objective for Hydropower Generation

The operation objective for hydropower generation mainly consists of maximizing the hydropower generation amount and maximizing the reliability of hydropower generation, which can be expressed mathematically as the following:

(1) Maximizing the hydropower generation amount

$$\max\left(\mathrm{EPow}_1\right)=\sum_{t=1}^{m}\mathrm{N}_t * t \tag{1.16}$$

where N_t is the hydropower generated output, $EPow_1$ is the total hydropower generation amount at the total operation period.

(2) Maximizing the reliability of hydropower generation

$$\max\left(\mathrm{EPow}_2\right)=\frac{\sum_{t=1}^{m}1\$\left(\mathrm{N}_t>\mathrm{N}_{pro}\right)}{m} \tag{1.17}$$

where N_{pro} is the hydropower generation guarantee output at the stage t, N_t is the actual hydropower generation at the stage t, $1\$(N_t > N_{std,\ flood})$ means if $N_t > N_{std,\ flood}$, the count of number is equals to 1; else the count of number is equals to 0. $\sum_{t=1}^{m} 1\$\left(N_t > N_{pro}\right)$ is the total periods of the actual hydropower generation over he hydropower generation guarantee output, $EPow_2$ is the reliability of hydropower generation.

4.1.4 The Operation Objective for Navigation

The operation objective for navigation includes two main objectives: maximizing the reliability of flow for navigation and maximizing the reliability of water depth for navigation. They can be expressed mathematically as the followings:

(1) Maximizing the reliability of flow for navigation

$$\max\left(\mathrm{Navi}_1\right)=\frac{\sum_{t=1}^{m}1\$\left(\mathrm{Q}_t>\mathrm{Q}_{pro,\ navi}\right)}{m} \tag{1.18}$$

where $Q_{pro,\ navi}$ is the river flow required for the navigation purpose at the stage t, Q_t is the river actual flow at the stage t, $1\$(Q_t > Q_{pro,\ navi})$ means if $Q_t > Q_{pro,\ navi}$, the count of number is equals to 1; else the count of number is equals to 0. $\sum_{t=1}^{m} 1\$\left(Q_t > Q_{pro,navi}\right)$ is the total periods of the actual river flow over the river flow required for the navigation purpose, $Navi_1$ is the reliability of flow for navigation.

(2) Maximizing the reliability of water depth for navigation

$$\max\left(Navi_2\right)=\frac{\sum_{t=1}^{m}1\$\left(H_t>H_{pro,\ navi}\right)}{m} \tag{1.19}$$

where $H_{pro,\ navi}$ is the river water depth required for the navigation purpose at the stage t, H_t is the river actual water depth at the stage t, $1\$(H_t > H_{pro,\ navi})$ means if $H_t > H_{pro,\ navi}$, the count of number is equals to 1; else the count of number is equals to 0. $\sum_{t=1}^{m}1\$\left(H_t>H_{pro,navi}\right)$ is the total periods of the actual river water depth over the river water depth required for the navigation purpose, $Navi_2$ is the reliability of water depth for navigation.

4.1.5 The Operation Objective for Ecology

The operation objective for ecology is composed of two main objectives: maximizing the reliability of suitable flow for ecology and maximizing the reliability of minimum flow for ecology. They can be expressed mathematically as the followings:

(1) Maximizing the reliability of suitable flow for ecology

$$\max\left(Eco_1\right)=\frac{\sum_{t=1}^{m}1\$\left(Q_t>Q_{pro,\ eco}\right)}{m} \tag{1.20}$$

where $Q_{pro,\ eco}$ is the river flow to satisfy the suitable ecology flow requirement at the stage t, Q_t is the river actual flow at the stage t, $1\$(Q_t > Q_{pro,\ eco})$ means if $Q_t > Q_{pro,\ eco}$, the count of number is equals to 1; else the count of number is equals to 0. $\sum_{t=1}^{m}1\$\left(Q_t>Q_{pro,eco}\right)$ is the total periods of the actual river flow over the river flow required for suitable ecology, Eco_1 is the reliability of suitable flow for ecology.

(2) Maximizing the reliability of minimum flow for ecology

$$\max\left(Eco_2\right)=\frac{\sum_{t=1}^{m}1\$\left(Q_t>Q_{min,\ eco}\right)}{m} \tag{1.21}$$

where $Q_{min,\ eco}$ is the river flow to satisfy the minimum ecology flow requirement at the stage t, Q_t is the river actual flow at the stage t, $1\$(Q_t > Q_{min,\ eco})$ means if $Q_t > Q_{min,\ eco}$, the count of number is equals to 1; else the count of number is equals to 0. $\sum_{t=1}^{m} 1\$\left(Q_t > Q_{min,eco}\right)$ is the total periods of the actual river flow over the minimum ecology flow requirement, Eco_2 is the reliability of minimum flow for ecology.

4.1.6 The Operation Objective for Sediment

The operation objective for sediment is mainly to satisfy the sediment discharge rate requirement. It can be expressed mathematically as the followings:

$$\max\left(Sed_1\right) = \frac{Sed_{out}}{Sed_{in}} \tag{1.22}$$

where Sed_{in} is the sediment amount into the reservoir at the stage t, Sed_{out} is the sediment amount out of the reservoir at the stage t; Sed_1 is the sediment discharge rate.

4.1.7 The Operation Objective for Environment

The operation objective for environment is composed of two main objectives: maximizing the achievement ratio of water quality and minimizing the eutrophia rate, which can be expressed mathematically respectively as the followings:

(1) Maximizing the achievement ratio of water quality

$$\max\left(WQ_1\right) = \frac{\sum_{t=1}^{m} 1\$\left(WQ_t < WQ_{std,\ wq}\right)}{m} \tag{1.23}$$

where $WQ_{std,\ wq}$ is water quality standard for some indexes, WQ_t is water quality index at the stage t, $1\$(WQ_t > WQ_{std,\ wq})$ means if $WQ_t > WQ_{std,\ wq}$, the count of number is equals to 1; else the count of number is equals to 0. $\sum_{t=1}^{m} 1\$\left(WQ_t < WQ_{std,\ wq}\right)$ is the total periods of water quality satisfies the water quality standard, WQ_1 is the achievement ratio of water quality.

(2) Minimizing the eutrophia rate

$$\max\left(WQ_2\right) = \frac{\sum_{t=1}^{m} 1\$\left(WEutro_t > WEutro_{std,\ wq}\right)}{m} \tag{1.24}$$

where $WEutro_{std,wq}$ is the eutrophia standard, $WEutro_t$ is the eutrophia index at the stage t, $1\$(WEutro_t > WEutro_{std,wq})$ means if $WEutro_t > WEutro_{std,wq}$, the count of number is equals to 1; else the count of number is equals to 0. $\sum_{t=1}^{m} 1\$(\text{WEutro}_t > \text{WEutro}_{\text{std, wq}})$ is the total periods of the water eutrophic level over the eutrophia standard, WQ_2 is the eutrophia rate.

4.2 Multi-Objective Optimization Algorithm

As a novel member of the family of swarm intelligence techniques, particle swarm optimization (PSO) has been widely extended to solve MOOPs [62–65]. The non-dominated sorting particle swarm optimization (NSPSO) algorithm [65] is one kind of multi-objective PSO algorithms (MOPSO), which combines single-objective PSO with the operations of comparison and sort based on non-domination.

For avoiding premature convergence of the NSPSO, an improved NSPSO algorithm (I-NSPSO) is developed by incorporating the multi-population mechanism into the NSPSO algorithm in this study. In order to investigate the algorithm effectiveness, two benchmark test problems are adopted to compare the I-NSPSO with the standard NSPSO. The I-NSPSO is also employed to optimize the proposed operating policy for the multi-reservoir system located at Taize River basin in China.

PSO algorithm [66] is one of the newest techniques within the family of EAs. As a population-based search algorithm, PSO uses the local and global search capabilities to find solutions of better quality based on the simulation of the social behavior of birds within a flock. Due to the simplicity and good efficiency that the algorithm presents for single-objective optimization, PSO has been widely extended to solve MOOPs [67]. The NSPSO, one of typical pareto-based approaches, is proposed by some researchers and its good performance for solving MOOPs has also been demonstrated [65].

Inspired by the algorithm of Shuffled Complex Evolution (SCE) [68], the idea of multi-population is adopted by many researchers to elevate the ability of EAs in finding the global optimum and avoiding premature convergence. For example, Chen and Chang [69] applied a real-coded multi-population genetic algorithm (GA) to multi-reservoir operation. Jalali et al. [70] proposed a special version of multi-colony algorithm to minimize the possibility of losing global optimum domain. Jiang et al. [71] presented an improved particle swarm optimization (IPSO), which owns several sub-swarms.

Based on the previous work, an improved NSPSO algorithm (I-NSPSO) is presented here by incorporating the multi-population mechanism into the NSPSO to improve its ability in finding the optimal trade-off curves or surfaces for MOOPs. The computational procedures of the I-NSPSO are illustrated below:

Step 1: Generate an initial population (population size $= N$) and velocity for each individual (agent or particle) in a feasible space; set the maximum speed $V_i^{\max}$ ($V_i^{\max} =$ its upper bound minus lower bound) for variables; determine the value of the inertia weight factor ω, the weighting factors c_1 and c_2. And get the initial multi-objective values for each particle, the initial particles are taken as the *pbest* temporarily.

Step 2: Sort the population based on the non-dominated and crowding distance ranking, save L non-dominated solutions as *gbest* in one external archive.

Step 3: Partition the main swarm A into p sub-swarms $\left(A^1, A^2, \cdots A^p\right)$, based one objective vector in order of increasing function value, each containing M points $(N = p \cdot M)$, such that: $A^k = \left\{X_j^k, f_j^k | X_j^k = X_{k+p\cdot(j-1)}, f_j^k = f_{k+p\cdot(j-1)}, j = 1, \ldots, m\right\}, k = 1, \ldots p$. The non-dominated solutions in the external archive as *gbest* are also divided into p parts for each sub-swarm.

Step 4: Evolve each sub-swarm A^k separately using the standard PSO. Select one solution randomly from the external archive *gbest* of sub-swarm A^k as the global best one for updating particles. Determine the *pbest* for each agent: if the new individual dominates the previous one, replace the *pbest* with the current individual; if the new individual is dominated by the previous one, keep the previous one as the *pbest*; if both of them are incomparable, choose one of them as the *pbest* randomly. Update and save Q non-dominated solutions as sub-swarm *gbest* in its internal archive $(Q = L/P)$.

Step 5: When update each sub-swarm up to the maximum iteration time, shuffle all the sub-swarms and their non-dominated solutions to obtain one new main swarm and L new non-dominated solutions.

Step 6: Combine the offspring and parent population of the global best individual to form extended population of $2L$. Compare the extended population based on the non-dominated and crowding distance ranking. Fill the new population of size L into the external archive *gbest*.

Step 7: Check the stopping criteria. If the stopping criteria are satisfied, stop. Otherwise, return to step 3.

The constraint handling mechanism proposed by Deb et al. [72] is adopted in this study. In this approach, a solution i is said to be a constrained-dominate solution j if any of the following conditions hold good: ① solution i is feasible and solution j is not, ② both solution i and solution j are infeasible, but solution i has a smaller overall constraint violation, ③ both solution i and solution j are feasible, but solution i dominates solution j.

In order to verify the efficiency of the I-NSPSO, two benchmark functions are performed, which were proposed by Deb [73] and used by Reddy and Kumar [61]. The objective functions, variable bounds and constraints of the test problems are summarized in Table 1.4. To evaluate the performance of the I-NSPSO, the standard NSPSO algorithm is employed to optimize the two benchmark functions for comparison. For the two algorithms, the parameter ω is decreasing from 0.9 to 0.4 linearly with the sub-swarm iteration times. The weighting factors c_1 and c_2 both equal 2.0. In the I-NSPSO algorithm, the number of sub-swarms p is 4 and the number of non-dominated solutions in the main-swarm external archive L is

Table 1.4 Test problems

Problem	Objective functions	Variable bounds	Constraints
BNH	*Minimize* $f_1(x) = 4x_1^2 + 4x_2^2$ $f_2(x) = (x_1 - 5)^2 + (x_2 - 5)^2$	$x_1 \in [0, 5]$ $x_2 \in [0, 3]$	$g_1(x) = (x_1 - 5)^2 + x_2^2 \leq 25$ $g_2(x) = (x_1 - 8)^2 + (x_2 + 3)^2 \geq 7.7$
KITA	*Maximize* $f_1(x) = -x_1^2 + x_2$ $f_2(x) = 0.5x_1 + x_2 + 1$	$x_1 \in [0, 3]$ $x_2 \in [6, 6.5]$	$g_1(x) = x_1/6 + x_2 - 6.5 \leq 0$ $g_2(x) = 0.5x_1 + x_2 - 7.5 \leq 0$ $g_3(x) = 5x_1 + x_2 - 30 \leq 0$

Table 1.5 Resulting statistics by the I-NSPSO and the NSPSO for test problems BNH and KITA

Test problem	Statistic	Set coverage metric (SC)		Spacing metric (SP)	
		SC(I-NSPSO, NSPSO)	SC(I-NSPSO, NSPSO)	I-NSPSO	NSPSO
BNH	Best	0.2153	0.1745	0.1900	0.3942
	Worst	0.1136	0.0854	0.4138	0.8140
	Mean	0.1724	0.1278	0.2622	0.6008
	Variance	0.0004	0.0006	0.0016	0.0028
KITA	Best	0.3060	0.2013	0.0360	0.0450
	Worst	0.1502	0.1103	0.1362	0.4192
	Mean	0.2135	0.1642	0.0835	0.2635
	Variance	0.0012	0.0018	0.0008	0.0015

40, which illustrate the number of the non-dominated solutions in the sub-swarm internal archive Q is 10. The population size (N) of the main swarm is 200 and the one of each sub-swarm is 50. The parameters in NSPSO algorithm have the same value as the ones of the main swarm in the I-NSPSO algorithm.

For the MOOPs, both of them are very important to guarantee the convergence to the Pareto optimal set and to maintain the diversity of solutions. Table 1.5 presents the best, worst, mean and variance values of the two performance metrics (SC and SP) obtained from 25 independent runs using the I-NSPSO and the NSPSO algorithms. The set coverage metric SC (A, B) provides a measure of how many solutions of B are dominated by A and vice versa. For example, the value SC (A, B) = 1 means that all solutions in B are weakly dominated by A, while SC (A, B) = 0 represents the situation when none of the solutions in B are weakly dominated by A. From Table 1.5, it can be seen that with respect to the SC metric, the performance of the I-NSPSO algorithm is better than the one of the NSPSO algorithm. This metric shows the better efficiency of the I-NSPSO in achieving convergence to true Pareto optimal fronts than the NSPSO. About the spacing metric (SP), the I-NSPSO gives smaller SP values for both the test problems than the NSPSO. The smaller SP indicates that the algorithm gives better distribution of solutions with good diversity. The obtained optimal trade-offs for both the test problems are shown in Fig. 1.23. Therefore, it can be concluded that the I-NSPSO algorithm has a good ability for solving the two constrained MOOPs.

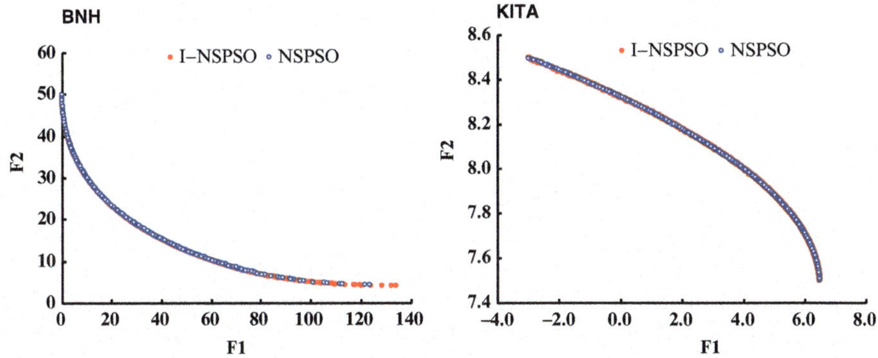

Fig. 1.23 Non-dominated solutions obtained by the I-NSPSO and the NSPSO for the test problems BNH and KITA

4.3 Multi-Objective Optimization Operation of Dan Jiangkou Reservoir for Water Transfer

Dan Jiangkou Reservoir provides water source for the middle line of South to North Water Transfer Project in China. At present, the reservoir is operated according to the original rule curves proposed during the reservoir design period. The amount of water transfer varies significantly within the year and presents large variation during the interannual period, which is not convenient for water resources management in the water intake area. For increasing the amount of water transfer to the recipient area and reducing the water spills of Dan Jiangkou Reservoir, the reservoir rule curves are needed to be updated again. Therefore, this study develops a multi-objective optimization model to derive new rule curves for satisfying the multi-objective requirements.

In the optimization model, there are three objectives to be taken into consideration, which includes maximizing the annual average amount of water supply, maximizing the annual minimum amount of water supply and maximizing the guarantee ratio of low flow. In this section, the water supply means the water transfer for the water recipient areas. Each item of the objectives in the optimization model is described mathematically from Eqs. (1.25) to (1.28).

$$Max\ W_{avg} = \frac{1}{Y}\sum_{y=1}^{Y}\sum_{t=1}^{T} W_{y,t}\Delta t \tag{1.25}$$

$$Max\ W_{\min} = \min\left(y, \sum_{t=1}^{T} W_{y,t}Dt\right) \tag{1.26}$$

$$Max\ P_W = \frac{\sum_{y=1}^{Y} \sum_{t=1}^{T} 1\$(W_{y,t} \geq W_{pro})}{Y \times T} \times 100\% \tag{1.27}$$

$$\begin{aligned} t &= 1, 2, \ldots, T \\ y &= 1, 2, \ldots, Y \end{aligned} \tag{1.28}$$

where: W_{avg} is the annual average amount of water supply, W_{min} is the annual minimum amount of water supply, $W_{y,t}$ is the amount of water supply at the period t in the year y, P_w is the guarantee ratio of water supply, Y is the year serial number, T is the computation serial period in each year. The constraints of the model consist of water balance equation, reservoir capacity, water transfer capacity of the canal. The decision variables are the rule curve positions and the water transfer flow at each operation zone. The optimization model is solved using the multi-objective optimization algorithm mentioned in the last section. The optimal rule curves obtained are shown in Fig. 1.24. The rule curves divide the reservoir capacity into four zones: the increment water supply zone, the normal water supply zone, the hedging zone 1 and the hedging zone 2. The number in each zone means the flow of water transfer from the reservoir at unit time.

The Pareto surface of the three objectives can be obtained by solving the multi-objective optimization model using the multi-objective optimization algorithm as given in Fig. 1.25. The relationship of every two objectives are projected in the two axes surface in Fig. 1.25. It can be found out that the original solution from the original rule curves is dominated by several non-dominated solutions derived by the multi-objective optimization algorithm. The rule curves shown in Fig. 1.24 corresponds to one of the non-dominated solutions in Figs. 1.25 and 1.26, which can help the decision maker choose the most suitable operation policy.

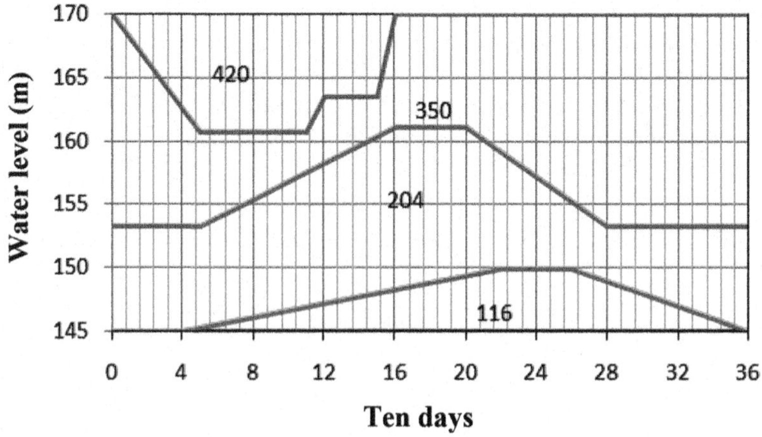

Fig. 1.24 The rule curves for Dan Jiangkou reservoir operation

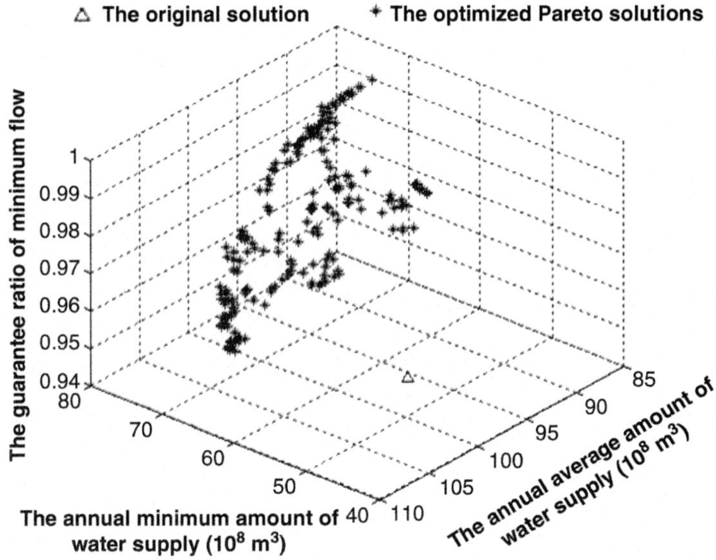

Fig. 1.25 The Pareto surface among the three operation objectives of Dan Jiangkou Reservoir

5 Multi-Reservoir Operation in Inter-Basin Water Transfer Project

The uneven distribution of water resources and imbalanced water demand in different regions make it inevitable to construct an inter-basin water transfer (IBWT) project across regional boundaries. Creation of storage and inter-basin transfer of water from surplus to deficit regions are rational options to overcome the problems caused by the mismatch of supply and demands, which can increase the resilience of the water system and decrease the risk of shortages [74].

Presently, the research on IBWT mainly focuses on optimal allocation of transferable water resources [75], alternative evaluation [76, 77], uncertainty analysis [78], Chen and Chang [79], strategic choice methodology in conflicts over water resources management by IBWT [80], hydrological impact [81] and inter-basin water transfer-supply model [82]. For example, Sadegh et al. [75] developed a new methodology based on crisp and fuzzy Shapley games for optimal allocation of inter basin water resources. Matete and Hassan [76] proposed a generalized analytical framework that can be applied to integrate environmental sustainability aspects into economic development planning in the case of exploiting water resources through IBWT. Li et al. [77] presented a new optimization method using fuzzy pattern recognition to appraise the water-supply decision schemes in inter-basin diversion systems. Dosi and Moretto [78] investigated the storage capacity and optimal guaranteed deliveries in IBWT, taking into account the uncertain nature of water surplus. Chen and Chang [79] used fuzzy sets for

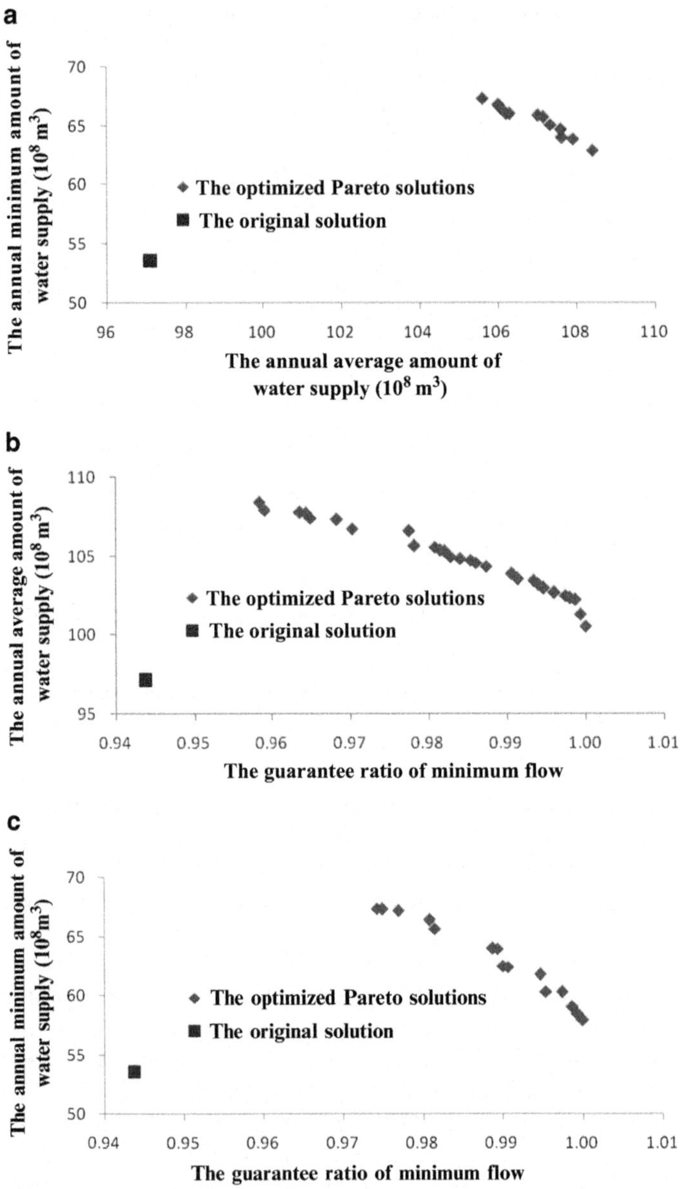

Fig. 1.26 The Pareto front between every two operation objectives of Dan Jiangkou Reservoir. (**a**) The relationship between the annual average and minimum amount of water supply. (**b**) The relationship between the annual average amount of water supply and the guarantee ratio of minimum flow. (**c**) The relationship between the annual minimum amount of water supply and the guarantee ratio of minimum flow

incorporating objective and subjective uncertainties to address the complexity in determining water resources redistribution alternatives in a trans-boundary channel-reservoir system. Carvalho and Magrini [80] analyzed the application of the strategic choice methodology in a dispute over transferring water between two river basins. Bonacci and Andric [81] described the hydrological changes of two rivers caused by IBWT and reservoir development. Xi et al. [82] developed a new inter-basin water transfer-supply and risk assessment model with consideration of rainfall forecast information.

As the most important facilities in IBWT project, reservoirs play an important role in storing and regulating water resources to meet certain requirements. The above review of previous work indicates that the interest of researchers on IBWT has spread widely throughout many respects, but there has been limited study on multi-reservoir operating policy in IBWT project, especially on the water transfer rule. In this section, a set of water transfer rule is proposed to direct the system manager under what condition to transfer water from the abundant to scare regions.

Regarding the reservoir operating rule for water supply, there has been much research and several types of reservoir operating rules have been proposed and discussed. Among these policies, the Standard Operating Policy (SOP) is a simple and the most often used operating policy. According to the SOP, reservoirs release as much water as they can provide to meet the target delivery. The SOP is the optimal operating policy with an objective to minimize the total deficit over the time horizon [83]. Besides, different forms of the Linear Decision Rule (LDR) are also applied widely in the practical operation of reservoirs. The LDR is formulated to assume the releases linearly related to storage and decision parameters and is usually optimized with linear programming [84]. Hedging rule, normally used for rationing the water supply during droughts, distributes deficits over a longer horizon to improve the efficiency of reservoir operation [54, 85]. During periods of drought, system managers would rather incur a sequence of smaller shortages in water supply than one potential catastrophic shortage [86]. Due to its good ability to deal with reservoir operation problem during droughts, hedging rule has attracted much attention of researchers [5–7, 9, 87]. In this section, hedging rule is adopted in the form of hedging rule curves for individual reservoirs in IBWT project to control their releases.

About multi-reservoir operation model, many advances in this area have been made during recent years. A lot of optimization methods are designed and applied to prevail over the high dimension, dynamic, nonlinear, multi-objective and stochastic characteristics of reservoir systems [14], which include implicit stochastic optimization, explicit stochastic optimization, real-time control with forecasting, and heuristic programming models. Increased application of heuristic programming to be linked directly with trusted simulation models is a great advantage. Fuzzy rule-based systems and neural networks may alleviate the difficulty in inferring operating policies from implicit stochastic optimization models. The detailed work and recent advancement on optimal operation of multi-reservoir system are scrutinized by Labadie [14].

Despite great advances on the study of reservoir operation, it can be observed, from the above review, that the problems of multi-reservoir water supply and water transfer in IBWT project have seldom been taken into consideration together. This may influence the utilization efficiency of water resources, because an improper water transfer will not only bring negative effect on the water supply of the reservoir(s) in water-exporting region but also can increase water spills of the reservoir(s) in water-importing region. Therefore, the water-supply rule in IBWT project should match up with the water transfer rule and both of them are ought to be considered at the same time.

For multi-reservoir operation problem in inter-basin water transfer-supply project, it involves decision makers at two distinct levels with a hierarchical relationship between them. The decision process involves two different decision makers, who represent the multi-reservoir system manager in charge of water transfer and the individual reservoir manager in charge of water supply, respectively. The system manager, which is at the upper level of the hierarchy, controls the distribution of water resources among water exporting and importing regions using a set of water transfer rule. The individual reservoir manager, at the lower level of the hierarchy, controls the water-supply process by hedging rule, which is influenced by the decision of the upper decision-maker. Both, in general, do not cooperate because of different optimization purposes. These characteristics make this problem unsuitable for modeling by standard mathematical programming. They are more likely to be modeled using bi-level programming (BLP), which has been proposed in the literature as an appropriate model for hierarchical decision processes with two non-cooperative decision makers, the leader at the upper level of the hierarchy and the follower at the lower level [88].

This section proposes a bi-level programming model for multi-reservoir operating policy in inter-basin water transfer-supply project. And a set of water transfer rule based on the storage of individual reservoir in the system is presented in this study. In this bi-level programming model, the leader wants to allocate trans-boundary water resources in accordance with the planned water transfer amount to satisfy water demand in every region and to reduce water spills of the system. The follower pursues the best water supply; meanwhile, the action of water transfer occurs. In other words, the objective of the leader is to minimize both the system water spills and the deviation of the actual transferred water from the water transfer target. The objective of the follower is to minimize water shortage or maximize the amount of water supply. The water transfer rule curves are decision variables of the leader in the hierarchical process, which determine the conditions to start water transfer or not. Besides, hedging rule curves are decision variables of the follower, which relate to some indexes reflecting water-supply efficiency. An improved particle swarm optimization algorithm (IPSO) proposed by Jiang et al. [71] is adopted in this section to solve the bi-level model. The East-to-West inter-basin water transfer project of Liaoning province in China is taken as a case study to verify the reasonability and efficiency of the proposed bi-level model and the water transfer rule.

5.1 Bi-Level Programming Model Theory

Decision-making in most real life problems fits within the framework of a leader–follower or Stackelberg game [89]. Such a game can be expressed mathematically by bi-level model, which has been proposed for dealing with decision process involving two decision makers with a hierarchical structure, the leader at the upper level and the follower at the lower level. Each decision maker controls a set of variables subject to a set of constraints and seeks to optimize his own objective function. Once the leader sets the value of his variables, the follower reacts by providing the value of his controlled variables and optimizes his objective function. In general, the leader can influence but cannot control the behaviors of the follower. The goal of the leader is to optimize his objective function and incorporate the reaction of the follower to the leader's course of action within the optimization scheme. In other words, the leader optimizes his objective function taking into account his own constraints and the reaction of the follower, who has the freedom of choosing his best decision [88]. General bi-level optimization model can be formulated as:

$$\begin{aligned} \min_{x} \quad & F(x, y) \\ s.t. \quad & G(x, y) \leq 0 \\ & \min_{y} \quad f(x, y) \\ & s.t. \quad g(x, y) \leq 0 \end{aligned} \tag{1.29}$$

where F = objective function of the upper-level decision maker (system manger); x = decision vector of the upper-level decision maker; G = constraint set of the upper-level decision vector; f = objective function of the lower-level decision maker; y = decision vector of the lower-level decision maker; g = constraint set of the lower-level decision vector.

It can be observed that many decision-making problems in real life can be described as Stackelberg game. Therefore, bi-level programming model has been widely used to deal with such practical problems as transportation control and management [90], production–distribution planning [88], pricing control [91] and aid in specification of surface and ground water related policies [92]. Among those problems, the bi-level model is studied most extensively in the area of transportation control and management. In the upper level of BLP model for transport network, the traffic planner makes decisions regarding management, control, and design to improve the performance of the system. In the lower level, the network users make choices with regard to rout, travel mode, origin and destination of their travel in response to the upper-level decision [90]. This type of bi-level traffic modeling and optimization problem has emerged as an important progress in handling transportation problems. Typical examples include traffic signal setting [93], optimal road capacity improvement [94], estimation of origin–destination matrices from traffic counts [95], ramp metering in freeway-arterial corridor [96], and optimization of road tolls [97].

In addition, Calvete et al. [88] addressed a hierarchical production–distribution planning problem, in which there were two different decision makers controlling the production and the distribution process, respectively. Marcotte et al. [91] developed a bi-level pricing model for a service firm, which explicitly took into account customer behavior as well as the underlying network topology of the problem. Bisschop et al. [92] applied two-level linear programming to aid in the specification of surface and ground water policies in Pakistan. In the model, the government played the role of the policy maker, while the farmers played the role of policy receivers. The government decided on surface water allocations, and set taxes or subsidies. The farmers, in turn, reacted to the setting of these policy instruments by using water (both surface and ground water) and choosing cropping patterns so as to maximize their own net income. Although bi-level model has been applied widely in practice, it has been seldom used to handle multi-reservoir operation problem in inter-basin water transfer-supply project.

5.2 *Bi-Level Model for Multi-Reservoir Operation in Inter-Basin Water Transfer Project*

In the proposed bi-level model for multi-reservoir operating policy, the upper level model optimizes the water transfer rule to distribute water resources between exporting and importing regions as the planned scheme and to minimize the total water spills of the multi-reservoir system. The lower level model optimizes the water supply rule to obtain the best water supply quality under the condition of water transfer. In order to describe the development process of the bi-level model, an inter-basin water transfer-supply project consisting of three reservoirs connected by water transfer pipelines is taken as an example, just like the one illustrated in Fig. 1.27.

5.2.1 The Upper Level Model

From the current literature on multi-reservoir operating policy, it is observed that there has been quite little research carried out on water transfer rule to direct the multi-reservoir system manger under what conditions to transfer water from the exporting reservoir to the importing reservoir. In this section, a set of water transfer rule is proposed based on the storage of each member reservoir in the system, shown schematically in Fig. 1.28. In this way, the inter-related dynamic water storage of each reservoir is taken as the main factor influencing the decision of water transfer when lack of future inflow information.

As shown in Fig. 1.28, the active storage of each reservoir between the maximum and minimum storages is divided into two parts: zone I and zone II. When the reservoir storage stays in zone I, it means that there is sufficient water in the

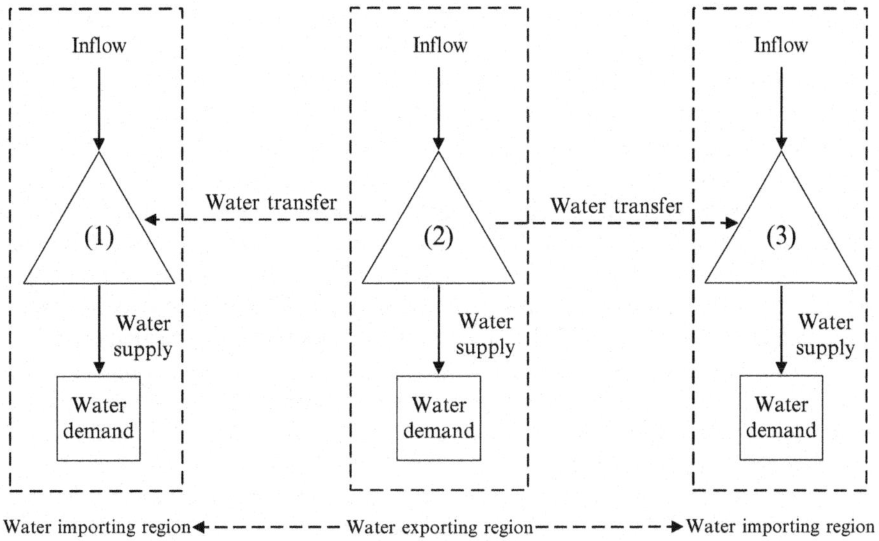

Fig. 1.27 The layout of an inter-basin water transfer-supply project

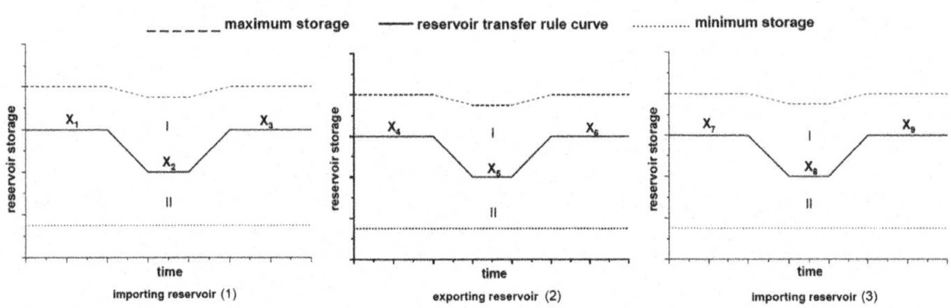

Fig. 1.28 The water transfer rule curve based on the storage of each member reservoir

reservoir. If the reservoir storage stays in zone II, it means that there is scare water in the reservoir. During the operation of the multi-reservoir system, there will be such 8 combinations occurring according to the three-reservoir storage as (I,I,I), (I,I,II), (I,II,I), (I,II,II), (II,I,I), (II,I,II), (II,II,I), (II,II,II). The detailed judgment procedures of water transfer are described as below. First, the storage of the exporting reservoir should be concerned. If the exporting reservoir storage stays in zone II, it means that there is not enough water to export. At this moment, the action of water transfer is stopped at all regardless of whatever storage in the importing reservoir so as to guarantee the water supply in water exporting region. This condition corresponds to 4 combinations of (II,I,I), (II,I,II), (II,II,I), (II,II,II).

If the exporting reservoir storage stays in zone I, it means the stored water in exporting reservoir is enough to be transferred into the importing region. Under this condition, the storage of the importing reservoir should be also paid an attention to. Because if the water level of the importing reservoir is quite high, the transferred water may produce a lot of water spills, which is obviously unreasonable. At present, the water transfer rule can be categorized into three conditions. First, if both the storages of the two importing reservoirs stay in zone I corresponding to (I,I, I), the action of water transfer is stopped. Second, if both the storages of the two importing reservoirs stay in zone II corresponding to (I,II,II), the action of water transfer is carried out. The transferred water amount from the exporting reservoir at this period is as much as the pipeline's transporting ability and it is divided into the two importing reservoirs according to an allocation ratio. Third, if the storage of one of the two importing reservoirs is in zone II and the other one is in zone I, corresponding to (I,I,II) and (I,II,I), the action of water transfer is carried out all the same. The transferred water amount from the exporting reservoir at this period is also as much as the pipeline's transporting ability and it is all transported into the importing reservoir whose storage is in zone II. The above procedure is illustrated schematically in Fig. 1.29.

For the upper model, it pursues to achieve a trans-boundary water resources allocation as the planned scheme by water transfer to satisfy water demand in every region and to minimize the total water spills of the multi-reservoir system. To realize the objective is by means of optimizing the positions of water transfer rule curves as shown in Fig. 1.28. The mathematical formulation of the upper level model is given by Eq. (1.30).

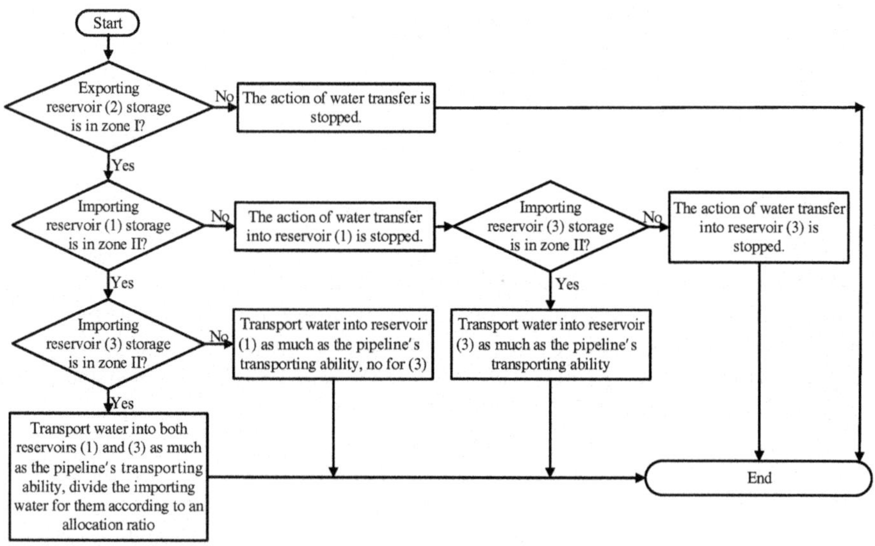

Fig. 1.29 The judgment flowchart of water transfer based on the proposed water transfer rule

$$\begin{aligned} & \min_{x} \quad F(x,y) = w_{DS} \cdot \sum_{i=1}^{m} |NDS_i - TNDS_i| + w_{SU} \cdot \sum_{i=1}^{m} SU_i \\ & s.t. \quad NDS_i = G(x,y), \ SU_i = g(x,y) \\ & \qquad ST_i^{\min} \le x_i \le ST_i^{\max}, \ ST_i^{\min} \le y_i \le ST_i^{\max} \\ & \qquad\qquad 0 \le DS \le DS_{\max} \\ & \qquad\qquad i = 1 \ldots m \end{aligned} \tag{1.30}$$

where x is the decision variable of the upper level model representing the position of water transfer rule curve during an operation period; y is the decision variable of the lower level model denoting the position of the Hedging rule curves in Fig. 1.30; both x and y are between the maximum and minimum storages; NDS_i and SU_i refer to the annual average transferred water amount and the water spill of reservoir i, which are related to water transfer rule and water supply rule and can be formulated as the function of x and y. In this study, the weighting approach is applied. For a given weight combination, single-objective optimization is used for optimization of the aggregated objective function. The weighting factors w_{DS} and w_{SU} can be obtained empirically.

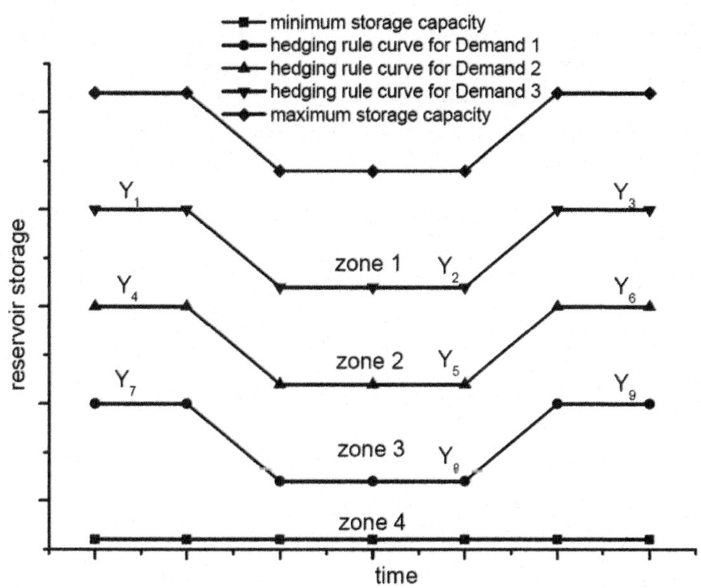

Fig. 1.30 The Hedging rule curves based on the storage of the member reservoir in the system

5.2.2 The Lower Level Model

In this section, the Hedging rule is adopted as the water supply operating rule in the lower level model, which has been discussed in different methods for reservoir operation. Srinivasan and Philipose [53] used hedging parameters, such as starting water availability, ending water availability and hedging factor (degree of hedging), to construct the hedging rules and evaluated effects on the reservoir performance indicators. Shih and ReVelle [54, 55] determined the trigger value for a continuous hedging rule and then for a discrete hedging rule, respectively. Neelakantan and Pundarikanthan [34] presented a simulation–optimization methodology using neural network and multiple hedging rules to improve reservoir operation performance. Tu et al. [56] considered a set of rule curves that are a function of the current storage level to trigger hedging for a multipurpose, multi-reservoir system.

In this study, the hedging rule based on the storage of each member reservoir consists of hedging rule curves and rationing factors for each water demand. Details of the hedging rule curves and its corresponding operating rule are illustrated in Fig. 1.30 and in Table 1.6. In previous works on hedging rule curves [56, 57], all planned water demand are met at the same level and are rationed at the same time when drought occurs. For single purpose of water supply operation, the water demand can be divided into various categories, such as irrigation, industry and domesticity. It should be noted that different kind of water demand requires different reliability and different degree of hedging in practice. In this study, different hedging rule curves and rationing factors are assigned to different kinds of water demand. When drought occurs, different types of water demand own different priority to get as much water as demand without rationing.

For the lower model, it optimizes the water supply rule to obtain the optimal water supply quality under the condition of water transfer. The mathematical formulation of the lower model is given by Eq. (1.31).

$$
\begin{aligned}
&\min_{y} f(x,y) = \sum_{i=1}^{m}\sum_{j=1}^{n} w_{ij}\cdot\left|Index_{ij} - Target_{ij}\right| \\
&s.t.\ \ Index_{ij} = k(x,y) \\
&\quad ST_i^{min} \le x_i \le ST_i^{max},\ ST_i^{min} \le y_i \le ST_i^{max} \\
&\quad i = 1\ldots m,\ j = 1\ldots n
\end{aligned}
\tag{1.31}
$$

Table 1.6 Water supply operating rule implied by the Hedging rule curves

	Water supply for each demand		
Reservoir storage	Demand 1 (D1)	Demand 2 (D2)	Demand 3 (D3)
Zone 1	D1	D2	D3
Zone 2	D1	D2	α3*D3
Zone 3	D1	α2*D2	α3*D3
Zone 4	α1*D1	α2*D2	α3*D3
Rationing factor	α1	α2	α3

where x and y have the same meaning as the ones in Eq. (1.30); $Index_{ij}$ refers to the water supply index for the water demand j of reservoir i, which can be water shortage index, water supply reliability or some other indexes; $Index_{ij}$ is the function of water transfer rule and water supply rule, which should get close to the target value $Target_{ij}$. The lower level objective function consists of water supply indexes of all the water demand in the multi-reservoir system, which also uses the weighting approach to combine these indexes.

5.2.3 Method Solution

Due to their structure, bi-level programs are non-convex and quite difficult to deal with and solve. Even bi-level problems in which all functions involved are linear are (strongly) NP-hard [98]. Exact approaches rang from studying the properties of the feasible region, to obtaining necessary and sufficient optimality conditions, replacing the lower level problem by its Karush–Kuhn–Tucker conditions, using penalty functions or using gradient methods. Most exact algorithms can only tackle relatively small problems, so meta-heuristic approaches have been widely applied for solving bilevel programming. For example, genetic algorithms, simulated annealing and tabu search are proposed or developed to solve bilevel programming [99–102]. Kuo and Huang [103] apply particle swarm optimization algorithm (PSO) for solving bi-level linear programming problem. However, PSO has premature convergence like other swarm intelligence methods, especially in complex multi-peak-search problems.

For solving the bi-level program proposed to model the inter-basin water transfer-supply problem, an improved particle swarm optimization (IPSO) by Jiang et al. [71] is adopted in this section. In IPSO, a population of points sampled randomly from the feasible space. Then the population is partitioned into several sub-swarms, each of which is made to evolve based on particle swarm optimization (PSO) algorithm. At periodic stages in the evolution, the entire population is shuffled, and then points are reassigned to sub-swarms to ensure information sharing. In this way, the ability of exploration and exploitation has been greatly elevated. The detailed IPSO strategy can be referred to the work of Jiang et al. [71], which is used to solve the inter-basin water transfer-supply bi-level programming problem in this section. The flowchart of solving bi-level model for multi-reservoir operating policy using IPSO is described schematically in Fig. 1.31.

5.3 East–West Water Transfer Project in Liaoning Province of China

5.3.1 Study Area

The reservoir system chosen for the application of the proposed bi-level model is Huanren, Qinghe and Baishi multi-reservoir system, which locate in Liaoning

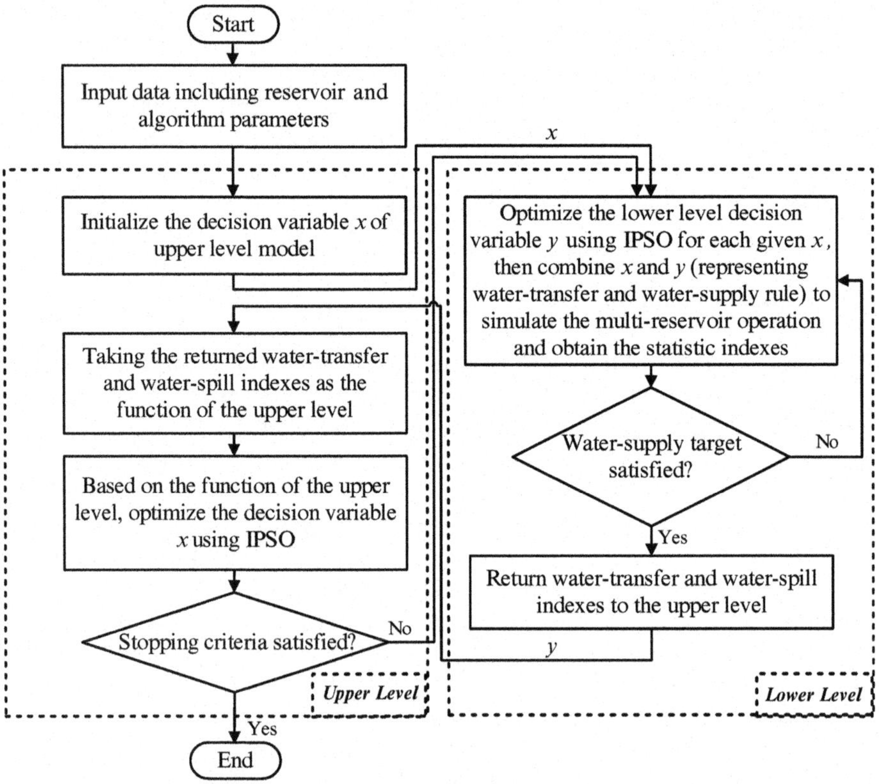

Fig. 1.31 The flowchart of solving bi-level model for multi-reservoir operating policy

province of Northeast China. As shown in Fig. 1.32, Huanren reservoir is situated in the eastern part of Liaoning province, Qinghe reservoir in central part, and Baishi reservoir in Liaoning western part.

Liaoning province covers an area of 145.9 thousand km^2 with an annual rainfall from 600 to 1100 mm, which has an extremely uneven distribution in space. The average amount of annual precipitation decreases from 1100 mm in east to 600 mm in west. In contrast, the population, industries, and agricultural areas mainly concentrate in the central and western parts. This situation can be reflected in Fig. 1.33. Comparing to Qinghe and Baishi reservoir, Huanren reservoir has the largest amount of annual average reservoir inflow and has the least amount of water demand to satisfy.

Creation of storage and inter-basin transfer of water from surplus to deficit regions is a rational option to overcome the problems caused by mismatch of supply and demand. Inter-basin transfer of water over long distances has been mooted as a long-term strategy for Liaoning province to meet the increasing water demand in water short areas. In the East-to-West inter-basin water transfer (EW-IBWT) project of Liaoning province, the abundant water in Huanren reservoir

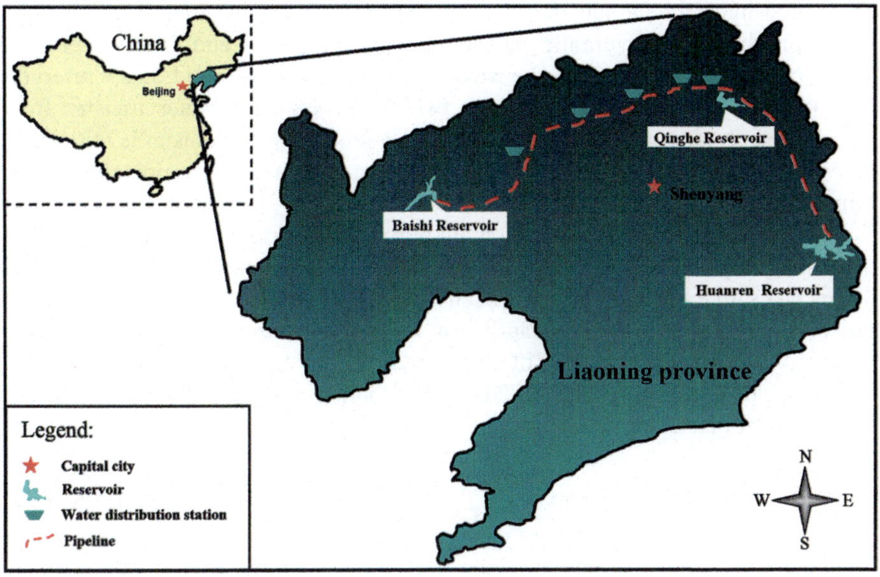

Fig. 1.32 The layout of the adopted reservoir system

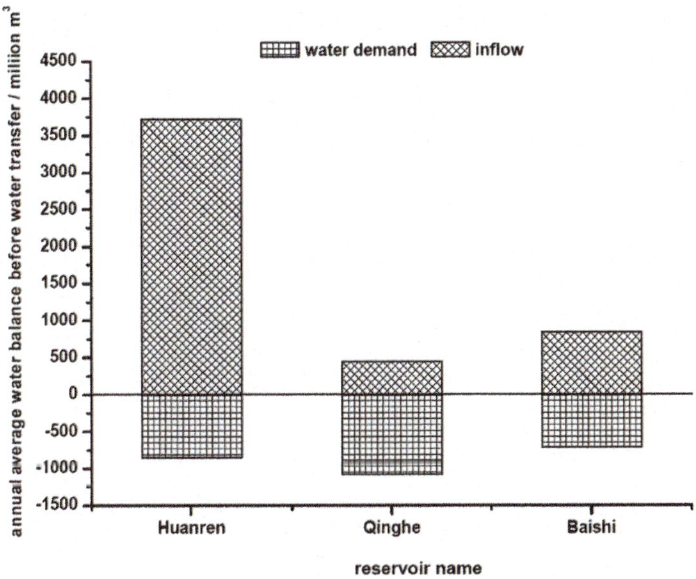

Fig. 1.33 The comparison of annual average reservoir inflow and water demand before water transfer

is diverted into Qinghe and Baishi reservoir separately using long distance pipelines. And there is no hydraulic connection between Qinghe and Baishi reservoirs. The allocation ratio of dividing the water into Qinghe reservoir and Baishi reservoir is 0.6:0.4 for this project when both two reservoirs need water transfer from Huanren reservoir at the same time. After the action of water transfer is initialized, the water supply of each reservoir in the system is also being in operation at the same time.

The reservoir characteristics and main purposes are presented in Table 1.7. The flood season mainly concentrates periods from July to September, during which the inflow takes up a large part of the annual inflow. The active storages of Qinghe and Baishi reservoir reduce significantly for the flood control during flood season. Inflow data for each reservoir of this system is collected as a whole series of hydrological record from 1951 to 2006, the monthly average inflow of which can be seen from Fig. 1.34.

Table 1.7 Reservoir characteristics

	Active storage (million m^3)			
Reservoir	Flood season	Drought season	Main purposes	Role in water transfer project
Huanren	1610.08	1610.08	I	Exporting water
Qinghe	409.13	573.13	I, P, R	Importing water
Baishi	523.08	625.00	I, P, R	Importing water

I Industry, *P* Paddy field, *R* Reed field

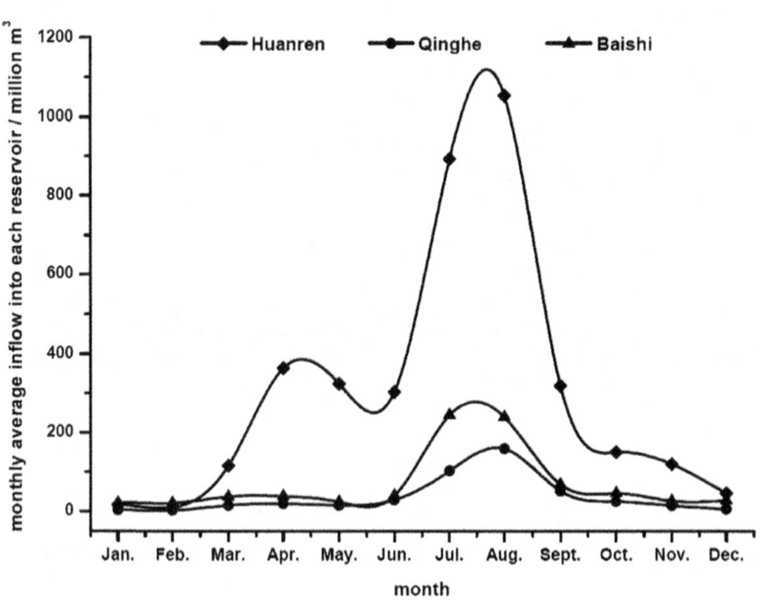

Fig. 1.34 The monthly average inflow of each reservoir in the system

In this section, a bi-level programming model is developed for the EW-IBWT project of Liaoning Province, in which the generalized shortage index (*GSI*) is adopted as the index to reflect water supply quality in the lower objective function. The *GSI* not only incorporates the basic shortage characteristics, but also emphasizes the consequential socioeconomic impacts of water shortage. It can be expressed as follows:

$$GSI = \frac{100}{N}\sum_{i=1}^{N}\left(\frac{DPD_i}{100 \times DY_i}\right)^k \tag{1.32}$$

$$DPD = \sum [\textit{Daily deficit rate}(\%) \times \textit{Number of days in a continuous deficit}] \tag{1.33}$$

where N = number of sample years; K = coefficient, usually taken as 2; DY_i = number of days in the ith year (365 or 366); and DPD_i = sum of all *DPD* in the ith year. In Eq. (1.33), $\sum$ represents the summation of the indicated values for all deficit events in the period under consideration. The daily deficit rate is defined as the ratio of the total deficit in a period to the designed water supply. The mathematic formulation of the bi-level model for the EW-IBWT project is presented in the Eq. (1.34).

$$\begin{aligned}
&\min_x \; F(x,y) = w_{DS} \cdot \sum_{i=1}^{m} |NDS_i - TNDS_i| + w_{SU} \cdot \sum_{i=1}^{m} SU_i \\
&\quad s.t. \; NDS_i = G(x,y),\; SU_i = g(x,y) \\
&\quad ST_i^{min} \le x_i \le ST_i^{max},\; ST_i^{min} \le y_i \le ST_i^{max} \\
&\quad 0 \le DS \le DS_{max} \\
&\quad i = 1 \ldots m \\
&\min_y \; f(x,y) = \sum_{i=1}^{m}\sum_{j=1}^{n} w_{ij} \cdot GSI_{ij} \\
&\quad s.t. \; GSI_{ij} = k(x,y) \\
&\quad ST_i^{min} \le x_i \le ST_i^{max},\; ST_i^{min} \le y_i \le ST_i^{max} \\
&\quad i = 1 \ldots m,\; j = 1 \ldots n
\end{aligned} \tag{1.34}$$

5.3.2 Results and Discussion

The development of regional society and economy requires a certain amount of available water resources to utilize. In order to verify the reasonability and validity of the proposed bi-level model, four scenarios of transferring water among

Table 1.8 The planned annual average transferred water amount among reservoirs/million m^3

Scenario	The amount of transferred water from Huanren reservoir	The amount of transferred water into Qinghe reservoir	The amount of transferred water into Baishi reservoir
1	1000	750	250
2	1000	650	350
3	1300	1050	250
4	1300	950	350

reservoirs are designed, which consider the future water demand of regional society development and economy growth.

As shown in Table 1.8, the annual average amount of transferred water from Huanren reservoir includes two alternatives of 1000 and 1300 million m^3 and the annual average amount of transferred water into Baishi reservoir also consists of two alternatives of 250 and 350 million m^3. Besides, the planned transferred water amount into Qinghe reservoir plus the one into Baishi reservoir is equal to the transferred water amount from Huanren reservoir. These alternatives in combination produce the four scenarios.

Under the four scenarios, the annual average water balance components of the multi-reservoir system vary along with the change of transferred water amount, which can be seen in Fig. 1.35. For Huanren reservoir, an increment of annual water amount transferred out, under scenarios 3 and 4, make the amount of water spills and water supply reduce comparing with the ones of scenarios 1 and 2. For Qinghe and Baishi reservoirs, an increment of annual water amount transferred in make their water supply and water spill get more. Although the water supply of Huanren reservoir gets less a little, the increment of its water transferred out is mainly from the significant reduction of water spill and the water supply amount of Qinghe and Baishi reservoirs has been raised significantly.

In Table 1.9, the generalized shortage indexes (GSI) of each water demand under different scenarios are listed in comparison to reflect the effect of water transfer on the water supply quality. A larger index of GSI means more frequencies of water shortage. It can be observed that the GSI of Huanren reservoir water supply for the downstream industry gets larger under the scenarios 3 and 4 comparing with the ones under scenarios 1 and 2. This relates directly to the amount increment of water transferred out from Huanren reservoir under scenarios 3 and 4. For each water demand of Qinghe and Baishi reservoirs, there is similar change about the GSI in common under these scenarios, which is that the GSI gets smaller when the water amount transferred into increases. Based on the changes of the GSI, it is concluded that the water supply quality of Qinghe and Baishi reservoirs can get much better with the increment of available water resources, in spite of a little impact on water supply of water exporting region.

From the content of the proposed water transfer rule, we can know that the higher position of water transfer rule curve of the importing reservoir can produce more importing chances and enhance the amount of importing water.

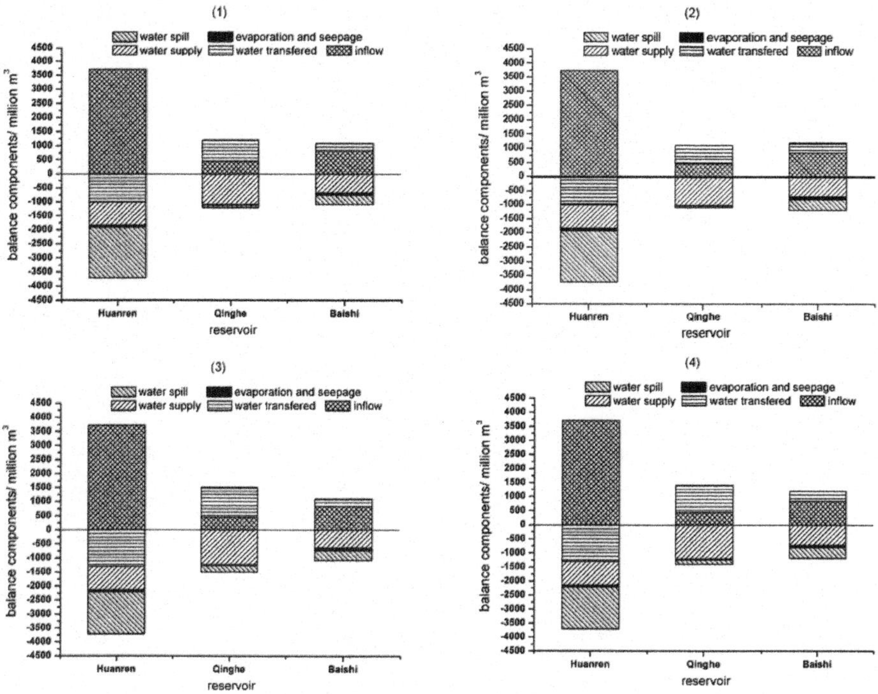

Fig. 1.35 The annual average water balance of the multi-reservoir system under different scenarios

Due to un-unique set of water transfer rule curve for one fixed water transfer target, we design an index to evaluate the position change of water transfer rule curve. As shown in Fig. 1.36, water transfer rule curve divide the reservoir active storage between minimum storage and maximum storage into two parts: zone I and zone II. The ratio of zone II area to the whole active storage area is just the index designed by us to delineate the position of water transfer rule curve. A larger area ratio of zone II represents a higher water transfer rule curve position. For Baishi reservoir, its water transfer rule curve position under scenario 2 becomes much higher comparing with the one under scenario 1. For Qinghe reservoir, its water transfer rule curve position under scenario 2 becomes a little lower comparing with the one under the scenario 1. The water transfer target adjustment of Qinghe and Baishi reservoirs between scenario 1 and scenario 2 is the main driving force of their water transfer rule curve position change. In other words, the position change of water transfer rule curve is a reasonable response to the adjustment of corresponding water transfer target. Under scenarios 3 and 4, the similar characteristic is also reflected for water transfer rule curves of Qinghe and Baishi reservoirs. For Huanren reservoir, there seems to be no significant change about its water transfer rule curve position under four scenarios. So its water transfer target adjustment may be

Table 1.9 The GSI comparison of each water demand under different scenarios

Scenario	Huanren	Qinghe			Baishi		
	Industry	Paddy field	Reed filed	Industry	Paddy field	Reed filed	Industry
1	9.42800E-07	8.66954E-05	8.40308E-05	9.66800E-07	8.09607E-04	2.72452E-04	4.97500E-06
2	8.40700E-07	1.83899E-04	1.43891E-04	9.16700E-07	3.24863E-04	1.20360E-04	4.42260E-06
3	3.11950E-06	4.64815E-05	4.69131E-05	3.05650E-06	1.09101E-03	3.18070E-04	2.01434E-05
4	3.58490E-06	7.41040E-05	7.56802E-05	3.07170E-06	2.91011E-04	1.03434E-04	4.97500E-06

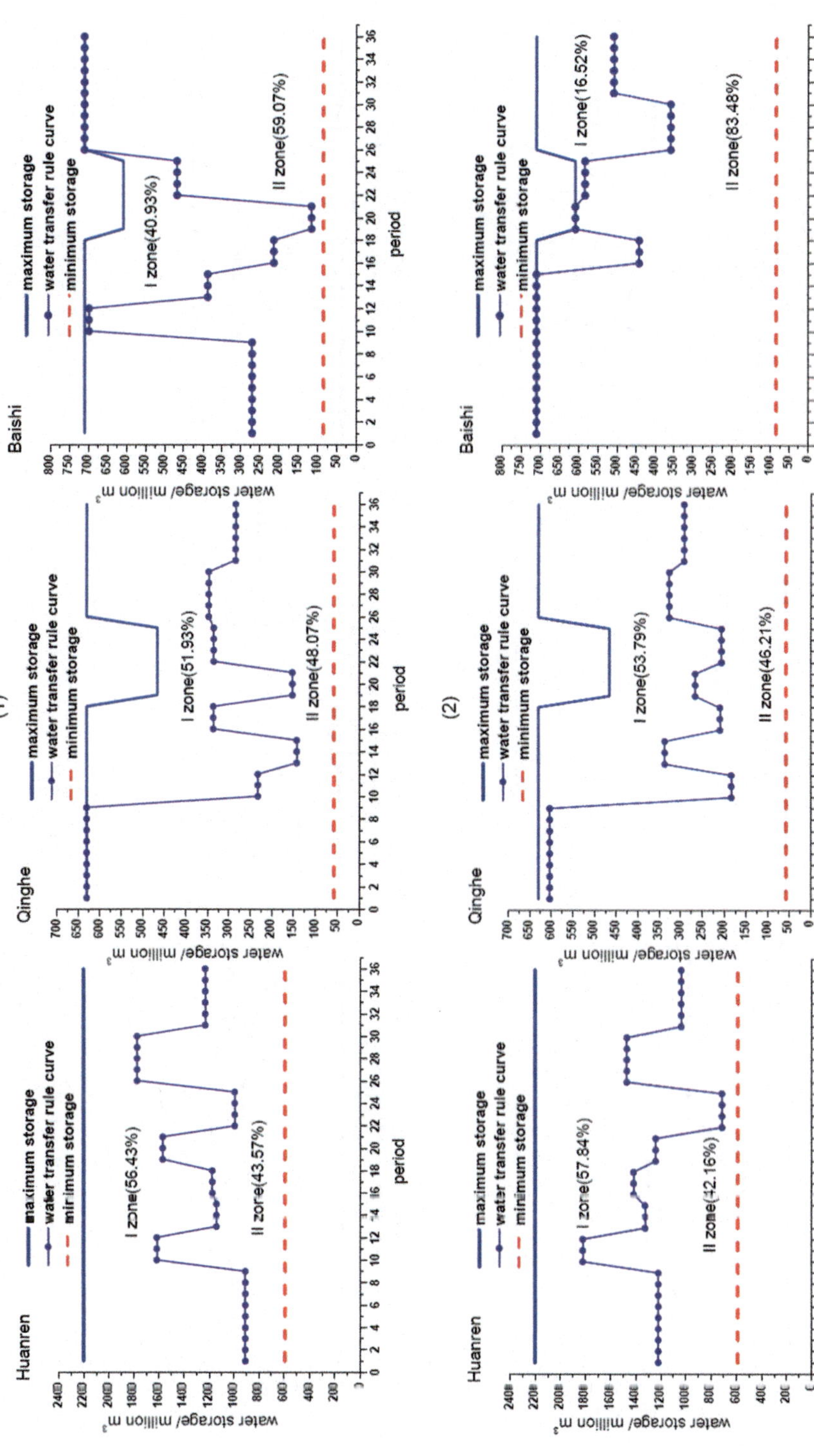

Fig. 1.36 The response of water transfer rule curves to different scenarios

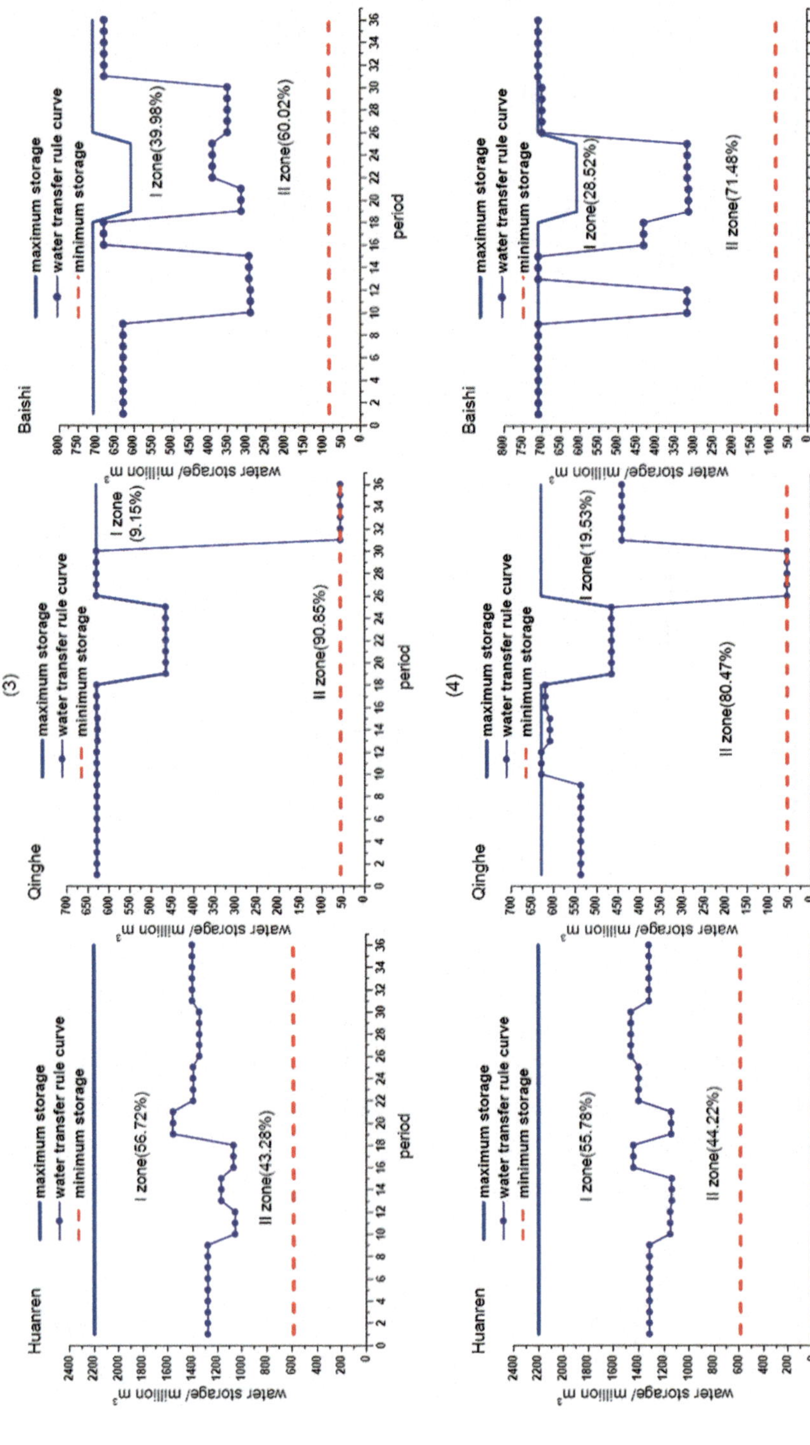

Fig. 1.36 (continued)

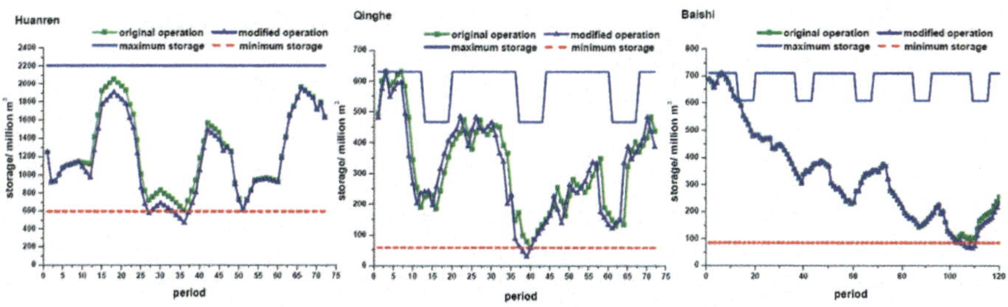

Fig. 1.37 The comparison of reservoir storage during drought period in original and modified operation

achieved by the position change of the importing reservoir water transfer rule curves. This also indicates the inter-related relation between the water exporting and importing reservoir water transfer rule curves.

In order to inquiry the effect of water transfer rule curve on reservoir storage in operation, we raise the flood season water transfer rule curve of Huanren reservoir under scenario 1 by 2 m and lower it by 2 m. Once it is lowered by 2 m, the water mount transferred out from Huanren reservoir gets more and the storage of Huanren reservoir during extreme drought periods falls down below the dead storage. In Fig. 1.37, the change of Huanren reservoir storage in original and modified operation is illustrated. When the flood season water transfer rule curve of Huanren reservoir is raised by 2 m, the water amount transferred out from Huanren reservoir becomes less. At this time, the storage of Qinghe and Baishi reservoirs during extreme drought period falls down below their dead storage, which is also shown in Fig. 1.37. As we know, the case that reservoir storage falls down below the dead storage is not allowable absolutely in the practical operation. The comparison of reservoir storage during extreme drought periods in original and modified operation indicates that the effect of water transfer rule curve on reservoir storage in operation is obviously and the original water transfer rule curve is optimal and reasonable.

6 Hydrology Forecast for Reservoir Operation

Various hydrological forecast products have been applied to real-time reservoir operation, including deterministic streamflow forecast (*DSF*), *DSF*-based probabilistic streamflow forecast (*pseudo-PSF*, *pPSF*), and ensemble or probabilistic streamflow forecast (denoted as *real-PSF*, *rPSF*). *DSF* represents forecast uncertainty in the form of deterministic forecast errors, *pPSF* a conditional distribution of forecast uncertainty for a given *DSF*, and *rPSF* a probabilistic uncertainty distribution. Based on previous studies on hydrology forecast for reservoir operation, this section attempts to model the dynamic evolution of uncertainties involved in the various forecast methods, explores their effect on real-time reservoir operation decisions and identifies the effective forecast horizon.

6.1 Effect of Inflow Forecast Uncertainty on Real-Time Reservoir Operation

6.1.1 Background

Advances in weather forecasting, hydrologic modeling, and hydro-climatic teleconnection relationships have significantly improved streamflow forecast precision and lead-time [104, 105] and provide great opportunities to improve the efficiency of water resources system operations [106, 107]. In recent years, forecast products, particularly long-term streamflow forecasts (with a lead-time longer than 15 days), have been applied to reservoir operation and water resources management [107, 108].

One important issue with implementing streamflow forecasts in real-time reservoir operation models is dealing with the uncertainty involved in streamflow forecast products. Although forecast uncertainty analysis has been one research focus in hydrology [109–111] there are comparatively less studies on the effect of forecast uncertainty on real-time reservoir operations [112, 113]. Deterministic or probabilistic streamflow forecast products are usually treated as ad hoc inputs for deterministic or stochastic reservoir operation models. That is to say, a deterministic forecast or a stochastic forecast represented by a number of scenarios is pre-designed for a specific reservoir operation problem for screening test, and no non-generalizable structure of the forecast error is endogenously involved in the operation analysis. Correspondingly, many previous studies on forecast and reservoir operation in the literature adopt a two-component approach, one provides ("recommends") a forecast scenario [104, 114, 115] as input to the other component [106, 108, 116] that dealing with forecast application. In general, such an approach suggests that forecast can always improve reservoir operation efficiency especially under extreme conditions.

This section aims at analyzing the effect of forecast uncertainty on real-time reservoir operations. As different forecast products, e.g., deterministic and probabilistic streamflow forecasts can exert different effects on real-time reservoir operation decisions in optimization and simulation models, this study will explicitly simulate the uncertainty in each of the streamflow forecasts examined and assess its effect on real-time reservoir operation decisions. Since the tool for such a purpose does not exist in the hydrologic literature, the Martingale Model of Forecasting Evolution (MMFE) [117, 118] used in supply chain management is introduced to quantify real-time streamflow forecast uncertainty and generate deterministic and probabilistic forecast products. Simulations based on standard operation policy (SOP), dynamic programming (DP), and stochastic dynamic programming (SDP) [13, 16] are adopted to determine release decisions for a hypothetical reservoir using synthetic streamflow forecasting products.

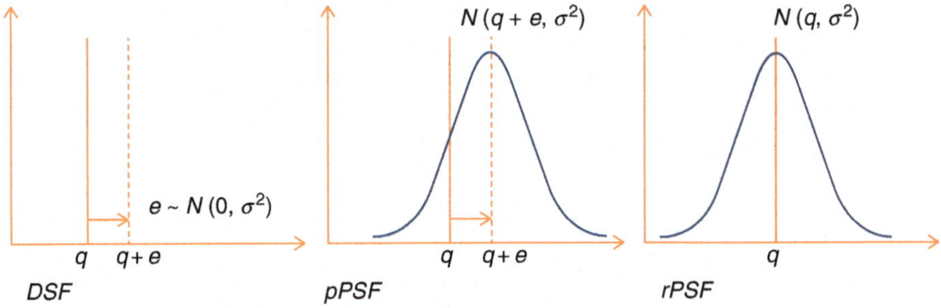

Fig. 1.38 Schematic of single-period streamflow forecast uncertainty

6.1.2 Streamflow Forecast Uncertainty Analysis Based on Martingale Model of Forecasting Evolution

Streamflow Forecast and Forecast Uncertainty

Both deterministic and probabilistic streamflow forecast products have been applied to real-time reservoir release decision making, as outlined in Fig. 1.38. Defining q as the actual streamflow and e as the forecast error, the relationship between deterministic streamflow forecast (*DSF*) and q can be interpreted by Eq. (1.35):

$$DSF = q + e \tag{1.35}$$

Equation (1.35) shows that the forecast uncertainty in *DSF* is characterized by a deterministic forecast error e. Usually, e is assumed to be stochastic and fit a normal distribution [105, 116]:

$$e \sim N(0, \sigma^2) \tag{1.36}$$

where σ^2 denotes the variance of e (i.e., uncertainty level)

Probabilistic streamflow forecasts (*PSF*) can be generated with two approaches. One involves treating the *PSF* as an empirical conditional distribution of forecast uncertainty for a given *DSF* (namely *Pseudo-PSF*, denoted as *pPSF* in this study) [116]. The premise of *pPSF* is that, since $q = DSF - e$ and $e \sim N(0, \sigma^2)$, the actual streamflow q fits a conditional normal distribution with mean *DSF* and variance σ^2.

$$pPSF \sim N(DSF, \sigma^2) = N(q + e, \sigma^2) \tag{1.37}$$

Equation (1.37) shows that the forecast uncertainty in *pPSF* depends on the deterministic forecast error e and the distribution of *pPSF* is conditional to the distribution of e.

The other approach for generating *PSF* takes a more rigorous way to handle forecast uncertainty, which is to characterize the streamflow forecast uncertainty by either the ensemble streamflow forecasting method [18, 112] or probabilistic streamflow forecasting methods [24, 119]. We denote this type of *PSF*, shown in Eq. (1.38), as a *real-PSF* (*rPSF*) to distinguish it from the *pseudo-PSF* (*pPSF*) presented in Eq. (1.37). Assuming a normal distribution for forecast uncertainty, *rPSF* can also be characterized with a normal distribution [120]:

$$rPSF \sim N\left(q, \sigma^2\right) \tag{1.38}$$

Equation (1.38) shows that the forecast uncertainty in *rPSF* is also represented by a probabilistic distribution form. This is different from Eq. (1.37), which contains a deterministic forecast error term as well as a probabilistic uncertainty term.

This study simplifies forecast uncertainty with the stationary Gaussian distribution assumption and characterizes the single period streamflow forecast uncertainty with σ^2 (the variance of e). In hydrology, σ^2 is closely related to popular hydrologic forecast evaluation criteria, such as the Nash–Sutcliffe efficiency coefficient (*NSE*) and Root Mean Square Error (*RMSE*) [104]. The calculation of *NSE* and *RMSE* are shown in Eqs. (1.39) and (1.40), respectively.

$$NSE = 1 - \frac{\sum_{i=1}^{M} (f_i - q_i)^2}{\sum_{i=1}^{M} (q_i - \overline{q})^2} \approx 1 - \frac{\sigma^2}{\overline{q}C_v} \tag{1.39}$$

$$RMSE = \sqrt{\frac{1}{M}\sum_{i=1}^{M} (f_i - q_i)^2} \approx \sigma \tag{1.40}$$

where M is the number of samples, C_v is the streamflow coefficient of variation, q_i is the streamflow, and f_i is the streamflow forecast. As can be ascertained from Eqs. (1.39) and (1.40), *NSE* measures the comparative level of forecast uncertainty to the streamflow standard deviation and represents the fraction of streamflow variability explained by the forecast while *RMSE* is a direct reflection of the forecast uncertainty itself.

The *PSF* evaluation criteria, e.g., the linear error in probability space (LEPS), the Brier score, mainly depend on the bias and dispersion of the forecasted streamflow distribution, of which σ^2 is an effective statistical indicator [104, 107, 114].

Martingale Model of Forecasting Evolution (MMFE)

In streamflow forecasts, denote H as the length of forecast lead time or forecast horizon, within which the streamflow is predictable with an available forecasting method. The streamflow forecasts can be represented by a vector:

$$F_{t,-} = \left[f_{t,t} \quad f_{1,t+1} \quad f_{t,t+2} \quad \cdots \quad f_{t,t+H} \right] \tag{1.41}$$

where $F_{t,-}$ is a vector denoting the forecast sequence made at period t; $f_{t,t+i}$ denotes the period t's forecast for the period $t+i$ streamflow. Denoting $\sigma_{t,t+i}$ as the uncertainty of $f_{t,t+i}$ and assuming (1) stationary forecast uncertainty (i.e., $\sigma_{t,t+i}$ does not change with t) and (2) a pre-determined ending time, two important properties of real-time streamflow forecasts hold (as shown in Fig. 1.39) [106, 114, 120]:

$$\sigma_{t,t} \leq \sigma_{t,t+1} \leq \sigma_{t,t+2} \leq \ldots \leq \sigma_{t,t+H} \tag{1.42}$$

$$\sigma_{t-H,t} \geq \sigma_{t-H+1,t} \geq \sigma_{t-H+2,t} \geq \ldots \geq \sigma_{t,t} \tag{1.43}$$

Equation (1.42) denotes that the uncertainty level of the streamflow forecast increases with the forecast lead time, which is intuitive since the longer the forecast lead time, the less reliable the forecast information is, as shown in the upper part of Fig. 1.39. Equation (1.43) represents a property that indicates the dynamic updating of the real-time streamflow forecast, i.e., when the forecast period moves towards the ending time, information becomes more reliable and the forecast uncertainty level decreases, as shown in the lower part of Fig. 1.39.

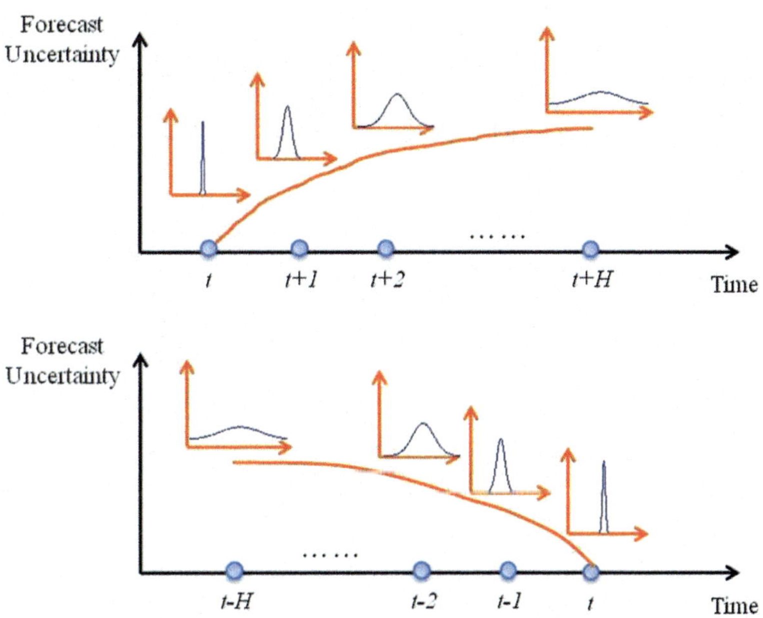

Fig. 1.39 Schematic diagram of the increase in forecast uncertainty with forecast lead-time

The MMFE model uses a decomposition approach to measure the uncertainty in each of the time periods within the forecast lead time (H):

$$\Delta F_{t,-} = \begin{bmatrix} \Delta f_{t,t} & \Delta f_{1,t+1} & \Delta f_{t,t+2} & \cdots & \Delta f_{t,t+H} \end{bmatrix} \tag{1.44}$$

where $\Delta F_{t,-}$ is a vector denoting the forecast update made at period t from the forecasts made at period $t-1$ and $\Delta f_{t,t+i}$ is the improvement of streamflow forecast at period $t+i$, and:

$$\Delta f_{t,t+i} = f_{t,t+i} - f_{t-1,t+i} \tag{1.45}$$

MMFE, which simulates the forecast improvement process, is based on the following four assumptions: (1) $F_{t,-}$ is an unbiased forecast for the future; (2) $\Delta f_{t,t+i}$ is uncorrelated with past forecast updates $\Delta f_{s,s+i}(s < t)$; (3) the forecast update $\Delta f_{t,t+i}$ forms a stationary stochastic process of t; and (4) the forecast update $\Delta f_{t,t+i}$ is normally distributed.

Under MMFE, the total forecast uncertainty can be characterized by the variance–covariance (VCV) matrix of $\Delta F_{t,-}$

$$VCV = \begin{bmatrix} \sigma_{0,0}^2 & \sigma_{0,1}^2 & \cdots & \sigma_{0,H}^2 \\ \sigma_{1,0}^2 & \sigma_{1,1}^2 & \cdots & \sigma_{1,H}^2 \\ \vdots & \vdots & \ddots & \vdots \\ \sigma_{H,0}^2 & \sigma_{H,1}^2 & \cdots & \sigma_{H,H}^2 \end{bmatrix}_{(H+1)\times(H+1)} \tag{1.46}$$

where $\sigma_{i,j}^2$ is the covariance between $\Delta f_{t,t+i}$ and $\Delta f_{t,t+j}$. Denoting $f_{t,t} = q_t$, with Eq. (1.45), $f_{t-i,t}$ can be expressed by:

$$f_{t-i,t} = q_t - \sum_{j=t-i+1}^{t} \Delta f_{j,t} \tag{1.47}$$

With Eq. (1.48) and the second assumption of MMFE, the forecast uncertainty level of $f_{t-i,t}$ can be calculated by:

$$\text{var}\left(q_t - f_{t-i,t}\right) = \sum_{j=0}^{i-1} \sigma_{j,j}^2 \tag{1.48}$$

Since $\text{var}\left(q_t - f_{t-i,t}\right)$ increases with i, MMFE naturally reflects some properties of stream flow forecasts, i.e., increased uncertainty with forecast lead-time and dynamic forecast updates.

It is important to note that MMFE is not a forecast model but rather a framework representing the dynamics of forecast updates [118, 121]. Due to its simplicity and effectiveness in illustrating the forecast uncertainty evolution processes, MMFE has been widely applied to operations research for quantifying the economic profits from forecast improvements [118] analyzing the optimality of supply chain management strategies [121, 122] determining the safety stock level in supply chain management [110], and supporting restocking decision making under forecast uncertainty [123].

Streamflow Forecast Uncertainty Analysis Based on (MMFE)

To use MMFE to model the uncertainty of streamflow forecasts, it is necessary to justify its assumptions, i.e. unbiasedness, non inter-period correlation, stationarity, and Gaussian distribution. Real-time streamflow forecasts are based on hydrologic model inputs, such as precipitation, temperature, and soil moisture. These inputs are updated at the beginning of each period with new weather forecasts and hydrologic observations (e.g., streamflow, soil moisture) to improve the preceding streamflow forecast. Since hydrologic model input errors are usually considered to be dominated by random factors rather than structural ones, the assumption of unbiasedness in MMFE (i.e., the structural error is negligible) has been widely adopted in hydrologic studies [112, 116].

The second assumption may be justified by the hypothetical problem setting in this study. As time moves forward to the prescribed ending period, the forecast lead time decreases and more information becomes available (Fig. 1.39). At the start of a new period, new information becomes available, which is not available for the previous periods. It is reasonable to assume that this new information is independent from the information that was previously available. Therefore, it can be assumed that the update to the streamflow forecast for a given period is independent of the updates in previous periods.

The third and fourth assumptions imply stationary and a Gaussian distribution of the uncertainty, respectively, which are common assumptions in hydrologic studies [116].

In MMFE, the *VCV* matrix of the linearly dependent components of $F_{t,-}$ in Eq. (1.46) plays a central role. Since the *VCV* matrix is positive semi-definite, it can be decomposed into the product of a matrix multiplied by its transpose through the Cholesky decomposition [124], i.e.,

$$VCV = VV^T \tag{1.49}$$

Denote $[X_1 \quad X_2 \quad \ldots \quad X_{H+1}]$ as a vector of $H+1$ independent standard normal variables and transposing it with matrix V^T:

$$[Y_1 \quad Y_2 \quad \ldots \quad Y_{H+1}] = [X_1 \quad X_2 \quad \ldots \quad X_{H+1}]V^T \tag{1.50}$$

Then, the generated vector of *Y* consists of normally distributed variables with a variance–covariance matrix equal to their original variance–covariance matrix, $VCV = VV^T$. Thus, Eq. (1.50) can be used for generating forecast errors:

$$\begin{bmatrix} \Delta f_{1,1} & \Delta f_{1,2} & \cdots & \Delta f_{1,H+1} \\ \Delta f_{2,2} & \Delta f_{2,3} & \cdots & \Delta f_{2,H+2} \\ \vdots & \vdots & \ddots & \vdots \\ \Delta f_{t,t} & \Delta f_{t,t+1} & \cdots & \Delta f_{t,H+t} \end{bmatrix} = \begin{bmatrix} x_{1,1} & x_{2,1} & \cdots & x_{H+1,1} \\ x_{1,2} & x_{2,2} & \cdots & x_{H+1,2} \\ \vdots & \vdots & \ddots & \vdots \\ x_{1,t} & x_{2,t} & \cdots & x_{H+1,t} \end{bmatrix} V^T \quad (1.51)$$

Assuming the actual streamflow sequence is known:

$$Q = [q_1 \quad q_1 \quad \ldots \quad q_N] \quad (1.52)$$

The deterministic streamflow forecast error can be expressed by:

$$e_{t-i,t} = q_t - f_{t-i,t} = \sum_{j=H+2-i}^{H+1} \Delta f_{t-H-1+j,t} = \sum_{j=0}^{i-1} \Delta f_{t-j,t} \quad (1.53)$$

where $e_{t-i,t}$ denotes the forecast error for period t streamflow in the forecast made during period $t-i$. The synthetic *DSF* forecast errors, e.g. $e_{1,5} = \sum_{i=2}^{5} \Delta f_{i,5}$, can then be generated through Eq. (1.53).

With the second assumption of MMFE, the variance of the forecast error $(e_{t-i,t})$ can be calculated by:

$$\operatorname{var}(e_{t-i,t}) = \sum_{j=0}^{i-1} \sigma_{j,j}^2 (i \geq 1) \quad (1.54)$$

Combining Eqs. (1.35), (1.37), and (1.38) with Eq. (1.53), Eq. (1.54), the *DSF*, *pDSF*, and *rDSF* made at period $t-i$ for period t streamflow can be explicitly expressed with the following equations:

$$DSF : DSF_{t-i,t} = q_t - e_{t-i,t} \left(e_{t-i,t} \sim N\left(0, \sum_{j=0}^{i=1} \sigma_{j,j}^2\right)\right) \quad (1.55)$$

$$pPSF : pPSF_{t-i,t} \sim N\left(q_t - e_{t-i,t}, \sum_{j=0}^{i-1} \sigma_{j,j}^2\right) \quad (1.56)$$

$$rPSF : rPSF_{t-i,t} \sim N\left(q_t, \sum_{j=0}^{i-1} \sigma_{j,j}^2\right) \quad (1.57)$$

Thus, using MMFE, *DSF*, *pPSF*, and *rPSF* can be synthetically generated with a common framework. For probabilistic forecasts (*pPSF* and *rPSF*), Eqs. (1.56) and

(1.57) depict the forecast uncertainty of period t without reflecting the correlation relationship between the uncertain terms expressed in Eqs. (1.44) and (1.46). To deal with this concern, this study adopts a scenario-based Monte-Carlo approach for forecast uncertainty analysis [17]. Then, with a deterministic or stochastic reservoir operation model, a framework for real-time reservoir release decisions can be established.

It is worthwhile to note that forecast uncertainty and forecast horizon are two important features of streamflow forecast and both can affect reservoir operation using the forecast [6, 113], as the forecast can be too uncertain if it is too long (i.e., it cannot reliably reflect inflow conditions) or too short to be applicable for supporting decision making. This study focuses on the effect of forecast uncertainty on real-time reservoir operation while the complicating effect of forecast uncertainty and forecast horizon will be analyzed in future work.

6.1.3 Effect of Streamflow Forecast Uncertainty on Reservoir Operation

The Hypothetical Reservoir System with Operation Strategies

Reservoir System

Besides forecast uncertainty, the efficiency of reservoir operations can also be affected by reservoir inflow variability, demand change, and reservoir capacity [112, 125]. To study these influential factors, the hypothetical reservoir operation model consists of four categories of parameters: forecast uncertainty, reservoir inflow, reservoir capacity, and the objective function.

(a) Forecast uncertainty: The forecast error standard deviation σ and the forecast error correlation ρ_{error} are introduced to characterize the streamflow forecast uncertainty [110, 111]. As shown in Eq. (1.58). The VCV matrix is simplified with the two forecast parameters:

$$\begin{bmatrix} \sigma^2 & \rho_{error}\sigma^2 & \cdots & 0 \\ \rho_{error}\sigma^2 & \rho_{error}\sigma^2 & \cdots & 0 \\ \vdots & \vdots & \ddots & \vdots \\ 0 & 0 & \cdots & \sigma^2 \end{bmatrix}_{(H+1)\times(H+1)} \tag{1.58}$$

σ represents the magnitude of uncertainty in the forecast. A higher σ value implies a greater forecast uncertainty. ρ_{error} reflects the temporal correlation relationship of the forecast uncertainty. In general, a negative ρ_{error} implies a lower amount of uncertainty in the total inflow, as the overestimated forecast errors are more likely to be balanced by the underestimated forecast errors; meanwhile, a positive ρ_{error} implies a higher degree of uncertainty in the total inflow forecast.

(b) Reservoir inflow: The reservoir inflow parameters include the mean, coefficient of variation, and the correlation coefficient of the streamflow, which are denoted as μ, C_v, and ρ_{flow}, respectively. A simplified Thomas–Fiering model is applied to generate the reservoir inflow sequences:

$$q_{t+1} = \mu + \rho_{flow}(q_t - \mu) + \sqrt{1 - \rho_{flow}^2}(\mu C_v)\delta \tag{1.59}$$

In Eq. (1.59), δ is a standard normal random number. The minimum streamflow is set to 0.4 so that 93 % of the generated streamflow sequences can be subsequently used in the MMFE streamflow forecast model when C_v is at its maximum value, i.e., 0.4.

(c) Reservoir capacity: The reservoir capacity (S) is represented by the active maximum storage, which is the difference between the maximum and the minimum storage

$$S = S_{\max} - S_{\min} \tag{1.60}$$

To avoid adverse effects of initial storage and end storage on reservoir operation decisions, the initial storage and end storage are set to half of S.

(d) Objective function: The objective function is defined as the sum of the single-period reservoir release utility [Eq. (1.61)] and is maximized in the DP and SDP formulations, in which the reservoir storage and inflow are discretized into intervals with a width of 0.01.

$$g_t = \left(\frac{D_t - D_{\min}}{D_{\max} - D_{\min}}\right)^{1/2} \tag{1.61}$$

where D_t is the beneficial release (excluding the reservoir spill DS_t) at time period t, while D_{max} and D_{min} represent the maximum and minimum beneficial releases, respectively. Equation (1.61) is concave with a decreasing marginal utility property [5, 125].

The parameters of the hypothetical reservoir operation model are summarized in Table 1.10. Each of the impact factors discussed above (as shown in Table 1.10) is assessed individually, i.e. adjusting the value of a given factor while holding the base values of all other parameters. Table 1.10 shows the range of values tested for each parameter. It is necessary to note that forecast uncertainty parameters [Eq. (1.58)] have already been specified with values in Table 1.10 for this hypothetical case study and the underlying assumption is that MMFE has already been validated before the policy simulation. For real-world application of MMFE, a validation step is needed.

Table 1.10 Parameters of the hypothetical reservoir system

Reservoir components	Parameters symbol	Type	Value range	Base value
Forecast uncertainty	σ	Variable	0.02–0.20	0.10
	ρ_{error}	Variable	−0.50–0.50	0
Reservoir inflow	μ	Constant	1	–
	C_v	Variable	0.05–0.40	0.30
	ρ_{flow}	Variable	0.4	–
Reservoir capacity	S	Variable	0.20–5.00	2.00
Reservoir utility	$D_{\min}$	Constant	0.4	–
	$D_{\max}$	Constant	1.2	–

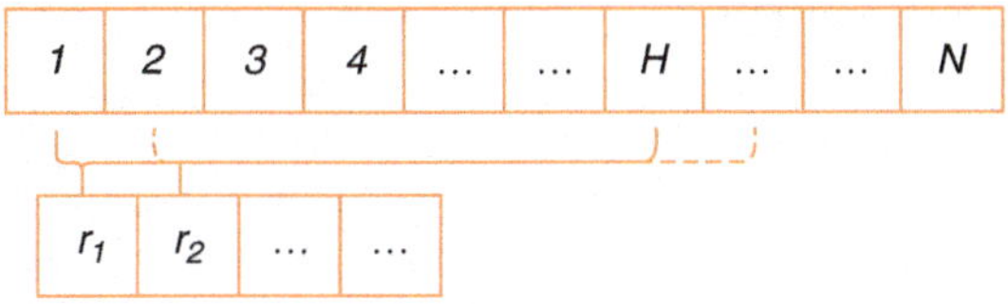

Fig. 1.40 Schematic of rolling horizon decision making in reservoir operation

Operation Strategies

The following generic procedures are used to model the hypothetical reservoir operation problem: (1) time series of streamflow Q during the N operation periods are generated using a flow synthesis model with given reservoir inflow statistics; (2) *DSF*, *pPSF*, and *rPSF* are generated with Q and MMFE using the predefined forecast uncertainty statistics; for *pPSF* and *rPSF*, 500 forecast error scenarios are generated to approximate the streamflow probability and state transition probability [17, 18]; (3) with the synthesized forecast products from (2), optimization models (DP and SDP) and the simulation model based on SOP are employed for reservoir operation analysis. For each parameter test, the numerical experiment is conducted with 100 randomly generated streamflow scenarios, and the mean value and standard deviation of the utility are computed using the 100 samples.

Decision horizon (DH, how long the generated decision is implemented), forecast horizon (FH, how long the inflow can be predicted), and operation horizon (OH, how long the reservoir operation is targeted) are important issues in reservoir operation. In our study, DH is set as 1 and FH is assumed to be the same as the length of OH (i.e., the lead time H is N periods at the beginning, N − 1 periods when decision moves to next period, and so on). The following procedures are undertaken for the modeling exercise: (1) reservoir operation decision is determined for each period with the streamflow forecast provided up to the end of the operation periods; (2) for the generated decision sequence, only the current period decision is treated as final; (3) decisions in future time periods will be updated period by period, i.e., at the beginning of the next period, the reservoir state is updated with inflow and release, and new release decision is made with updated forecast (i.e., rolling horizon decision making, see Fig. 1.40). This process is repeated from period 1 to N (N is set as 6 in this study).

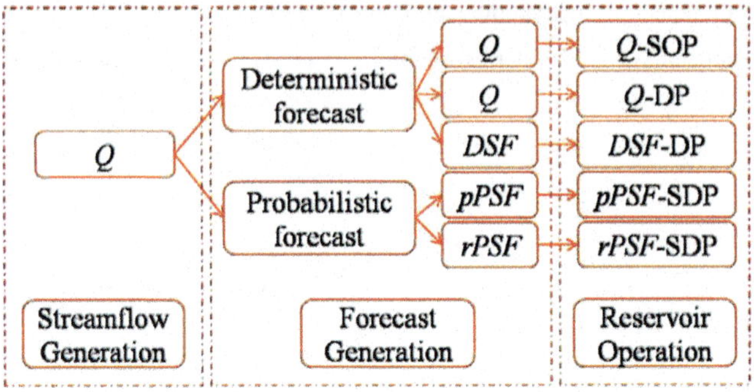

Fig. 1.41 Procedures of modeling exercise for testing the various forecasts with reservoir operation models

This study undertakes a finite horizon specified with the ending storage, which is set equal to the initial storage for this theoretical study. Five operation scenarios, shown in the last column of Fig. 1.41, are examined. The optimization models of dynamic programming (DP) and stochastic dynamic programming (SDP) are utilized to generate the operation decisions with the streamflow forecast. The perfect forecast, Q, and DSF are implemented through DP while the probabilistic forecast scenarios ($pPSF$ and $rPSF$) are implemented through SDP. These results are compared to a simulation model of standard operation policy (SOP) using Q.

Effect Analysis of Factors Relating to Streamflow and Reservoir

The Role of Forecast Uncertainty for Reservoir Operation

With the base parameter values, effects of different values of σ and ρ_{error} are assessed. Figures 1.42 and 1.43 show the effect of σ and ρ_{error} on the utility level of the reservoir operation, respectively. The mean value and standard deviation of the reservoir operation improvement (in terms of utility increase) with deterministic forecasts (Q-DP and DSF-DP) and probabilistic forecasts ($pPSF$-SDP and $rPSF$-SDP) from Q-SOP are compared in Fig. 1.42. As can be seen from the upper part of Fig. 1.42, a threshold level exists in the $rPSF$-SDP performance. With a medium uncertainty level $\sigma < 0.1$ ($\sigma = 0.1$ is about one third of the stream flow standard deviation $\mu C_v = 0.3$), the operation of $rPSF$-SDP is similar to Q-DP (i.e., the optimal reservoir release decision) in terms of the mean utility improvement from Q-SOP. Beyond this uncertainty level $\sigma > 0.1$, $rPSF$-SDP has a decreasing trend with the increase of forecast uncertainty level. The performances of DSF-DP and $pPSF$-SDP are similar with a declining trend in the forecast uncertainty level. In terms of the standard deviation of the utility improvement, Q-DP and $rPSF$-SDP both exhibit a lower variation while DSF-DP and $pPSF$-SDP show a higher variation. In general, $rPSF$ performs superiorly to $pPSF$ in terms of improving the real-time reservoir operation, which suggests that merely carrying out an

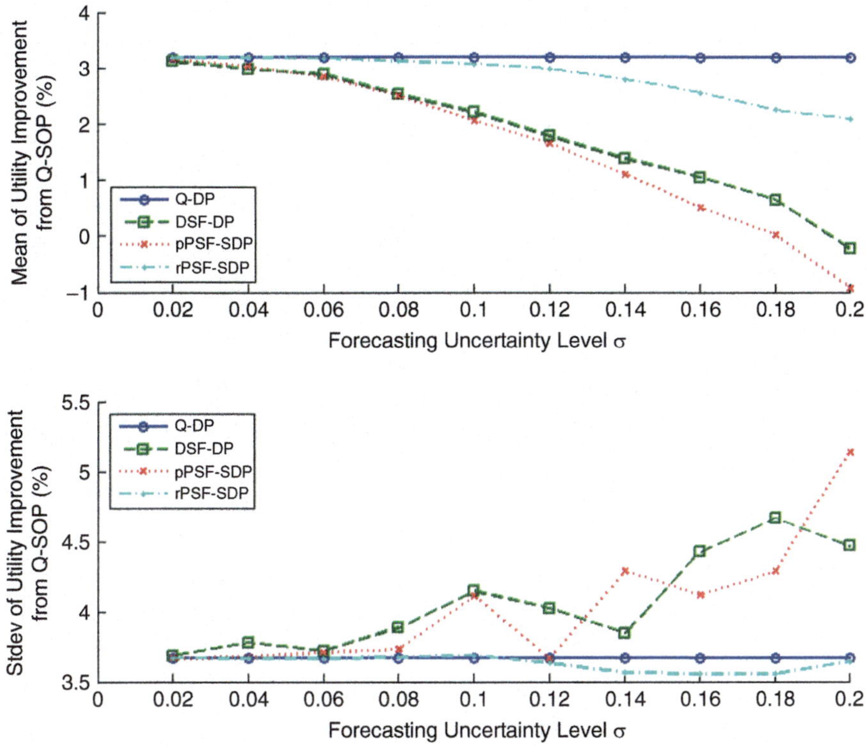

Fig. 1.42 Relationship between reservoir operation efficiency improvement from SOP and streamflow forecast uncertainty level

empirical uncertainty analysis based on *DSF* is not as efficient as an ensemble streamflow forecast.

Figure 1.43 shows the reservoir performances under different forecast uncertainty correlations varying between −0.5 and 0.5. As with the impact of uncertainty levels (Fig. 1.42), both *Q*-DP and *rPSF*-SDP perform similarly, which further illustrates the robustness of reservoir operation under *rPSF* with respect to uncertainty correlation. *DSF*-DP and *pPSF*-SDP both show worse performances than *Q*-DP and *rPSF*-SDP. However, the performance of *pPSF*-SDP is more stable with different ρ_{error} levels, while the mean performance of *DSF*-DP shows a slightly declining trend with ρ_{error}, as shown in Fig. 1.43.

One characteristic of the probabilistic streamflow forecast is its explicit probabilistic representation of future low and high flow conditions, which is important in decision risk analysis. In reservoir operation practice, hedging, which means slightly reducing the current water supply to mitigate future water shortages, is an important real-time reservoir operation practice. As forecast uncertainty increases, it becomes more beneficial to adopt hedging to avoid large shortages. Comparing the first period reservoir release reduction under *DSF*-DP, *pPSF*-SDP, and *rPSF*-

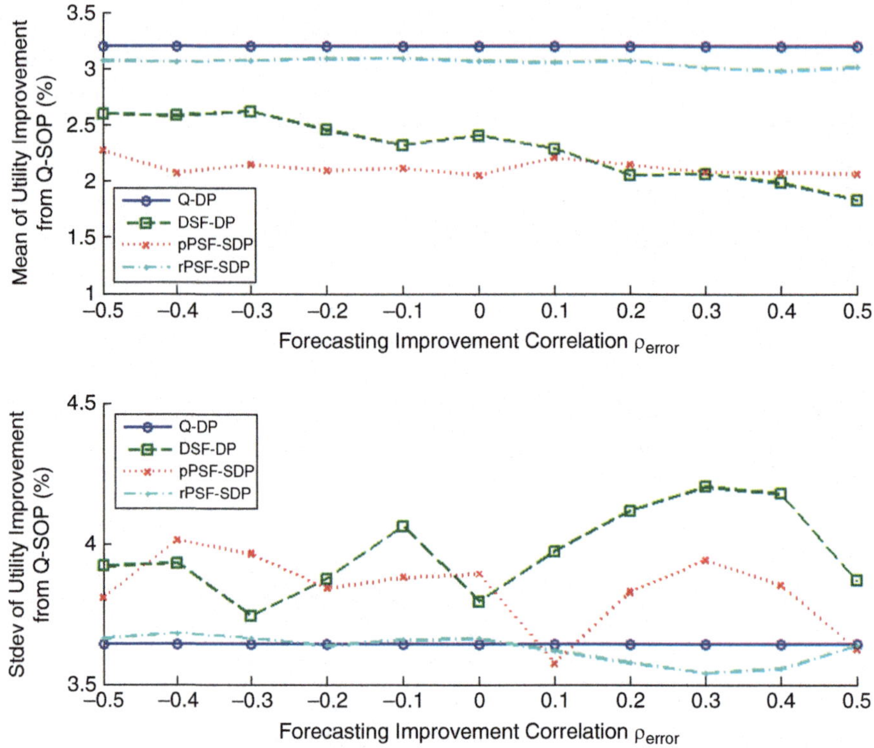

Fig. 1.43 Relationship between reservoir operation efficiency improvement from SOP and correlation of forecast errors

SDP to that under the perfect forecast Q-DP (i.e., the optimal reservoir operation without forecast uncertainty), the hedging effects of both $pPSF$-SDP and $rPSF$-SDP exhibit an increasing trend with the increase of the uncertainty level. On the other hand, *DSF-DP* shows no significant hedging effect (as shown in Fig. 1.43), which illustrates the effectiveness of adopting probabilistic streamflow forecasts to represent the future risks. Meanwhile, although the hedging trends under $pPSF$ and $rPSF$ are similar, there are differences between $pPSF$-SDP and $rPSF$-SDP in terms of utility improvement from Q-SOP (as shown in Figs. 1.42 and 1.43). The reason can be that the $pPSF$-SDP operation hedges against both the deterministic forecast error and the random forecast uncertainty. Since the magnitude of the deterministic forecast error is approximate to that of the forecast uncertainty (denoted by the standard deviation of the deterministic forecast error, as shown in Eqs. (1.35), (1.36), and (1.37)), the benefit of hedging is not as significant in $pPSF$-SDP as $rPSF$-SDP. Also, the hedging effect of $pPSF$-SDP tends to be more variable than that of $rPSF$-SDP.

Effect of Streamflow Variability

A reservoir is built to regulate natural streamflow variability and to maintain a reliable utility from natural streamflow. The coefficient of variation C_v, which is defined as the ratio of the streamflow standard deviation over the mean value, is commonly used to characterize the inter-period streamflow variability. Figure 1.44 displays the effect of C_v on reservoir operation performances under the various forms of forecast uncertainty.

Figure 1.45 illustrates that, with the increase of C_v, the utility improvements relative to SOP under all the optimized solutions with deterministic or probabilistic forecasts tend to increase. This generally implies that the more variable the streamflow is, the more valuable the forecast is for improving reservoir operation efficiency. Meanwhile, reservoir operation under *rPSF* shows robustness with a high uncertainty level (comparable to the natural variability), for example, when $C_v = 0.1$ (i.e., the forecast uncertainty is comparable to the streamflow variability $\mu C_v = \sigma$), about 50 % of the reservoir operations under *DSF*-DP and about 80 % of the reservoir operations under *pPSF*-SDP are inferior to *Q*-SOP, while *rPSF*-SDP shows a performance similar to *Q*-DP and better than *DSF*-DP.

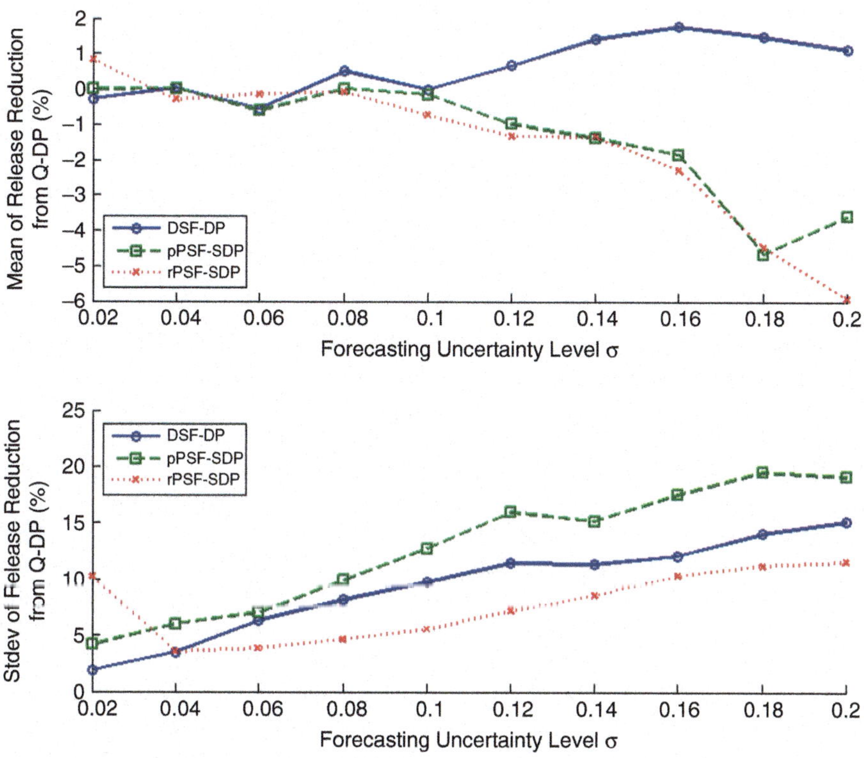

Fig. 1.44 Hedging effects resulting from application of streamflow forecasts to reservoir operation

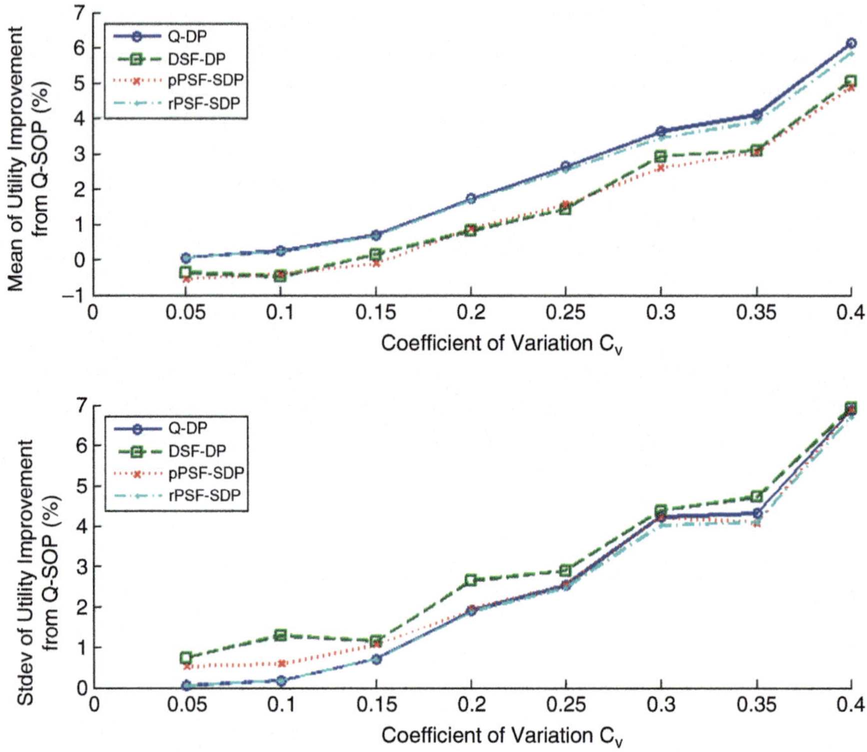

Fig. 1.45 Effect of streamflow variability on the application of streamflow forecasts to reservoir operation

Effect of Reservoir Capacity

The effect of reservoir capacity is studied by varying reservoir capacity S from 20 to 500 % of the mean inflow [μ in Eq. (1.59)], of which the results are shown in Fig. 1.46. With respect to mean utility improvement, the DSF-DP and rPSF-SDP perform similarly. DSF-DP performs more poorly than pPSF-SDP when the storage is small, and gradually improves and approaches the performance of pPSF-SDP as the reservoir storage becomes larger. This is different from the above comparisons between DSF-DP and pPSF-SDP, where the DSF-DP performs similarly to pPSF-SDP under various forecast uncertainty and streamflow variability levels. Note that pPSF differs from DSF because it includes an empirical uncertainty analysis that addresses the risk induced by forecast error. Thus the poor performance of DSF-DP compared to pPSF-SDP when the reservoir storage is small implies that small reservoirs are more sensitive to forecast uncertainties. Standard deviation values of utility improvements show a similar performance with the increase of reservoir storage, except that the DSF-DP has a larger standard deviation when the reservoir storage is small. This also suggests that the DSF-guided reservoir operation is vulnerable to forecast errors when the reservoir is small.

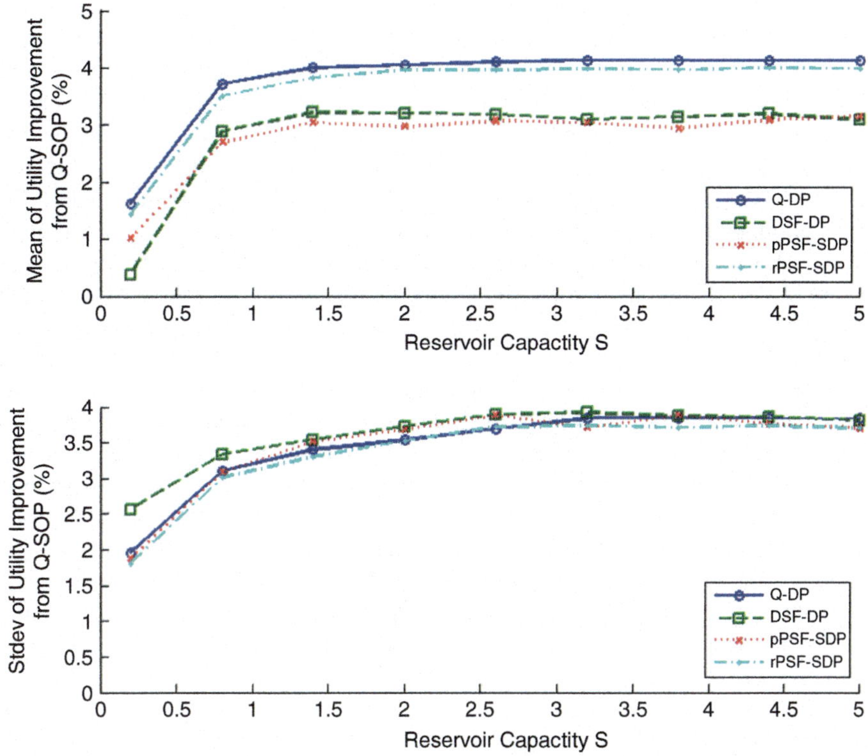

Fig. 1.46 Effect of reservoir capacity on the application of streamflow forecasts to reservoir operation

6.2 *Identifying Effective Forecast Horizon for Real-Time Reservoir Operation*

6.2.1 Background

Reservoirs are built to regulate streamflow variability and to meet operation objectives, such as water supply, irrigation, flood control, and hydropower generation. Conventionally, reservoir water release decisions are determined by prescribed operation rules. These rules are based on historical streamflow records, which specify reservoir release under certain storage and inflow conditions. However, in real-time reservoir operation, operation rules cannot effectively utilize streamflow forecasts [106]. In recent years, various optimization and simulation models have been developed to exploit streamflow forecasts for reservoir decision making [126, 127]. As advances in weather forecasting, hydrologic modeling, and hydro-climatic teleconnections have significantly reduced streamflow forecast uncertainty and prolonged the forecast horizon, streamflow forecasts are now a more promising tool for improving reservoir operation efficiency [18, 103, 114].

Applications of streamflow forecasts to reservoir operation are constrained by their limited length and a meaningful forecast horizon (*FH*) is usually shorter than the reservoir operation horizon. For example, reservoir flood control operations may last several months, while a streamflow forecast is only available several weeks in advance. The practical length of a streamflow forecast is also limited by the complicating relationship between forecast uncertainty (*FU*) and forecast horizon (*FH*), i.e., the longer the forecast horizon, the more complete the information for decision making, but the larger the forecast uncertainty [105, 112, 113]. Although the complicating effect of *FU* and *FH* is interesting, reservoir operation studies incorporating streamflow forecast do not explicitly analyze the effects. The current study is then motivated to explore explicit relationships between operation decision optimality and the *FU* and *FH* characteristics.

In previous reservoir operation studies, the importance of reducing *FU* has been illustrated by using both an operational forecast with actual reservoir systems [103, 114] and a synthetic forecast with hypothetical reservoir systems [8, 111, 124]. Meanwhile, most studies assumed that the available *FH* was as long as the operation horizon and did not consider the complicating effect of *FU* and *FH*. However, a few previous publications have shed light on the issue. By employing a Kalman filter forecast technique, Simonovic and Burn [113] illustrated the existence of an empirical *FH* threshold, and argued that a *FH* longer than the threshold would not contribute to reservoir performance. More recently, You and Cai [6] presented both a theoretical and a numerical framework to determine the optimal forecast horizon for a given decision horizon, which is defined as the initial periods in which decisions are not affected by forecast data beyond the forecast horizon. Furthermore, You and Cai [7] addressed the dual problem of You and Cai [6], i.e., how far the actual decision is away from the optimal one under a given forecast with a limited horizon. Following You and Cai [6] and You and Cai [7] the current study develops metrics to evaluate reservoir release decisions under a limited forecast and procedures to analyze the effect of *FU* and *FH* and to identify *EFH*. To conduct the analysis, a fixed ending storage is specified for a reservoir operation problem. With this realistic setting, a monotonic relationship between reservoir release and ending storage is derived with a concave objective function. With the monotonic relationship, metrics to evaluate reservoir release decisions under a limited forecast can be calculated. Following that numerical experiments based on synthetic forecasts are designed to analyze the complicating effect of *FH* and *FU* and to illustrate the procedures to identify the an effective forecast horizon (*EFH*), beyond which *FH* increase will not contribute to reservoir operation efficiency.

6.2.2 Evaluation Metrics of Limited Inflow Forecast for Reservoir Operation

This section first discusses FU and FH with a particular reservoir operation optimization model, and then provides evaluation metrics of a limited inflow forecast for reservoir operation.

Problem Formulation

Consider a reservoir operation problem with an operation horizon of *N* periods and denote the variables as follows:

i	the index of time periods;
s_i	reservoir storage at the end of period i;
q_i	reservoir inflow at period i;
x_i	forecast of period i reservoir inflow;
r_i	period i reservoir release decision;
$f\_i()$	period i reservoir utility function;
$\underline{s}$	the minimum reservoir storage;
$\overline{s}$	the maximum reservoir storage;
$\underline{r}$	the minimum reservoir release;
$\overline{r}$	the maximum reservoir release;
d	discount ratio of reservoir utility;
l	loss ratio of reservoir storage;
s_O	the initial reservoir storage;
s_T	the target storage at the end of reservoir operation horizon (N);
s'_T	the target storage at the end of reservoir inflow forecast horizon (H).

Based on the aforementioned variables and the selection of reservoir release r_i as decision variable, the multiple-period reservoir operation optimization model can be formulated as follows:

$$\max \quad \sum_{i=1}^{N} \frac{1}{(1+d)^{i-1}} f_i(r_i) \qquad (1.1)$$

$$s.t. \begin{cases} (1-l)s_{i-1} + q_i - r_i = s_i (i = 1, \ldots, N) & (1.2) \\ \underline{s} \le s_i \le \overline{s} & (1.3) \\ \underline{r} \le r_i \le \overline{r} & (1.4) \\ s_N = s_T & (1.5) \end{cases} \qquad (1.62)$$

Equation (1.62), (1.1) denotes the reservoir operation objective and it is defined as the sum of discounted reservoir utility, which is a function of reservoir release r_i and assumed concave (i.e., diminishing marginal utility or $f''(r_i) \le 0$, for example, willingness-to-pay for one more unit of water is high in drought periods and it decreases as water scarcity abates) in this section [5, 128]; (1.2) is the water balance equation which illustrates the conservation of mass between reservoir storage, inflow and release; (1.3) is the storage capacity constraint; (1.4) is the release capacity constraint; and (1.5) is the ending storage constraint.

Reservoir operation optimization shown in Eq. (1.62) is an ideal case and the underlying assumption is that reservoir inflows through the operation horizon *N* are perfectly known (i.e., no limitation on *FH* and *FU*). In real-world reservoir operation, the inflow information is available only within a limited horizon *H* [128, 129] and involves uncertainties [112, 124, 126]:

$$x_i = q_i + \varepsilon_i \tag{1.63}$$

where ε_i is a random variable indicating forecast uncertainty of period i ($i \leq H$) reservoir inflow.

To bridge the gap between the operation horizon *N* and the forecast horizon *H*, real-world reservoir operation typically employs a rolling horizon approach (Fig. 1.47), i.e., (1) making the release decision for the decision horizon with a limited forecast; (2) implementing the current release decision; (3) move to the next period and repeat (1) and (2) with updated inflow forecast and reservoir storage until the end of operation horizon [8, 112, 130].

In the rolling horizon approach, decision horizon (*DH*, how long the generated decision is implemented), forecast horizon (*FH*, how long the inflow can be predicted), and operation horizon (*OH*, how long the reservoir operation is targeted) are important issues (Fig. 1.47). As the inflow forecast can update period by period in real-time reservoir operation, *DH* is usually set as 1 (i.e., only the current period decision is treated as final and decisions in future periods will be updated with the new forecast). *FH* depends on the forecast technology and *OH* is set equal to *FH*, although the original *OH* [Eq. (1.62)] is determined by inflow variability and reservoir characteristics (e.g., a seasonal reservoir has an *OH* of several months and an annual reservoir has an *OH* of 1 year).

The reservoir optimization operation model for rolling horizon decision making can be formulated as follows:

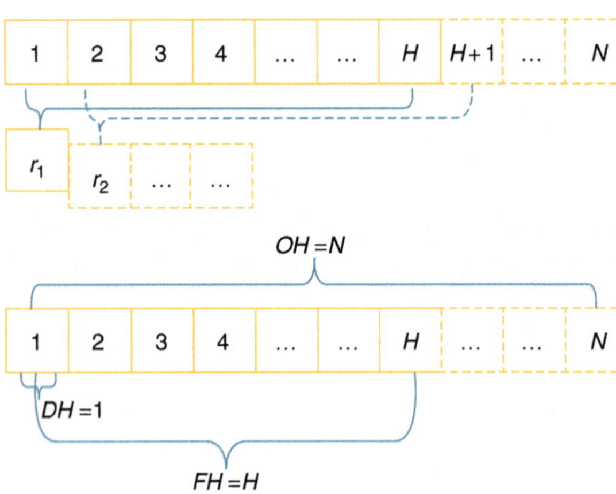

Fig. 1.47 Schematic of reservoir operation rolling horizon decision making and its operation horizon, forecast horizon, and decision horizon

$$\max \quad \sum_{i=1}^{H} \frac{1}{(1+d)^{i-1}} f_i(r_i)$$
$$s.t. \quad \begin{cases} (1-l)s_{i-1} + x_i - r_i = s_i (i = 1, \ldots, H) \\ \underline{s} \le s_i \le \overline{s} \\ \underline{r} \le r_i \le \overline{r} \\ s_H = s'_T \end{cases} \tag{1.64}$$

Equation (1.64) represents a practical reservoir operation model, which is different from Eq. (1.62) in: (1) x_i contains forecast uncertainty; (2) the operation horizon is H; (3) the user-specified ending storage is s'_T (which is differentiated from s_T used in Eq. (1.62), the target end storage). With the forecast updated and s'_T specified in each period, the rolling horizon approach can be employed to determine real-time reservoir releases.

The practical model [Eq. (1.64)] can generate a release sequence $[r'_1, r'_2, \ldots, r'_H]$ within *FH* while only the current decision r'_1 is implemented (*DH* is 1) and subsequent decisions will be obtained by re-running the practical model with an updated inflow forecast through the rolling horizon process. For reservoir operation applications, the rolling horizon approach generally exhibits superior performance to the climatology scenario based stochastic approach, especially in extreme hydrologic conditions [115, 130]. For the comparability of decision making through the rolling horizon process, this section focuses on one single period and investigates the gap between r'_1 [the local optimal decision with a limited forecast, Eq. (1.64)] and r^*_1 [the global optimal decision with a perfect forecast, Eq. (1.62)].

Monotonicity Property and Reservoir Release Bounds

One direct question from the rolling horizon approach is how the generated local optimal decision can approximate the global optimal decision, i.e., the gap between r^*_1 and r'_1. Since s'_T indicates the trade-off between water-use within *FH* and water-use beyond *FH* and affects r'_1 [i.e., ending storage effect, Eq. (1.64)], another question is what effect s'_T exerts on the gap between r'_1 and r^*_1. This section addresses the second question first and illustrates a monotonic relationship between s'_T and r'_1:

Theorem *Given a pre-determined forecast horizon FH and inflow forecast* $[x_1 \quad x_2 \quad \ldots \quad x_H]$, *if the reservoir release utility function* $f_i()$ *exhibits a diminishing marginal utility property (i.e., concavity and* $f''_i() < 0$*), then* r'_1 *underlying a given* s'_T *will not increase if* s'_T *increases.*

This theorem illustrates a monotonic ending storage s'_T effect on first period decision r'_1. For rolling horizon reservoir operation, this monotonic relationship generally suggests that the current release decision r'_1 will not increase if we set a higher s'_T (i.e., to save more water for periods beyond FH). It is important to note that the monotonicity dependence relationship has been studied in supply chain management [131, 132]; this section introduces the relationship to reservoir operation and extends it by considering the effects of utility discount and storage loss.

A direct corollary of the monotonicity theorem is that: r'_1 is bounded by $r'_{1,u}$ (upper bound, defined as r'_1 corresponding to $s'_T = \underline{s}$) and $r'_{1,l}$ (lower bound, defined as r'_1 corresponding to $s'_T = \overline{s}$). In general, $r'_{1,u}$ implies the release decision under the most optimistic expectation of future streamflow, as total inflow plus initial storage are scheduled to be used up within FH. Alternately, $r'_{1,l}$ implies the release decision under the most pessimistic expectation of future streamflow and that storage and inflow are planned to be saved as much as possible for periods beyond FH. For real-time reservoir operation, when practical values of $\underline{s}$ and $\overline{s}$ are given, these two releases bounds $r'_{1,u}$ and $r'_{1,l}$ can be used to determine the range of reservoir release decisions under a given streamflow forecast (Fig. 1.48).

Based on the monotonicity property of r'_1 and the two release bounds $r'_{1,u}$ and $r'_{1,l}$, we can analyze the gap between r^*_1 and r'_1 (i.e., the optimality of reservoir operation decision with a limited forecast). The gap between r^*_1 and r'_1 is bounded and can be represented by the gaps between r^*_1 and $r'_{1,u}$ and between r^*_1 and $r'_{1,l}$. Three error bound indices are derived:

$$EBR = r'_{1,u} - r'_{1,l} \tag{1.65}$$

$$EBU = r'_{1,u} - r^*_1 \tag{1.66}$$

$$EBL = r^*_1 - r'_{1,l} \tag{1.67}$$

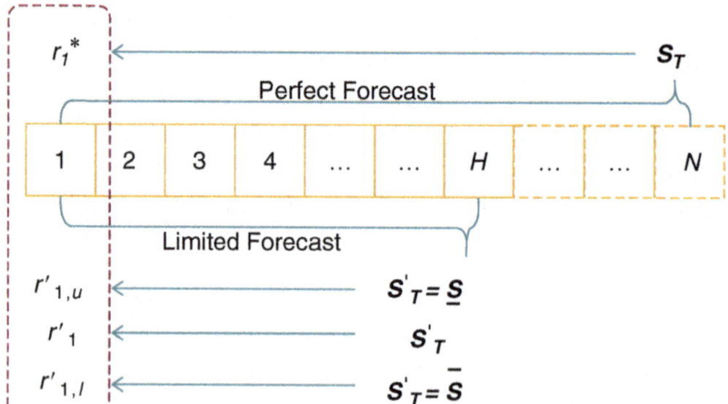

Fig. 1.48 The optimal first period release decision r*1 of the ideal model and the upper and lower bound for the first period release decision r′1 of the practical model

EBR (called error bound range) is the difference between $r'_{1,u}$ and $r'_{1,l}$, or rather, the variation in the range of r'_1. *EBU* (called upper error bound) and *EBL* (called lower error bound) indicate the error bound of the most optimistic and pessimistic expectation of future streamflow, respectively.

For these three indices, *EBR* is applicable to real-time reservoir operation for the determination of the release range. As r^*_1 is obtained through perfect information of reservoir inflow (not possible in real-time reservoir operation), *EBU* and *EBL* are retrospective analysis indices with the assumption that the inflow process has already been realized and release decisions $r'_{1,u}$ and $r'_{1,l}$ were made prior to inflow realization. This section is, thus, a retrospective analysis of real-time reservoir operation and applies *EBR*, *EBU* and *EBL* to diagnosing the optimality of release decision under a limited forecast. In the following sections, this section will investigate *EBR*, *EBU* and *EBL* by varying *FH*, *FU*, streamflow variability, and reservoir characteristics. Considering the difficulty of obtaining an analytic optimal solution for the multiple period reservoir optimization operation model [Eqs. (1.62) and (1.65)], numerical experiments with a hypothetical reservoir are adopted in the subsequent analysis.

6.2.3 A Real-World Case Study to Detect Effective Forecast Horizon

The analysis described above is applied to a real-world case study—the Danjiangkou Reservoir, one of the major reservoirs for water transfer in the central route of the South–North Water Transfer Project (SNWTP) in China. The retrospective analysis period is selected as June 1 to September 30 in 2008 (122 days).

The case study focuses on the water supply function of the reservoir and the objective function is chosen as to minimize the shortage index:

$$\min \quad SI = \sum_{i=1}^{122} \left(\frac{TS_i}{TD_i} \right)^2 \tag{1.68}$$

in which i is the time index ($i = 1, .. 122$); TD_i is the water demand in period i; TS_i is the water shortage defined as below

$$TS_i = \max(0, TD_i - r_i) \tag{1.69}$$

The constraints are similar to that in Eqs. (1.62) and (1.64). s and $\bar{s}$ are set as $121.0*10^8$ m^3 and $146.6*10^8$ m^3, respectively; the release constraint is not considered. Since the reservoir also serves for flood control, as well as water supply, during the study period, the ending storage s'_T is set as the minimum storage $(\underline{s})$ for flood control, 12.1 billion m^3. When s'_T is set as the minimum following the flooding control regulation and r_1' is the maximum, only *EBU* described in Sect. 3 is analyzed for the case study.

The auto-regressive forecast uncertainty is incorporated into the retrospective analysis in a rolling time window. In each day, a deterministic streamflow forecast within the *FH* is generated and it is assumed that the forecast error τ_i fits a gamma distribution with the mean of $(1-\gamma_i)\mu_i$ and the variance of $\left(1-\gamma_i^2\right)\mu^2 C_v^2$ [115]. It is assumed that the model determined release at the current day is always implemented, and the reservoir storage is updated to next day. In the numerical experiments, *FH* is varied from 3 days to 15 days and three δ levels ($\delta = 0.005, 0.05,$ and 0.10) are tested. The results that illustrate the relationships between *FU*, *FH* and average shortage index are presented in the lower part of Fig. 1.49.

As can be seen from Fig. 1.49, at the same *FH* level, the lower the *FU*, the better the reservoir performance; meanwhile, at the same *FU* level, if there is no uncertainty ($\delta = 0.00$), the longer *FH*, the better the reservoir performance; while if *FU* exists, to prolong *FH* beyond a certain threshold (i.e., *EFH*) will even decline the reservoir performance. *EFH* = 12 corresponding to $\delta = 0.05$ and *EFH* = 7 when $\delta = 0.10$, which verifies that a larger *FU* ends with a shorter *EFH*.

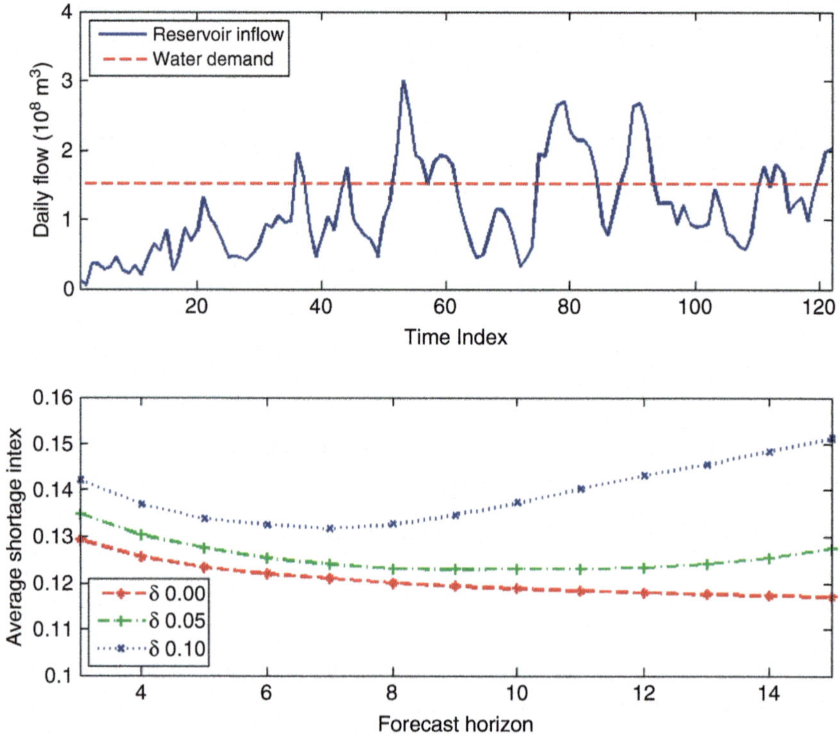

Fig. 1.49 Danjiangkou reservoir water supply operation evaluation under different FU and FH conditions

6.3 *Generalized Marginal Model of the Uncertainty Evolution of Inflow Forecasts*

6.3.1 Background

Streamflow forecasts provide useful information about the future. Forecasts are especially important for predicting extreme hydrologic events and are used to guide management decisions on water resource systems [134, 135]. Advances in weather forecasting, hydrologic modeling, and hydro-climatic teleconnections have improved the ability to forecast streamflows [136–138]. As a result, streamflow forecasts have been extensively applied in water resource management. The applications usually take a two-component approach. One paper develops innovative models for streamflow forecasting, whereas the other proposes novel optimization models to incorporate the forecast into decision making. For example, Carpenter and Georgakakos [136] generated an ensemble streamflow forecast that considered both atmospheric forcing and hydrologic model uncertainty. Yao and Georgakakos [107] then developed forecast-management schemes with operation rules and optimization models. Mauer and Lettenmaier [114] assessed the seasonal streamflow predictability of the Mississippi River basin. Based on these data, Mauer and Lettenmaier [107] evaluated the value of seasonal streamflow forecast to guide the Missouri River main-stem reservoir operation. Ajami et al. [139] proposed an integrated Bayesian uncertainty estimator to account for input, parameter, and model structural uncertainty in hydrologic prediction, after which they demonstrated the importance of considering hydrological uncertainty in sustainable water resource management [127].

Uncertainty is an inherent and important characteristic of streamflow forecasting. In both real-world and hypothetical studies focusing on the applications of streamflow forecasting, uncertainty has been identified as the major influencing factor of the value of the forecast [8, 104, 127]. Real-world studies that aim to develop decision support systems for a targeted river basin generally address forecast uncertainty using advanced forecast techniques (e.g., ensemble forecasts) and optimization (or simulation) models [107, 128, 134]. Hypothetical studies typically use synthesized forecast uncertainty based on certain assumptions, e.g., unbiasedness, and Gaussian distributions [8, 140, 141]. Testing the validity of these assumptions is an important issue for this type of hypothetical study. In this investigation, we use real-world forecast data and perform statistical tests on assumptions of forecast uncertainty.

Forecast uncertainty evolves in real-time because streamflow forecasts are dynamically updated. On one hand, the uncertainties of forecasts for future periods become larger as forecast lead-time increases. On the other hand, the uncertainties of forecasts for a certain time period decrease over time as more hydrologic information becomes available. Heath and Jackson [118] proposed a martingale model of forecast evolution (MMFE) to formulate the uncertainty evolution of demand forecasts in supply chain management. Zhao et al. [8] applied MMFE to model deterministic, ensemble, and probabilistic streamflow forecasts and illustrated that ensemble and probabilistic forecasts are more effective than a deterministic forecast. MMFE formulates the total forecast uncertainty of a streamflow in one future period as the sum of forecast improvements in the intermediate periods. This study tests the assumptions of MMFE, i.e., single-period forecast improvements are unbiased, Gaussian distributed, temporally independent, and stationary. Furthermore, this study proposes a generalized martingale model GMMFE to address cases wherein the assumptions are violated.

6.3.2 Methodology

Streamflow forecasts are updated in real-time. At the beginning of one time period, forecasts of streamflow in the subsequent time periods are made based on currently available hydrologic information. As time progresses and as more hydrologic information becomes available, the forecasts are dynamically updated. This section introduces the MMFE, which describes this dynamic forecast-updating process.

Mathematical Formulation of Uncertainty Evolution

$f_{s,t}$ is denoted as the forecast made at period s for the streamflow at period t (s must be less than or equal to t). The forecasts made at period s form a vector $F_{s,-}$ comprising $f_{s,s+i}$ ($i = 0, 1, \ldots, h$; h denotes the forecast horizon) with lead time ranging from 0 to h periods, i.e.,

$$F_{s,-} = \left[f_{s,s} \quad f_{s,s+1} \quad \cdots \quad f_{s,s+h} \right] \tag{1.70}$$

In subsequent periods $s+1$, $s+2$, ..., $F_{s+1,-}$, $F_{s+2,-}$, ..., are made. A schematic of the rolling horizon process of a real-time streamflow forecast with a forecast horizon h (h is set as 4 periods for example) is given at the upper part of Fig. 1.50.

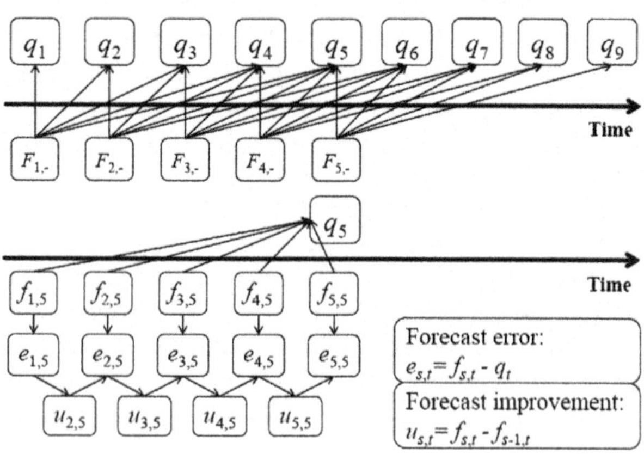

Fig. 1.50 Schematic of uncertainty evolution in real-time streamflow forecasting

On one hand, $F_{s,-}$ contains multiple forecasts of streamflow in the subsequent h periods. On the other hand, streamflow q_t at period t corresponds to multiple forecasts made at the precedent periods, as shown in the lower part of Fig. 1.50. The $h+1$ forecasts $f_{t-i,t}$ ($i = h, \ldots, 1, 0$) corresponding to q_t form a vector $F_{-,t}$ ($F_{-,t}$ is differentiated from $F_{s,-}$).

$$F_{-,t} = \begin{bmatrix} f_{t-h,t} & f_{t-h+1,t} & \cdots & f_{t,t} \end{bmatrix} \tag{1.71}$$

The relationship among $f_{s,t}$ (the estimated value), q_t (the real value), and $e_{s,t}$ (the forecast error) are additive [8, 118, 141].

$$e_{s,t} = f_{s,t} - q_t \tag{1.72}$$

The forecast errors of $F_{-,t}$ also form a vector $E_{-,t}$

$$E_{-,t} = \begin{bmatrix} e_{t-h,t} & e_{t-h+1,t} & \ldots & e_{t,t} \end{bmatrix} \tag{1.73}$$

With $e_{s,t}$ ($s = t-h, t-h+1, \ldots, t$), forecast improvement $u_{s,t}$ can be defined as the difference between the forecast errors of two consecutive periods

$$\begin{aligned} u_{s,t} &= e_{s,t} - e_{s-1,t} \\ &= \left(f_{s,t} - q_t\right) - \left(f_{s-1,t} - q_t\right) \\ &= f_{s,t} - f_{s-1,t} \end{aligned} \tag{1.74}$$

As shown in Eq. (1.74) and Fig. 1.50, $u_{s,t}$ represents the improvement in $f_{s,t}$ (the period s forecast of q_{-t}) from $f_{s-1,t}$ (the previous period's forecast of q_t). We have a total of h updates of $u_{s,t}$ ($s = t{-}h+1, t{-}h+2, \ldots, t$) for q_t, which correspond to the $h+1$ elements in $E_{-,t}$ and $F_{-,t}$.

Assuming that the observation (denoted as $f_{t,t}$) at the current period is perfect,

$$f_{t,t} = q_t \tag{1.75}$$

The relationship between $e_{s,t}$ and $u_{s,t}$ can be formulated as follows:

$$\begin{cases} e_{t,t} = 0 \\ e_{t-1,t} = e_{t,t} - u_{t,t} = -u_{t,t} \\ e_{t-2,t} = e_{t-1,t} - u_{t-1,t} = -u_{t,t} - u_{t-1,t} \\ \ldots \\ e_{t-h+1,t} = e_{t-h+2,t} - u_{t-h+2,t} = -\sum\limits_{i=0}^{h-2} u_{t-i,t} \\ e_{t-h,t} = e_{t-h+1,t} - u_{t-h+1,t} = -\sum\limits_{i=0}^{h-1} u_{t-i,t} \end{cases} \tag{1.76}$$

By incorporating Eq. (1.76) into Eq. (1.72), forecast $f_{s,t}$ can be formulated with q_t and $u_{s,t}$ ($s = t\text{–}h+1, t\text{–}h+2, \ldots, 0$)

$$\begin{cases} f_{t,t} = q_t \\ f_{t-1,t} = q_t - u_{t,t} \\ f_{t-2,t} = q_t - u_{t,t} - u_{t-1,t} \\ \ldots \\ f_{t-h+1,t} = q_t - \sum_{i=0}^{h-2} u_{t-i,t} \\ f_{t-h,t} = q_t - \sum_{i=0}^{h-1} u_{t-i,t} \end{cases} \tag{1.77}$$

The equation also indicates that

$$\begin{cases} f_{t,t} = f_{t-1,t} + u_{t,t} \\ f_{t-1,t} = f_{t-2,t} + u_{t-1,t} \\ \ldots \\ f_{t-h+1,t} = f_{t-h,t} + u_{t-h+1,t} \end{cases} \tag{1.78}$$

which implies that $f_{s,t}$ continues to be improved by $u_{s,t}$ as s increases from $t-h$ to t, as shown in Fig. 1.50.

Generalized Martingale Model

The MMFE model deals with unbiased Gaussian samples and synthesizes the unbiased Gaussian forecast improvements based on the variance–covariance matrix [8, 118]. However, the non-Gaussian properties of forecast improvements hinder the application of MMFE in real-world studies. To bridge this gap, this study integrates the NQT method, which can convert biased non-Gaussian distributed variables into unbiased Gaussian variables [119, 142], with the conventional MMFE model. This new model GMMFE comprises three steps: NQT, MMFE, and inverse-NQT, as shown in Fig. 1.51.

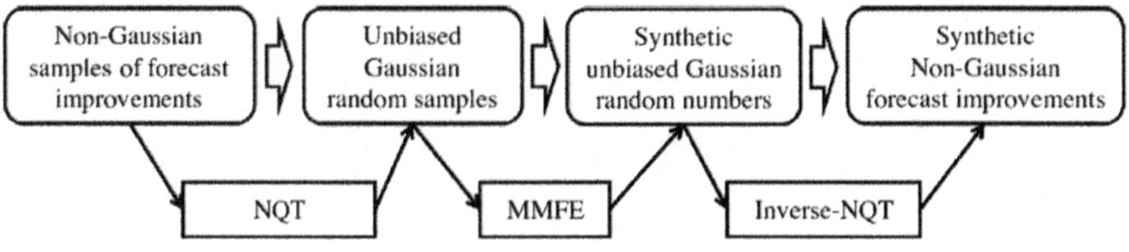

Fig. 1.51 Flowchart of the generalized marginal model GMMFE

(1) Step 1 is the transformation of variables by NQT. $u_{s,s+i-1}$ ($i = 1, 2, \ldots, h$) are converted into standard Gaussian random variables $u'_{s,s+i-1}$ ($i = 1, 2, \ldots, h$) based on

$$u'_{s,s+i-1} = CDF^{-1}_{Gaussian}(CDF_i(u_{s,s+i-1})) \tag{1.79}$$

where $CDF_i()$ is the CDF of $u_{s,s+i-1}$, and $CDF^{-1}_{Gaussian}()$ is the inverse of the CDF of standard Gaussian distribution. Equation (1.79) comprises two sub-steps. First, $CDF_i()$ transforms $u_{s,s+i-1}$ into the corresponding cumulative probability, which is uniformly distributed between 0 and 1. Thereafter, $CDF^{-1}_{Gaussian}()$ converts the cumulative probabilities into the standard Gaussian random $u'_{s,s+i-1}$. The principle behind NQT is similar to that of the quantile–quantile plot [143] in that the two random variables are matched based on the values of their cumulative probability.
(2) Step 2 is the application of MMFE to the transformed variables. The variance–covariance of $u'_{s,s+i-1}$ ($i = 1, 2, \ldots, h$) is calculated, after which u'_i ($i = 1, 2, \ldots, h$) is generated based on the Cholesky decomposition of the variance–covariance matrix. Both u'_i ($i = 1, 2, \ldots, h$) and $u'_{s,s+i-1}$ ($i = 1, 2, \ldots, h$) are standard Gaussian random variables that have the same cross correlation relationships.
(3) Step 3 is the inverse transformation of the variables by NQT. u'_i ($i = 1, 2, \ldots, h$) and forecast improvements u_i ($i = 1, 2, \ldots, h$) are generated with the inverse application of NQT

$$u_i = CDF^{-1}_i\left(CDF_{Gaussian}\left(u'_i\right)\right). \tag{1.80}$$

In Eq. (1.80), $CDF_{Gaussian}()$ converts u'_i into the corresponding cumulative probability, whereas $CDF^{-1}_i()$ converts the cumulative probabilities into new forecast updates u_i ($i = 1, 2, \ldots, h$). Notably, $CDF^{-1}_i()$ plays an important role in determining the statistical distribution of u_i ($i = 1, 2, \ldots, h$). For example, the substitution of $CDF^{-1}_i()$ with the inverse of the CDFs of other distributions, e.g., $N(\mu_i, \sigma^2_i)$, enables u_i to fit the Gaussian distribution with mean μ_i and standard deviation σ_i.

When applying GMMFE to real-world cases, these three steps are needed, and special attention should be focused on fitting the CDF for the given samples of $u_{s,s+i-1}$ ($i = 1, 2, \ldots, h$). However, the first step of NQT for handling given samples is unnecessary for hypothetical studies. On the other hand, the two steps of MMFE and inverse-NQT are needed. The variance–covariance matrix should be set for Step 2 to account for the dependence relationships [118, 144]. $CDF^{-1}_i()$ ($i = 1, 2, \ldots, h$) should be specified for Step 3 to determine the statistical distribution of forecast improvements. In a simple case where forecast improvements in each period [Eq. (1.79)] are assumed to be independent, only Step 3 is needed, and u_i ($i = 1, 2, \ldots, h$) can be individually simulated. The streamflow forecast can then be synthesized based on Eqs. (1.70)–(1.78).

Simulation of Streamflow Forecast

The GMMFE model is applied to generate synthetic streamflow forecast improvements for the TGR during the main flood season of 2008 (Fig. 1.51). The CDF of us, s + i − 1 (i = 1, 2, . . ., h) samples are estimated by the non-parametric kernel density function [145]. To examine the effects of the assumptions of unbiasedness and Gaussianity, three cases are designed:

(1) In the case UG, $CDF_i^{-1}()$ in Eq. (1.40) is substituted by the inverse of the CDF of $N(0, \sigma_i^2)$, and ui fits unbiased Gaussian distribution;
(2) In the case BG, $CDF_i^{-1}()$ is substituted by the inverse of the CDF of $N(\mu_i, \sigma_i^2)$, and ui fits the biased Gaussian distribution; and
(3) In the case NG, $CDF_i^{-1}()$ is represented by the inverse of the kernel cumulative density function, and ui fits the non-Gaussian distribution.

In the cases of UG and BG, μ_i and σ_i denote the mean and the standard deviation of the us, s + i − 1 (i = 1, 2, . . ., h; h = 4) samples of the main flood season in 2008, respectively.

The GMMFE model generates forecast improvements at each period [Eq. (1.79)] and simulates streamflow forecast [Eqs. (1.76) and (1.77)]. We run 100 Monte-Carlo simulations with GMMFE and derive the forecast errors of synthetic forecast with a lead-time of 1 d to 4 d for the three cases. The mean, standard deviation (stdev), and the coefficient of skewness (Cs) of the forecast errors are presented by boxplots, as shown in Fig. 1.52. The boxplots illustrate the median as a central mark, the 25th and 75th percentiles as edges, the range of data points as whiskers, and the outliers as plus signs. For comparison, the statistics of the forecast errors of the TGR streamflow forecast records are represented by the circles linked by a line in Fig. 1.52.

The effect of the violation of the unbiasedness assumption is detected by the mean value in the case UG, which shows that this assumption results in the underestimation of the forecast error by approximately 1000 m^3/s when the lead time is 4 d. The effect of the violation of the Gaussianity assumption is illustrated by the Cs values in the cases of UG and BG, which indicates that this assumption results in the underestimation of the skewness of the forecast errors because Gaussian distribution is symmetric. The case of NG considers both unbiased and non-Gaussian forecast uncertainty, which results in statistics of synthetic forecast errors similar to actual values. In summary, the three cases demonstrate the effectiveness and generality of the GMMFE model in simulating evolution of forecast uncertainties.

6.3.3 Effect of Forecast Uncertainty Distribution on Reservoir Operation

Rolling-Horizon Reservoir Operation

Reservoir operation utilizes a rolling-horizon process to incorporate the dynamically updated streamflow forecast into decision-making [134, 146]. This study

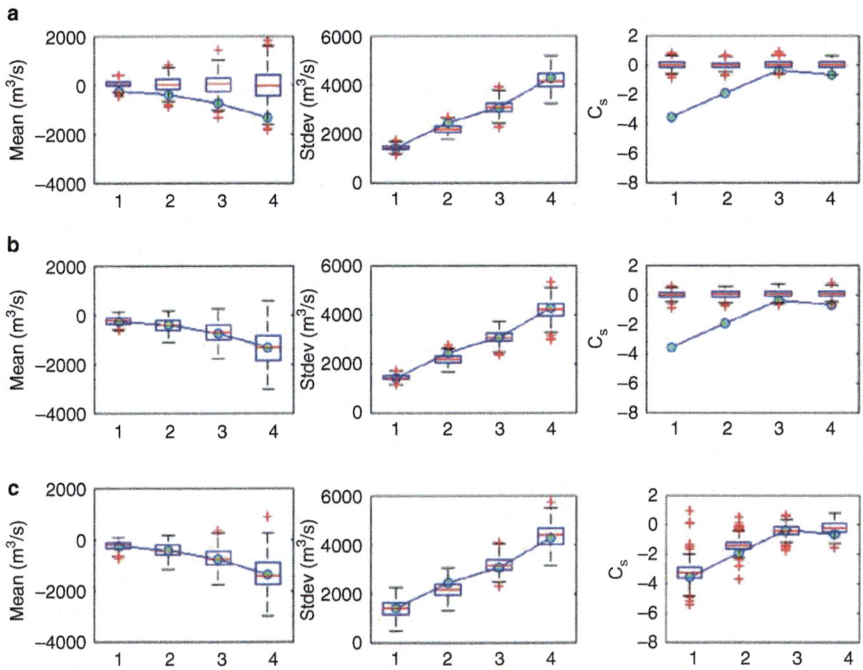

Fig. 1.52 Distribution of the means, standard deviations, and coefficients of skewness of the observed and simulated forecast errors

employs a hypothetical reservoir and synthesizes streamflow forecasts for reservoir operation. The reservoir operation model aims to maximize total utility, i.e.,

$$\max \quad \sum_{i=1}^{n} g_i(r_i)$$
$$s.t. \quad \begin{cases} s_i + x_i - r_i = s_{i+1}(i = 1, \ldots, n) \\ s_{\min} \leq s_i \leq s_{\max}(i = 1, \ldots, n) \\ s_1 = s_{ini} \\ s_{n+1} = s_{end} \\ r_i \geq r_{\min} \end{cases} \tag{1.81}$$

where i is the index of time periods ranging from 1 (the current period) to n (the operation horizon); s_i denotes the reservoir storage at the beginning of period i; x_i and r_i represent the period i's streamflow forecast and release decision respectively; $s_{\min}$ and $s_{\max}$ are the minimum and the maximum of reservoir storage, respectively; s_{ini} and s_{end} denote initial and ending storage, respectively; and $r_{\min}$ is the lower bound of reservoir release.

In the reservoir operation with a study horizon of T periods, the rolling-horizon process comprises the following steps:

(1) Release decisions are made based on the forecasts made at the current period s

$$[x_1 \quad x_2 \quad \ldots \quad x_n] = [f_{s,s} \quad f_{s,s+1} \quad \cdots \quad f_{s,T}] \tag{1.82}$$

(2) The current release decision r_1 is implemented, and r_1 is saved as r_s.
(3) The next period $s+1$ is considered, and the initial storage and streamflow forecast are updated:

$$s_{ini} = s_1 + q_s - r_s \tag{1.83}$$

$$[x_1 \quad x_2 \quad \ldots \quad x_n] = [f_{s+1,s+1} \quad f_{s+1,s+2} \quad \cdots \quad f_{s+1,T}] \tag{1.84}$$

These three steps are repeated until the end of the study horizon T. The operation horizon n in Eq. (1.75) notably reduces from T to 1 as s progresses from 1 to T. In each period, the current release decision is saved, and the single-period utility is evaluated. The total utility is the sum of all single-period utilities. Finally, the total utility of the rolling-horizon reservoir operation can be compared with that of the baseline scenario, which is defined as a case without any reservoir regulation. The utility improvement indicates the value of forecast-based reservoir operation.

Experiment Setting

The experiment is set up based on the reservoir operation model given by Zhao et al. [8]. The reservoir takes a concave utility function, i.e., r_i exhibits a diminishing marginal utility (for instance, the case of water supply operation)

$$g_i(r_i) = \begin{cases} \sqrt{\dfrac{r_i - r_{\min}}{r_{\max} - r_{\min}}} & (r_{\max} \geq r_i \geq r_{\min}) \\ 1 & (r_i > r_{\max}) \end{cases} \tag{1.85}$$

The parameters of the reservoir are as follows: $s_{\min} = 0, r_{\min} = 0.2, r_{\max} = 1.2$, and $s_{ini} = s_{end} = \frac{s_{\max}}{2}$. The study horizon T is set as six periods. Two scenarios are set for $s_{\max}$, i.e., 0.5 and 2.0, examining the effects of reservoir storage capacity [146, 147].

The reservoir inflow is generated using a simplified Thomas–Fiering model, i.e.,

$$q_{t+1} = \mu + \rho_{flow}(q_t - \mu) + \sqrt{1 - \rho_{flow}^2}(\mu C_v)\delta \tag{1.86}$$

where q_t denotes streamflow in period t; δ is a standard Gaussian random number for Monte-Carlo simulation; and the parameters are set as $\mu = 1$, $\rho_{flow} = 0.4$, and $C_v = 0.3$.

The streamflow forecast is generated by GMMFE. To simplify the formulation, $u_{s,t}$ ($t=s, \ldots, T$) are treated as independent and identically distributed random variables, and $u_{s,t}$ is simulated individually. The skewedness in forecast errors is considered by setting the positively and negatively skewed distributions for non-Gaussian forecast uncertainty. Four cases are designed:

1) In the case of UG, $u_{s,t}$ is from an unbiased Gaussian distribution with a mean of 0 and a stdev of σ

$$u_{s,t} = \sigma\delta, \delta \sim N\left(0, 1^2\right) \tag{1.87}$$

2) In the case of BG, $u_{s,t}$ is from a biased and Gaussian distribution with a mean of Δ and a stdev of σ

$$u_{s,t} = \Delta + \sigma\delta, \delta \sim N\left(0, 1^2\right) \tag{1.88}$$

3) In the case of NGn, $u_{s,t}$ is from a negatively-skewed log-normal distribution with a mean of Δ and a stdev of σ

$$u_{s,t} = \Delta + \sigma(2 - \delta), \delta \sim Logn\left(2, 1^2\right) \tag{1.89}$$

4) In the case of NGp, $u_{s,t}$ is from a positively-skewed log-normal distribution with a mean of Δ and a stdev of σ

$$u_{s,t} = \Delta + \sigma(\delta - 2), \delta \sim Logn\left(2, 1^2\right) \tag{1.90}$$

In Eqs. (1.89, 1.90), δ is from the log-normal distribution with a mean of 2 and a stdev of 1.

By fixing Δ as 0.05 and varying σ from 0.02 to 0.20, this study generates streamflow forecasts for the four defined uncertainty distributions using GMMFE. The forecasts are incorporated into the rolling horizon reservoir operation. The effects of forecast uncertainty distribution on reservoir operation among the four cases are analyzed by comparing the utility improvements from the baseline case without any reservoir regulation.

Result Analysis

This study conducts 100 Monte-Carlo experiments for each σ value and evaluates the utility improvements for the four cases of forecast uncertainty. Figure 1.53 presents the utility improvements when reservoir capacity $s_{\max}$ is 0.50. The applications of streamflow facilitate utility improvements in comparison with the baseline case. The mean of the utility improvements decreases, and the stdev of utility improvements tends to slightly increase with increasing σ. Comparing the UG case with the three other cases, the presence of Δ in the BG, NGn, and NGp cases

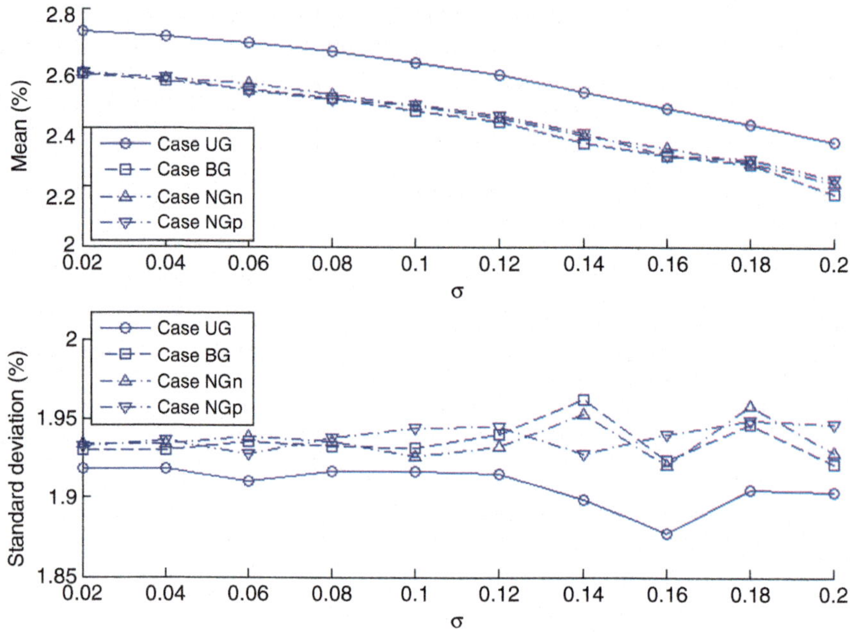

Fig. 1.53 Relationship of forecast uncertainty with the mean and standard deviation of utility improvements (smax = 0.50)

reduces the mean utility improvements and contributes to the increase in standard deviation.

Figure 1.53 further presents the utility improvements when reservoir capacity $s_{max} = 2.0$. Compared with Fig. 1.53, the applications of streamflow forecast bring about greater but also more variable utility improvements. In the BG, NGn and NGp cases, there are marginal differences in terms of the mean of utility improvements. Moreover, there are minimal differences in the stdev of the utility improvements when σ is small, but major differences in stdev are observed as σ increases. The stdev of the utility improvements exhibits the most rapid increase in the NGp case (biased and positively-skewed forecast uncertainty distribution), followed by BG (biased and Gaussian distribution), NGn (biased and negatively-skewed distribution), and UG (unbiased and Gaussian distribution) cases. A larger reservoir can regulate streamflow at a longer timeframe and exploit forecasts with a longer lead time [147]. However, a streamflow forecast with a longer lead time involves greater uncertainty. The effects of the non-Gaussian forecast uncertainty on the reservoir operation are greater in Fig. 1.53 than in Fig. 1.54. This finding implies that more attention should be focused on the non-Gaussian characteristics of forecast uncertainty in the operation of larger reservoirs. Given that unbiased-Gaussian distributions are often simply assumed for forecast uncertainties in real cases, Figs. 1.53 and 1.54 suggest that this assumption results in the overestimation of the utilities from the applications of the streamflow forecast if the actual uncertainties are not

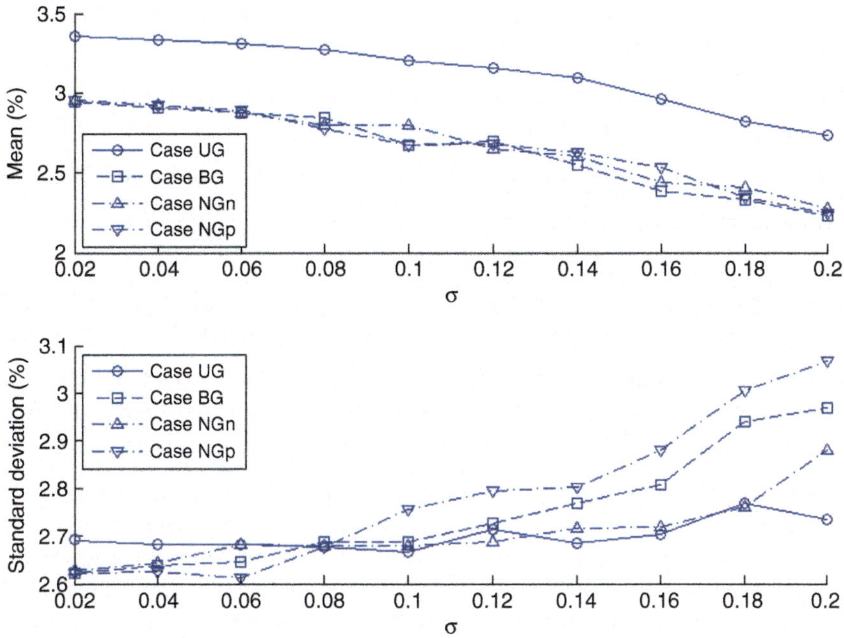

Fig. 1.54 Relationship of forecast uncertainty with the mean and standard deviation of utility improvements (smax = 2.00)

unbiased-Gaussian distributed. Therefore, more attention should be paid to evaluating the statistical distribution of forecast uncertainty before using streamflow forecasts.

Furthermore, the benefit from streamflow forecasts in reservoir operations may vary considerably because of the dependence on a number of factors, e.g., hydrological characteristics, objective functions, and physical constraints [8, 106, 147]. Thus, the effect of forecast uncertainty distribution on reservoir operations may also vary on a case-by-case basis. However, properly estimating the forecast uncertainty distribution is evidently necessary before utilization, and the GMMFE model provides a general tool to address this issue.

Glossary of Reservoir System Operation Terms

Dead Storage The volume in a reservoir below the lowest controllable level.

Flood Control Storage Storage of water in reservoirs to abate flood damage.

Flood The inundation of a normally dry area caused by high flow, or overflow of water in an established watercourse, such as a river, stream, or drainage ditch; or ponding of water at or near the point where the rain fell.

Gaging Station A particular site on a watercourse where systematic observations of stage and/or flow are measured.

Head The difference between the pool height and tailwater height. Usually expressed in feet of head, or in lbs./sq. inch.

Head Loss The decrease in total head caused by friction, entrance and exit losses.

Hydrograph A graph showing the water level (stage), discharge, or other property of a river volume with respect to time.

Hydrologic Model A conceptual or physically-based procedure for numerically simulating a process or processes which occur in a watershed.

Hydrologic Equation The water inventory equation (Inflow = Outflow + Change in Storage) which expresses the basic principle that during a given time interval the total inflow to an area must equal the total outflow plus the net change in storage.

Hydrology The applied science concerned with the waters of the earth, their occurrences, distribution, and circulation through the unending hydrologic cycle of: precipitation, consequent runoff, infiltration, and storage; eventual evaporation; and so forth. It is concerned with the physical and chemical reaction of water with the rest of the earth, and its relation to the life of the earth.

Inactive Storage Capacity The portion of capacity below which the reservoir is not normally drawn, and which is provided for sedimentation, recreation, fish and wildlife, aesthetic reasons, or for the creation of a minimum controlled operational or power head in compliance with operating agreements or restrictions.

Interbasin Water Transfer The physical transfer of water from one watershed to another.

Maximum Spillway Discharge Spillway discharge (cfs) when reservoir is at maximum designed water surface elevation.

Multi-reservoir System More than one reservoir connected with some hydraulic relationship or some operation objective in common, which becomes a system. Multi-reservoir system operation can obtain more benefits than surplus of the ones from individual operation of all member reservoir.

Multipurpose Reservoir A reservoir constructed and equipped to provide storage and release of water for two or more purposes such as flood control, power development, navigation, irrigation, recreation, pollution abatement, domestic water supply, etc.

Municipal Use of Water The various uses of water in developed urban areas, including domestic use, industrial use, street sprinkling, fire protection, etc.

Normal Year A year during which the precipitation or stream flow approximates the average for a long period of record.

Outflow Channel A natural stream channel which transports reservoir releases.

Outlet An opening through which water can be freely discharged from a reservoir.

Precipitation As used in hydrology, precipitation is the discharge of water, in a liquid or solid state, out of the atmosphere, generally onto a land or water surface. It is the common process by which atmospheric water becomes surface, or subsurface water. The term "precipitation" is also commonly used to designate the quantity of water that is precipitated. Precipitation includes rainfall, snow, hail, and sleet, and is therefore a more general term than rainfall.

Reservoir A man-made facility for the storage, regulation and controlled release of water.

Reservoir Inflow The amount of water entering a reservoir expressed in acre-feet per day or cubic feet per second.

Reservoir Operation Management of water release from reservoir to satisfy some specific objective.

Reservoir Regulation (or Operating) Procedure Operating procedures that govern reservoir storage and releases.

Reservoir Surface Area The area covered by a reservoir when filled to a specified level.

Reservoir Volume The volume of a reservoir when filled to normal pool or water level.

Seepage The interstitial movement of water that may take place through a dam, its foundation, or abutments.

Spillway A structure over or through which excess or flood flows are discharged. If the flow is controlled by gates, it is a controlled spillway, if the elevation of the spillway crest is the only control, it is an uncontrolled spillway.

Spillway Crest The elevation of the highest point of a spillway.

Storage Equation The equation for the conservation of mass.

Stream Gage A site along a stream where the stage (water level) is read either by eye or measured with recording equipment.

Stream flow Water flowing in the stream channel. It is often used interchangeably with discharge.

Surcharge Capacity The volume of a reservoir between the maximum water surface elevation for which the dam is designed and the crest of an uncontrolled spillway, or the normal full-pool elevation of the reservoir with the crest gates in the normal closed position.

Total Gross Reservoir Capacity The total amount of storage capacity available in a reservoir for all purposes from the streambed to the normal water or normal water or normal pool surface level. It does not include surcharge, but does include dead storage.

Watershed Land area from which water drains toward a common watercourse in a natural basin.

Water Year The time period form October 1 through September 30.

References

1. Robert, M. H., Jared, L. C., & Charles, S. R. (1977). Gains from joint operation of multiple reservoir systems. *Water Resources Research, 13*(2), 239–245.
2. Lund, J. (2004). Discussion of "Optimal Operation of Multi-reservoir Systems: State-of-the-Art Review" by John W. Labadie. *Journal of Water Resources Planning and Management, 130*(2), 93–111.
3. Lund, J. R., & Guzman, J. (1999). Derived operating rules for reservoirs in series or in parallel. *Journal of Water Resources Planning and Management, 125*(3), 143–153.
4. Lund, J. R. (2000). Derived power production and energy drawdown rules for reservoirs. *Journal of Water Resources Planning and Management, 126*(2), 108–111.
5. Draper, A. J., & Lund, J. R. (2004). Optimal hedging and carryover storage value. *Journal of Water Resources Planning and Management, 130*(1), 83–87.
6. You, J. Y., & Cai, X. M. (2008). Hedging rule for reservoir operations: 1. A theoretical analysis. *Water Resources Research, 44*, W01415. doi:10.1029/2006WR005481.
7. You, J. Y., & Cai, X. M. (2008). Hedging rule for reservoir operations: 2. A numerical model. *Water Resources Research, 44*(1), W01416. doi:10.1029/2006WR005482.
8. Zhao, J. S., Cai, X. M., & Wang, Z. J. (2011). Optimality conditions for a two-stage reservoir operation problem. *Water Resources Research, 47*(8). doi:10.1029/2010WR009971.
9. Shiau, J.-T. (2011). Analytical optimal hedging with explicit incorporation of reservoir release and carryover storage targets. *Water Resources Research, 47*, W01515. doi:10.1029/2010WR009166.
10. Celeste, A. B., & Billib, M. (2009). Evaluation of stochastic reservoir operation optimization models. *Advances in Water Resources, 32*, 1429–1443.
11. Simonovic, S. P. (1992). Reservoir systems analysis closing gap between theory and practice. *Journal of Water Resources Planning and Management, 118*(3), 262–280.
12. Yeh, W. W. G. (1985). Reservoir management and operation models: A state-of-the-art review. *Water Resources Research, 21*(12), 1797–1818.
13. Wurbs, R. A. (1993). Reservoir system simulation and optimization models. *Journal of Water Resources Planning and Management, 119*(4), 455–472.
14. Labadie, J. W. (2004). Optimal operation of multireservoir systems: State-of-the-art review. *Journal of Water Resources Planning and Management-ASCE, 130*, 93–111.
15. Rani, D., & Moreira, M. M. (2010). Simulation–optimization modeling: A survey and potential application in reservoir systems operation. *Water Resources Management, 24*, 1107–1138.
16. Braga, B. P. F., Yen, W. W.-G., Becker, L., & Barros, M. T. L. (1991). Stochastic optimization of multiple-reservoir system operation. *Journal of Water Resources Planning and Management-ASCE, 117*(4), 471–481.
17. Kelman, J., Stedinger, J. R., Cooper, L. A., Hsu, E., & Yuan, S.-Q. (1990). Sampling stochastic dynamic-programming applied to reservoir operation. *Water Resources Research, 26*(3), 447–454.

18. Faber, B. A., & Stedinger, J. R. (2001). Reservoir optimization using sampling SDP with ensemble streamflow prediction (ESP) forecasts. *Journal of Hydrology, 249*(1–4), 113–133.
19. Kim, T., & Heo, J.-H. (2000). Application of implicit stochastic optimization in the Han River basin. In *Proceedings of the Fourth International Conference on Hydroscience and -engineering*. Seoul, Korea: Korean Society of Water Resources.
20. Zhao, T. T. G., Cai, X. M., Lei, X. H., & Wang, H. (2012). Improved dynamic programming for reservoir operation optimization with a concave objective function. *Journal of Water Resources Planning and Management, 138*(6), 590–596.
21. Arunkumar, S., & Yeh, W. W.-G. (1973). *Probabilistic models in the design and operation of a multipurpose reservoir system. Contribution 144*. California Water Resources Center, University of California, Davis.
22. Wang, Y. C., Yoshitani, J., & Fukami, K. (2005). Stochastic multiobjective optimization of reservoirs in parallel. *Hydrological Processes, 19*(18), 3551–3567.
23. Young, G. K. (1967). Finding reservoir operating rules. *Journal of the Hydraulics Division, ASCE, 93*(HY6), 297–321.
24. Karamouz, M., & Houck, M. H. (1982). Annual and monthly reservoir operating rules generated by deterministic optimization. *Water Resources Research, 18*(5), 1337–1344.
25. Willis, R., Finney, B. A., & Chu, W.-S. (1984). Monte Carlo optimization for reservoir operation. *Water Resources Research, 20*(9), 1177–1182.
26. Chandramouli, V., & Raman, H. (2001). Multireservoir modeling with dynamic programming and neural networks. *Journal of Water Resources Planning and Management, 127*(2), 89–98.
27. Jain, S. K., Das, A., & Srivastava, D. K. (1999). Application of ANN for reservoir inflow prediction and operation. *Journal of Water Resources Planning and Management, 125*(5), 263–271.
28. Raman, H., & Chandramouli, V. (1996). Deriving a general operating policy for reservoirs using neural network. *Journal of Water Resources Planning and Management, 122*(5), 342–347.
29. Chandramouli, V., & Deka, P. (2005). Neural network based decision support model for optimal reservoir operation. *Water Resources Management, 19*(4), 447–464.
30. Mousavi, S. J., Ponnambalam, K., & Karray, F. (2005). Reservoir operation using a dynamic programming fuzzy rule-based approach. *Water Resources Management, 19*(5), 655–672.
31. Panigrahi, D. P., & Mujumdar, P. P. (2000). Reservoir operation modelling with fuzzy logic. *Water Resources Management, 14*(2), 89–109.
32. Russell, S. O., & Campbell, P. F. (1996). Reservoir operating rules with fuzzy programming. *Journal of Water Resources Planning and Management, 122*(3), 165–170.
33. Cancelliere, A., Giuliano, G., Ancarani, A., & Rossi, G. (2002). A neural networks approach for deriving irrigation reservoir operating rules. *Water Resources Management, 16*(1), 71–88.
34. Neelakantan, T. R., & Pundarikanthan, N. V. (1999). Hedging rule optimization for water supply reservoirs system. *Water Resources Management, 13*(6), 409–426.
35. Koutsoyiannis, D., & Economou, A. (2003). Evaluation of the parameterization–simulation–optimization approach for the control of reservoir systems. *Water Resources Research, 39*(6), 1170. doi:10.1029/2003WR002148.
36. Nalbantis, I., & Koutsoyiannis, D. (1997). A parametric rule for planning and management of multiple reservoir systems. *Water Resources Research, 33*(9), 2165–2177.
37. Tung, C., Hsu, S., Liu, C.-M., & Li, J. S. (2003). Application of the genetic algorithm for optimizing operation rules of the LiYuTan reservoir in Taiwan. *Journal of the American Water Resources Association, 39*(3), 649–657.
38. Momtahen, S., & Dariane, A. B. (2007). Direct search approaches using genetic algorithms for optimization of water reservoir operating policies. *Journal of Water Resources Planning and Management, 133*(3), 202–209.

39. Kangrang, A., Compliew, S., & Chaiyapoom, W. (2009). Heuristic algorithm with simulation model for searching optimal reservoir rule curves. *American Journal of Applied Sciences, 6* (2), 263–267.
40. Jia, Y., Wang, H., & Yan, D. (2006). Distributed model of hydrological cycle system in Heihe River basin I. Model development and verification. *Journal of Hydraulic Engineering, 37*(5), 534–542 (in Chinese).
41. Jia, Y.-W., Wang, H., Wang, J.-H., Luo, X.-Y., Zhou, Z.-H., & Yan, D.-H. (2005). Development and verification of a distributed hydrological model for the Yellow River Basin. *Journal of Natural Resources, 20*(2), 300–308.
42. Lei, X., Jia, Y., Jiang, Y., & Tian, Y. (2009). Parameter optimization of WEP model and its application to the upstream of Han River. *Journal of Hydraulic Engineering, 40*(12), 112–119 (in Chinese).
43. Lei, X., Yu, T., & Jia, Y. (2010). Global parameter sensitivity analysis of WEP Model and concerned application in Hanjiang river upstream. *Journal of China Hydrology, 30*(6), 12–18 (in Chinese).
44. Lei, X. H., Yu, T., Jiang, Y. Z., & Luo, X. Y. (2009). General catchment delineation method and its application into the middle route project of China's South-to-North water diversion. *The HIKE Transactions*, 23–29.
45. Lei, X. H., Zhou, Z. H., & Ding, X. Y. (2009). Study on the treatment of boundary river and coast line in the watershed subdivision of distributed hydrological model. *Journal of China Hydrology, 29*(6), 1–5 (in Chinese).
46. Lei, X. H., Wang, H. C., & Tian, Y. (2009). Sub-watershed delineation for the taken area of south-to-north water diversion project. *South-to-North Water Transfers and Water Science & Technology, 7*(3), 10–14 (in Chinese).
47. Chen, L. (2003). Real coded genetic algorithm optimization of long term reservoir operation. *Journal of the American Water Resources Association, 39*(5), 1157–1165.
48. Oliveira, R., & Loucks, D. P. (1997). Operating rules for multireservoir systems. *Water Resources Research, 33*(4), 839–852.
49. Ngo, L. L., Madsen, H., & Rosbjerg, D. (2007). Simulation and optimization modeling approach for operation of the Hoa Binh reservoir, Vietnam. *Journal of Hydrology, 336* (3–4), 269–281.
50. Reddy, M. J., & Kumar, D. N. (2007). Optimal reservoir operation for irrigation of multiple crops using elitist-mutated particle swarm optimization. *Hydrological Sciences Journal, 52* (4), 686–701.
51. Chang, L.-C., & Chang, F.-J. (2009). Multi-objective evolutionary algorithm for operating parallel reservoir system. *Journal of Hydrology, 377*(1–2), 12–20.
52. Wilhite, D. A. (1993). The enigma of drought. In D. A. Wilhite (Ed.), *Drought assessment, management and planning: Theory and case studies* (pp. 3–15). Kluwer Academic.
53. Srinivasan, K., & Philipose, M. C. (1996). Evaluation and selection of hedging policies using stochastic reservoir simulation. *Water Resources Management, 10*(3), 163–188.
54. Shih, J. S., & ReVelle, C. (1994). Water supply operations during drought: Continuous hedging rule. *Journal of Water Resources Planning and Management, 120*(5), 613–629.
55. Shih, J. S., & ReVelle, C. (1995). Water supply operations during drought: A discrete hedging rule. *European Journal of Operational Research, 82*, 163–175.
56. Tu, M. Y., Hsu, N. S., & Yeh, W. W. G. (2003). Optimization of reservoir management and operation with hedging rules. *Journal of Water Resources Planning and Management, 129* (2), 86–97.
57. Tu, M. Y., Hsu, N. S., Tsai, F. T. C., & Yeh, W. W. G. (2008). Optimization of hedging rules for reservoir operations. *Journal of Water Resources Planning and Management, 134*(1), 3–13.
58. Arvanitidis, N. V., & Rosing, J. (1970). Composite representation of a multi-reservoir hydroelectric power system. *IEEE Transactions on Power Apparatus and Systems, 89*(2), 319–325.

59. Brandão, J. L. B. (2010). Performance of the equivalent reservoir modeling technique for multi-reservoir hydropower systems. *Water Resources Management, 24*, 3101–3114.
60. Reddy, M. J., & Kumar, D. N. (2006). Optimal reservoir operation using multi-objective evolutionary algorithm. *Water Resources Management, 20*, 861–878.
61. Reddy, M. J., & Kumar, D. N. (2007). Multi-objective particle swarm optimization for generating optimal trade-offs in reservoir operation. *Hydrological Processes, 21*, 2897–2909.
62. Hu, X., & Eberhart, R. (2002). Multiobjective optimization using dynamic neighborhood particle swarm optimization. In *Congress on Evolutionary Computation* (Vol. 2, pp. 1677–1681). Piscataway, NJ.
63. Fieldsend, J. E., & Singh, S. (2002). A multi-objective algorithm based upon particle swarm optimization, an efficient data structure and turbulence. In *Proceedings of the 2002 U.K. Workshop on Computational Intelligence* (pp. 37–44). Birmingham, UK.
64. Hui, X., Eberhart, R. C., & Shi, Y. (2003). Particle swarm with extended memory for multiobjective optimization. In *IEEE Swarm Intelligence Symposium Proceedings* (pp. 193–197). Indianapolis, IN, USA.
65. Liu, Y. (2009). Automatic calibration of a rainfall-runoff model using a fast and elitist multi-objective particle swarm algorithm. *Expert Systems with Applications, 36*, 9533–9538.
66. Kennedy, J., & Eberhart, R. (1995). Particle swarm optimization. In *Proceedings of the IEEE International Conference on Neural Networks: 1942–1945*.
67. Margarita, R. S., & Carlos, A. C. (2006). Multi-objective particle swarm optimizers: A survey of the state-of-the-art. *International Journal of Computational Intelligence Research, 2*(3), 287–308.
68. Duan, Q. Y., Sorooshian, S., & Gupta, V. K. (1992). Effective and efficient global optimization for conceptual rainfall-runoff models. *Water Resources Research, 28*, 1015–1031.
69. Chen, L., & Chang, F. J. (2007). Applying a real-coded multi-population genetic algorithm to multi-reservoir operation. *Hydrological Processes, 21*, 688–698.
70. Jalali, M. R., Afshar, A., & Marrino, M. A. (2007). Multi-colony ant algorithm for continuous multi-reservoir operation optimization problem. *Water Resources Management, 21*, 1429–1447.
71. Jiang, Y., Hu, T. S., Huang, C. C., & Wu, X. N. (2007). An improved particle swarm optimization algorithm. *Applied Mathematics and Computation, 193*, 231–239.
72. Deb, K., Pratap, A., Agarwal, S., & Meyarivan, T. (2002). A fast and elitist multi-objective genetic algorithm: NSGA-II. *IEEE Transactions on Evolutionary Computation, 6*, 182–197.
73. Deb, K. (2001). *Multi-objective optimization using evolutionary algorithms*. Chichester, UK: Wiley.
74. Jain, S., Agarwal, P., & Singh, V. (2007). Hydrology and water resources of India. *Water Science and Technology Library, 57*(4), 1065–1109.
75. Sadegh, M., Mahjouri, H., & Kerachian, R. (2010). Optimal inter-basin water allocation using crisp and fuzzy Shapley games. *Water Resources Management, 24*, 2291–2310.
76. Matete, M., & Hassan, R. (2006). Integrated ecological economic accounting approach to evaluation of inter-basin water transfers: An application to the Lesotho Highlands Water Project. *Ecological Economics, 60*, 246–259.
77. Li, X. S., Wang, B. D., Rajeshwar, M., Ashish, S., & Wang, G. L. (2009). Considering of trends in evaluating inter-basin water transfer alternatives within a fuzzy decision making framework. *Water Resources Management, 23*, 3207–3220.
78. Dosi, C., & Moretto, M. (1994). Inter-basin water transfers under uncertainty: Storage capacity and optimal guaranteed deliveries. *Environmental and Resource Economics, 4*, 331–352.
79. Chen, H. W., & Chang, N. B. (2010). Using fuzzy operators to address the complexity in decision making of water resources redistribution in two neighboring river basins. *Advances in Water Resources, 33*, 652–666.
80. Carvalho, R. C. D., & Magrini, A. (2006). Conflicts over water resource management in Brazil: A case study of inter-basin transfers. *Water Resources Management, 20*, 193–213.
81. Bonacci, O., & Andric, I. (2010). Impact of an inter-basin water transfer and reservoir operation on a karst open stream flow hydrological regime: An example from the Dinaric karst (Croatia). *Hydrological Processes, 24*, 3852–3863.

82. Xi, S. F., Wang, B. D., Liang, G. H., Li, X. S., & Lou, L. L. (2010). Inter-basin water transfer-supply model and risk analysis with consideration of rainfall forecast information. *Science China Technological Sciences, 53*(12), 3316–3323.
83. Stedinger, J. R. (1984). The performance of LDR models for preliminary design and reservoir operation. *Water Resources Research, 20*(2), 215–224.
84. ReVelle, C., Joeres, E., & Kirbt, W. (1969). The linear decision rule in reservoir management and design. I. Development of the stochastic model. *Water Resources Research, 5*(4), 767–777.
85. Neelakantan, T. R., & Pundarikanthan, N. V. (2000). Neural network based simulation–optimization model for reservoir operation. *Journal of Water Resources Planning and Management, 126*(2), 57–64.
86. Lund, J. R., & Reed, R. U. (1995). Drought water rationing and transferable rations. *Journal of Water Resources Planning and Management, 121*(6), 429–437.
87. Bayazit, M., & Unal, N. E. (1990). Effects of hedging on reservoir performance. *Water Resources Research, 26*(4), 713–719.
88. Calvete, H. I., Gale, C., & Oliveros, M. J. (2011). Bilevel model for production-distribution planning solved by using ant colony optimization. *Computers & Operations Research, 38*, 320–327.
89. Stackelberg, H. (1952). *The theory of the market economy*. New York, Oxford: Oxford University Press.
90. Yang, H., & Bell, M. G. H. (2001). Transport bilevel programming problems: Recent methodological advances. *Transportation Research Part B, 35*, 1–4.
91. Marcotte, P., Savard, G., & Zhu, D. L. (2009). Mathematical structure of a bilevel strategic pricing model. *European Journal of Operational Research, 193*, 552–566.
92. Bisschop, J., Candler, W., Duloy, J. H., & O'mara, G. T. (1982). The Indus basin model: A special application of two-level linear programming. *Mathematical Programming Study, 20*, 30–38.
93. Allsop, R. E. (1974). Some possibilities for using traffic control to influence trip destinations and route choice. In *Proceedings of the Sixth International Symposium on Transportation and Traffic Theory* (pp. 345–374). Amsterdam: Elsevier.
94. Abdulaal, M., & LeBlanc, L. J. (1979). Continuous equilibrium network design models. *Transportation Research Part B, 13*, 19–32.
95. Yang, H., Sasaki, T., Iida, Y., & Asakura, Y. (1992). Estimation of origin–destination matrices from link traffic counts on congested networks. *Transportation Research Part B, 26*, 1–18.
96. Yang, H., & Yagar, S. (1994). Traffic assignment and traffic control in general freeway-arterial corridor systems. *Transportation Research Part B, 28*, 463–486.
97. Yang, H., & Bell, M. G. H. (1997). Traffic restraint, road pricing and network equilibrium. *Transportation Research Part B, 31*, 303–314.
98. Hansen, P., Jaumard, B., & Savard, G. (1992). New branch-and-bound rules for linear bilevel programming. *SIAM Journal on Scientific and Statistical Computing, 13*, 1194–1217.
99. Calvete, H. I., Gale, C., & Mateo, P. (2008). A new approach for solving linear bilevel problems using genetic algorithms. *European Journal of Operational Research, 188*(1), 14–28.
100. Hejazi, S. R., Memariani, A., Jahanshahloo, G., & Sepehri, M. M. (2002). Linear bilevel programming solution by genetic algorithm. *Computers and Operations Research, 29*, 1913–1925.
101. Sahin, K. H., & Ciric, A. R. (1998). A dual temperature simulated annealing approach for solving bilevel programming problems. *Computers and Chemical Engineering, 23*, 11–25.
102. Rajesh, J., Gupta, K., Kusumakar, H. S., Jayaraman, V. K., & Kulkarni, B. D. (2003). A tabu search based approach for solving a class of bilevel programming problems in chemical engineering. *Journal of Heuristics, 9*, 307–319.
103. Kuo, R. J., & Huang, C. C. (2009). Application of particle swarm optimization algorithm for solving bi-level linear programming problem. *Computers and Mathematics with Applications, 58*, 678–685.
104. McCollor, D., & Stull, R. (2008). Hydrometeorological short-range ensemble forecasts in complex terrain. Part I: Meteorological evaluation. *Weather and Forecasting, 23*(4), 533–556.

105. Sankarasubramanian, A., Lall, U., Devineni, N., & Espinueva, S. (2009). The role of monthly updated climate forecasts in improving intraseasonal water allocation. *Journal of Applied Meteorology and Climatology, 48*(7), 1464–1482.
106. Maurer, E. P., & Lettenmaier, D. P. (2004). Potential effects of long-lead hydrologic predictability on Missouri River main-stem reservoirs. *Journal of Climate, 17*(1), 174–186.
107. Yao, H., & Georgakakos, A. (2001). Assessment of Folsom Lake response to historical and potential future climate scenarios 2. Reservoir management. *Journal of Hydrology, 249*, 176–196.
108. McCollor, D., & Stull, R. (2008). Hydrometeorological short-range ensemble forecasts in complex terrain. Part II: Economic evaluation. *Weather and Forecasting, 23*(4), 557–574.
109. Krzysztofowicz, R. (1999). Bayesian theory of probabilistic forecasting via deterministic hydrologic model. *Water Resources Research, 35*, 2739–2750.
110. Schaefli, B., Talamba, D. B., & Musy, A. (2007). Quantifying hydrological modeling errors through a mixture of normal distributions. *Journal of Hydrology, 332*, 303–315.
111. Schoups, G., & Vrugt, J. A. (2009). A formal likelihood function for parameter and predictive inference of hydrologic models with correlated, heteroscedastic, and non-Gaussian errors. *Water Resources Research, 46*, W10531. doi:10.1029/2009WR008933.
112. Georgakakos, K. P., & Graham, N. E. (2008). Potential benefits of seasonal inflow prediction uncertainty for reservoir release decisions. *Journal of Applied Meteorology and Climatology, 47*, 1297–1321.
113. Simonovic, S. P., & Burn, D. H. (1989). An improved methodology for short-term operation of a single multipurpose reservoir. *Water Resources Research, 25*(1), 1–8.
114. Maurer, E. P., & Lettenmaier, D. P. (2003). Predictability of seasonal runoff in the Mississippi River basin. *Journal of Geophysical Research-Atmospheres, 108*(D16).
115. Sankarasubramanian, A., Lall, U., Filho, F. A. S., & Sharma, A. (2009). Improved water allocation utilizing probabilistic climate forecasts: Short-term water contracts in a risk management framework. *Water Resources Research, 45*(W11409).
116. Sankarasubramanian, A., Lall, U., & Espinueva, S. (2009). Role of retrospective forecasts of GCMs forced with persisted SST anomalies in operational streamflow forecasts development. *Journal of Hydrometeorology, 9*(2), 212–227.
117. Graves, S. C. (1986). A tactical planning-model for a job shop. *Operational Research, 34*, 522–533.
118. Heath, D. C., & Jackson, P. L. (1994). Modeling the evolution of demand forecasts with application to safety stock analysis in production distribution-systems. *IIE Transactions, 26* (3), 17–30.
119. Krzysztofowicz, R. (1997). Transformation and normalization of variates with specified distributions. *Journal of Hydrology, 197*(1–4), 286–292.
120. Lettenmaier, D. P. (1984). Synthetic streamflow forecast generation. *Journal of Hydraulic Engineering-ASCE, 110*, 277–289.
121. Iida, T., & Zipkin, P. H. (2006). Approximate solutions of a dynamic forecast-inventory model. *Manufacturing & Service Operations Management, 8*, 407–425.
122. Wang, T., & Toktay, B. L. (2008). Inventory management with advance demand information and flexible delivery. *Management Science, 54*, 716–732.
123. Wang, Y. M., & Tomlin, B. (2009). To wait or not to wait: Optimal ordering under lead time uncertainty and forecast updating. *Naval Research Logistics, 56*, 766–779.
124. Arteaga, F., & Ferrer, A. (2010). How to simulate normal data sets with the desired correlation structure. *Chemometrics and Intelligent Laboratory Systems, 101*, 38–42.
125. Graham, N. E., & Georgakakos, K. P. (2010). Toward understanding the value of climate information for multiobjective reservoir management under present and future climate and demand scenarios. *Journal of Applied Meteorology and Climatology, 49*(4), 557–573.
126. Loucks, D. P., Stedinger, J. R., & Haith, D. A. (1981). *Water resources systems planning and analysis*. Englewood Cliffs, NJ: Prentice-Hall, Inc.
127. Ajami, N. K., Hornberger, G. M., & Sunding, D. L. (2008). Sustainable water resource management under hydrological uncertainty. *Water Resources Research, 44*(11). doi:10.1029/2007WR006736.
128. Valeriano, O. C. S., Koike, T., Yang, K., Graf, T., Li, X., Wang, L., et al. (2010). Decision support for dam release during floods using a distributed biosphere hydrological model driven

by quantitative precipitation forecasts. *Water Resources Research, 46*. doi:10.1029/2010WR009502.
129. You, J. Y., & Cai, X. (2008). Determining forecast and decision horizons for reservoir operations under hedging policies. *Water Resources Research, 44*(11). doi:10.1029/2008WR006978.
130. You, J. Y. (2008). *Hedging rule for reservoir operation: How much, when and how long to hedge*. PhD Dissertation.
131. Martinez, L., & Soares, S. (2002). Comparison between closed-loop and partial open-loop feedback control policies in long term hydrothermal scheduling. *IEEE Transactions on Power Systems, 17*(2), 330–336.
132. Veinott, A. F. (1964). Production planning with convex costs – A parametric study. *Management Science, 10*(3), 441–460.
133. Huang, W. C., & Hsieh, C. L. (2010). Real-time reservoir flood operation during typhoon attacks. *Water Resources Research, 46*.
134. Alemu, E. T., Palmer, R. N., Polebitski, A., & Meaker, B. (2011). Decision support system for optimizing reservoir operations using ensemble stream flow predictions. *Journal of Water Resources Planning and Management-ASCE, 137*(1), 72–82.
135. Boucher, M. A., Tremblay, D., Delorme, L., Perreault, L., & Anctil, F. (2012). Hydro-economic assessment of hydrological forecasting systems. *Journal of Hydrology, 416*, 133–144.
136. Carpenter, T. M., & Georgakakos, K. P. (2001). Assessment of Folsom lake response to historical and potential future climate scenarios: 1. Forecasting. *Journal of Hydrology, 249* (1–4), 148–175.
137. Wei, W. W., & Watkins, D. W. (2011). Data mining methods for hydroclimatic forecasting. *Advances in Water Resources, 34*(11), 1390–1400.
138. Wood, E. F., Roundy, J. K., Troy, T. J., van Beek, L. P. H., Bierkens, M. F. P. & Blyth, E. (2011). Hyperresolution global land surface modeling: Meeting a grand challenge for monitoring Earth's terrestrial water. *Water Resources Research, 47*. doi:10.1029/2010WR010090.
139. Ajami, N. K., Duan, Q. Y., & Sorooshian, S. (2007). An integrated hydrologic Bayesian multi model combination framework: Confronting input, parameter, and model structural uncertainty in hydrologic prediction. *Water Resources Research, 43*(1). doi:10.1029/2005WR004745.
140. Li, X. A., Guo, S. L., Liu, P., & Chen, G. Y. (2010). Dynamic control of flood limited water level for reservoir operation by considering inflow uncertainty. *Journal of Hydrology, 391* (1–2), 126–134.
141. Weigel, A. P., Liniger, M. A., & Appenzeller, C. (2008). Can multi-model combination really enhance the prediction skill of probabilistic ensemble forecasts. *Quarterly Journal of the Royal Meteorological Society, 134*(630), 241–260.
142. Montanari, A. (2005). Deseasonalisation of hydrological time series through the normal quantile transform. *Journal of Hydrology, 313*(3–4), 274–282.
143. Tan, W. D., Gan, F. F., & Chang, T. C. (2004). Using normal quantile plot to select an appropriate transformation to achieve normality. *Computational Statistics and Data Analysis, 45*(3), 609–619.
144. Ilich, N., & Despotovic, J. (2007). A simple method for effective multi-site generation of stochastic hydrologic time series. *Stochastic Environmental Research and Risk Assessment*. doi:10.1007/s00477-007-0113-6.
145. Sharma, A., Tarboton, D. G., & Lall, U. (1997). Streamflow simulation: A nonparametric approach. *Water Resources Research, 33*(2), 291–308.
146. Zhao, T. T. G., Yang, D. W., Cai, X. M., Zhao, J. S., & Wang, H. (2012). Identifying effective forecast horizon for real-time reservoir operation under a limited inflow forecast. *Water Resources Research, 48*, W01540. doi:10.1029/2011WR010623.
147. Vogel, R. M., & Stedinger, J. R. (1987). Generalized storage reliability yield relationships. *Journal of Hydrology, 89*(3–4), 303–327.
148. Wang, L. K., & Yang, C. T. (Eds.). (2014). *Modern water resources engineering*. New York: Humana Press, Springer Science + Business Media. 866 pp.

Chapter 2
Management of Aquifer Systems Connected to Streams Using Semi-Analytical Models

Domenico Baù and Azzah Salah El-Din Hassan

Contents

D. Baù, Ph.D. (✉)
Department of Civil and Structural Engineering, University of Sheffield, Sheffield, S3 7HQ, United Kingdom
e-mail: d.bau@sheffield.ac.uk

A.S.E.-D. Hassan, M.S.
Department of Geology and Geophysics, Texas A&M University, College Station, Texas 77843-3115, United States of America
e-mail: az_a_kurdi@hotmail.com

L.K. Wang, C.T. Yang, and M.-H.S. Wang (eds.), *Advances in Water Resources Management*, Handbook of Environmental Engineering, Volume 16,
DOI 10.1007/978-3-319-22924-9_2

Abstract In this chapter, we present a suite of analytically-based algorithms that can be used for simulating the effect of groundwater pumping and recharge in alluvial aquifers hydraulically connected to a stream. These algorithms are developed by systematically applying the principle of linear superposition of effects to the analytical solutions for the aquifer drawdown and the stream depletion rate derived by Theis (*AGU Transactions 16*(2), 519–524, 1935) and Glover and Balmer (*AGU Transactions, 35*(3), 468–470, 1954). These analytical models hypothesize the aquifer as confined, constant-thickness, homogeneous and isotropic, and semi-infinite, i.e. limited by a rectilinear boundary representing the stream. We extend these solutions to the cases in which the aquifer is laterally finite and bounded by either a no-flow boundary, representing for example the physical boundary of the alluvium, or a constant-head boundary, representing for example another stream or irrigation canal. In addition, we consider also effects of the cyclic operation of wells, which occurs when groundwater is extracted to support irrigation in the growing season, and recharged for stream augmentation during the cold seasons. These algorithms are used to implement a computationally efficient simulation-optimization framework that can be used to support the planning and management of groundwater resources in conditions where the impact of well pumping on stream flows must be minimized. Given their analytical basis, which relies upon highly idealistic assumptions on the hydrogeological structure of the aquifer-stream system, the use of these planning tools is not suitable for detailed simulations and predictions, but can be recommended for gaining insight into the qualitative behavior of the alluvial system, to conduct first-hand screening calculations and risk analyses.

Keywords Aquifer drawdown • Groundwater recharge • Stream depletion • Cyclic well operation • Analytical solution • Superposition of effects • Linear optimization

Nomenclature

a	Well-stream distance, L
A	Cross-section area, L/T
b	Aquifer saturated thickness, L
g	Gravity acceleration, L/T^2
h	Water level, or hydraulic (piezometric) head, L
h_{max}	Maximum allowed hydraulic head, L
h_{min}	Minimum allowed hydraulic head, L
h_o	Initial hydraulic head, L
k	Permeability, L^2
K	Hydraulic conductivity, L/T
n_{ow}	Number of operating wells, /
n_{mw}	Number of monitoring wells, /
p	Water pressure relative to atmospheric, M/L/T^2

q	Darcy's velocity or specific discharge (flow rate per unit cross-section area), L/T
q'	Aquifer flow rate per unit width, L^2/T
Q	Total flow rate, L^3/T
$Q_{Available}$	Total available recharge rate, L^3/T
Q_{Demand}	Total demand rate, L^3/T
Q_E	Well extraction rate, L^3/T
Q_I	Well injection rate, L^3/T
Q_r	Stream depletion rate, L^3/T
Q_{ratio}	Stream depletion rate ratio, /
r	Radial distance, L
r_I	Distance between observation point and imaginary well, L
s	Head drawdown, L
S	Aquifer storativity, /
S_s	Specific elastic storage, 1/L
S_y	Specific yield, /
t	Time, T
T	Aquifer transmissivity, L^2/T
u	Boltzmann unitless variable, /
V_r	Stream depletion volume, L^3
V_{ratio}	Stream depletion volume ratio, /
$\mathrm{V}_{r(ex)}$	Cumulative stream depletion volume, L^3
$\mathrm{V}_{r(in)}$	Cumulative stream accretion volume, L^3
w	Aquifer width (distance between stream and lateral boundary), L
W	Theis well function, /
W_B	Well function for a semi-infinite bounded aquifer, /
W_{BB}	Well function for a finite constant-width aquifer, /
W_C	Well function for a cycling pumping operation, /
W_{CSI}	Well function for cycling pumping operation in a semi-infinite aquifer, /
W_{FC}	Well function for cycling pumping operation in a finite constant-width aquifer, /
W_N	Well function for a semi-infinite aquifer delimited by a no-flow boundary, /
W_{NR}	Well function for a finite constant-width aquifer comprised between a no-flow boundary and a recharge boundary, /
W_R	Well function for a semi-infinite aquifer delimited by a recharge boundary, /
W_{RR}	Well function for a finite aquifer comprised between two recharge boundaries, /
x	West-to-east coordinate, L
y	South-to-north coordinate, L
Δt_E	Time interval over which an extraction well is operated, T
Δt_I	Time interval over which an injection well is operated, T
Δt_{on}	Time interval over which a well is activated, T

Δt_{off}	Time interval over which a well is inactive, T
ρ	Water density, M/L^3
μ	Water dynamic viscosity, M/L/T
τ_E	Starting time for extraction well operation, T
τ_I	Starting time for injection well operation, T

1 Introduction

In water management, conjunctive use is defined as the combined use of surface and water resources in order to maximize the global net benefit of users. In Colorado, conjunctive use is of paramount importance for shallow aquifers situated in the South Platte and in the Arkansas River basins. Indeed, these aquifers, which were formed by alluvial deposition processes, are hydraulically connected to the surface network of streams and irrigation ditches, such that consumptive use of subsurface water inevitably affects surface water regimes and vice versa.

In the Western US, water use is mostly regulated under the Doctrine of Prior Appropriation, which gives senior water rights to users with earlier appropriation dates (first in time, first in right). Each year, senior users have the right to use water according to their full allocation, if available, over "junior" users, who can exert their water rights only if they do not impinge on water rights that are senior to thier's.

Historically, surface water rights (stream direct flow, reservoirs) were fully allocated by the 1920s. Well water rights were adjudicated much later concurrently with the development of high capacity turbine pumps, as it became evident that the use of groundwater had an impact on stream flows. As a result, groundwater users are generally junior to surface users and, in times of drought, might not receive their full allocation.

It is widely acknowledged that the regulations established by the Doctrine of Prior Appropriation impose strong limitations to the conjunctive use of water resources since groundwater users can damage senior water rights. Given these limitations, the management of groundwater resources in Colorado is often thought of as the identification of pumping schemes that meet irrigation demands while minimizing the impact on stream flows.

The earliest fundamental models for assessing the impact of well pumping on water levels and stream flows in aquifers hydraulically connected to streams were derived analytically by Theis [1] and Glover and Balmer [2]. These models hypothesize the aquifer as constant-thickness, homogeneous and isotropic, and semi-infinite, i.e. limited by a rectilinear boundary representing the stream. Jenkins [3] proposed the stream depletion factor (SDF) method to extend the Glover model to finite heterogeneous aquifers bounded by meandering streams.

Most often, the applicability of these models to realistic conditions is limited by the simplifying hypotheses on which they rely, which make them valid only under much idealized conditions. For real-world scenarios, the use of numerical models, such as USGS's MODFLOW [4, 5] is strongly preferable. However, given that these models can be computationally very expensive, the use of close-form analytical solutions remains attractive for analyzing the dynamics of aquifer-stream interaction, screening calculations and overall risk assessment. This explains why models based upon the Glover solution [2] or the SDF method [3] are still broadly used by regulatory agencies to estimate stream depletion due to groundwater pumping and evaluate plans for stream augmentation by groundwater recharge.

Analytical models were originally developed for semi-infinite aquifers under constant-rate pumping from a single well. In practice, alluvial aquifers are always finite and delimited between the stream and another boundary, which can be modeled as impervious, where the alluvium terminates, or as a recharge boundary, where the aquifer is in hydraulic contact, for example, with an irrigation ditch. In addition, every year, irrigation pumping occurs from several wells, which are operated only during the growing season with generally variable pumping rates.

In these situations, both Theis and Glover's models can be extended to account for the effect of boundaries and multiple wells with time-varying pumping rates. Indeed, since these models are linear, there exists a direct time-dependent proportionality between well pumping rate, head drawdown, stream depletion rate and stream depletion volume. Therefore, superposition of effects and temporal convolution methods [6] can be applied to estimate the impact on water level and stream flow of well fields, no-flow or constant-head boundaries parallel to the stream, and time-varying pumping rates.

Based on these ideas, we have developed a suite of semi-analytical models that rely upon Theis and Glover's analytical solutions, to simulate and plot the water level spatial-temporal distributions and the times series of stream depletion rate and stream depletion volume due to a well field in which each well operates cyclically over prescribed on-off sequences.

These models are developed for three basic alluvial aquifer conditions: (1) semi-infinite aquifer in hydraulic contact with a stream; (2) finite-width aquifer in hydraulic contact with a stream on one side and delimited by a no-flow physical boundary on another; (3) finite-width aquifer in hydraulic contact with a stream on one side and another stream, or an irrigation ditch, on the other side. These models are implemented in MATLAB and, together with explanation notes and user manuals, are included in the Appendices 1–5 to this chapter.

2 Saturated Groundwater Flow Theory

2.1 Darcy's Law

The equations governing groundwater flow in saturated porous media rely on Darcy's law, an empirical law stating that the rate of flow through a porous medium is proportional to the cross-section area and the energy loss, and inversely proportional to the distance between start and end points in the direction of flow [7]. This law is expressed as:

$$Q = A \cdot q = -A \cdot K \cdot \frac{\Delta h}{l} \tag{2.1}$$

where: Q is the flow rate (L^3/T); A is the cross-section area (L^2); q is Darcy's velocity (or specific discharge) (L/T), which represents the volume flux per unit bulk area per unit time; K is the hydraulic conductivity (L/T); Δh is the variation of piezometric head h (L); and l is the distance (L). The hydraulic or piezometric head h represents the energy per unit weight of water at any point in the aquifer, and is given by the following equation:

$$h = z + \frac{p}{\rho \cdot g} \tag{2.2}$$

where: z is the elevation (L); p is the water pressure relative to the atmospheric pressure ($ML^{-1}\ T^{-2}$); g is the gravity acceleration (L/T^2); and ρ is the density of water (M/L^3).

The hydraulic conductivity K is a hydrogeological parameter combining both the fluid and the porous medium properties, which represents the ability of the medium to conduct water [8]. The hydraulic conductivity is given by [9]:

$$K = \frac{\rho \cdot g}{\mu} \cdot k \tag{2.3}$$

where: μ is the dynamic viscosity (M/L/T) of the fluid and k is the intrinsic permeability (L^2). In a three dimensional reference system, the Darcy's law is written as:

$$\boldsymbol{q} = \begin{bmatrix} q_x \\ q_y \\ q_z \end{bmatrix} = -\boldsymbol{K} \cdot \nabla h = -\begin{bmatrix} K_{xx} & K_{xy} & K_{xz} \\ K_{yx} & K_{yy} & K_{yz} \\ K_{zx} & K_{zy} & K_{zz} \end{bmatrix} \cdot \begin{bmatrix} \frac{\partial h}{\partial x} \\ \frac{\partial h}{\partial y} \\ \frac{\partial h}{\partial z} \end{bmatrix} \tag{2.4}$$

where: $\boldsymbol{q}$ is the specific discharge vector, ∇ is the differential operator and ∇h is the gradient vector of the piezometric head. The matrix $\boldsymbol{K}$ is the hydraulic conductivity tensor, which becomes "diagonal" when the coordinate axes x, y, and z are set collinearly with the principal directions of the hydraulic conductivity [10].

With respect to its hydraulic conductivity properties, an aquifer is homogenous if $\boldsymbol{K}$ is independent of position (or uniform in space), or it is heterogeneous otherwise. An aquifer is isotropic if its hydraulic conductivity is independent of direction ($K_{xx} = K_{yy} = K_{zz}$). Vice versa, an aquifer is anisotropic if its hydraulic conductivity depends on direction ($K_{xx} \neq K_{yy}$ or $K_{xx} \neq K_{zz}$ or $K_{yy} \neq K_{zz}$). In condition of isotropy, the hydraulic conductivity tensor $\boldsymbol{K}$ is identified with a single scalar coefficient K, generically referred to as hydraulic conductivity.

2.2 *Equation of Continuity*

The continuity equation represents the mass balance for a fluid in a closed system and, in a three-dimensional form, it is written as:

$$\nabla[\boldsymbol{K} \cdot \nabla h] + f = S_s \cdot \frac{\partial h}{\partial t} \tag{2.5}$$

where f (1/T) represents generic source/sink terms (e.g., forcing terms for water extraction, injection, or recharge) and S_s is the specific elastic storage (1/L) (the volume of water released from storage per unit volume of the aquifer per unit decline in pressure head). In the case of a diagonal hydraulic conductivity tensor, Eq. (2.5) becomes:

$$\frac{\partial}{\partial x}\left(K_{xx} \cdot \frac{\partial h}{\partial x}\right) + \frac{\partial}{\partial y}\left(K_{yy} \cdot \frac{\partial h}{\partial y}\right) + \frac{\partial}{\partial z}\left(K_{zz} \cdot \frac{\partial h}{\partial z}\right) + f = S_s \cdot \frac{\partial h}{\partial t} \tag{2.6}$$

Equation (2.6) applies to heterogeneous and anisotropic confined aquifers. It can be modified to reflect other conditions for the aquifer hydraulic conductivities. For example, for homogenous and anisotropic aquifers, Eq. (2.6) takes on the form:

$$K_{xx} \cdot \frac{\partial^2 h}{\partial x^2} + K_{yy} \cdot \frac{\partial^2 h}{\partial y^2} + K_{zz} \cdot \frac{\partial^2 h}{\partial z^2} + f = S_s \cdot \frac{\partial h}{\partial t} \tag{2.7}$$

For homogenous and isotropic aquifers, Eq. (2.6) is further simplified to the form:

$$K \cdot \left(\frac{\partial^2 h}{\partial x^2} + \frac{\partial^2 h}{\partial y^2} + \frac{\partial^2 h}{\partial z^2}\right) + f = S_s \cdot \frac{\partial h}{\partial t} \tag{2.8}$$

The integration of the continuity Equation (2.5) requires assigning initial and boundary conditions. The initial conditions necessitate prescribing the hydraulic head h_o distribution throughout the aquifer domain before the changes made by the external influences (e.g., operating wells) applied, which is written as:

$$h(x, y, z; 0) = h_0(x, y, z) \ \forall (x, y, z) \in \Omega \tag{2.9}$$

where Ω is the aquifer domain. Three types of boundary conditions are generally imposed: Dirichlet's, Neumann's and Cauchy's. As described by Willis and Yeh [10],

Neumann's and Cauchy's boundary conditions mathematically represent a head or a flow/flux state along the aquifer boundary. Dirichlet boundary conditions are used when the hydraulic head is known at any time over a given portion Γ_D of the domain boundary:

$$h(x, y, z; t) = h_D(x, y, z; t) \ \ \forall (x, y, z) \in \Gamma_D \ \ \forall t > 0 \tag{2.10}$$

where h_D is the boundary-head time-dependent function. Neumann boundary conditions are prescribed when the flow across a portion Γ_N of the aquifer boundary is known at any time:

$$-\boldsymbol{K} \cdot \nabla h \cdot \boldsymbol{n} = g_N(x, y, z; t) \ \ \forall (x, y, z) \in \Gamma_N \ \ \forall t > 0 \tag{2.11}$$

where g_N (1/T) is the flux across the boundary and $\boldsymbol{n}$ is the unit vector normal to the boundary. Cauchy boundary conditions consist of a linear combination of Dirichlet and Neumann conditions, imposed over a prescribed portion Γ_C of the aquifer domain:

$$\delta_D \cdot h + \delta_N \cdot \boldsymbol{K} \cdot \nabla h \cdot \boldsymbol{n} = g_C(x, y, z; t) \ \ \forall (x, y, z) \in \Gamma_C \ \ \forall t > 0 \tag{2.12}$$

where δ_D and δ_N are coefficients and g_C is the so-called Cauchy potential function. The differential equations governing the flow in aquifers can be solved analytically only under highly simplified assumptions for the aquifer setting (e.g. [1, 11, 12]), which limits the application of these methods to only very ideal conditions. For realistic systems, the solution to these equations is preferably obtained numerically using finite-difference or finite-element model, such as USGS's MODFLOW [4] and SUTRA [13]. The development of numerical models often requires, however, the availability of large subsurface datasets and intensive efforts for implementation and computational time. Analytical and semi-analytical methods (e.g. [3, 14]) may offer a more valid alternative to numerical models for screening calculations and risk analyses.

3 Fundamental Analytical Solutions

3.1 The Theis Model

Theis [1] derived the solution to Eq. (2.5) for an infinite-extent, horizontal, constant-thickness, homogeneous and isotropic confined aquifer under the effect of a radial unsteady flow caused by a fully penetrating well, located at the center of the aquifer and operating with a constant pumping rate Q. The initial conditions under which Theis's solution is solved require the initial head h_0 to be uniform over the aquifer domain. Boundary conditions state that the hydraulic head may remain undisturbed and equal to the initial head h_0 at infinite distance from the well at any

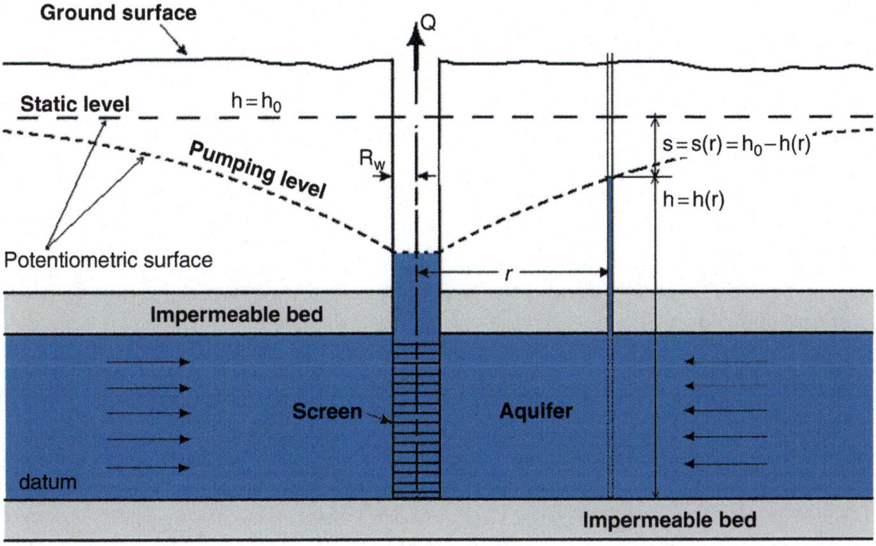

Fig. 2.1 Ideal setting for the Theis solution (1935), consisting of a laterally-infinite, constant-thickness, horizontal, homogeneous and isotropic confined aquifer affected by the operation of a single pumping well

time during pumping operations. According to Theis's assumptions, the spatial dependency on head is reduced to the two horizontal coordinates x and y, whereas the vertical coordinate z is dropped given the condition of ideal horizontal flow.

Figure 2.1 shows the features of the perfectly confined aquifer setting addressed by Theis [1], with a fully penetrating pumping well. The dashed lines represent the position of the potentiometric surface initially and during pumping.

When studying the effect of radial flow, the saturated flow equation is written in terms of aquifer drawdown instead of hydraulic head. The aquifer drawdown (L) represents the reduction of the piezometric with respect to its initial state:

$$s(x, y; t) = h_0 - h(x, y; t) \tag{2.13}$$

Under the conditions of radial, the partial differential equation (PDE) (2.5) governing unsteady flow in Theis's confined aquifer may be simplified as follows:

$$\frac{1}{r} \cdot \frac{\partial}{\partial r}\left(r \cdot \frac{\partial s}{\partial r}\right) + f = \frac{\partial^2 s}{\partial r^2} + \frac{1}{r} \cdot \frac{\partial s}{\partial r} + f = \frac{S}{T} \cdot \frac{\partial s}{\partial t} \tag{2.14}$$

In Eq. (2.14), r is the distance between the operating well, located at the origin of the Cartesian system and the observation point (x, y) $\left(r = \sqrt{x^2 + y^2}\right)$, T is the aquifer transmissivity (L^2/T) (the rate at which water is transmitted through a unit width of aquifer under a unit hydraulic gradient), and S is the elastic storage coefficient or storativity (/) (the volume of water released from a column of a unit cross-section area per unit pressure-head decline). Note that the function f is zero everywhere, expect that at the well location, that is, $r = 0$.

The initial conditions associated with the PDE (2.14) require the drawdown at time $t = 0$ to be zero everywhere:

$$s(r;0) = 0 \ \forall r \tag{2.15}$$

The prescribed boundary conditions require a zero drawdown at infinite distance from the pumping well:

$$\lim_{r \to \infty} s(r;t) = 0 \ \forall t > 0 \tag{2.16}$$

Note that Eq. (2.16) is a Dirichlet boundary condition. In addition, the following Neumann condition must be imposed in proximity of the pumping well location:

$$\lim_{r \to 0} r \cdot \frac{\partial s}{\partial r} = -\frac{Q}{2 \cdot \pi \cdot T} \forall t > 0 \tag{2.17}$$

In practice, condition (2.17) is derived by applying the equation of continuity together with the Darcy's law across the lateral surface of a cylinder of infinitesimal radius centered on the well, from where a constant flow rate Q is drawn. Theis's solution is derived by introducing the unitless variable:

$$u(r,t) = \frac{S}{4 \cdot T} \cdot \frac{r^2}{t} \tag{2.18}$$

known as the Boltzmann variable. Using this variable, the first and second derivatives of the drawdown s with respect to r in Eq. (2.14) can be calculated using the chain rule:

$$\frac{\partial s}{\partial r} = \frac{ds}{du} \cdot \frac{\partial u}{\partial r} = \frac{ds}{du} \cdot \frac{2 \cdot u}{r} \tag{2.19}$$

$$\frac{\partial^2 s}{\partial r^2} = \frac{\partial}{\partial r}\left(\frac{ds}{du} \cdot \frac{2 \cdot u}{r}\right) = \frac{ds}{du} \cdot \frac{\partial}{\partial r}\left(\frac{2 \cdot u}{r}\right) + \frac{2 \cdot u}{r} \cdot \frac{\partial}{\partial r}\left(\frac{ds}{du}\right) \tag{2.20}$$

Equation (2.20) can be further developed as follows:

$$\frac{\partial^2 s}{\partial r^2} = \frac{ds}{du} \cdot \frac{2 \cdot u}{r^2} + \left(\frac{2 \cdot u}{r}\right)^2 \cdot \frac{d^2 s}{du^2} \tag{2.21}$$

The derivative of s with respect to t is obtained as:

$$\frac{\partial s}{\partial t} = \frac{ds}{du} \cdot \frac{\partial u}{\partial t} = -\frac{ds}{du} \cdot \frac{u}{t} \tag{2.22}$$

Substituting Eqs. (2.19), (2.21) and (2.22) into PDE (2.14) gives:

$$\left(\frac{2 \cdot u}{r}\right)^2 \cdot \frac{d^2 s}{du^2} + 2 \cdot \left(\frac{2 \cdot u}{r^2}\right) \cdot \frac{ds}{du} = -\frac{S}{T} \cdot \frac{ds}{du} \cdot \frac{u}{t} \tag{2.23}$$

By dividing both sides of Eq. (2.23) by $2 \cdot \frac{u}{r^2}$ and rearranging its terms, the following ordinary differential equation (ODE) is obtained:

$$\frac{d^2 s}{du^2} + \left(1 + \frac{1}{u}\right) \cdot \frac{ds}{du} = 0 \tag{2.24}$$

Based on Eq. (2.15), the initial conditions for the ODE are:

$$t \to 0 \Rightarrow u \to \infty : \quad s(u \to \infty) = 0 \tag{2.25}$$

Based on Eqs. (2.16) and (2.17) the boundary conditions for the ODE (2.24) are, respectively:

$$r \to \infty \Rightarrow u \to \infty : \quad s(u \to \infty) = 0 \tag{2.26}$$

$$\begin{gathered} \lim_{r \to 0} r \cdot \frac{\partial s}{\partial r} = \lim_{r \to 0} r \cdot \frac{ds}{du} \cdot \frac{2 \cdot u}{r} = \lim_{u \to 0} 2 \cdot u \cdot \frac{ds}{du} = -\frac{Q}{2 \cdot \pi \cdot T} \\ \Downarrow \\ \lim_{u \to 0} u \cdot \frac{ds}{du} = -\frac{Q}{4 \cdot \pi \cdot T} \end{gathered} \tag{2.27}$$

Assuming $= \frac{ds}{du}$, the integration of the ODE (2.24) proceeds as follows:

$$\begin{gathered} \frac{dP}{du} + \left(1 + \frac{1}{u}\right) \cdot P = 0 \Rightarrow \frac{dP}{P} + \left(1 + \frac{1}{u}\right) \cdot du = 0 \\ \Downarrow \\ \int \frac{dP}{P} = -\int \left(1 + \frac{1}{u}\right) \cdot du \Rightarrow \ln P = -u - \ln u + c \\ \Downarrow \\ P = \mathrm{e}^{-u - \ln u + c} = \mathrm{e}^{c} \cdot \frac{\mathrm{e}^{-u}}{u} \end{gathered} \tag{2.28}$$

After introducing the constant $c' = \mathrm{e}^c$, Eq. (2.28) provides:

$$P = \frac{ds}{du} = c' \cdot \frac{\mathrm{e}^{-u}}{u} \tag{2.29}$$

Equation (2.29) can be integrated by separation of variables:

$$ds = c' \cdot \frac{e^{-u}}{u} \cdot du \Rightarrow \int_{s(u)}^{s(u \to \infty)} ds = c' \cdot \int_{u}^{\infty} \frac{e^{-u'}}{u'} \cdot du' \tag{2.30}$$

From which:

$$s(\infty) - s(u) = c' \cdot \int_{u}^{\infty} \frac{e^{-u'}}{u'} \cdot du' \tag{2.31}$$

Based on conditions (2.25) and (2.26), Eq. (2.31) becomes:

$$s(u) = -c' \cdot \int_{u}^{\infty} \frac{e^{-u'}}{u'} \cdot du' \tag{2.32}$$

The constant c' can be obtained by deriving both hand sides of Eq. (2.32) with respect to u, and imposing conditions (2.26) and (2.27):

$$\begin{aligned} \frac{ds}{du} &= -\left[c' \cdot \frac{e^{-u'}}{u'}\right]_{u}^{\infty} = 0 + c' \cdot \frac{e^{-u}}{u} \\ u \cdot \frac{ds}{du} &= c' \cdot e^{-u} \\ \lim_{u \to 0} u \cdot \frac{ds}{du} &= \lim_{u \to 0} c' \cdot e^{-u} = c' = -\frac{Q}{4 \cdot \pi \cdot T} \end{aligned} \tag{2.33}$$

From Eqs. (2.32) and (2.33), the drawdown function is thus:

$$s(u) = \frac{Q}{4 \cdot \pi \cdot T} \cdot \int_{u}^{\infty} \frac{e^{-u'}}{u'} \cdot du' \tag{2.34}$$

Equation (2.34) represents the drawdown distribution in time and space due to a single well operating at a constant rate Q in a homogenous, isotropic, horizontal, constant thickness confined aquifer. The exponential integral in Eq. (2.34) is known as the Theis well function:

$$W(u) = \int_{u}^{\infty} \frac{e^{-u'}}{u'} \cdot du' \tag{2.35}$$

Note that in Eq. (2.34), drawdown is positive if Q is positive, that is, if water is extracted from the aquifer. Here, an opposite sign rule is adopted, so that Q is positive if injected and negative if extracted. Accordingly with this assumption, the Theis Equation (2.34) is rewritten as:

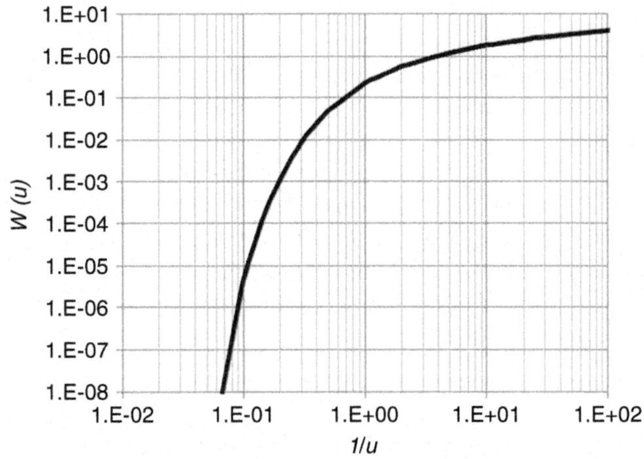

Fig. 2.2 Profile of the Theis well function vs. the inverse of the Boltzmann variable u

$$s(r,t) = -\frac{Q}{4 \cdot \pi \cdot T} \cdot W\left(\frac{S}{4 \cdot T} \cdot \frac{r^2}{t}\right) \tag{2.36}$$

Figure 2.2 shows the profile of the exponential well function W with respect to the inverse of u. Note that, in Eq. (2.36) the drawdown $s(r,t)$ increases if the distance r decreases and the time t increases. With respect to the transmissivity T, it can be shown that the drawdown cone becomes shallower but wider if T is increased, and vice versa. With respect to the storativity S, it can be shown that the drawdown cone becomes both shallower and narrower if S is increased, and vice versa.

From a numerical standpoint, the calculation of the Eq. (2.35) can be achieved using the series expansion $-0.5772 - ln(u) - \sum_{i=1}^{\infty} (-1)^i \cdot \frac{u^i}{i \cdot i!}$ if u is less than or equal to one (1). Indeed, within these conditions the terms within the sum become infinitesimal for relatively low values of the index i, so that the series converges rather quickly. However, if u is greater than one this method is prone to numerical errors, thus a better solution is achieved by using a numerical technique for the direct calculation of the exponential integral in Eq. (2.35). In MATLAB, the well function $W(u)$ is given by the special function `expint(x)`.

3.2 *Superposition of Effects: Aquifer Drawdown*

The Theis [1] solution (Eq. 2.36) can be extended to dealing with conditions where the constant pumping rate assumption and the single operating well assumption are not met. In order to remove these assumptions the principle of superposition is introduced.

The principle of superposition of effects or solutions in physics states that the total response of a linear system governed by linear differential equations can be evaluated as the sum of individual, elementary, linear responses in space and time caused by multiple source/sink terms [6]. Given that the solution of the ground water flow problem involves satisfying initial and boundary conditions, to be able to apply the principle of superposition these conditions have to be linear as well. In the case of the Theis equation, the principle of superposition may be used in the calculation of the drawdown at a certain point in space and time due to the effect of: (a) a well field, that is, a set of pumping wells operating simultaneously; (b) a single operating well with varying operation rate; and (c) no-flow and recharge boundaries that render the aquifer semi-infinite.

a) **Well Field**. In this case, head drawdowns or build-ups in the aquifer occur as a response to spatially distributed operating wells. In the two dimensional extent of the aquifer, the drawdown at any observation point (x, y) and time t due to a well field with a number n_{ow} of operating wells, each one with a rate Q_i and operation starting time $t_i (i = 1, 2, .., n_{ow})$ can be calculated as:

$$s(x, y, t) = -\sum_{i=1}^{n_{ow}} \frac{Q_i}{4 \cdot \pi \cdot T} \cdot W\left(\frac{S}{4 \cdot T} \cdot \frac{r_i^2}{t - t_i}\right) \tag{2.37}$$

where r_i is the distance of the observation point (x, y) from well i, located at coordinates (x_i, y_i), which is calculated as:

$$r_i = \sqrt{(x - x_i)^2 + (y - y_i)^2} \tag{2.38}$$

Given that the drawdown at any point in the aquifer is affected by pumping from well i only if the well is activated at the time of observation, each of the terms at the right-hand side Eq. (2.37) must be accounted for only if $t > t_i$ and not otherwise.

An application example of Eq. (2.37) is given in Fig. 2.3, which shows the contour line plot of the drawdown distribution, obtained at a time $t = 120$ d (d) in an aquifer with the hydrogeological parameters given in Table 2.1. The wells locations, operation rates, starting times and operation periods are listed in Table 2.2. The plot in Fig. 2.3 is obtained with a MATLAB code that implements Eq. (2.37). This code, called, `Drawdown2D.m` is provided in Appendix 1.

b) **Time-varying operation rates**. Since the Theis solution considers only constant pumping rates, the principle of superposition can be used to deal with cases that do not meet this condition. In these cases, the operation rate is treated as a time-dependent function, and the drawdown (L) is computed using the following "convolution" integral:

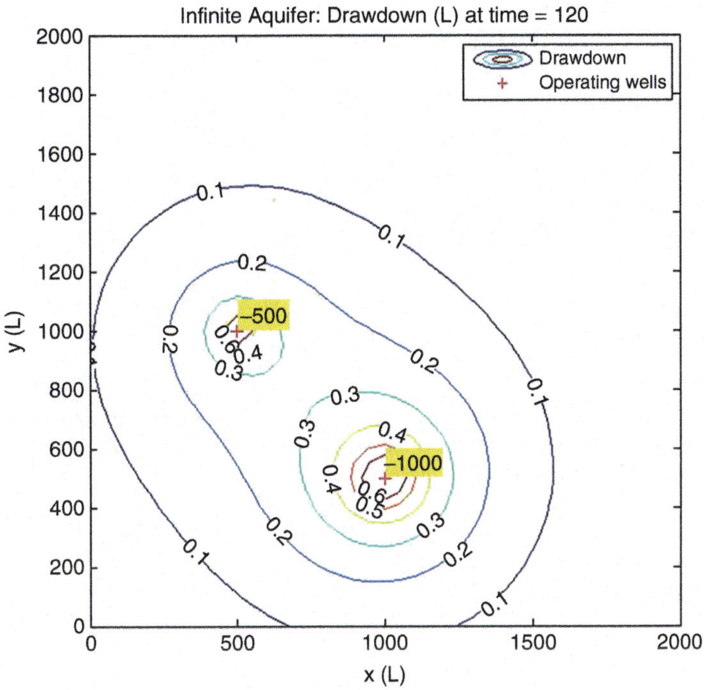

Fig. 2.3 Drawdown contour representation for two wells located at (500, 1000) and (1000, 500) (m), operating with a constant rates $-500\ m^3/d$ and $-1000\ m^3/d$, respectively, for a time $t = 120$ d

Table 2.1 Example aquifer hydrogeological parameters

h_0 (m)	$T = K \cdot h_0$ (m^2/d)	S (/)
30	648	0.2

Table 2.2 Example well field data

Well	Location (x_i, y_i) (m)	Operation starting time t_i (d)	Operation period Δt_{on} (d)	Operation rate Q_i (m^3/d)
1	(500, 1000)	0	180	−500
2	(1000, 500)	60	120	−1000

$$s(r,t) = -\frac{1}{4 \cdot \pi \cdot T} \cdot \int_0^t \frac{\partial Q(\tau)}{\partial \tau} \cdot W\left(\frac{S}{4 \cdot T} \cdot \frac{r_i^2}{t-\tau}\right) \cdot d\tau \tag{2.39}$$

where τ (T) is the generic instant within the well operation time interval [0, t]. A MATLAB code called `Drawdown_VS_Time.m` that calculates Eq. (2.39) is provided in Appendix 1. Figure 2.4a shows an example of time-varying well operation rate. Figure 2.4b shows the corresponding drawdown profile obtained at an observation point located at a distance r equal to 150 m from the pumping well, using the MATLAB code mentioned above. The aquifer parameters used in this example are given in Table 2.1.

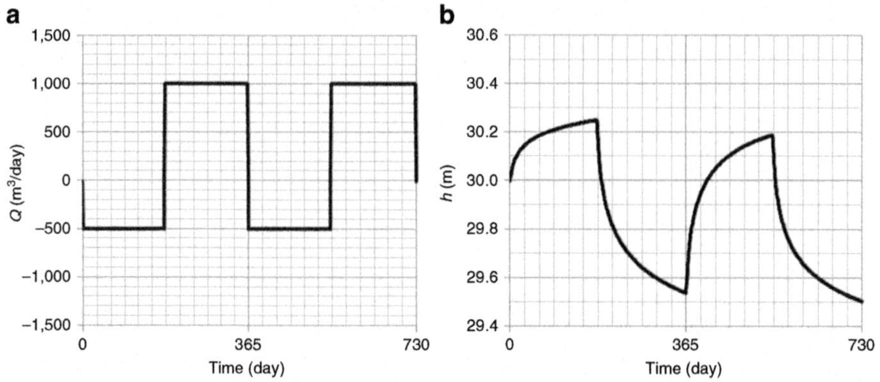

Fig. 2.4 (**a**) Example of time-varying rates of an extracting well and (**b**) resulting drawdown time series in an observation point located at distance $r = 140$ m from the operating well. Table 2.1 reports the aquifer parameters used in this example

Equations (2.13), (2.37) and (2.39) can be combined together to provide a general expression for the hydraulic head in an aquifer subject to pumping from a well field with time-dependent pumping rates:

$$h(x,y,t) = h_0 + \frac{1}{4\cdot\pi\cdot T}\cdot\sum_{i=1}^{n_{ow}}\left[\int_0^t \frac{\partial Q_i(\tau_i)}{\partial \tau_i}\cdot W\left(\frac{S}{4\cdot T}\cdot\frac{r_i^2}{t-\tau_i}\right)\cdot d\tau_i\right] \tag{2.40}$$

c) **Effect of recharge and no-flow boundaries**. A recharge boundary is considered as a boundary subject to a Dirichlet constant-head condition, where the drawdown is maintained constant and equal to zero over time. Conversely, a no-flow boundary consists of an impervious boundary across which groundwater cannot flow. The Theis Equation (2.36) can be extended to dealing with cases in which these boundaries are rectilinear.

To create a mathematically equivalent condition for either a recharge boundary or a no-flow boundary and restore the infinite aquifer condition, an "image" well may be introduced to the system, located at a point symmetrical to the real well with respect to the boundary. In the case of recharge boundary, the image well performs simultaneously the opposite type of operation of the actual well with the same rate in order to keep the state of zero drawdown at the boundary. In the case of a no-flow boundary, the image well performs the same type of operation to create zero constant flux condition on the impermeable boundary line [8].

Figure 2.5 shows the layouts for a semi-infinite aquifer with the actual and image wells, in the cases of a recharge boundary (subpanel a) and a no-flow boundary (subpanel b). Correspondingly, subpanels (c) and (d) illustrate the heads levels due to the operation of each well and the resulting combined head levels.

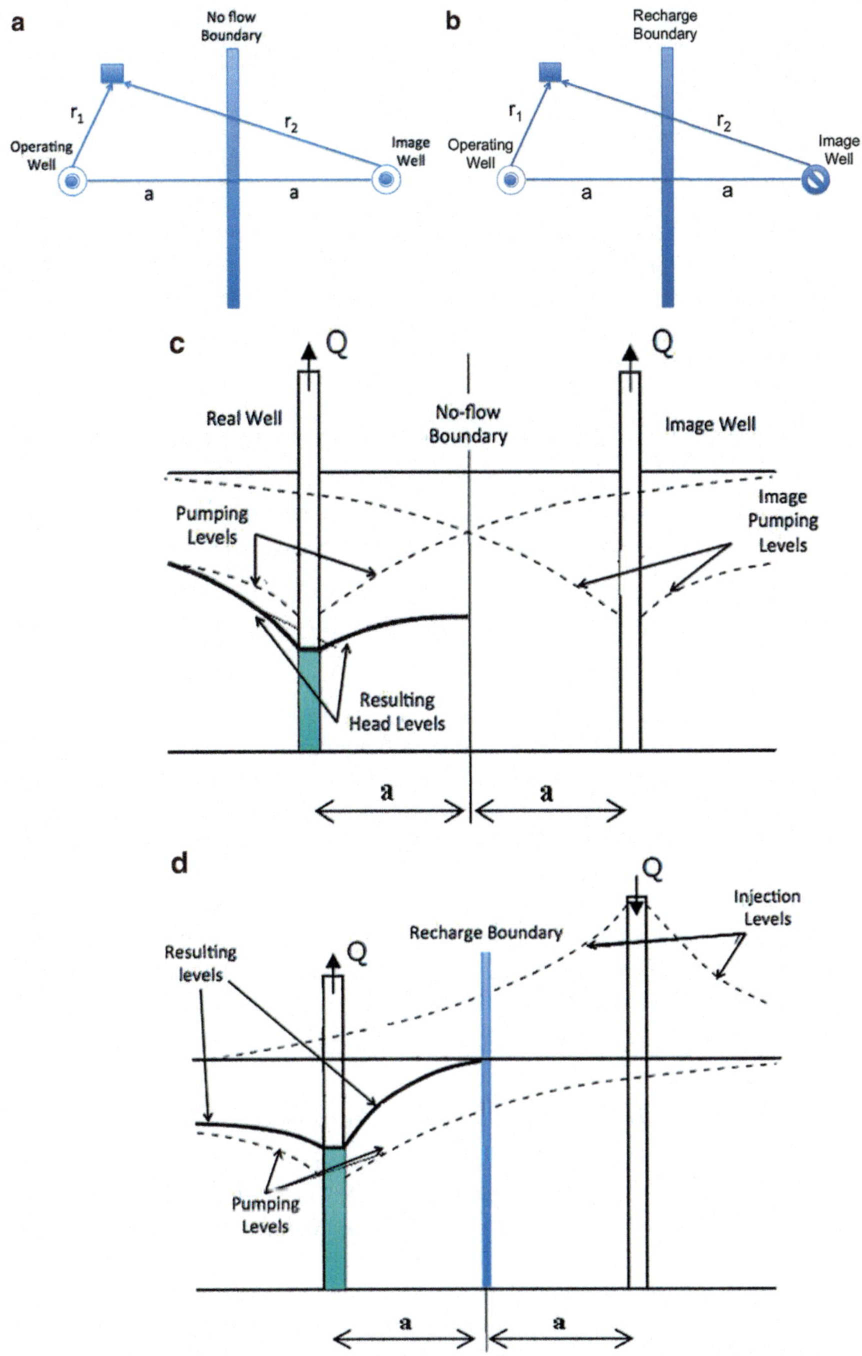

Fig. 2.5 Image-well layouts for semi-infinite aquifers delimited by (**a**) a recharge boundary and (**b**) a no-flow boundary. Subpanels (**c**) and (**d**) show the head profiles corresponding to layouts (**a**) and (**b**), respectively

The general drawdown equation due to a single well operating in proximity of a recharge boundary is thus given by the sum of the effects of the real well and the image well:

$$s(x,y,t) = -Q \cdot \frac{W[u(r_1, t - t_i)] - W[u(r_2, t - t_i)]}{4 \cdot \pi \cdot T} = -Q \cdot W_R(r_1, r_2, t, t_i) \quad (2.41)$$

where r_1 and r_2 (L) represent the distances of the observation point (x, y) from the real and the image wells, respectively (Fig. 2.5a). The well functions for the operating well and the image well are both calculated using Eq. (2.37).

The general drawdown equation due to a single well operating in proximity a no-flow boundary is the following:

$$s(x,y,t) = -Q \cdot \frac{W[u(r_1, t - t_i)] + W[u(r_2, t - t_i)]}{4 \cdot \pi \cdot T} = -Q \cdot W_N(r_1, r_2, t, t_i) \quad (2.42)$$

The drawdown distribution due to a well field operating in an aquifer delimited by either a recharge or a no-flow rectilinear boundary is obtained as:

$$s(x,y,t) = -\sum_{i=1}^{n_{ow}} Q_i \cdot W_B(r_{i,1}, r_{i,2}, t, t_i) \quad (2.43)$$

where the "boundary" well function W_B is equal to W_R(Eq. 2.41) for a recharge boundary, or W_N (Eq. 2.42) for a no-flow boundary.

Figure 2.6 shows the drawdown distribution at time $t = 120$ di n a semi-infinite aquifer subject to extraction from two wells at the coordinates (500, 1000) and (1000, 500) (m). In subpanel 2.6a, a recharge boundary is located at $y = 0$ (the x-axis), whereas in subpanel 2.6b a no-flow boundary is present at $y = 0$. Aquifer parameters used in these scenarios are given in Table 2.1. Detailed information about the extracting wells (schedule and pumping rates) is given in Table 2.2. These plots are obtained using a MATLAB code called `Drawdown2D.m` provided in Appendix 1.

3.3 *The Theis Model in Unconfined Aquifers*

Unconfined aquifers differ from confined aquifers in that their upper boundary, known as "water table", constitutes a free surface boundary, at which the relative pressure is equal to zero. In an unconfined aquifer, the change in the water storage occurs as a response to drainage or recharge of the pores within the cone of depression. Different from confined aquifers, the saturated thickness of unconfined aquifers changes with time.

In unconfined aquifers, solution to the saturated ground water flow PDE (2.5) is particularly challenging because of the unknown location of the water table, which

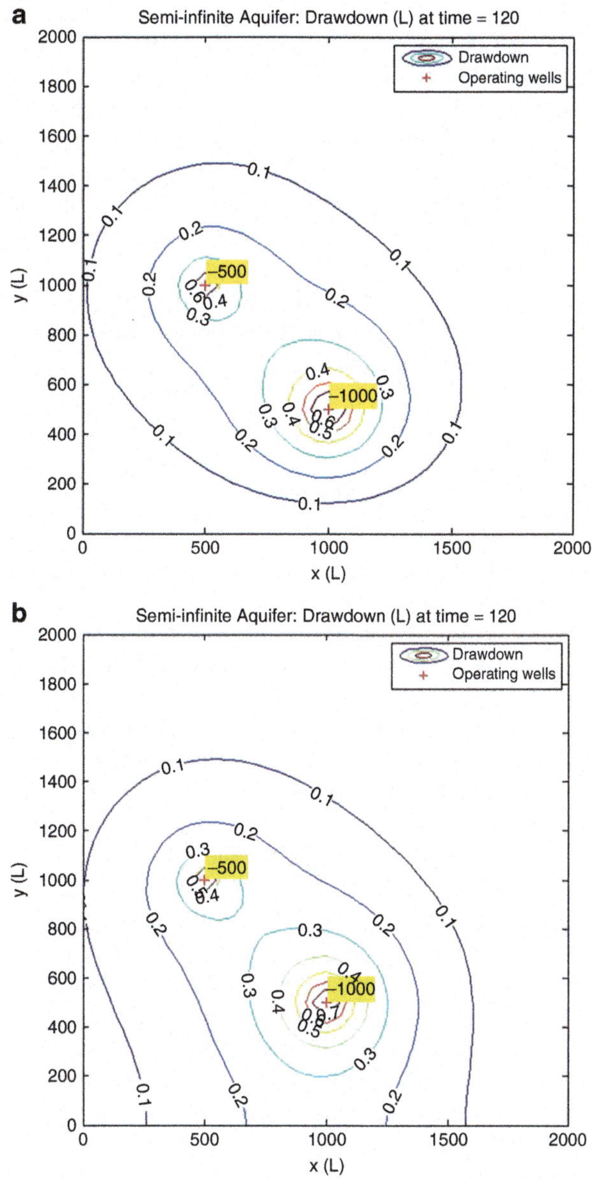

Fig. 2.6 Drawdown contours at $t = 120$ d due to two extracting wells operating at locations (500, 1000) and (1000, 500) (m), respectively, in semi-infinite aquifer with a recharge boundary (subpanel **a**) and with a no-flow boundary (subpanel **b**). Extraction rates are $-500\ \mathrm{m^3/d}$ for the first operating well, and $-1000\ \mathrm{m^3/d}$ for the second operating well

is required as a known boundary condition. In practice, the Theis solution can still be extended to model flow in unconfined aquifers under the assumption of prevalent horizontal flow (Dupuit approximation). Polubarinova-Kochina [15] shows that this assumption is sufficiently accurate if the aquifer drawdown is small compared to the initial saturated thickness of the aquifer. Under this condition the aquifer transmissivity T(L^2/T) and the aquifer storativity S(/),which truly depend on the saturated thickness of the aquifer, can be assumed as constants, and the vertical component of the fluid velocities may be neglected in relation to their horizontal component. However, corrections should be applied to the Theis solution in vicinity of operating wells, where the vertical component of the pore velocity may be significant and the occurring drawdown is typically large.

It is worth noting that storativity values for unconfined aquifers are practically equal to the specific yield, S_Y (the unitless ratio of the volume of water added or removed directly from the saturated zone of the aquifer to the resulting change in the volume of aquifer below water). In this respect, the storativity value of 0.2 given in Table 2.1 and used in the example application presented so far is truly appropriate for unconfined aquifers.

3.4 The Glover Model

Glover and Balmer's [2] solution was developed to compute the effects of depletion/accretion due to an operating well on a stream hydraulically connected to the aquifer. In this situation, the stream may be seen as a recharge boundary, which constitutes a constant-head boundary condition for the semi-infinite aquifer. The solution was derived based on the approximation the proposed by Theis [16] to evaluate stream depletion by integrating the Darcy's flux (in terms of drawdown) per unit width of the stream over the entire length of the stream. The Glover and Balmer [2] solution is still widely used in ground-water/surface-water conjunctive management.

The assumptions underlying the Glover model [2] are the following: (a) the aquifer is semi-infinite, horizontal, homogeneous and isotropic; (b) stream and aquifer are initially at equilibrium (the initial head h_0 in aquifer is constant and the same as the stage level in the river); (c) the aquifer transmissivity and storativity are uniform and constant over time (the saturated thickness does not change significantly); (d) the stream stage remains constant over time; (e) the stream forms a straight line and fully penetrates the aquifer (flow is horizontal); and (f) the stream is perfectly connected to the aquifer (no resistance to flow is caused by fine sediments at the streambed).

Note that a significant part of these assumptions were made to derive the Theis solution. The Glover model [2] is developed for the case of a single well operating at a constant extraction rate Q and starting time $t=0$, located at the origin of the Cartesian system in an aquifer with transmissivity T and storativity S. The stream is represented by the straight line located at $x = a$. Figure 2.7 shows a schematic of the aquifer-stream layout for Glover's model.

Let $q'_x(a, y, t)$ be flow rate per unit width (L^2/T) along the x direction on a generic point located on the stream line at $x = a$. Using Darcy's law this flow rate can be expressed as:

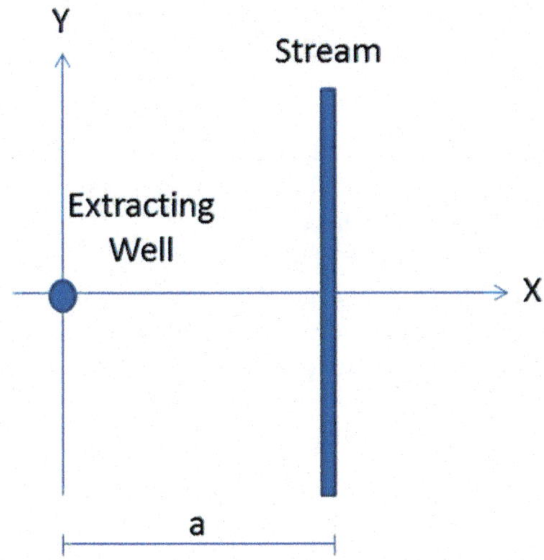

Fig. 2.7 Aquifer-stream system layout used to derive the Glover equation

$$q'_x(a,y,t) = T \cdot \frac{\partial s}{\partial x}(a,y,t) \tag{2.44}$$

Rewriting Eq. (2.18) in the Cartesian system and using the chain rule, the partial derivative at the right hand side of Eq. (2.44) is:

$$\frac{\partial s}{\partial x}(a,y,t) = \frac{\partial s}{\partial u}(a,y,t) \cdot \frac{\partial u}{\partial x}(a,y,t) \tag{2.45}$$

where (see Eq. 2.34):

$$\begin{aligned}\frac{\partial s}{\partial u}(a,y,t) &= \frac{\partial\left[\frac{Q}{4\cdot\pi\cdot T}\cdot\int_u^{\infty}\frac{\mathrm{e}^{-u'}}{u'}\cdot du'\right]}{\partial u} = \frac{Q}{4\cdot\pi\cdot T}\cdot\left[\frac{\mathrm{e}^{-u'}}{u'}\right]_u^{\infty} = \\ &= -\frac{Q}{4\cdot\pi\cdot T}\cdot\frac{\mathrm{e}^{-u}}{u} = -\frac{Q}{4\cdot\pi\cdot T}\cdot\frac{\mathrm{e}^{-\frac{S}{4\cdot T}\cdot\frac{a^2+y^2}{t}}}{\frac{S}{4\cdot T}\cdot\frac{a^2+y^2}{t}}\end{aligned} \tag{2.46}$$

and

$$\frac{\partial u}{\partial x}(a,y,t) = \frac{\partial\left[\frac{S}{4\cdot T}\cdot\frac{x^2+y^2}{t}\right]}{\partial x}(a,y,t) = \frac{S}{4\cdot T}\cdot\frac{2\cdot \mathrm{a}}{t} \tag{2.47}$$

Substitution of Eqs. (2.46) and (2.47) into Eq. (2.44) yields:

$$q'_x(a,y,t) = -T \cdot \frac{Q}{4\cdot\pi\cdot T} \cdot \frac{e^{-\frac{S}{4\cdot T}\cdot\frac{a^2+y^2}{t}}}{\frac{S}{4\cdot T}\cdot\frac{a^2+y^2}{t}} \cdot \frac{S}{4\cdot T} \cdot \frac{2\cdot \mathrm{a}}{t} = -\frac{Q}{2\cdot\pi} \cdot \frac{e^{-\frac{S}{4\cdot T}\cdot\frac{a^2+y^2}{t}}}{a^2+y^2} \cdot \mathrm{a} \quad (2.48)$$

Note that, in Eq. (2.48), q'_x is negative if both Q and a are positive. Indeed, in that case groundwater flow will have a negative horizontal component, that is, opposite to the sense of the x-axis.

The stream depletion rate can be obtained by spatial integration of q'_x along the stream line $x = a$ and using the method of images presented in Sect. 2.2, given that the stream provides a discharge per unit length equal to two times that given by Eq. (2.48). At any given time t he total stream depletion rate is thus given by:

$$Q_r(a,t) = \int_{-\infty}^{+\infty} q'_x(a,y,t)\cdot dy = -\frac{Q}{\pi} \cdot \int_{-\infty}^{+\infty} \frac{e^{-\frac{S}{4\cdot T}\cdot\frac{a^2+y^2}{t}}}{a^2+y^2} \cdot \mathrm{a}\cdot dy \quad (2.49)$$

In Eq. (2.49), the stream depletion rate is positive if the well withdraws water from the aquifer ($Q_r > 0$ if $Q < 0$) and negative otherwise. To calculate the integral in Eq. (2.49), the fraction P_r between the stream depletion rate Q_r and the pumping rate Q is considered, and the parameter $\alpha = S/(4\cdot T\cdot t)$ is substituted in Eq. (2.49):

$$P_r(a,t) = \frac{Q_r(a,t)}{Q} = -\frac{1}{\pi} \cdot \int_{-\infty}^{+\infty} \frac{e^{-\alpha\cdot\frac{a^2+y^2}{t}}}{a^2+y^2} \cdot \mathrm{a}\cdot dy \quad (2.50)$$

The integral at the right-hand side of Eq. (2.50) may be calculated by observing the following:

$$\begin{aligned}
\frac{\partial P_r}{\partial \alpha} &= -\frac{1}{\pi} \cdot \frac{\partial\left[\int_{-\infty}^{+\infty} \frac{e^{-\alpha\cdot(a^2+y^2)}}{a^2+y^2} \cdot \mathrm{a}\cdot dy\right]}{\partial\alpha} = -\frac{1}{\pi} \cdot \int_{-\infty}^{+\infty} \frac{\partial\left[\frac{e^{-\alpha\cdot(a^2+y^2)}}{a^2+y^2} \cdot \mathrm{a}\right]}{\partial\alpha} \cdot dy \\
&= \frac{1}{\pi} \cdot \int_{-\infty}^{+\infty} \frac{e^{-\alpha\cdot(a^2+y^2)}}{a^2+y^2} \cdot \mathrm{a}\cdot (a^2+y^2)\cdot dy = \frac{a}{\pi} \cdot \int_{-\infty}^{+\infty} e^{-\alpha\cdot(a^2+y^2)} \cdot dy \\
&= \frac{a}{\pi} \cdot e^{-\alpha\cdot a^2} \int_{-\infty}^{+\infty} e^{-\alpha\cdot y^2} \cdot dy
\end{aligned} \quad (2.51)$$

The integral at the far right-hand side of Eq. (2.51) may be solved by introducing the variable $z = \sqrt{\alpha}\cdot y$, where $dz = \sqrt{\alpha}\cdot dy$ and using the Gauss integral $\int_{-\infty}^{+\infty} e^{-z^2}$ $\cdot dz = \sqrt{\pi}$:

$$\int_{-\infty}^{+\infty} e^{-\alpha\cdot y^2} \cdot dy = \frac{1}{\sqrt{\alpha}} \cdot \int_{-\infty}^{+\infty} e^{-z^2} \cdot dz = \sqrt{\frac{\pi}{\alpha}} \quad (2.52)$$

Thus, Eq. (2.51) becomes:

$$\frac{\partial P_r}{\partial \alpha} = \frac{a}{\sqrt{\pi \cdot \alpha}} \cdot \mathrm{e}^{-\alpha \cdot a^2} \tag{2.53}$$

Note that the integration with respect to y has been performed in Eq. (2.53). Equation (2.51) is now integrated with respect to α to obtain P_r (Eq. 2.50). To do so, the variable $\chi = \sqrt{\alpha}$ is introduced (note that $d\chi = \frac{d\alpha}{2 \cdot \sqrt{\alpha}}$):

$$\begin{aligned} P_r &= \int \frac{\partial P_r}{\partial \alpha} \cdot d\alpha = \int \frac{a}{\sqrt{\pi \cdot \alpha}} \cdot \mathrm{e}^{-\alpha \cdot a^2} \cdot d\alpha = \frac{a}{\sqrt{\pi}} \int \frac{\mathrm{e}^{-\alpha \cdot a^2}}{\sqrt{\alpha}} \cdot d\alpha \\ &= \frac{a}{\sqrt{\pi}} \int \frac{\mathrm{e}^{-(a \cdot \chi)^2}}{\chi} \cdot 2 \cdot \chi \cdot d\chi = \frac{2}{\sqrt{\pi}} \int \mathrm{e}^{-(a \cdot \chi)^2} \cdot d(a \cdot \chi) \end{aligned} \tag{2.54}$$

Since the error function is defined as $\mathrm{erf}(v) = \frac{2}{\sqrt{\pi}} \cdot \int_0^v \mathrm{e}^{-v'^2} \cdot dv'$, Eq. (2.54) may be rewritten as:

$$P_r = \frac{2}{\sqrt{\pi}} \int \mathrm{e}^{-(a \cdot \chi)^2} \cdot d(a \cdot \chi) = \mathrm{erf}(a \cdot \chi) + c'' = \mathrm{erf}\left(\sqrt{\alpha} \cdot a\right) + c'' \tag{2.55}$$

where c'' is a constant of integration that may be calculated from the condition: $P_r(\alpha \to \infty) = 0$ (note that $\alpha \to \infty$ corresponds to $t \to 0$). Since the error function is such that $\mathrm{erf}(v \to \infty) = \frac{2}{\sqrt{\pi}} \int_0^\infty \mathrm{e}^{-v^2} dv$, the constant c'' is determined as follows:

$$P_r(\alpha \to \infty) = \lim_{\alpha \to \infty} \mathrm{erf}\left(\sqrt{\alpha} \cdot a\right) + c'' = 1 + \mathrm{c}'' = 0 \tag{2.56}$$

Equation (2.56) proves that $\mathrm{C}'' = -1$. The function P_r is thus determined to be equal to:

$$\begin{aligned} P_r(a, t) &= -1 + \mathrm{erf}\left(\sqrt{\alpha} \cdot a\right) = -\mathrm{erfc}\left(\sqrt{\alpha} \cdot a\right) = -\mathrm{erfc}\left(\sqrt{\frac{S}{4 \cdot T \cdot t}} \cdot a\right) \\ &= -\mathrm{erfc}\left(\sqrt{\frac{S}{4 \cdot T} \cdot \frac{a^2}{t}}\right) \end{aligned} \tag{2.57}$$

From Eq. (2.50), the stream depletion flow rate is thus obtained as:

$$Q_r(a, t) = -Q \cdot \mathrm{erfc}\left(\sqrt{\frac{S}{4 \cdot T} \cdot \frac{a^2}{t}}\right) \tag{2.58}$$

Equation (2.58) constitutes the Glover model [2] and can be integrated over time to obtain an equation for the cumulative stream depletion volume (2.14):

$$V_r(a,t) = -Q \cdot t \cdot \left[\left(1 + \frac{S}{2 \cdot T} \cdot \frac{a^2}{t}\right) \cdot \text{erfc}\left(\sqrt{\frac{S}{4 \cdot T} \cdot \frac{a^2}{t}}\right) - \frac{2}{\sqrt{\pi}} \cdot \sqrt{\frac{S}{4 \cdot T} \cdot \frac{a^2}{t}} \cdot \exp\left(-\frac{S}{4 \cdot T} \cdot \frac{a^2}{t}\right)\right] \tag{2.59}$$

Equations (2.58) and (2.59) can be rearranged to obtain the ratio between the stream depletion rate and the well pumping rate at time t for a well that started pumping a generic time τ ($t > \tau$):

$$Q_{ratio}(a,t,\tau) = \frac{Q_r(a,t,\tau)}{Q} = -\text{erfc}\left(\sqrt{\frac{S}{4 \cdot T} \cdot \frac{a^2}{t-\tau}}\right) \tag{2.60}$$

Likewise, the ratio between the stream depletion volume and the volume of groundwater pumped by the well is given by:

$$\begin{aligned} V_{ratio}(a,t,\tau) &= \frac{V_r(a,t,\tau)}{Q \cdot (t-\tau)} \\ &= -\left[\left(1 + \frac{S}{2 \cdot T} \cdot \frac{a^2}{t-\tau}\right) \cdot \text{erfc}\left(\sqrt{\frac{S}{4 \cdot T} \cdot \frac{a^2}{t-\tau}}\right)\right. \\ &\quad \left. - \frac{2}{\sqrt{\pi}} \cdot \sqrt{\frac{S}{4 \cdot T} \cdot \frac{a^2}{t-\tau}} \cdot \exp\left(-\frac{S}{4 \cdot T} \cdot \frac{a^2}{t-\tau}\right)\right] \end{aligned} \tag{2.61}$$

A MATLAB code called `Glover.m` that calculates and plots Eqs. (2.60 and 2.61) vs. time is provided in Appendix 2, along with instructions and examples for its use. Figure 2.8 shows the profiles of the stream depletion rate ratio (represented by the solid line) and the stream depletion volume ratio (represented by the dashed line) as a function of time, respectively, due to a well located 1000 m away from the stream, for an indefinite period of continuous well operation. These profiles are obtained using the `Glover.m` code. Aquifer properties used to obtain these figures are presented in Table 2.1. Figure 2.8 shows that both functions Q_{ratio} and V_{ratio} tend asymptotically to a steady-state value equal to 1.

3.5 Superposition of Effects: Stream Depletion

Similar to the Theis solution [1], Eqs. (2.58) and (2.59) indicate that both Q_r and V_r are linearly proportional to the well pumping rate Q [2]. In practice, the principle of superposition of solution can be applied to the Glover model [2] to calculate stream

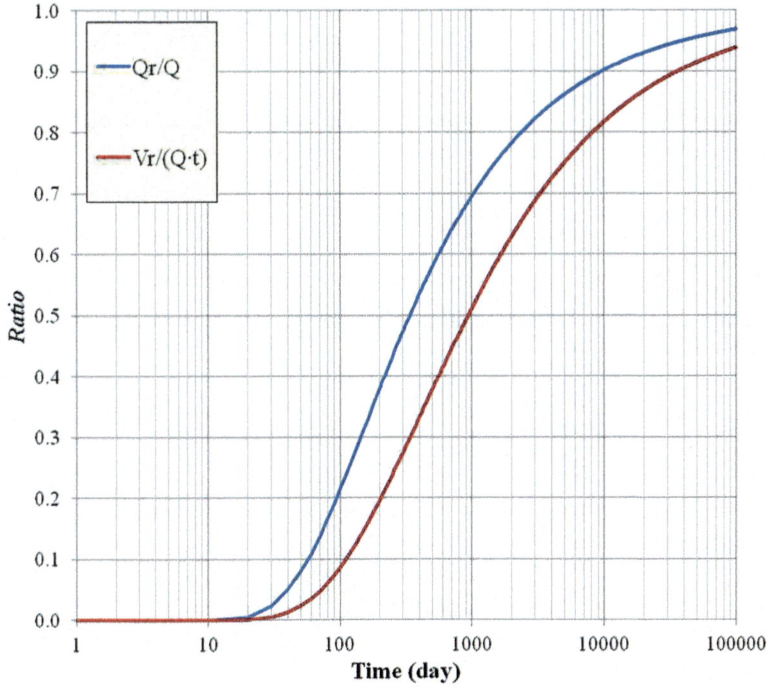

Fig. 2.8 Profiles of the stream depletion rate (*solid line*) and stream depletion volume (*dashed line*) ratios vs. time due to a well operating continuously and located at a distance of 1000 m from the stream

depletion rates and volumes from a generic well field with time varying pumping rates. The stream depletion rate due to a well field can be calculated as:

$$Q_r(t) = \sum_{i=1}^{n_{ow}} Q_i \cdot Q_{ratio}(a_i, t, \tau_i) \tag{2.62}$$

Note that in the sum of Equation (2.62) the Q_{ratio} function for well i is accounted for only if $t > \tau_i$. The stream depletion rate due to a single well with time-varying pumping rates is calculated as:

$$Q_r(t) = \int_0^t \frac{\partial Q}{\partial \tau} \cdot Q_{ratio}(a, t, \tau) \cdot d\tau \tag{2.63}$$

The stream depletion rate due to a well field with each well having time-varying pumping rates is obtained combining Equation (2.62) and (2.63):

$$Q_r(t) = \sum_{i=1}^{n_{ow}} \int_0^t \frac{\partial Q_i}{\partial \tau_i} \cdot Q_{ratio}(a_i, t, \tau_i) \cdot d\tau_i \quad (2.64)$$

Equations similar to (2.62–2.64) can be obtained for the stream depletion volume. In the case of a well field, the stream depletion volume is given by:

$$V_r(t) = \sum_{i=1}^{n_{ow}} Q_i \cdot (t - \tau_i) \cdot V_{ratio}(a_i, t, \tau_i) \quad (2.65)$$

The stream depletion volume due to a single operating well with time-varying pumping rate is obtained as:

$$V_r(t) = \int_0^t \frac{\partial Q}{\partial \tau} \cdot (t - \tau) \cdot V_{ratio}(a, t, \tau) \cdot d\tau \quad (2.66)$$

The stream depletion volume in a well field with a time varying pumping rates is calculated as:

$$V_r(t) = \sum_{i=1}^{n_{ow}} \int_0^t \frac{\partial Q_i}{\partial \tau_i} \cdot (t - \tau_i) \cdot V_{ratio}(a_i, t, \tau_i) \cdot d\tau_i \quad (2.67)$$

Using the stream depletion equations listed above, one can study the ratio between stream depletion volume during pumping operations and after pumping has ceased. Let us consider, for example, the case of five operating wells in an alluvial aquifer. The aquifer is in perfect hydraulic contact with a stream located at $x = 0$ (the y-axis), all wells are extracting with a constant operation rate of $-1000\ m^3/d$ for 180 d and are shut off afterwards. The second column of Table 2.3 displays the distance between the operating wells and the stream.

A simulation is performed for a total time of 5 years (1825 d), where the wells are activated individually and their impact on stream, in terms of depletion volume, is calculated during pumping and after the well is shut off. The results of this simulation are summarized in Table 2.3.

Table 2.3 Example of wells operating in proximity of a stream

Well	Stream distance a (m)	$V_{r,1}$ (m^3) (at 180 d—at pumping shut off)	$V_{r,2}$ (m^3) (at 1825 d—4.5 years after pumping shut off)	$\frac{V_{r,1}}{V_{r,2}}$ (/)
1	200	131,843	171,264	0.77
2	800	45,647	145,378	0.31
3	1200	20,071	128,697	0.16
4	2000	2866	97,702	0.03
5	5000	0.031	23,079	0.00

Simulation results are reported for stream depletion volumes during operation, and at the end of the simulation period

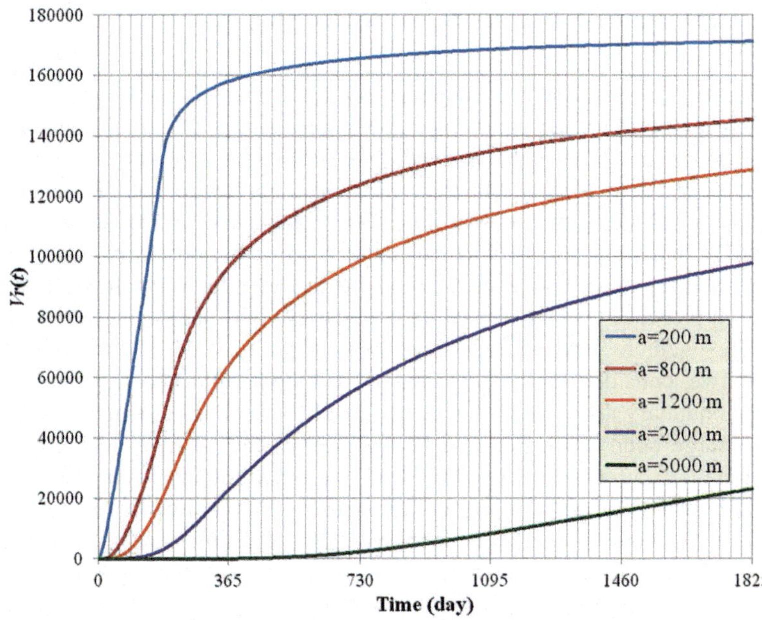

Fig. 2.9 Profiles of the stream-depletion volume vs. time for the five well configurations shown in Table 2.3. *Each curve* represents the well-stream distance as shown in the legend

It is interesting to note that, since the stream is the only recharge source for the aquifer, it provides a continuous supply even after pumping ceases. However, the ratio between the stream depletion volume during and after well operation varies significantly, from 0.77 to about 0, as the distance of the well to the stream is increased, and all the pumped volume is eventually extracted completely from the stream. This effect is similar to that observed in Fig. 2.8, where the stream depletion volume ratio reaches an asymptotic value of 1.0. The total time needed to restore the initial condition of equilibrium aquifer-stream is theoretically infinite. In practical terms, both this time and the volume of water extracted directly from the stream during well operations depend largely on the distance between the operating well and the stream.

Figure 2.9 shows the V_{ratio}-vs.-time profiles for the five wells considered in the example presented above. These profiles show that at larger times the ratio between the stream depletion volume and the total pumped water tends to be 1, which is in agreement with the conclusion that the extracted amount of water comes completely from the stream.

3.6 Pumping in a Bounded Finite Aquifer

In this section, an aquifer is considered "finite" if characterized by a finite areal extension due to the presence of two parallel boundaries in a 2-dimensional domain. The aquifer has a width w equal to the distance between the two boundaries, and an operating well is located within this width. The application of the principle of superposition of solutions provides an effective approach for extending the Theis and Glover equations in order to calculate the distribution of drawdown in the aquifer, and the impact on the stream flow due to well operations. In particular, the concurrent presence of the two parallel boundaries yields an infinite series number of "image" wells located "outside" the aquifer domain.

Here, we consider two configurations: (a) an aquifer bounded between a recharge boundary and a no-flow boundary; and (b) an aquifer comprised between two precharge boundaries. Figure 2.10 shows practical examples of these two. In Fig. 2.10a the aquifer is comprised between a no-flow boundary represented by the physical boundary of an alluvial shallow aquifer (the aquifer terminates due the presence of an outcropping bedrock formation), and a boundary represented by a stream that recharges the aquifer. In Fig. 2.10b the aquifer is bounded between an irrigation ditch and a stream. In the latter example, both boundaries can be conceptualized as recharge, constant-head boundaries.

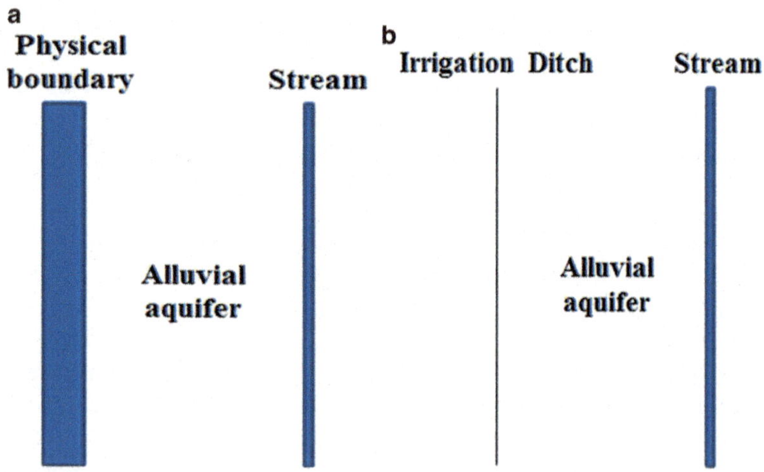

Fig. 2.10 Examples of finite constant-width aquifers. In case (**a**), the aquifer is delimited by a stream, which acts as a recharge boundary, and by its own physical boundary, which constitutes a no-flow boundary. In case (**b**) the aquifer is comprised between two recharge boundaries, such as an irrigation ditch and a stream

3.6.1 Superposition of Effects: Drawdown

Finite Aquifer Comprised Between a No-Flow Boundary and a Recharge Boundary

Let us study first the case of a finite aquifer comprised between a no-flow boundary and a recharge boundary. Figure 2.11 shows the well layout scheme derived by applying the method of images and accounting for the combined effect of the two boundaries.

As shown in Fig. 2.11, the real well, denoted as $i = 0$, calls for two image wells, one across each of the two boundaries. Each of these two image wells is in turn going to require a new image across the other boundary. This process eventually generates an infinite series of image wells across the two boundaries of the finite aquifer. Note that the wells are situated symmetrically with respect to the no-flow boundary. On one side of this boundary, the wells are grouped into couples ordered sequentially according to the index j ($j = 1, 2, 3, ..$). Each couple is characterized by wells with flow rates that are opposite in sign. If j is an odd integer the first well is extracting and the second well is injecting, whereas if j is an even integer the first well is injecting and the second well is extracting. In Fig. 2.11, the coordinates of wells in group j, having indices $i = 2 \cdot j - 2$ and $i = 2 \cdot j - 1$ are given by:

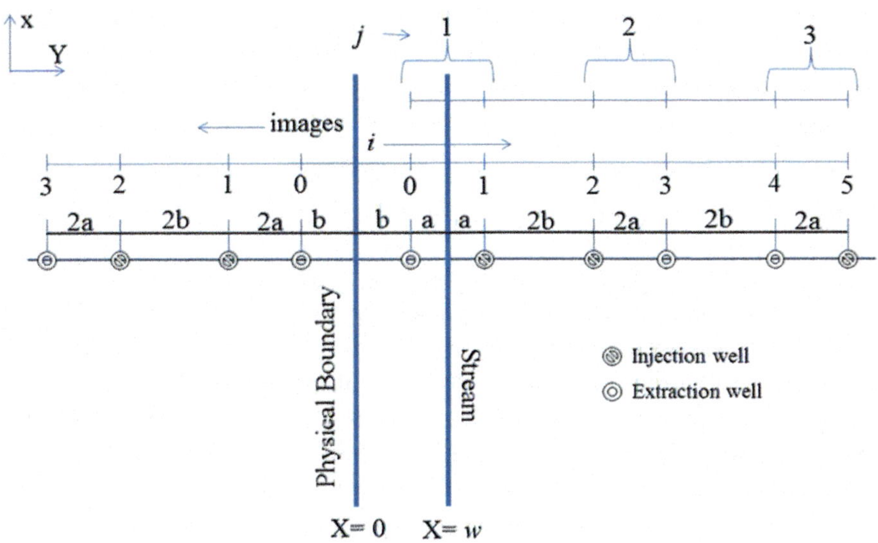

Fig. 2.11 Distribution of image wells with their operation type for an well operating within an aquifer bounded between a no-flow boundary, such as the aquifer physical boundary, and a constant-head boundary, such as a stream

$$\begin{aligned} x_{2\cdot j-2} &= (2\cdot j-1)\cdot w-a \\ x_{2\cdot j-1} &= (2\cdot j-1)\cdot w+a \end{aligned} \tag{2.68}$$

and the coordinates of the image wells opposite with respect to the axis of symmetry are:

$$\begin{aligned} x_{I,2\cdot j-2} &= -x_{2\cdot j-2} \\ x_{I,\,2\cdot j-1} &= -x_{2\cdot j-1} \end{aligned} \tag{2.69}$$

All of the wells in the series have the same y coordinate, regardless of i. Based on the wells layout given in Fig. 2.11, the drawdown at a generic location (x, y) and time t, due to a single well operating continuously with a constant rate Q, starting at time τ, in a bounded aquifer characterized by transmissivity T and storativity S, is thus given by:

$$\begin{aligned} s(x,y,t) &= -Q\cdot\left\{\sum_{j=1}^{\infty}\frac{(-1)^{j+1}}{4\cdot\pi\cdot T}\cdot\left[W\left(u\left(r_{2\cdot j-2},t-\tau\right)\right)-W\left(u\left(r_{2\cdot j-1},t-\tau\right)\right)\right.\right. \\ &\quad \left.\left.+W\left(u\left(r_{I,2\cdot j-2},t-\tau\right)\right)-W\left(u\left(r_{I,2\cdot j-1},t-\tau\right)\right)\right]\right\} \\ &= -Q\cdot W_{NR}\left(w,x_w,y_{w,},x,y,\tau,t\right) \end{aligned} \tag{2.70}$$

Each term of the sum in Eq. (2.70) represents an equivalent well function equal to the sum of the four functions of well group j and its image, forming the contribution to drawdown in a bounded aquifer between a no-flow boundary (N) and a recharge boundary (R). $r_{2\cdot j-2}$ and $r_{2\cdot j-1}$ are the Euclidean distances (Eq. 2.17) between the operating well locations $\left(x_{2\cdot j-2}, y_{2\cdot j-2}\right)$ and $\left(x_{2\cdot j-1}, y_{2\cdot j-1}\right)$ and the observation point (x, y). Similarly, $r_{I,2\cdot j-2}$ and $r_{I,2\cdot j-1}$ are the Euclidean distances between the image well locations $\left(x_{I,2\cdot j-2}, y_{I,2\cdot j-2}\right)$ and $\left(x_{I,2\cdot j-1}, y_{I,2\cdot j-1}\right)$ and the observation point (x, y). The well functions in Eq. (2.70) are calculated using the Theis well function (Eq. 2.35). Obviously, Eq. (2.70) applies only if $t > \tau$. Such condition applies to all well functions introduced in this section.

At the right-hand side of Eq. (2.70), $W_{NR}(w, x_w, y_w, x, y, \tau, t)$ represents the equivalent well function for a well located at coordinates (x_w, y_w) within an aquifer delimited between a no-flow boundary and a constant-head boundary (NR). This equation can be generalized to calculate the drawdown at a generic location (x, y) and time t due to a well field with constant-rate pumping wells:

$$s(x,y,t) = -\sum_{i=1}^{n_{ow}} Q_i\cdot W_{NR}\left(x_{w,i}, y_{w,i}, x, y, \tau_i, t\right) \tag{2.71}$$

Equations (2.70) and (2.71) are implemented in a MATLAB code called `BCYC_Drawdown2D.m` provided in Appendix 3. Figure 2.12a, b show the

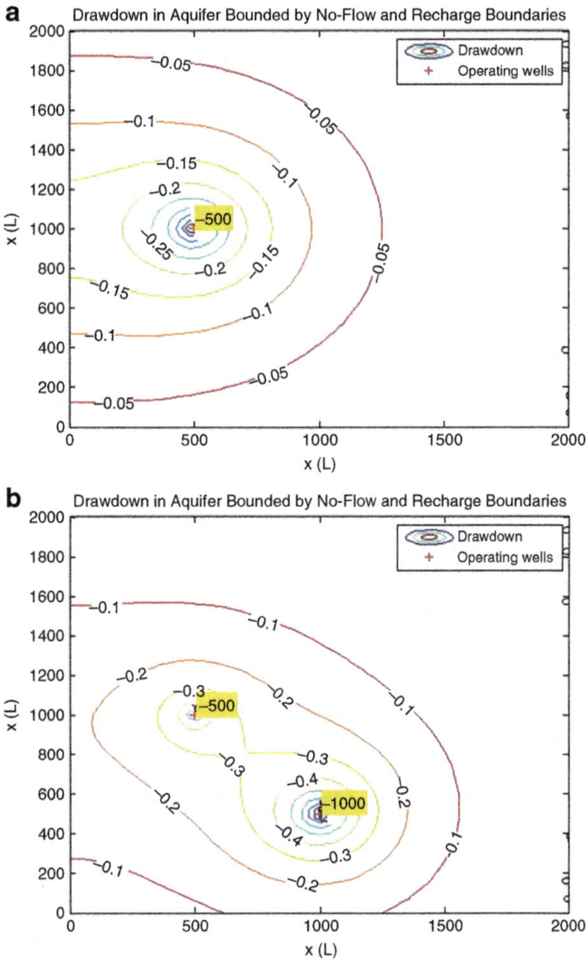

Fig. 2.12 Drawdown fields at $t = 120$ d in a finite aquifer delimited by a no-flow boundary at $x = 0$ and a recharge boundary at $x = 2000$ m, and subject to (**a**) a single continuously extracting well and (**b**) two extracting wells. Well locations and extraction rates are labeled in each subpanel. Other well data are provided in Table 2.2

drawdown distributions obtained using this model, at a time equal to 120 d due to a single well, and due to a two-well well field, respectively. Aquifer properties are indicated in Table 2.1. The no-flow boundary is located on the y-axis, whereas the recharge boundary is located at $x = 2000$ m (the aquifer width w is 2000 m). Well locations and extraction rates are labeled in each subpanel. Other well data are provided in Table 2.2.

Finite Aquifer Limited Between Two Recharge Boundaries

Let us now consider a finite aquifer limited between two recharge boundaries. Figure 2.13 presents the well layout scheme created using the method of images to model the concurrent presence of the two boundaries.

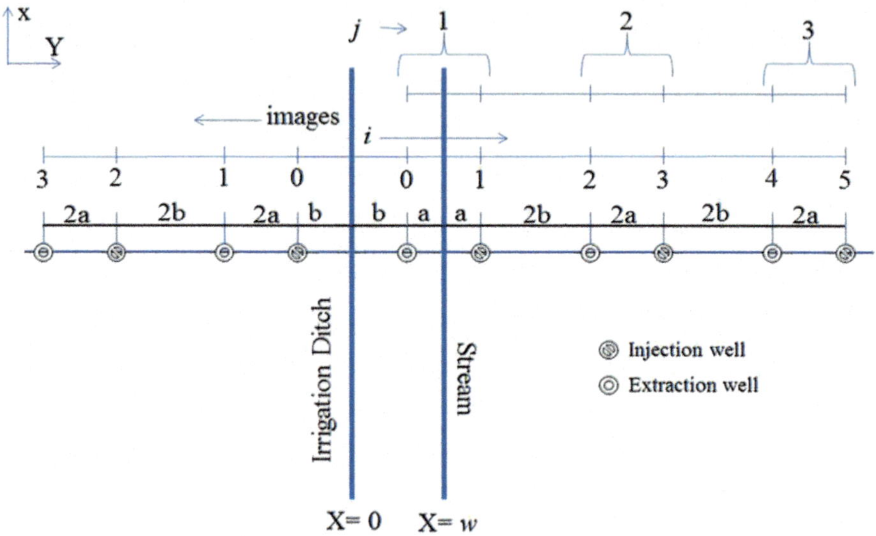

Fig. 2.13 Distribution of image wells with their operation type for a well operating within an aquifer bounded between two recharge boundaries, such as an irrigation ditch and a stream

In this case, the no-flow boundary in Fig. 2.11 is substituted by a constant-head boundary representing, for example, the presence of an irrigation ditch. This boundary represents the symmetry axis for the well layout. The image wells created across the axis of symmetry will have opposite operation types with respect to their corresponding wells across a no-flow boundary. This condition causes the two wells in the generic well group j to have the same order of operation types for all groups, that is, extraction for the first well and injection for the second. Well coordinates are obtained using Eqs. (2.68) and (2.69) and all wells distances to the observation point (x, y) are the same as in the former case. The drawdown general equation for the configuration in Fig. 2.13 is as follows:

$$\begin{aligned} s(x,y,t) &= -Q\cdot\left\{\sum_{j=1}^{\infty}\frac{1}{4\cdot\pi\cdot T}\cdot\left[W\left(u\left(r_{2\cdot j-2},t-\tau\right)\right)-W\left(u\left(r_{2\cdot j-1},t-\tau\right)\right)\right.\right.\\ &\quad\left.\left.-W\left(u\left(r_{I,2\cdot j-2},t-\tau\right)\right)+W\left(u\left(r_{I,2\cdot j-1},t-\tau\right)\right)\right]\right\}\\ &= -Q\cdot W_{RR}(x_w,y_w,x,y,\tau,t) \end{aligned} \quad (2.72)$$

where $W_{RR}(x_w, y_w, x, y, \tau, t)$ is the equivalent well function for a well located at coordinates (x_w, y_w) within an aquifer delimited between two constant-head recharge boundaries (RR). Consequently, the drawdown due to a well field of m wells is calculated by:

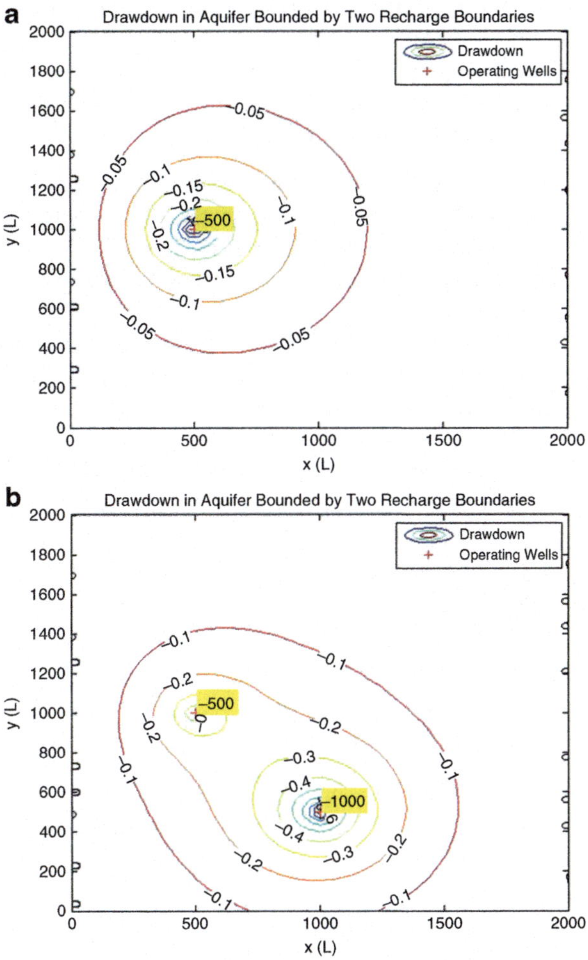

Fig. 2.14 Drawdown fields at $t = 120$ d in a finite aquifer delimited by two recharge boundaries at $x = 0$ and at $x = 2000$ m, and subject to (**a**) a single continuously extracting well and (**b**) two extracting wells. Well locations and extraction rates are labeled in each subpanel. Other well data are provided in Table 2.2

$$s(x, y, t) = -\sum_{i=1}^{n_{ow}} Q_i \cdot W_{RR}\left(x_{w,i}, y_{w,i}, x, y, \tau_i, t\right) \tag{2.73}$$

Figure 2.14a shows the drawdown distribution due to a single continuously extracting well, and Fig. 2.14b shows the drawdown distributions due to two extracting wells. Both distributions are plotted at time equal to 120 d using the MATLAB code `BCYC_Drawdown2D.m` provided in Appendix 3. Aquifer properties are given in Table 2.1. The irrigation ditch is located on the y-axis and the stream is at $x = 2000$ m (the aquifer width is 2000 m). The labels in the figures indicate well locations and extraction rates. Table 2.2 gives detailed well data.

3.6.2 Superposition of Effects: Stream Depletion

To develop stream depletion general equations for a finite aquifer bounded between two parallel boundaries, the principle of superposition of effects is applied in the same manner described in Sect. 3.6.1.

(a) First, a finite aquifer comprised between a no-flow boundary and a recharge boundary is considered. Figure 2.15 shows the layout scheme for the image wells of an operating well located at $x = w - a = b$. In this case, the no-flow boundary represents symmetry axis for the well layout, and it is located on the y-axis.

As shown in Fig. 2.15, imaginary wells located on the left side of the symmetry axis are grouped in couples indexed as "Glover groups" ($g = 1, 2, \ldots$). The wells in the group g have an inverse order (operation wise) with respect to wells in the group $g + 1$. Wells coordinates are obtained using equations similar to Eqs. (2.68) and (2.69). The total stream depletion rate can be obtained by applying superposition of solutions for the well system represented in Fig. 2.15:

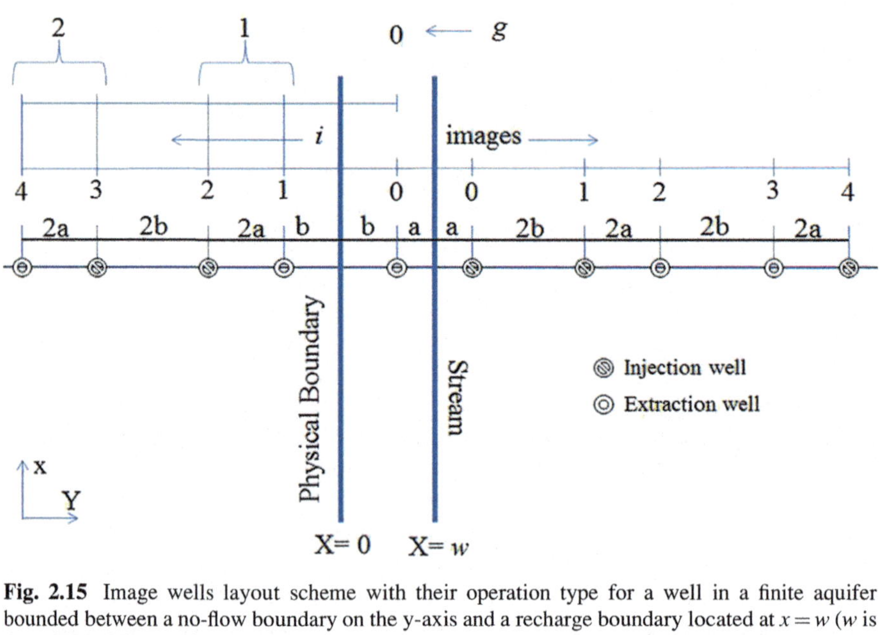

Fig. 2.15 Image wells layout scheme with their operation type for a well in a finite aquifer bounded between a no-flow boundary on the y-axis and a recharge boundary located at $x = w$ (w is the width of the aquifer)

$$Q_r(w,a,t,\tau) = Q \cdot \left\{ Q_{ratio}(a,t,\tau) + \sum_{g=1}^{\infty} (-1)^{g+1} \cdot [Q_{ratio}(2 \cdot g \cdot w - a, t, \tau) - Q_{ratio}(2 \cdot g \cdot w + a, t, \tau)] \right\} = Q \cdot BNQ_{ratio}(w,a,t,\tau) \tag{2.74}$$

Analogously, the stream depletion volume is obtained as:

$$V_r(w,a,t,\tau) = Q \cdot (t-\tau) \cdot \left\{ V_{ratio}(a,t,\tau) + \sum_{g=1}^{\infty} (-1)^{g+1} \cdot [V_{ratio}(2 \cdot g \cdot w - a, t, \tau) - V_{ratio}(2 \cdot g \cdot w + a, t, \tau)] \right\} = Q \cdot BNV_{ratio}(w,a,t,\tau) \tag{2.75}$$

In Eqs. (2.74) and (2.75), BNQ_{ratio} and BNV_{ratio} represent, respectively, the stream depletion rate ratio and the stream depletion volume ratio for a stream in a bounded aquifer with a no-flow boundary as the secondary boundary. Equations (2.74) and (2.75) are implemented in a MATLAB code called `BGlover.m` provided in Appendix 4.

Figure 2.16 shows the profiles of the stream depletion rate ratio and the stream depletion volume ratio vs. time, due to a well pumping continuously in an aquifer bounded between a no-flow boundary and a stream. The aquifer width is 2000 m, and the well is located 500 m away from the stream. These profiles are obtained using the MATLAB code `BGlover.m`. Aquifer properties used to develop the profiles in Fig. 2.16 are listed in Table 2.1.

Figure 2.17 shows the comparison between the stream depletion volumes in a semi-infinite aquifer (blue line) and in a 2000 m wide finite aquifer bounded by a no-flow boundary (red line). In both cases, the stream-well distance is 500 m. The stream depletion volume ratio in the finite aquifer (red line) tends to reach "steady state" conditions (when the ratio equals 1.0) faster than its equivalent in the semi-infinite aquifer (blue line). This is due to the limited lateral extent of the system in the former case, which prevents the expansion of the depression cone in the direction of the no-flow boundary and forces it to reach the stream earlier. Note that the two profiles are practically indistinguishable until the time when the depression cone reaches the no-flow boundary in the case of the finite aquifer.

The stream depletion general equations for the case of a well field are obtained by applying the principle of superposition. The stream depletion rate is given by:

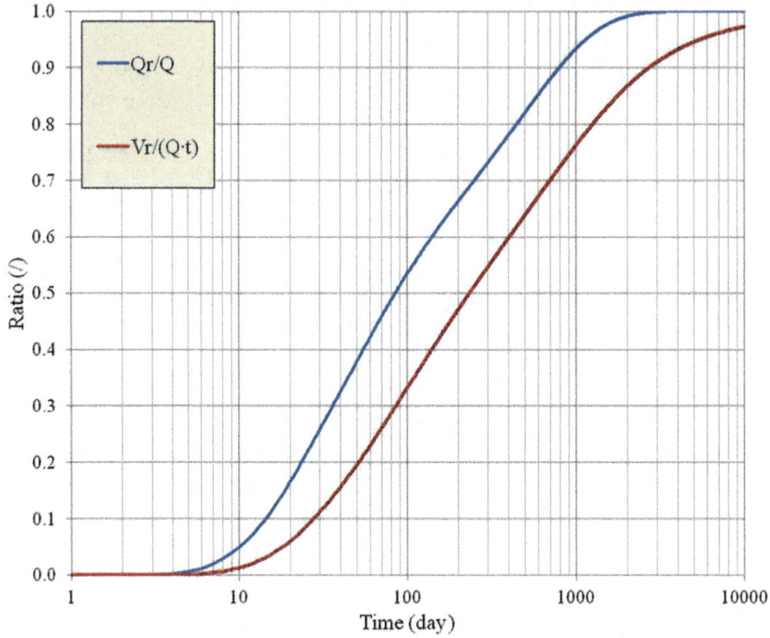

Fig. 2.16 Profiles of the stream depletion rate (*solid line*) and stream depletion volume (*dashed line*) ratios vs. unitless time due to a well operating continuously in a 2000 m wide finite aquifer bounded between a no-flow boundary and a stream. The well operates continuously and is located at a distance of 500 m from the stream

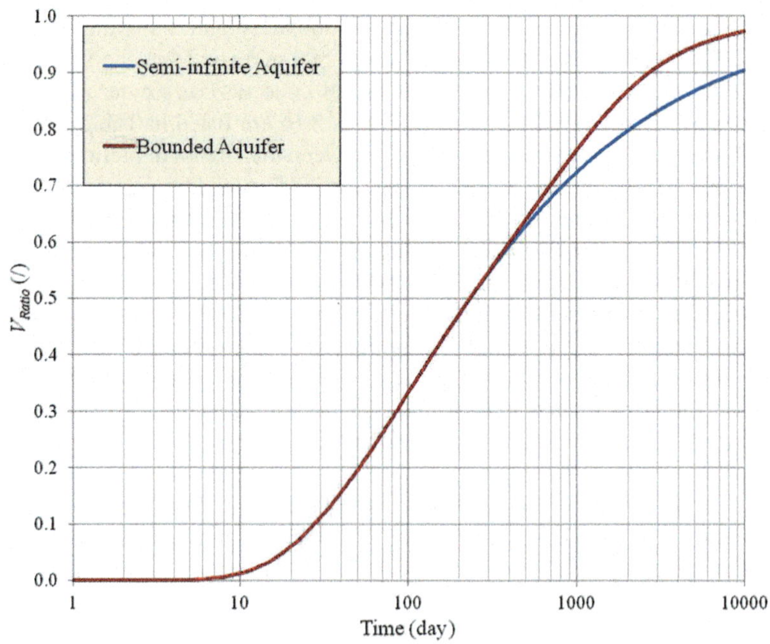

Fig. 2.17 Comparison between the stream depletion volume in a semi-infinite aquifer bounded by a stream (*dashed line*) and a 2000 m wide finite aquifer bounded by a no-flow boundary and a stream (*solid line*). The pumping well is located 500 m away from the stream

$$Q_r(t) = \sum_{i=1}^{n_{ow}} Q_i \cdot BNQ_{ratio}(w, a_i, t, \tau_i) \tag{2.76}$$

The stream depletion volume is calculated as:

$$V_r(t) = \sum_{i=1}^{n_{ow}} Q_i \cdot (t - \tau_i) \cdot BNV_{ratio}(w, a_i, t, \tau_i) \tag{2.77}$$

In the sum in Eqs. (2.76) and (2.77), the generic well i is accounted for only if $t > \tau_i$.

(b) Let us now consider a finite aquifer comprised between two recharge boundaries, such as a stream and an irrigation ditch. The application of the principle of superposition of effects to stream depletion due to image wells produces the well layout shown in Fig. 2.18. Glover groups similar to those created for the configuration of Fig. 2.15 are used here. Note that, in Fig. 2.15, the wellorder (injection-extraction) does not change from one group to the next since the boundary representing the axis of symmetry is a recharge boundary.

Given the well layout presented in Fig. 2.18, the stream depletion rate due to a single operating well is:

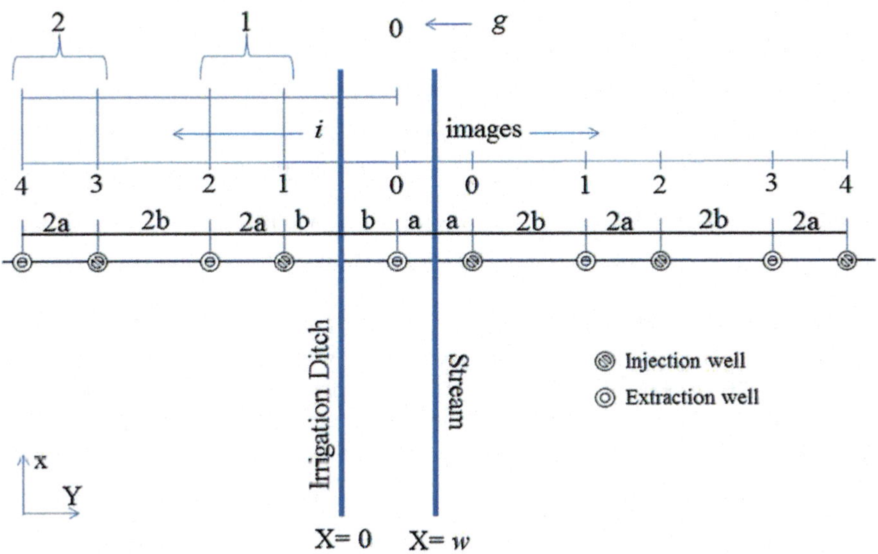

Fig. 2.18 Distribution of image wells with their operation type for an well operating within an aquifer bounded between two constant-head boundaries, such as a stream and an irrigation ditch

$$Q_r(w,a,t,\tau) = Q \cdot \left\{ Q_{ratio}(a,t,\tau) + \sum_{g=1}^{\infty} [-Q_{ratio}(2 \cdot g \cdot w - a, t, \tau) + Q_{ratio}(2 \cdot g \cdot w + a, t, \tau)] \right\} = Q \cdot BRQ_{ratio}(w,a,t,\tau) \quad (2.78)$$

Correspondingly, the volume of stream depletion is:

$$V_r(w,a,t,\tau) = Q \cdot (t-\tau) \cdot \left\{ V_{ratio}(a,t,\tau) + \sum_{g=1}^{\infty} [-V_{ratio}(2 \cdot g \cdot w - a, t, \tau) + V_{ratio}(2 \cdot g \cdot w + a, t, \tau)] \right\} = Q \cdot BRV_{ratio}(w,a,t,\tau) \quad (2.79)$$

where BRQ_{ratio} is the stream depletion ratio function, and BRV_{ratio} is the stream depletion volume ratio function, in an aquifer bounded between two recharge boundaries.

Equations (2.78) and (2.79) are valid for $t > \tau$. The BRQ_{ratio} and BRV_{ratio} functions are implemented in MATLAB routines provided in the code `BGlover.m` included in Appendix 4. Figure 2.19 shows the profiles for the stream depletion rate ratio and the stream depletion volume ratio vs. time, due to a single well pumping continuously. The well is located 500 m away from the stream in a 2000 m wide aquifer. These profiles are obtained using the previously mentioned MATLAB code `BGlover.m`. Aquifer properties used to develop these plots are listed in Table 2.1.

Figure 2.20 shows the stream depletion rate and volume ratio profiles with unit less time for a well operating in a semi-infinite aquifer bounded by a stream (dashed line) and in a 2000-m wide finite aquifer bounded between a an irrigation ditch and a stream (solid line). In both cases, the well is located 500 m away from the stream and pumping at constant rate. Aquifer properties used to develop these plot are listed in Table 2.1.

Note that, in the bounded aquifer configuration considered here, the secondary recharge boundary (the irrigation ditch) represents a supply source for the aquifer in addition to primary boundary (the stream). This explains why in Figs. 2.19 and 2.20 both the stream depletion rate and volume ratios tend to a "steady-state" value less than one. The actual pumped volume tends asymptotically to equalize the sum of the stream depletion volumes from both boundaries.

In the case of a well field, the total stream depletion rate is given by:

$$Q_r(t) = \sum_{i=1}^{n_{ow}} Q_i \cdot BRQ_{ratio}(w, a_i, t, \tau_i) \quad (2.80)$$

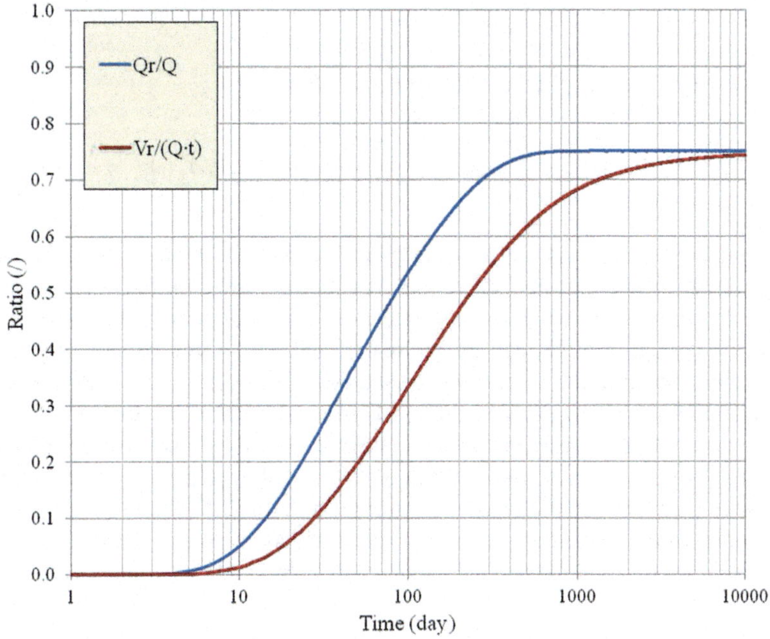

Fig. 2.19 Stream depletion rate ratio (*solid line*) and stream depletion volume ratio (*dashed line*) for a well operating continuously in a 2000-m wide finite aquifer bounded between an irrigation ditch and a stream. The well-stream distance is 500 m

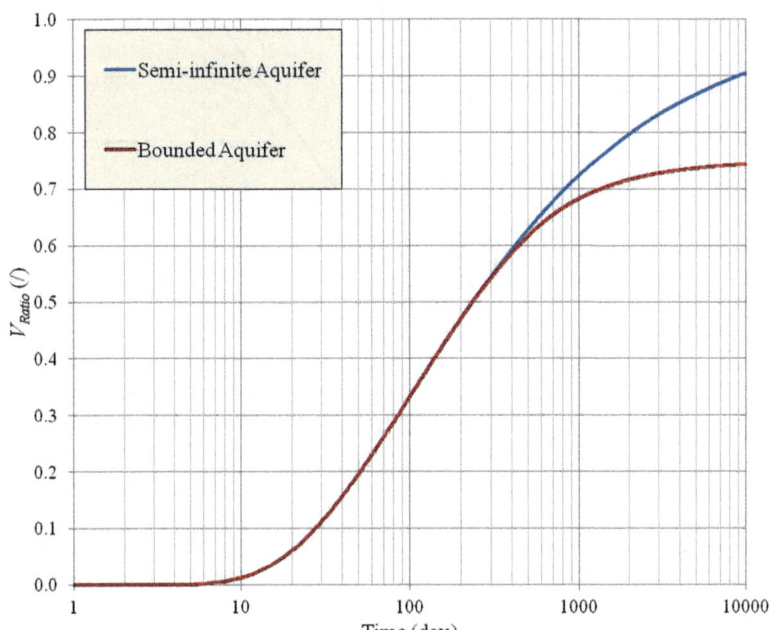

Fig. 2.20 Comparison between the stream depletion volume in a semi-infinite aquifer bounded by a stream (*dashed line*) and a 2000-m wide finite aquifer bounded by a stream and an irrigation ditch (*solid line*). The pumping well is located 500 m away from the stream

Likewise, the stream depletion volume is calculated as:

$$V_r(t) = \sum_{i=1}^{n_{ow}} Q_i \cdot (t - \tau_i) \cdot BRV_{ratio}(w, a_i, t, \tau_i) \tag{2.81}$$

Note that, Eqs. (2.80) and (2.81) are used to estimate the impact of well pumping on the stream, which, in Fig. 2.17 is represented by the straight line $x = w$. The rates and volumes of depletion produced on the secondary recharge boundary, that is, the irrigation ditch, can be calculated using the same equations presented above, after changing the stream-well distances to $b_i = w - a_i$ instead of a_i.

Figure 2.21 shows the profile profiles of stream depletion volume ratio functions vs. time for a well operating in a finite aquifer delimited by a no-flow boundary and a recharge boundary (blue line), and in a finite aquifer delimited by two recharge boundaries (red line). In both cases the aquifer width is 2000 m, the well is located at 500 m from the stream and operates continuously at constant rate. The aquifer properties used to develop this plot are listed in Table 2.1. The comparison between

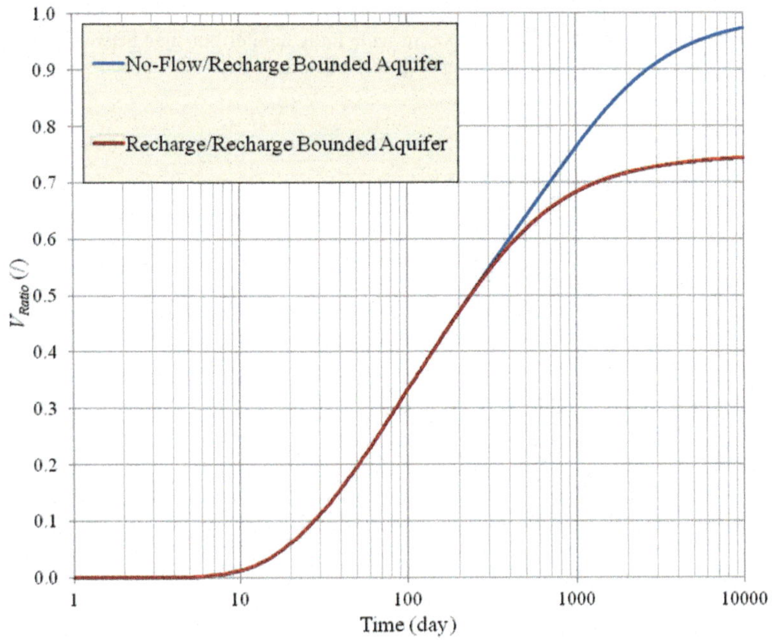

Fig. 2.21 Stream depletion volume ratio vs. unit less for a well operating in a finite aquifer bounded between a no-flow boundary and a stream (*solid line*) and in a finite aquifer bounded between a an irrigation ditch and a stream (*dashed line*). The aquifer is 2000-m wide. The well-stream distance is 500 m

the two profiles demonstrate that a much larger volume is drawn for the stream when the aquifer is bounded on the opposite side by a no-flow boundary than when the aquifer is bounded on the opposite side by another recharge boundary. Indeed in the former case, the no-flow boundary cannot supply any recharge and all the water extracted by the well must eventually come from the stream, which constitutes the only source of recharge available.

3.7 Cyclical Operation of Wells

In this section, we consider the case of a periodic operation of wells. This situation is quite typical, particularly when alluvial groundwater is being used seasonally for irrigation purposes. Figure 2.22 illustrates the cyclic operation rate profile over time for a single well. The well operates at a constant rate Q over a given period Δt_{on} and is shut off during a period Δt_{off}. The length of the full cyclic period Δt_{off} is given by the sum $\Delta t_{on} + \Delta t_{off}$.

As previously mentioned, continuous operation is one of the limitations of the Theis model (Eq. 2.34) and Glover model (Eq. 2.58). The principle of superposition may be used to overcome this obstacle. Whereby there is a variation of the pumping rate, one can simulate the activation of a new imaginary well, located at the same position of the real well with a rate equal to the change of the pumping. For example, for the cyclic operation depicted in Fig. 2.22, at any given time t, the effect occurring on the system during the generic interval of time Δt_{on} within the generic operation cycle $i(i = 1, 2, 3, \ldots)$ can be simulated by two wells: the first well starts operating continuously at time $\tau_i = (i - 1) \cdot \Delta t$ with a rate Q, and the second wells starts operating continuously at time $\tau_i = (i - 1) \cdot \Delta t + \Delta t_{on}$ with a rate $-Q$. At any generic time t, the number of full periods that have been completed is equal to:

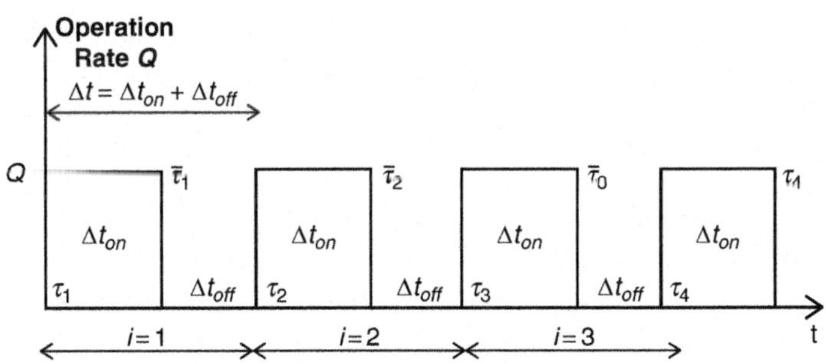

Fig. 2.22 Schedule plan for a well operating cyclically. The well operates with a constant rate Q over a given interval Δt_{on} starting at time τ_1, and is shut off at time $\overline{\tau}_1 = \tau_1 + \Delta t_{on}$. At time $\tau_2 = \overline{\tau}_1 + \Delta t_{off}$, the well is turned on again, and the cycle is repeated

$$n = \text{int}(t/\Delta t) \tag{2.82}$$

where int(·) represents the integer part of the real number (·).

3.7.1 Superposition of Effects: Drawdown

The drawdown general equation for the cyclic operation of a well is developed using the principle of superposition for the effects of the n couples of wells operating during n full cycles, plus an extra term accounting for the well operation during the current cycle. The latter has a different expression depending if t falls within the period Δt_{on} or within the period Δt_{off}. In a laterally infinite aquifer the resulting drawdown is thus given by the flowing equation:

$$\begin{aligned} s(r,t) &= -\frac{Q}{4\cdot\pi\cdot T}\cdot\left[\sum_{i=1}^{n}\{W[u(r,t-\tau_i)] - W[u(r,t-\overline{\tau}_i)]\}\right. \\ &\quad \left. +\begin{Bmatrix} W[u(r,t-\tau_{n+1})] & \text{if } n\cdot\Delta t < t \le n\cdot\Delta t + \Delta t_{on} \\ W[u(r,t-\tau_{n+1})] - W[u(r,t-\overline{\tau}_{n+1})] & \text{if } n\cdot\Delta t + \Delta t_{on} < t \le (n+1)\cdot\Delta t \end{Bmatrix}\right] \\ &= -Q\cdot W_C(r,\Delta t,\Delta t_{on},\tau_1,t) \end{aligned} \tag{2.83}$$

where $W_C(r, \Delta t, \Delta t_{on}, \tau_1, t)$ is the overall well function, representing the accumulated response of the system to the operation of the wells couples during n full cycles plus an extra term accounting for the well operation during the most current period. Note that Eq. (2.83) is valid only if $t > \tau_1$, otherwise the well function W_C, and thus the drawdown s, is equal to zero. The drawdown general equation for a system of wells operating cyclically with the same period Δt is the following:

$$s(x,y,t) = -\sum_{i=1}^{n_{ow}} Q_i\cdot W_C(r_i,\Delta t,\Delta t_{on,i},\tau_{1,i},t) \tag{2.84}$$

In Eq. (2.84), r_i is the distance between the generic well i and the observation point (x, y). Each well has a generic operation starting time $\tau_{1,i}$. Figure 2.23 shows the hydraulic head profile time series over a 10-year period, calculated at two observation points located at a distance r equals 15 m (red line) and r equals 150 m (blue line). Over a period $\Delta t = 365$ d (1 year) the well operates at a rate of-500 m^3/d for 180 d (Δt_{on}) and is turned off for the remaining 185 d. Pumping starts at time $t = 0$. The profiles are obtained using the MATLAB code called `Drawdown_VS_Time.m` included in Appendix 1. Aquifer properties for these examples are given in Table 2.1.

Figure 2.24a, b show the drawdown spatial distributions, calculated using the MATLAB code called `Drawdown2D.m` included in Appendix 1, at time $t = 100$ d, and $t = 200$ d, respectively, for a well operating cyclically, for a period of 180 d every year. The drawdown distributions shown in Fig. 2.24c, d are for a two-well

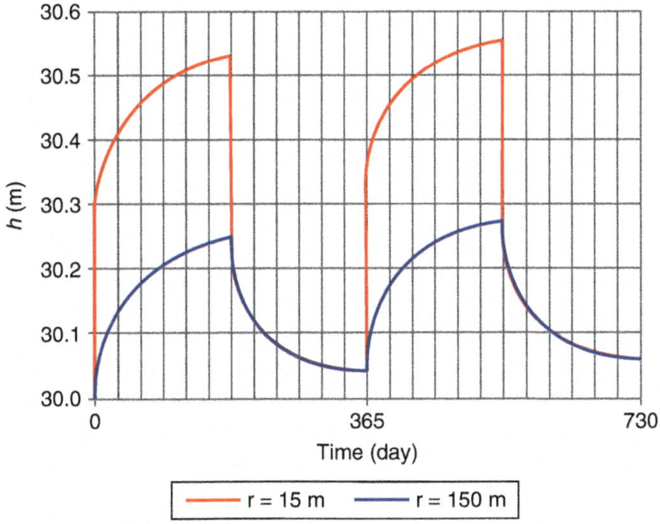

Fig. 2.23 Drawdown time series for two observation points distant 15 m (red *line*) and 150 m (*blue line*) from a well operating cyclically over a period of 10 years. The well operates at a rate of $-500\ \mathrm{m^3/d}$ during 180 d in each cycle with a cycle length equal to 365 d

well field for the same times as in subpanels (a) and (b). These "snapshot" times are selected during and after the operation period, respectively. Welllocations and operation rates are labeled in each plot. Table 2.1 shows the aquifer properties used in these examples, and Table 2.2 contains detail well field data.

In the case of semi-infinite aquifers bounded by either a recharge or a no-flow boundary, an equation similar to (2.84) may be applied, with the well function W_C replaced by the following well function:

$$\begin{aligned}
&W_{CSI}(r, r_I, \Delta t, \Delta t_{on}, \tau_1, t) \\
&\quad = \frac{1}{4 \cdot \pi \cdot T} \cdot \\
&\quad \left[\sum_{i=1}^{n} \left\{ W_{R/N}[u(r, r_I, t-\tau_i)] - W_{R/N}[u(r, r_I, t-\overline{\tau}_i)] \right\} \right. \\
&\quad \left. + \begin{cases} W_{R/N}[u(r, r_I, t-\tau_{n+1})] & \text{if } n \cdot \Delta t < t \leq n \cdot \Delta t + \Delta t_{on} \\ W_{R/N}[u(r, r_I, t-\tau_{n+1})] - W_{R/N}[u(r, r_I, t-\overline{\tau}_{n+1})] & \text{if } n \cdot \Delta t + \Delta t_{on} < t \leq (n+1) \cdot \Delta t \end{cases} \right]
\end{aligned} \tag{2.83}$$

where $W_{R/N}$ represents either the well function W_R (Eq. 2.41) in the case of a semi-infinite aquifer delimited by a recharge boundary, or the well function W_N (Eq. 2.42) in the case of a semi-infinite aquifer delimited by a no-flow boundary. r and r_I are the distances of the observation point (x, y) from the real well and its image across the boundary, respectively.

The drawdown equation for a well field in a semi-infinite aquifer, where each well operates cyclically, is thus obtained by superposition of solutions as follows:

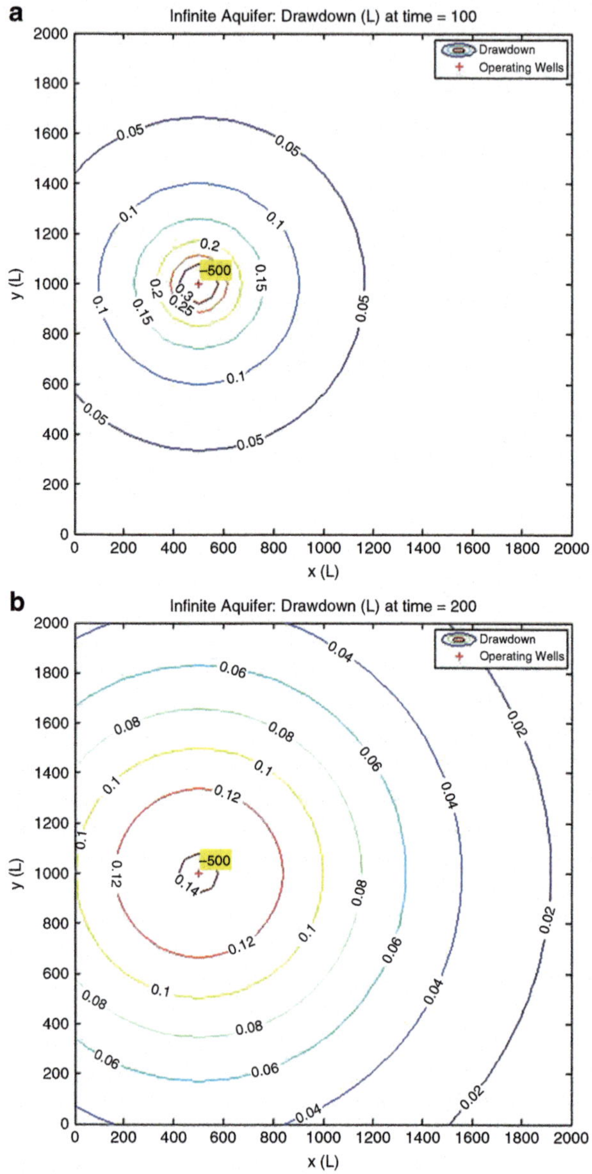

Fig. 2.24 Drawdown distributions due to a single well operating cyclically at a rate of $-500\ m^3/d$ for 180 d per year at times (**a**) t = 100 d and (**b**) t = 200 d. Subpanels (**c**) and (**d**) show the corresponding drawdown distributions due to a two-well well field. Wells locations and pumping rates are labeled in each subpanel

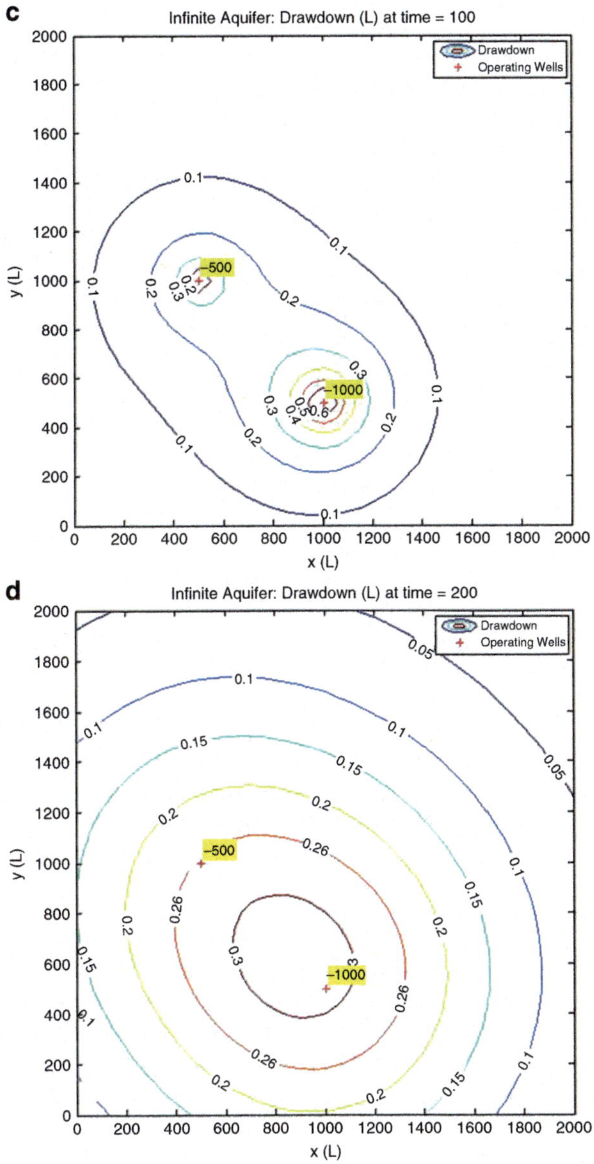

Fig. 2.24 (continued)

$$s(x, y, t) = -\sum_{i=1}^{n_{ow}} Q_i \cdot W_C(r_i, r_{I,i}, \Delta t, \Delta t_{on,i}, \tau_{1,i}, t) \tag{2.84}$$

Figure 2.25a shows the hydraulic head vs. time profiles over 10 years, plotted using the MATLAB code `Drawdown_VS_Time.m` (see Appendix 1), for two

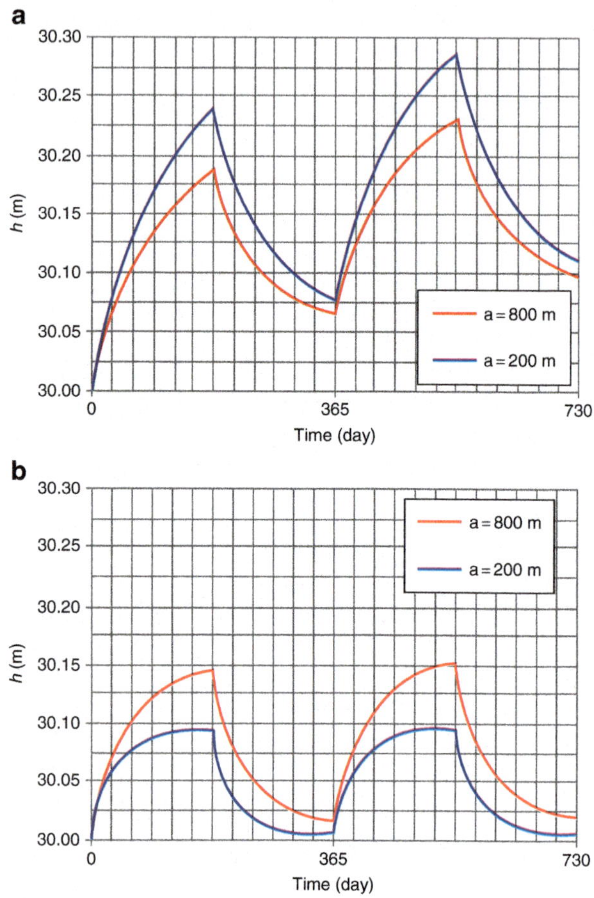

Fig. 2.25 Drawdown time series in a semi-infinite aquifer with (**a**) a recharge boundary, and (**b**) a no-flow boundary. Two observation points are considered, located at 200 m (*blue line*) and 800 m (*red line*) from the boundary. Both observation points are at a distance of 425 m from the operating well, which is located 500 m away from the boundary. The well operates at a rate of $-500\ m^3/d$ for 180 d each year

observation points in a semi-infinite aquifer located at distances of 800 m (red line) and 200 m (blue line) from a no-flow boundary, respectively. The two observation points are at the same distance $r = 300$ m from a well that operates cyclically for 180 d every year ($\Delta t = 365$ d) with a pumping rate of $-500\ m^3/d$. The well is located 500 m away from the recharge boundary.

Similarly, Fig. 2.25b shows the hydraulic head vs. time profiles, for the same setting as in subpanel (a), with the only difference that the semi-infinite aquifer is delimited by a recharge boundary.

Figure 2.26a, b show the drawdown distributions obtained using the MATLAB code `Drawdown2D.m` (Appendix 1) at time t = 200 d, due to two extracting wells, in a semi-infinite aquifer with, respectively, (a) a no-flow boundary and (b) a recharge boundary, both located on the *x*-axis. The aquifer properties are given in Table 2.1, whereas the extraction rates and operation schedule of the well field are described in Table 2.2.

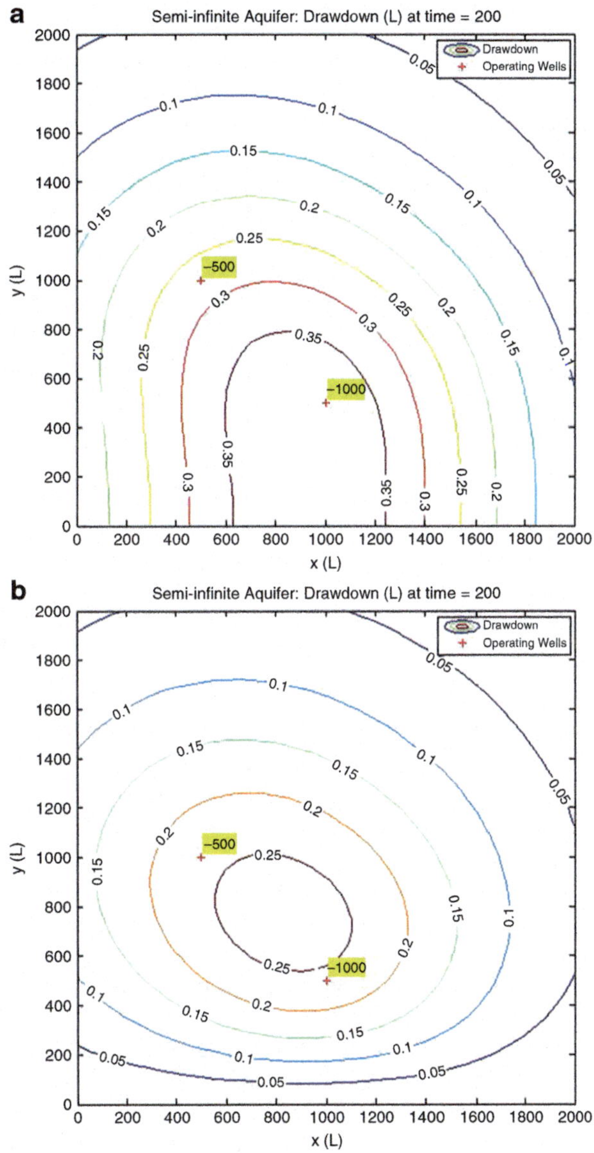

Fig. 2.26 Spatial distributions of drawdown occurring at time $t = 200$ d, during the cyclic extraction period of a two-well well field in (**a**) in a semi-infinite aquifer with a recharge boundary (**b**) in a semi-infinite aquifer with a no-flow boundary. Well field data are listed in Table 2.2

3.7.2 Superposition of Effects: Stream Depletion

Equations similar to those derived in Sect. 3.7.1 can be obtained to assess the impact on stream depletion due to the cyclic operation of wells in an aquifer hydraulically connected to a stream. Similar to Eq. (2.83), the stream depletion rate Q_r and the stream depletion volume V_r are given by the two following equations:

$$Q_r(a, \Delta t, \Delta t_{on}, \tau_1, t) = Q \cdot Q_{C,ratio}(a, \Delta t, \Delta t_{on}, \tau_1, t) \tag{2.85}$$

$$V_r(a, \Delta t, \Delta t_{on}, \tau_1, t) = Q \cdot V_{C,ratio}(a, \Delta t, \Delta t_{on}, \tau_1, t) \tag{2.86}$$

where:

$$\begin{aligned} &Q_{C,ratio}(a, \Delta t, \Delta t_{on}, \tau_1, t) \\ &= \sum_{i=1}^{n} [Q_{ratio}(a, t - \tau_i) - Q_{ratio}(a, t - \overline{\tau}_i)] \cdot \\ &\quad + \begin{cases} Q_{ratio}(a, t - \tau_{n+1}) & \text{if } n \cdot \Delta t < t \le n \cdot \Delta t + \Delta t_{on} \\ Q_{ratio}(a, t - \tau_{n+1}) - Q_{ratio}(a, t - \overline{\tau}_{n+1}) & \text{if } n \cdot \Delta t + \Delta t_{on} < t \le (n+1) \cdot \Delta t \end{cases} \end{aligned} \tag{2.87}$$

and

$$\begin{aligned} &V_{C,ratio}(a, \Delta t, \Delta t_{on}, \tau_1, t) \\ &= \sum_{i=1}^{n} [(t - \tau_i) \cdot V_{ratio}(a, t - \tau_i) - (t - \overline{\tau}_i) \cdot V_{ratio}(a, t - \overline{\tau}_i)] \\ &\quad + \begin{cases} (t - \tau_{n+1}) \cdot V_{ratio}(a, t - \tau_{n+1}) & \text{if } n \cdot \Delta t < t \le n \cdot \Delta t + \Delta t_{on} \\ (t - \tau_{n+1}) \cdot V_{ratio}(a, t - \tau_{n+1}) - (t - \overline{\tau}_{n+1}) \cdot V_{ratio}(a, t - \overline{\tau}_{n+1}) & \text{if } n \cdot \Delta t + \Delta t_{on} < t \le (n+1) \cdot \Delta t \end{cases} \end{aligned} \tag{2.88}$$

Equations (2.87) and (2.88) can be generalized to calculate the stream depletion rate and the stream depletion volume due to a system of cyclically operating wells:

$$Q_r(t) = \sum_{i=1}^{n_{ow}} Q_i \cdot Q_{C,ratio}(a_i, \Delta t, \Delta t_{on,i}, \tau_{1,i}, t) \tag{2.89}$$

$$V_r(t) = \sum_{i=1}^{n_{ow}} Q_i \cdot V_{C,ratio}(a_i, \Delta t, \Delta t_{on,i}, \tau_{1,i}, t) \tag{2.90}$$

Note that in the sums indicated in Eqs. (2.89) and (2.90) only those wells such that $t > \tau_{1,i}$ must be accounted for.

Equations (2.87–2.90) are implemented in MATLAB code called `Glover.m` included in Appendix 2. Figure 2.27a, b show the time series for the stream depletion rate and the stream depletion volume, obtained using the above mentioned code, for the cyclical operation of a well located at a distance of 500 m from the stream. The well operates cyclically with a rate of -500 m^3/d for a period of 180 d over an annual cycle ($\Delta t = 365$ d) starting at time $t = 0$. Aquifer properties are those listed in Table 2.1.

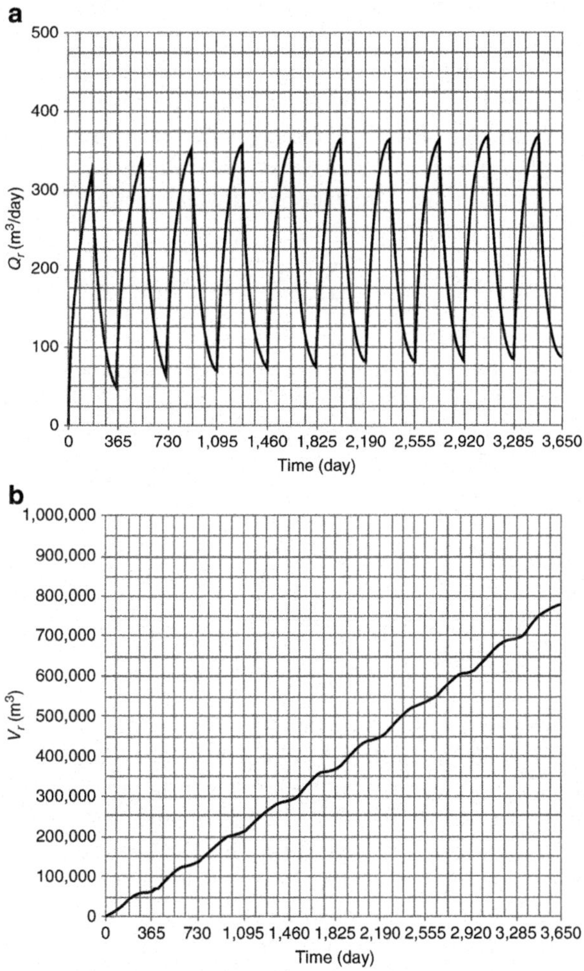

Fig. 2.27 Time series for (**a**) the stream depletion rate and (**b**) the stream depletion volume due to the cyclical operation of a well located at a distance of 500 m from the stream. The well is operating cyclically with a rate of -500 m^3/d for a period of 180 d every year

3.8 *Cyclic Operation of Wells in a Finite Aquifer*

This case addresses two limitations of the Theis and Glover solutions, that is, the condition of infinite areal extension of the aquifer and the condition of continuous constant-rate operation of the well. Once again, superposition of solutions is the approach used to remove these hypotheses. The response of the system is thus obtained by assuming the presence of two groups of wells: the first group represents image wells that simulate the presence of the aquifer boundaries; the second group represents the imaginary wells that simulate the cyclic well operation.

3.8.1 Superposition of Effects: Drawdown

In a constant-width finite aquifer characterized by one of the two configurations presented in Figs. 2.11 and 2.13, the drawdown at a generic point (x, y) produced by a system of cyclically operating wells, each with an operation schedule such as that graphed in Fig. 2.22, can be derived using an equation similar to Eq. (2.84), with the well function W_{CSI} replaced by a well function W_{FC}:

$$s(x, y, t) = -\sum_{i=1}^{n_{ow}} Q_i \cdot W_{FC}\left(x_{w,i}, y_{w,i}, x, y, \Delta t, \Delta t_{on,i}, \tau_{1,i}, t\right) \quad (2.91)$$

In Eq. (2.91), the well function W_{FC} is the overall well function for a well operating cyclically in a finite aquifer. Similar to Eq. (2.84), this function is calculated as:

$$\begin{aligned} & W_{FC}\left(x_{w,i}, y_{w,i}, x, y, \Delta t, \Delta t_{on,i}, \tau_{1,i}, t\right) \\ & = \frac{1}{4 \cdot \pi \cdot T} \cdot \\ & \left[\sum_{j=1}^{n} \left\{ W_{BB}\left(x_{w,i}, y_{w,i}, x, y, \tau_{i,j}, t\right) - W_{BB}\left(x_{w,i}, y_{w,i}, x, y, \bar{\tau}_{i,j}, t\right) \right\} \right. \\ & \left. + \begin{cases} W_{BB}\left(x_{w,i}, y_{w,i}, x, y, \tau_{i,n+1}, t\right) & \text{if } n \cdot \Delta t < t \leq n \cdot \Delta t + \Delta t_{on} \\ W_{BB}\left(x_{w,i}, y_{w,i}, x, y, \tau_{i,n+1}, t\right) - W_{BB}\left(x_{w,i}, y_{w,i}, x, y, \bar{\tau}_{i,n+1}, t\right) & \text{if } n \cdot \Delta t + \Delta t_{on} < t \leq (n+1) \cdot \Delta t \end{cases} \right] \end{aligned} \quad (2.92)$$

where W_{BB} is the well function W_{NR} (Eq. 2.70) if the aquifer is delimited by a no-flow and a constant-head boundary, or the well function W_{RR} (Eq. 2.72) if the aquifer is comprised between two constant-head boundaries.

Equation (2.92) is implemented in the MATLAB code called `Drawdown_VS_Time.m` provided in Appendix 1. This code is used to obtain the following results. Figure 2.28a shows the hydraulic headtime series due for a cyclically operating well located at a distance of 1000 m from the recharge boundary in a 2000-m wide aquifer bounded between a no-flow boundary and a constant-head stream. The aquifer parameters are given in Table 2.1. The well is operated at a constant rate of -1000 m^3/d for a period of 180 d every year. The profiles are obtained for two observation points located at distances 1500 m (red-line) and 500 m (dashed line) from the stream. These observation points are 400 and 600 m away from the operating well, respectively. Figure 2.28b shows the same setting as in subpanel (a), for the case of a finite aquifer comprised between two constant-head boundaries, such as irrigation ditch and a stream.

Figure 2.29 shows the drawdown distributions at $t = 200$ d due to two wells operating cyclically and simultaneously for 180 d per year, in a finite aquifer with a constant width of 2000 m. In Fig. 2.29a the aquifer is delimited by no-flow boundary, located on the y-axis and a stream. In Fig. 2.29b the aquifer is delimited by two recharge boundaries, such an irrigation ditch, located at the y-axis, and a stream. Wells locations and extraction rates are labeled in each plot. Other welldata are given in Table 2.2. Aquifer properties are indicated in Table 2.1.

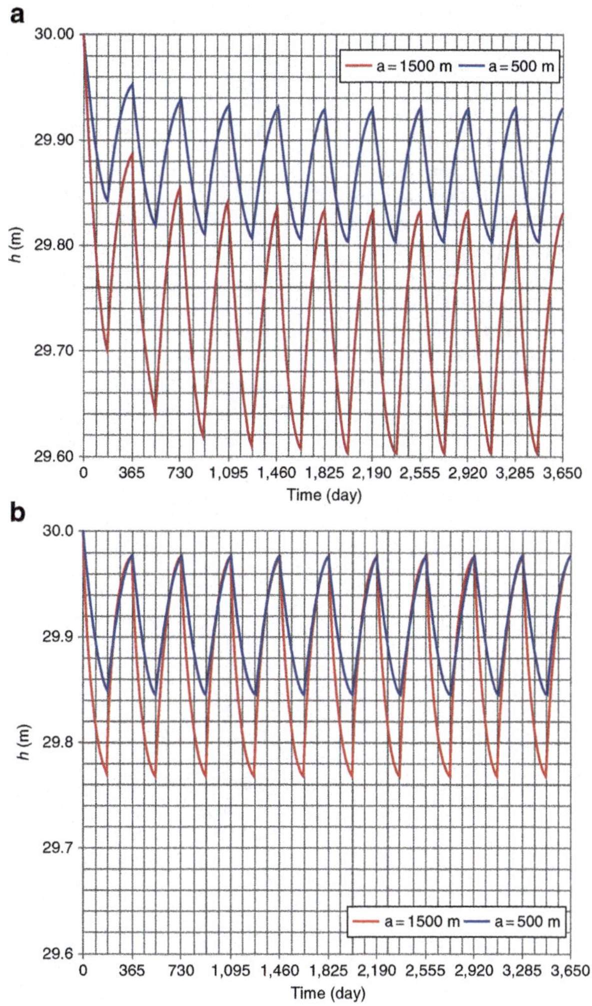

Fig. 2.28 Hydraulic head time series due to a well operating cyclically in a finite aquifer with a constant width of 2000 m comprised between (**a**) a no-flow and a recharge boundary, and (**b**) tow recharge boundaries. The profiles are obtained at two observation points located 1500 m (*red line*) and 500 m (*blue line*) away from the stream boundary. The well is operated cyclically at $-1000\ m^3/d$ for 180 d per year. The two observation points are at a distance of 400 and 600 m from the operating well, respectively

3.8.2 Superposition of Effects: Stream depletion

Based on superposition of effects, the depletion rate from a stream constituting a boundary for a finite-width aquifer caused by a cyclically operating well may be expressed as:

$$Q_r(a, w, \Delta t, \Delta t_{on}, \tau_1, t) = Q \cdot Q_{FC,ratio}(a, w, \Delta t, \Delta t_{on}, \tau_1, t) \qquad (2.93)$$

where $Q_{FC,ratio}$ is the accumulated sum of the stream depletion rate ratio functions for a cyclically operating well. This function is calculated for a stream hydraulically connected to an aquifer bounded between the stream and another boundary of either no-flow type or recharge type. This function is calculated as:

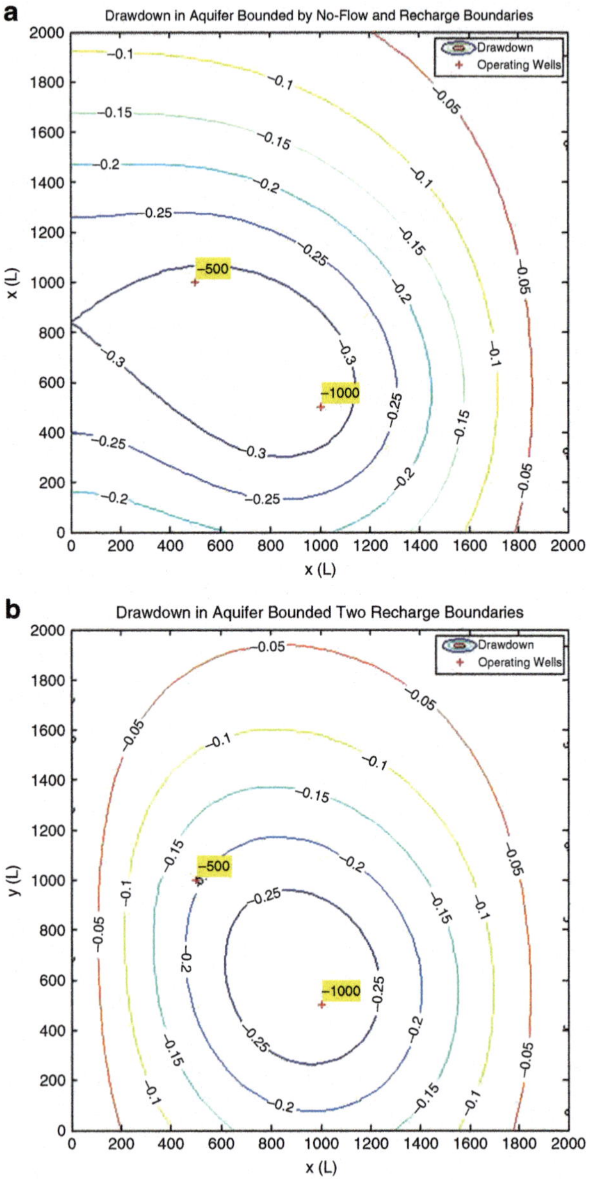

Fig. 2.29 Drawdown distributions in a constant width finite aquifer ($w = 2000$ m) bounded between (**a**) a no-flow boundary, located at the y-axis, and a stream, and (**b**) an irrigation ditch, located at the y-axis, and a stream. Distributions are at time $t = 200$ d, due to the operation of two wells (see Table 2.2) for 180 d very year. Operation rates and wells locations are labeled in each subpanel

$$
\begin{aligned}
&Q_{FC,ratio}(a, w, \Delta t, \Delta t_{on}, \tau_1, t)\\
&= \sum_{i=1}^{n} [BQ_{ratio}(a, w, t, \tau_i) - BQ_{ratio}(a, w, t, \overline{\tau}_i)]\\
&\quad + \begin{cases} BQ_{ratio}(a, t, \tau_{n+1}) & \text{if } n \cdot \Delta t < t \le n \cdot \Delta t + \Delta t_{on} \\ BQ_{ratio}(a, w, t, \tau_{n+1}) - BQ_{ratio}(a, w, t, \overline{\tau}_{n+1}) & \text{if } n \cdot \Delta t + \Delta t_{on} < t \le (n+1) \cdot \Delta t \end{cases}
\end{aligned} \tag{2.94}
$$

The function BQ_{ratio} in Eq. (2.94) is given by the function BNQ_{ratio} (Eq. 2.76) if the aquifer is limited, in addition to the stream, by a no-flow boundary, or by the function BRQ_{ratio} (Eq. 2.78) if the aquifer is limited by another constant head boundary. Similar to the stream depletion rate, the stream depletion volume is obtained as:

$$
V_r(a, w, \Delta t, \Delta t_{on}, \tau_1, t) = Q \cdot V_{FC,ratio}(a, w, \Delta t, \Delta t_{on}, \tau_1, t) \tag{2.95}
$$

where $V_{FC,\,ratio}$ is the overall stream depletion volume ratio function representing the response to a well operating cyclically in a finite aquifer, bounded between a constant-head stream and another boundary of either no-flow type or recharge type. This function is calculated as:

$$
\begin{aligned}
&V_{FC,ratio}(a, w, \Delta t, \Delta t_{on}, \tau_1, t)\\
&= \sum_{i=1}^{n} [(t - \tau_i) \cdot BV_{ratio}(a, w, t, \tau_i) - (t - \overline{\tau}_i) \cdot BV_{ratio}(a, w, t, \overline{\tau}_i)]\\
&\quad + \begin{cases} (t - \tau_{n+1}) \cdot BV_{ratio}(a, w, t, \tau_{n+1}) & \text{if } n \cdot \Delta t < t \le n \cdot \Delta t + \Delta t_{on} \\ (t - \tau_{n+1}) \cdot BV_{ratio}(a, w, t, \tau_{n+1}) - (t - \overline{\tau}_{n+1}) \cdot BV_{ratio}(a, w, t, \overline{\tau}_{n+1}) & \text{if } n \cdot \Delta t + \Delta t_{on} < t \le (n+1) \cdot \Delta t \end{cases}
\end{aligned} \tag{2.96}
$$

The function BV_{ratio} in Eq. (2.96) is given by the function BNV_{ratio} (Eq. 2.77) if the aquifer is delimited by a no-flow boundary, or by the function BRV_{ratio} (Eq. 2.79) if the aquifer is limited by another constant head boundary.

In the case of well field, the total stream depletion rate is given by:

$$
Q_r(t) = \sum_{i=1}^{n_{ow}} Q_i \cdot Q_{FC,ratio}(a_i, w, \Delta t, \Delta t_{on,i}, \tau_{1,i}, t) \tag{2.97}
$$

Likewise, the stream depletion volume is calculated as:

$$
V_r(t) = \sum_{i=1}^{n_{ow}} Q_i \cdot V_{FC,ratio}(a_i, w, \Delta t, \Delta t_{on,i}, \tau_{1,i}, t) \tag{2.98}
$$

Equations (2.93–2.98) have been implemented in MATLAB code called `BGlover.m` provided in Appendix 4. This code is used to obtain the following results. Figure 2.30a, b shows the time series for the stream depletion rate and the stream depletion volume, respectively, obtained for a well operating cyclically in a

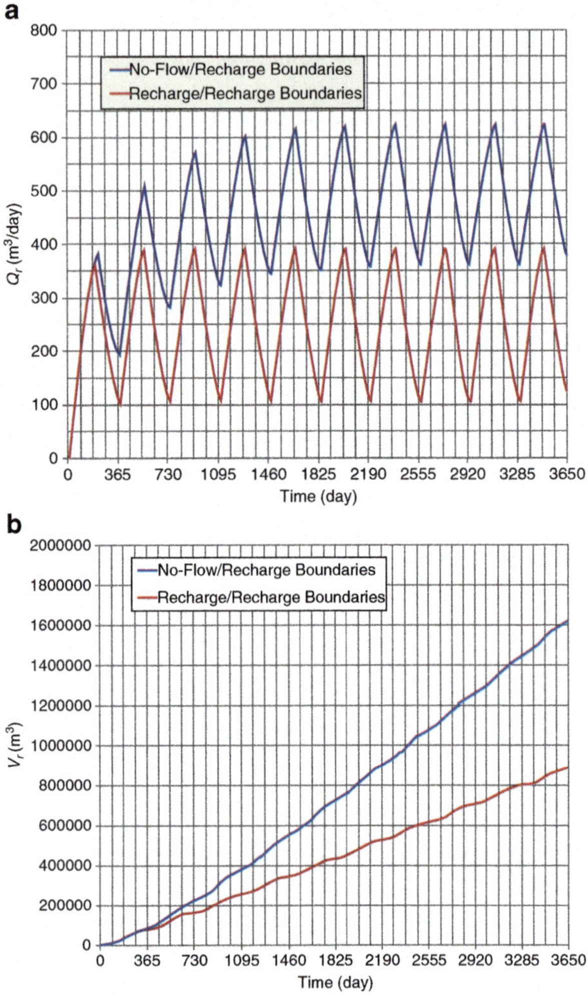

Fig. 2.30 (**a**) Stream depletion rate and (**b**) stream depletion volume vs. time obtained for a well operating cyclically in a constant-width finite aquifer ($w = 2000$ m) bounded between the stream under study and either a no flow boundary (*blue lines*), or another recharge boundary (*red lines*). The operating well is located at 1000 m from the stream and is operated cyclically at a rate of -1000 m^3/d for 180 d per year

2000-m wide finite aquifer comprised between the stream and either a no flow boundary (blue lines) or another recharge boundary (red lines). The operating well is located 1000 m away from the stream and operated at a rate -1000 m^3/d for 180 d every year.

4 Groundwater Management

The conjunctive management of water deals with the coordinated combined consumptive use of surface water and ground water resources, in order to efficiently meet the demands during times of water deficiency as well as availability. It is subject to laws regulating the water use, such as the prior appropriation system (also known as "priority doctrine") widely practiced in the western US.

The phrase "first in time, first in right" describes the doctrine of prior appropriation, according to which, water users with earlier appropriation decrees or "senior right holders" have a superior right in full water allocation before "junior right holders", who can get their water quote only if that does not impact availability for senior users. According to prior appropriation laws, well water users are junior right holders since historically they were granted use rights much later than when surface water use rights were fully allocated. Therefore, the water conjunctive management becomes a complex task when applied to a hydraulically connected stream/aquifer system under the prior appropriation system. Consequently, groundwater can be pumped from the aquifer for junior use (e.g. irrigation) under the condition of maintaining the minimal effect on the senior rights of the surface water [17].

An example of the application of such a system is the non-tributary and not non-tributary aquifers within the South-Platte river basins in northern Colorado. Pumping of these aquifers is permitted for overlaying landowners at a rate of 1 % a year to avoid affecting the connected surface water up until 100 years; otherwise, groundwater pumping is allowed provided that the stream is recharged with an amount of water equivalent to that extracted [18].

4.1 *Optimization of Groundwater Use*

This section addresses the groundwater management problem of an agricultural land irrigated by extracting water from the underlying aquifer, which is hydraulically connected to a stream. Given the fact that the stream will be compensating for the volume extracted from the aquifer, pumping groundwater is a junior act, which potentially affects the senior water right on the surface water. Therefore the management goal is to minimize the impact of water pumping on stream flows while satisfying the required irrigation needs. Withdrawing water from aquifer storage, not only affects the stream but also the hydraulic head levels in the aquifer. If the aquifer is over pumped it may not recover properly and will eventually fall short of providing required quantity of water, adding another constraint on the desired objective. Satisfying these conditions while meeting water demand is possible by replacing the extracted amount of groundwater back to the aquifer to keep heads levels slightly unchanged and to replenish the stream. In principle, this process may be carried out through [19–23]: (a) aquifer storage and recovery (ASR), where each operating well is provided with a pump able to extract water during periods of water need for irrigation and inject water when surface water is available for storage; (b) aquifer pumping and artificial recharge (APR), where, after being pumped, the aquifer is recharged with surface water at prescribed locations.

Both ASR and APR management problems can be formulated in a mathematical framework, as optimization problems with the objective of minimizing the total depletion/accretion of the stream caused by both pumping (extraction) and recharge

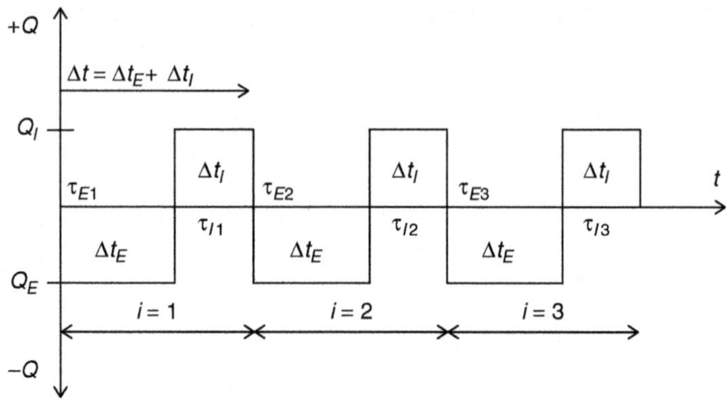

Fig. 2.31 Single-well operation schedule for aquifer storage and recovery (ASR) management

(injection), with constraints that represent the requirements of irrigation demand, the availability of water to inject in the aquifer, the maximum and minimum allowed aquifer head levels, and the maximum and minimum values of well operation rates established by well capacities.

Before proceeding to the description of the optimization problem setting, we have to distinguish the groundwater management formulations to be considered into: (a) single well operation, on which ASR relies; and (b) dual well operation, on which APR relies.

In single well operation, each well is activated in an operation mode in which it is set to perform one operation type during a certain period of time (e.g., extraction during the growing season), and reverse it for the rest of cycle period (injection during the off season). Given this assumption, operating wells have a cyclic operation schedule similar to that presented in Fig. 2.31. Each operating well extracts with a rate Q_E starting at time τ_E for a period Δt_E, and subsequently injects with a rate Q_I starting at time $\tau_I = \tau_E + \Delta t_E$ for a period Δt_I. As introduced in Sect. 3.7, the length of the single operation cycle Δt equals the sum $\Delta t_I + \Delta t_E$.

In dual well operation, as used in APR, there are two different well groups. These groups consist of pumping wells and injection wells, or generic recharge facilities. The two groups may be operating during generic periods, which may or may not be overlapping. Figure 2.32 shows an example of the schedule plan for two cyclically operating wells. The first well extracts with a rate Q_E starting at time τ_E for a period Δt_E, whereas the "second" well injects at time τ_I for a period Δt_I at a rate Q_I. Since these wells are independent, there is no relation between Δt_E, Δt_I and Δt. Generally, there may be parts of the (annual) operation cycle when neither pumping nor recharge is occurring.

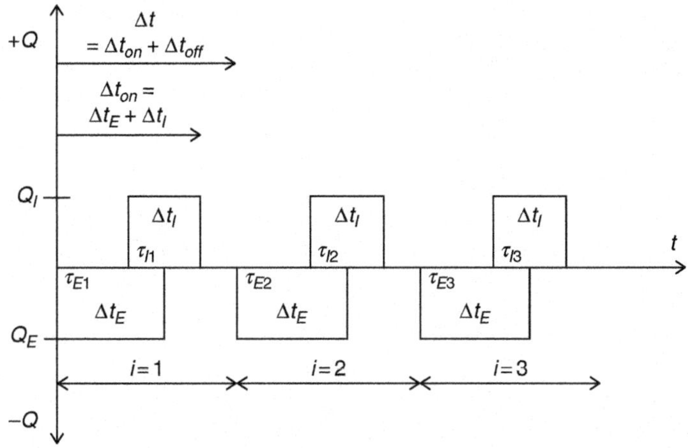

Fig. 2.32 Single-well operation schedule for aquifer pumping and recharge (APR) management

4.2 Linear Optimization Approach

The linear semi-analytical models presented in Sect. 2 for assessing the stream depletion/accretion and the aquifer drawdown can be applied to simulate the ground water management problems presented above, and solve them using linear programming techniques. In this case, the "independent" decision variables of the problem consist of the pumping rate, Q, at a number of prescribed well locations. The solution of a linear optimization problem requires expressing the objective function and the constraints as follows:

$$\text{Optimize}\{\mathbf{c}^{\mathrm{T}} \cdot \mathbf{Q}\} \tag{2.99}$$

$$\text{Subject to}: \ \mathbf{A} \cdot \mathbf{Q} \leq \mathbf{b} \tag{2.100}$$

where **Q** represents the decision variable (operation rate) vector, **b** and **c** are vectors of known coefficients and **A** is a matrix of known coefficients.

Management Objective. Equation (2.86), which estimates the effect on stream flow due to a cyclically operating well in an aquifer hydraulically connected to such a stream, can be applied to estimate the stream depletion volume, $\mathrm{V}_{r(ex)}$, or the stream accretion volume $\mathrm{V}_{r(in)}$ over a given time horizon t:

$$\begin{array}{ll} \mathrm{V}_{r(ex)}(t) = Q \cdot V_{C,ratio}(a, \Delta t, \Delta t_E, \tau_E, t) & if\, Q = Q_E < 0 \\ \mathrm{V}_{r(in)}(t) = Q \cdot V_{C,ratio}(a, \Delta t, \Delta t_I, \tau_I, t) & if\, Q = Q_I > 0 \end{array} \tag{2.101}$$

When planning the use of groundwater under prior appropriation rules, the optimization objective is to minimize the sum of the effects on the stream, so that there is a minimum recharge volume loss to the stream, as well as a minimum depletion volume from it. This objective can be expressed in terms of minimizing the absolute value of the total volume of stream depletion, $\mathrm{V}_r(t)$, over the investigated time horizon:

$$\begin{aligned} \mathrm{Min}|\mathrm{V}_r(t)| = \mathrm{Min}\Bigg| & \sum_{i=1}^{n_E} Q_i \cdot V_{C,ratio}(a_i, \Delta t, \Delta t_{E,i}, \tau_{E,i}, t) \\ & + \sum_{i=n_E+1}^{n_E+n_I} Q_i \cdot V_{C,ratio}(a_i, \Delta t, \Delta t_{I,i}, \tau_{I,i}, t)\Bigg| \end{aligned} \tag{2.102}$$

where n_E is the number well of extracting wells, and n_I is the number well of injecting wells. An ideal solution to the objective (2.102) is one that yields $\mathrm{V}_r(t) = 0$, which signifies that well operations have no net impact on stream flow at the selected time t.

This formulation of the optimization problem has a complication brought up by introducing an absolute value operator in the objective function, which causes a loss of linearity, such that the optimization problem cannot be tackled using a linear optimization method. This complication, however, can be by-passed by substituting the original objective function (2.102) with another objective function $\overline{\mathrm{V}}_r(t)$ equal to the absolute value of $\mathrm{V}_r(t)$:

$$\mathrm{Min}|\mathrm{V}_r(t)| = \mathrm{Min}\{\overline{\mathrm{V}}_r(t)\} \tag{2.103}$$

and adding the two following linear constraints:

$$\mathrm{V}_r(t) \le \overline{\mathrm{V}}_r(t) \tag{2.104}$$

$$-\mathrm{V}_r(t) \le \overline{\mathrm{V}}_r(t) \tag{2.105}$$

The constraints (2.104) and (2.105) have the effect of forcing $\overline{\mathrm{V}}_r(t)$ to equal $\mathrm{V}_r(t)$ upon being minimized, so that the objective function (2.103) is equivalent to the original objective function (2.102). Note that in this formulation $\overline{\mathrm{V}}_r(t)$ acts as both objective function and additional decision variable. Since this problem statement does not contain the absolute value operator, it can be solved using linear programming.

The vector **Q** of decision variables (see problem statement 2.99 and 2.100) for a generic system made up by n_E extraction wells, and n_I injection wells can be thus expressed as:

$$\mathbf{Q} = \left[Q_1, Q_2, .., Q_{n_E}; Q_{n_E+1}, Q_{n_E+2}, .., Q_{n_E+n_I}; \overline{\mathrm{V}}_r(t)\right] \tag{2.106}$$

where $Q_1, .., Q_{n_E}$ are extraction rate (negative) values and $Q_{n_E+1}, .., Q_{n_E+n_I}$ are injection rate (positive) values. The objective function (2.103) can thus be rewritten using vector notation as:

$$\mathrm{Min}\{\overline{\mathrm{V}}_r(t)\} = \mathrm{Min}\{\mathbf{I}_{n_E+n_I+1} \cdot \mathbf{Q}\} \tag{2.107}$$

where $\mathbf{I}_{n_E+n_I+1}$ is a $1 \times (n_E + n_I + 1)$ row vector, whose coefficients are all equal to zero, except the last one which equals 1.

Management Constraints. The objective function (2.107) must be optimized under a number of constraints on: (a) operation rates; (b) hydraulic head values at prescribed control points; and (c) irrigation demand and recharge availability.

(a) Operation rate constraints need to be imposed based on minimum (maximum extraction) and maximum (maximum injection) flow rate values. For each generic pumping well $i\,(i = 1, 2, .., n_E)$ the flow rate Q_i must be such that:

$$Q_{E,i,min} \le Q_i \le 0 \tag{2.108}$$

where $Q_{E,i,min}$ represents the maximum extraction rate at which the well can be operated. Similarly, for each injection unit, which may be either an injection well or a recharge facility, the injection rate $Q_i(i = n_E + 1, n_E + 2, .., n_E + n_I)$ must be such that:

$$0 \le Q_i \le Q_{I,i,max} \tag{2.109}$$

where $Q_{I,i,max}$ is the maximum injection rate at which the well or recharge unit can be operated. Using a matrix-vector notation, the constraints (2.108) and (2.109) may be rewritten as follows:

$$\mathbf{A}_\mathrm{L} \cdot \mathbf{Q} = \begin{bmatrix} 1 & 0 & \cdots & 0 & 0 & 0 & \cdots & 0 & 0 \\ -1 & 1 & \cdots & 0 & 0 & 0 & \cdots & 0 & 0 \\ 0 & -1 & \cdots & 0 & 0 & 0 & \cdots & 0 & 0 \\ \cdots & \cdots & \cdots & \cdots & \cdots & \cdots & \cdots & \cdots & \cdots \\ 0 & 0 & \cdots & 1 & 0 & 0 & \cdots & 0 & 0 \\ 0 & 0 & \cdots & -1 & 1 & 0 & \cdots & 0 & 0 \\ \cdots & \cdots & \cdots & \cdots & \cdots & \cdots & \cdots & \cdots & \cdots \\ 0 & 0 & \cdots & 0 & 0 & 0 & \cdots & 1 & 0 \\ 0 & 0 & \cdots & 0 & 0 & 0 & \cdots & -1 & 0 \end{bmatrix} \cdot \begin{bmatrix} Q_1 \\ Q_2 \\ \vdots \\ Q_{n_E} \\ Q_{n_E+1} \\ Q_{n_E+2} \\ \vdots \\ Q_{n_E+n_I} \\ \overline{\mathrm{V}}_r(t) \end{bmatrix} \le \begin{bmatrix} 0 \\ -Q_{E,1,min} \\ \vdots \\ -Q_{E,n_E,min} \\ -Q_{I,n_E+1,max} \\ 0 \\ \vdots \\ -Q_{n_E+n_I,max} \\ 0 \end{bmatrix} = \mathbf{b}_\mathrm{L} \tag{2.110}$$

where the matrix $\mathbf{A}_\mathrm{L}$ has size $2\,(n_E + n_I) \times (n_E + n_I + 1)$ and the vector $\mathbf{b}_\mathrm{L}$ has a size $2\,(n_E + n_I) \times 1$.

(b) <u>Hydraulic head constraints</u> require maximum and minimum allowable heads not to be exceeded at a number n_{mw} of prescribed control points in the aquifer, where monitoring wells are located. At the generic monitoring well $j\,(j = 1, 2, .., n_{mw})$ the hydraulic head at a given time t_j can be calculated by combining Eqs. (2.13) and (2.84):

$$h\left(x_j, y_j, t_j\right) = h_0 + \sum_{i=1}^{n_E} Q_i \cdot W_{CSI}\left(r_{ij}, r_{I,ij}, \Delta t, \Delta t_{E,i}, \tau_i, t\right) + \sum_{i=n_E+1}^{n_E+n_I} Q_i \cdot W_{CSI}\left(r_{ij}, r_{I,ij}, \Delta t, \Delta t_{I,i}, \tau_i, t\right) \quad (2.111)$$

where r_{ij} and $r_{I,ij}$ denote the distances of the generic monitoring well j from the generic pumping well i and its image across the stream, respectively. In Eq. (2.111), the coefficients W_{CSI} are calculated using Eq. (2.83) and will be hereafter denoted as β_{ij}. Constraints on hydraulic head at the generic monitoring well $j\,(j = 1, 2, .., n_{mw})$ are thus expressed as:

$$h_0 - h_{j,min} \leq \sum_{i=1}^{n_E+n_I} \beta_{i,j} \cdot Q_i \leq h_{j,max} - h_0 \quad (2.112)$$

where $h_{j,min}$ and $h_{j,max}$ are the minimum and maximum hydraulic head values allowed at the monitoring well, respectively. Using matrix-vector notation, hydraulic head constraints at the n_{mw} monitoring wells can be expressed as:

$$\mathbf{A}_{\mathrm{H}} \cdot \mathbf{Q} = \begin{bmatrix} \beta_{1,1} & \beta_{1,2} & \cdots & \beta_{1,n_E} & \beta_{1,n_E+1} & \beta_{1,n_E+2} & \cdots & \beta_{1,n_E+n_I} & 0 \\ -\beta_{1,1} & -\beta_{1,2} & \cdots & -\beta_{1,n_E} & -\beta_{1,n_E+1} & -\beta_{1,n_E+2} & \cdots & -\beta_{1,n_E+n_I} & 0 \\ \vdots & \vdots & \cdots & \vdots & \vdots & \vdots & \cdots & \vdots & \vdots \\ \beta_{n_{mw},1} & \beta_{n_{mw},2} & \cdots & \beta_{n_{mw},n_E} & \beta_{n_{mw},n_E+1} & \beta_{n_{mw},n_E+2} & \cdots & \beta_{n_{mw},n_E+n_I} & 0 \\ -\beta_{n_{mw},1} & -\beta_{n_{mw},2} & \cdots & -\beta_{n_{mw},n_E} & -\beta_{n_{mw},n_E+1} & -\beta_{n_{mw},n_E+2} & \cdots & -\beta_{n_{mw},n_E+n_I} & 0 \end{bmatrix} \cdot \begin{bmatrix} Q_1 \\ Q_2 \\ \vdots \\ Q_{n_E} \\ Q_{n_E+1} \\ Q_{n_E+2} \\ \vdots \\ Q_{n_E+n_I} \\ \overline{V}_r(t) \end{bmatrix} \leq \begin{bmatrix} h_{1,max} - h_0 \\ h_0 - h_{1,min} \\ \vdots \\ h_{n_{mw},max} - h_0 \\ h_0 - h_{n_{mw},min} \end{bmatrix} = \mathbf{b}_{\mathrm{H}} \quad (2.113)$$

where $\mathbf{A}_{\mathrm{H}}$ is a $2 \cdot n_{mw} \times (n_E + n_I + 1)$ matrix and $\mathbf{b}_{\mathrm{H}}$ is a column vector of size $2 \cdot n_{mw} \times 1$. It is worth noting that the index j identifies a control point where the head value is checked at a given time. If at the same monitoring well, heads must be checked at different times, then additional constraints are to be added.

(c) Irrigation demand and recharge availability constraints require that during the pumping season the total sum of the (negative) extraction rates is less than or equal to the total (negative) demand rate Q_{Demand}:

$$\sum_{i=1}^{n_E} Q_i \leq Q_{Demand} \tag{2.114}$$

At the same time, during the period in which surface water is made available for aquifer recharge, the total sum of injection well rates must be less than or equal to total available recharge rate $Q_{Available}$:

$$\sum_{i=n_E+1}^{n_E+n_I} Q_i \leq Q_{Available} \tag{2.115}$$

Using matrix-vector notation, the irrigation demand and recharge availability constraints can be expressed as follows:

$$\mathbf{A}_{\mathrm{DA}} \cdot \mathbf{Q} = \begin{bmatrix} 1 & 1 & \cdots & 1 & 0 & 0 & \cdots & 0 & 0 \\ 0 & 0 & \cdots & 0 & 1 & 1 & \cdots & 1 & 0 \end{bmatrix} \cdot \begin{bmatrix} Q_1 \\ Q_2 \\ \vdots \\ Q_{n_E} \\ Q_{n_E+1} \\ Q_{n_E+2} \\ \vdots \\ Q_{n_E+n_I} \\ \overline{V}_r(t) \end{bmatrix} \leq \begin{bmatrix} Q_{Demand} \\ Q_{Available} \end{bmatrix} = \mathbf{b}_{\mathrm{DA}} \tag{2.116}$$

where $\mathbf{A}_{\mathrm{DA}}$ is a $2 \times (n_E + n_I + 1)$ matrix and $\mathbf{b}_{\mathrm{DA}}$ is matrix and is a 2×1 column vector.

(d) Two additional constraints are necessary to prescribe the inequalities (2.104) and (2.105) introduced in order to remove the absolute value from the objective function (2.103). Inequalities (2.104) and (2.105) can thus be rewritten, respectively, as:

$$\sum_{i=1}^{n_E+n_I} Q_i \cdot V_{C,ratio}(a_i, \Delta t, \Delta t_i, \tau_i, t) \leq \overline{V}_r(t) \tag{2.117}$$

$$-\sum_{i=1}^{n_E+n_I} Q_i \cdot V_{C,ratio}(a_i, \Delta t, \Delta t_i, \tau_i, t) \leq \overline{V}_r(t) \tag{2.118}$$

In inequalities (2.117) and (2.118), the coefficients $V_{C,\,ratio}$ are calculated using Eq. (2.86) and will be hereafter denoted as α_i. Using matrix-vector notation, the two constraints (2.117) and (2.118) can thus be rewritten as:

$$\mathbf{A}_{\mathrm{OF}} \cdot \mathbf{Q} = \begin{bmatrix} \alpha_1 & \alpha_2 & \cdots & \alpha_{n_E} & \alpha_{n_E+1} & \alpha_{n_E+2} & \cdots & \alpha_{n_E+n_I} & -1 \\ -\alpha_1 & -\alpha_2 & \cdots & -\alpha_{n_E} & -\alpha_{n_E+1} & -\alpha_{n_E+2} & \cdots & -\alpha_{n_E+n_I} & -1 \end{bmatrix} \cdot \begin{bmatrix} Q_1 \\ Q_2 \\ \vdots \\ Q_{n_E} \\ Q_{n_E+1} \\ Q_{n_E+2} \\ \vdots \\ Q_{n_E+n_I} \\ \overline{\mathrm{V}}_r(t) \end{bmatrix} \le \begin{bmatrix} 0 \\ 0 \end{bmatrix} = \varnothing \tag{2.119}$$

where the matrix $\mathbf{A}_{\mathrm{OF}}$ has a size $2 \times (n_E + n_I + 1)$, and $\varnothing$ is the 2×1 null vector.

The linear optimization problem (2.99) and (2.100) into which the groundwater management is formulated can thus be structured as follows:

$$\mathrm{Min}\left\{\mathbf{c}^{\mathrm{T}} \cdot \mathbf{Q}\right\} \qquad \text{Subject to}: \ \mathbf{A} \cdot \mathbf{Q} \le \mathbf{b} \tag{2.120}$$

where:

$$\mathbf{c}^{\mathrm{T}} = \mathbf{I}_{n_E+n_I+1} \tag{2.121}$$

$$\mathbf{A} = \begin{bmatrix} \mathbf{A}_{\mathrm{L}} & \mathbf{A}_{\mathrm{H}} & \mathbf{A}_{\mathrm{DA}} & \mathbf{A}_{\mathrm{OF}} \end{bmatrix}^{\mathrm{T}} \tag{2.122}$$

$$\mathbf{b} = \begin{bmatrix} \mathbf{b}_{\mathrm{L}} & \mathbf{b}_{\mathrm{H}} & \mathbf{b}_{\mathrm{DA}} & \varnothing \end{bmatrix}^{\mathrm{T}} \tag{2.123}$$

The constraint matrix $\mathbf{A}$ has size $2 \cdot (n_E + n_I + n_{mw} + 2) \times (n_E + n_I + 1)$. The vector $\mathbf{b}$ has size $2 \cdot (n_E + n_I + n_{mw} + 2) \times 1$.

A formally identical linear optimization formulation can be developed to minimize stream depletion/accretion from a stream representing a boundary in a constant-width finite aquifer bounded on the opposite side by either a no-flow boundary or a recharge boundary. In this case, the $V_{C,\,ratio}$ coefficients that define the vector $\mathbf{c}$ and the matrix $\mathbf{A}_{\mathrm{OF}}$ must be replaced by the coefficients $V_{FC,ratio}$ given by Eq. (2.95). Correspondingly, the W_{CSI} coefficients that define the matrix $\mathbf{A}_{\mathrm{H}}$ must be replaced by the coefficients W_{FC} given by Eq. (2.92).

A MATLAB code called `LP_APR_BNR.m` has been implemented to solve the groundwater management problem formulated above. This code relies upon the highly efficient linear-programming (LP) routine "`linprog`" [24–26] available in MATLAB libraries, and is included in Appendix 5. Because of the computational efficiency of semi-analytical models, these LP problems may be solved at a relatively low computational cost, which allows for conducting extensive analyses of connected stream-aquifer systems, thus achieving improved insight into key aspects of groundwater management.

The following section presents the application of the linear optimization model described above to finite aquifers comprised between a no-flow boundary and a stream. The examples considered here are simulations of APR groundwater management. It is worth mentioning that, even though the semi-analytical models presented in Sect. 2 are here applied to both extraction and injection wells, the recharge is normally achieved by surface infiltration ponds, to which the Theis and the Glover model are not directly applicable. In this respect, Molden et al. [27] have proposed an approach to calculate recharge volumes due to surface infiltration, which could be used to improve the results of the developed semi-analytical models. However, it is reasonable to assume that the Theis-Glover derived models can still provide acceptable results, particularly within the limits of approximation that make the semi-analytical approach conceptually adequate for screening and proof-of-concept calculations.

4.3 Groundwater Management Problem Setting

Let us consider the 8000 m × 6000 m portion of an alluvial aquifer limited by a stream and a no-flow physical boundary (Fig. 2.33). The aquifer parameters are listed in Table 2.4.

The groundwater management problem requires providing groundwater for irrigation during a 4-month growing season, from April 1 to July 31, corresponding to the cumulative consumptive use of 1 m over a cultivated area of 30 % of the total area of the aquifer, and recharging the aquifer with an equivalent amount of surface water to offset potential stream over pumping. This consumptive use is representative of corn crop type. In practice, the aquifer is used to provide water in the amount of 1.2×10^7 m^3/year (~1000 aft/year).

A proposed solution to potential stream depletion is to acquire an equivalent amount of surface water to recharge the aquifer in a manner that offsets surface water withdrawal. Recharge is assumed to occur every year for 180 d, starting October 1.

Given a number of potential or pre-existing well locations and recharge areas (Fig. 2.33), the groundwater management goal is to identify the spatial distribution of pumping wells and aquifer recharge ponds that minimize the absolute value of the stream depletion volume over an operation period of 10 years.

Assuming the bottom of the alluvial aquifer as the datum, the initial head in the aquifer equals its average saturated thickness h_0 of 30 m (Table 2.4). In order to smooth out the variation of water levels, constraints are imposed such that h cannot be below prescribed minimum and maximum levels, h_{min} and h_{max}, at the control monitoring wells shown in Fig. 2.33. It is worth pointing out that, in practice, the water level variations will be significantly larger nearby pumping wells or recharge areas. Table 2.5 shows the three water level constraint scenarios hypothesized in

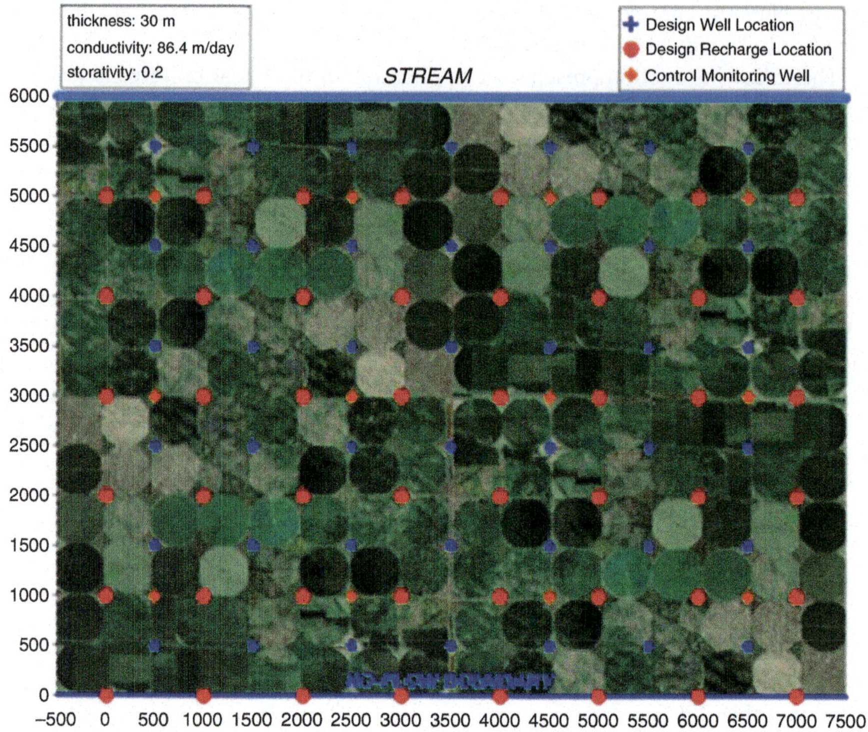

Fig. 2.33 Stream-aquifer setting. The map includes also the location of candidate groundwater wells, recharge units, and available monitoring wells

Table 2.4 Aquifer properties

h_0 (m)	K (m/d)	S (/)
30	86.4	0.2

these examples. With respect to the "base" Scenario 1, in Scenario 2 water level constraints are more stringent, whereas in Scenario 3, they are more relaxed.

Constraints are also prescribed on the irrigation demand Q_{Demand} and water availability for recharge $Q_{Availability}$. Table 2.6 lists the hypothesized values of the groundwater demand and recharge availability. The annual recharge availability is assumed to be equal to the annual irrigation demand in Scenarios 1 and 2 (see also Table 2.5), whereas in Scenario 3 recharge availability is assumed to be about 85 % of the irrigation demand.

Technical constraints are finally imposed to the maximum pumping capacity of injection wells and the maximum injection rate of recharge areas, which are both set equal to 5000 m^3/d (~2 cfs, cubic feet per second).

The LP algorithm previously introduced is used to obtain the optimal well and recharge layouts for the three scenarios described in Tables 2.5 and 2.6. These optimal layouts are depicted in Figs. 2.34, 2.35, and 2.36. Table 2.6 summarizes of

Table 2.5 Water level constraints hypothesized in the APR management problem

Scenario	h_{min} (m)	h_{max} (m)
1	29	31
2	29.5	30.5
3	25	31

Table 2.6 Water demand and availability hypothesized in the APR management problem

Scenario	Demand	Availability
1, 2	1.2×10^7 m^3/year (1×10^5 m^3/d)	1.2×10^7 m^3/year (0.85×10^5 m^3/d)
3	1.2×10^7 m^3/year (1×10^5 m^3/d)	1.0×10^7 m^3/year (0.71×10^5 m^3/d)
Operation period	120 d (March 15–July 15)	140 d (October 1–March 1)

the total daily recharge and extraction rate and 10-year net stream depletion volume calculated for the Scenarios 1–3.

Figure 2.34 shows the optimal layout identified in Scenario 1. During their respective operation periods (Table 2.6), all candidate wells and recharge units are "activated" at the rates noted above each marker in Fig. 2.34. In this scenario, the net volume of stream depletion over 10 years is equal to zero (Table 2.7), which indicates that all irrigation demand is truly met by extracting groundwater in equal amount to the aquifer recharge. In Fig. 2.34, it is interesting to observe that ground water extraction rates are lower in proximity of the stream, in order to minimize stream depletion, as well as in proximity of the no-flow boundary of the aquifer, in order to minimize aquifer drawdown. Conversely, recharge rates are progressively decreasing away from the stream, which suggests that a significant portion of recharged water is being used for stream augmentation.

With respect to Scenario 1, in Scenario 2 requires the variation of water level with respect to the initial value $h_0 = 30$ m (Table 2.4) at all monitoring wells (orange markers in Fig. 2.33) not to exceed 0.5 m in magnitude (Table 2.5). The optimal layout for Scenario 2 is shown in Fig. 2.35.

As indicated in Table 2.6, under Scenario 2 the stream is depleted of about 5.9×10^6 m^3 (~4800 aft) of water over 10 years, even though enough recharge capacity is available to satisfy the irrigation groundwater demand (Table 2.6). Comparison between the optimal layouts in Scenarios 1 and 2 (Figs. 2.34 and 2.35) reveals the dramatic change in pumping and recharge strategies that is required under more stringent water level constraints. Figure 2.35 shows that irrigation demands is met using about 20 high capacity pumping wells. At least five of these wells are positioned closer to the stream, where water levels are less sensitive to pumping due to the presence of the constant-head boundary. These wells are ultimately "responsible" for long-term stream depletion (Table 2.7). In addition, a few high capacity recharge units are located along the stream, to mitigate the impact of pumping on stream depletion, and along the physical boundary, to reduce aquifer drawdown as imposed by water level constraints. Scenario 2 provides a clear demonstration that the need to minimize water level

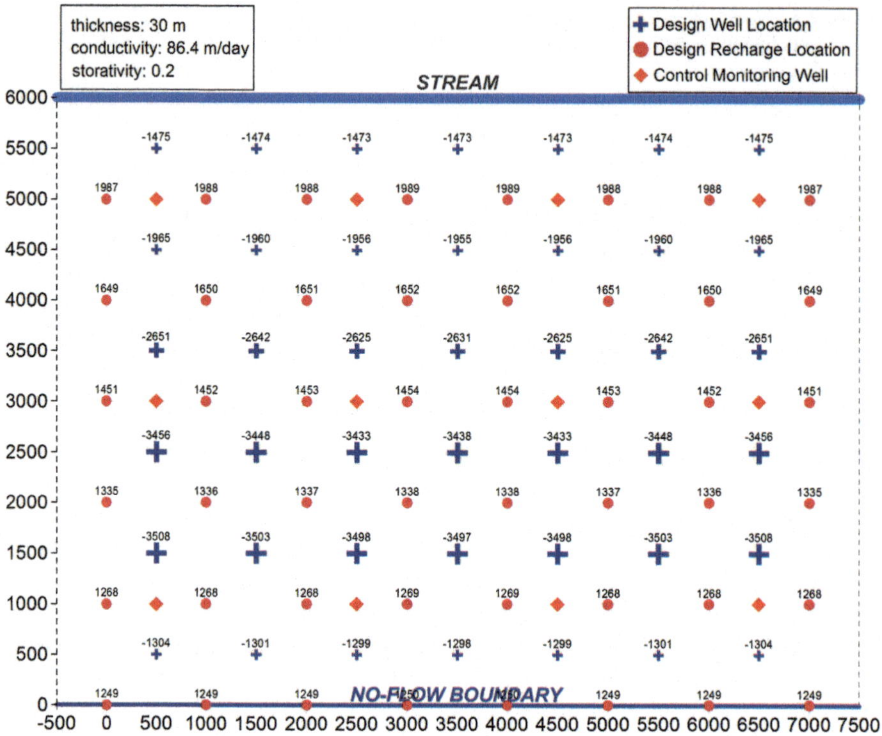

Fig. 2.34 Optimal APR layout in Scenario 1

variations competes directly competing with the need of minimizing stream depletion volumes.

With respect to Scenario 1, in Scenario 3 the maximum allowed water level h_{max} remains the same, whereas the minimum allowed water level h_{min} is lower. In addition, the recharge capacity is reduced to 85 % of the irrigation demand. Figure 2.36 shows the optimal well-recharge layout for Scenario 3.

As shown in Table 2.7, the net volume of stream depletion over 10 years is equal to zero even though recharge capacity is less than groundwater demand. Such result is opposite to that observed in Scenario 2, where the recharge capacity equaled the groundwater demand, yet the stream depletion volume was significant. This is due to the fact that, in Scenario 3, a significant portion of the irrigation demand is supplied directly from the aquifer storage, however at the expense of a larger decrease in water levels, as allowed by the lower value of h_{min} (Table 2.5). Accordingly, Fig. 2.36 shows that extraction wells are concentrated in the lower half of the aquifer, since this minimizes the impact of pumping on stream depletion. Even though pumping in proximity of the no-higher produces higher drawdown, this is allowed because of the more relaxed constraint on water levels. On the other hand, aquifer recharge occurs mostly in its uppermost portion along the stream. In

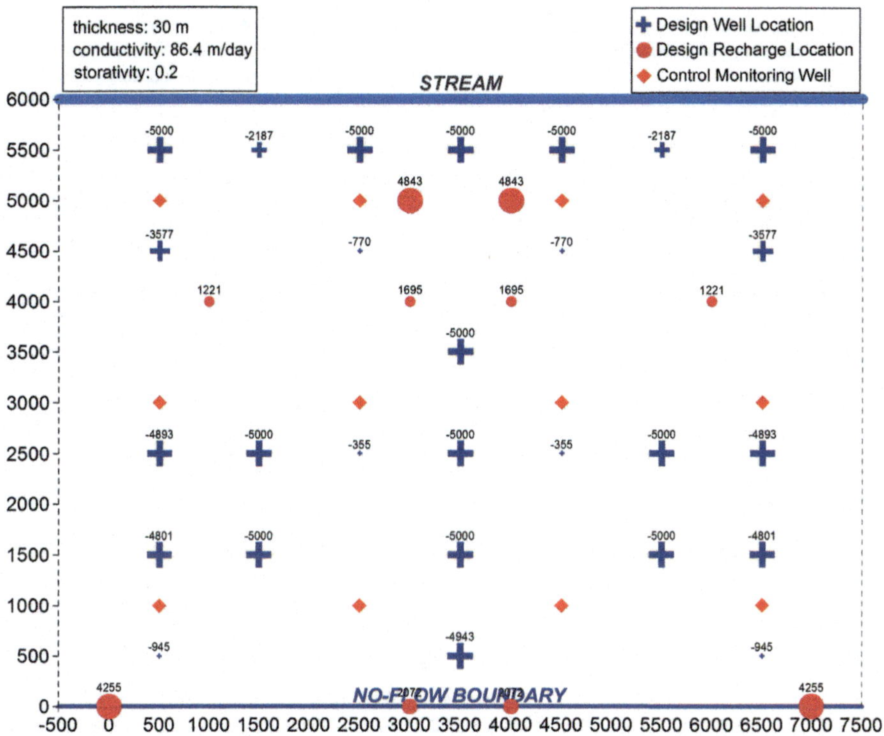

Fig. 2.35 Optimal APR layout in Scenario 2

this case, recharge units serve to the purpose of "shielding" the stream from aquifer pumping, and are activated towards the end of the irrigation season (Table 2.6), when the cone of depression from aquifer pumping is likely to be reaching the stream.

Finally, Fig. 2.37a, b display the time series for the total stream depletion rate and the total cumulative stream depletion volume in Scenarios 1, 2, and 3, respectively. It is worth noting that the spatial distribution of recharge areas with respect to regions where pumping is concentrated as a strong impact on the stream depletion volume profiles. For example, in both Scenarios 1 and 3 the volume of stream depletion at the end of the 10-year time horizon is equal to zero. However, while in Scenario 1 the stream remains depleted for the whole period and recovers only at the very end, in Scenario 3, where stronger recharge occurs in proximity of the stream (Fig. 2.36), the stream is augmented for most of the time, except at towards the end of the simulated period, when the effect of groundwater pumping from the lower portion of the aquifer finally impacts the stream.

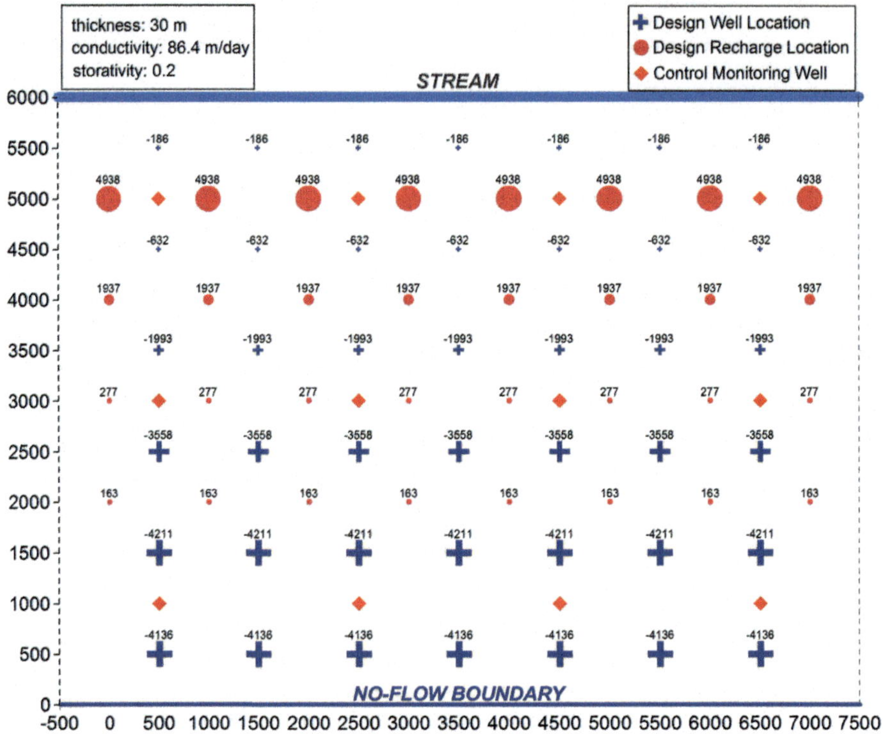

Fig. 2.36 Optimal APR layout in Scenario 3

Table 2.7 Summary of the recharge and extraction rate and net stream depletion volume estimated for the Scenarios described in Tables 2.4, and 2.5

Scenario	Recharge (m^3/d) (October 1–March 1)	Extraction (m^3/d) (March 15–July 15)	10-year Net stream depletion volume (m^3)
1	7.16×10^4	1.00×10^5	~0
2	6.82×10^4	1.00×10^5	5.86×10^6
3	5.85×10^4	1.03×10^5	~0

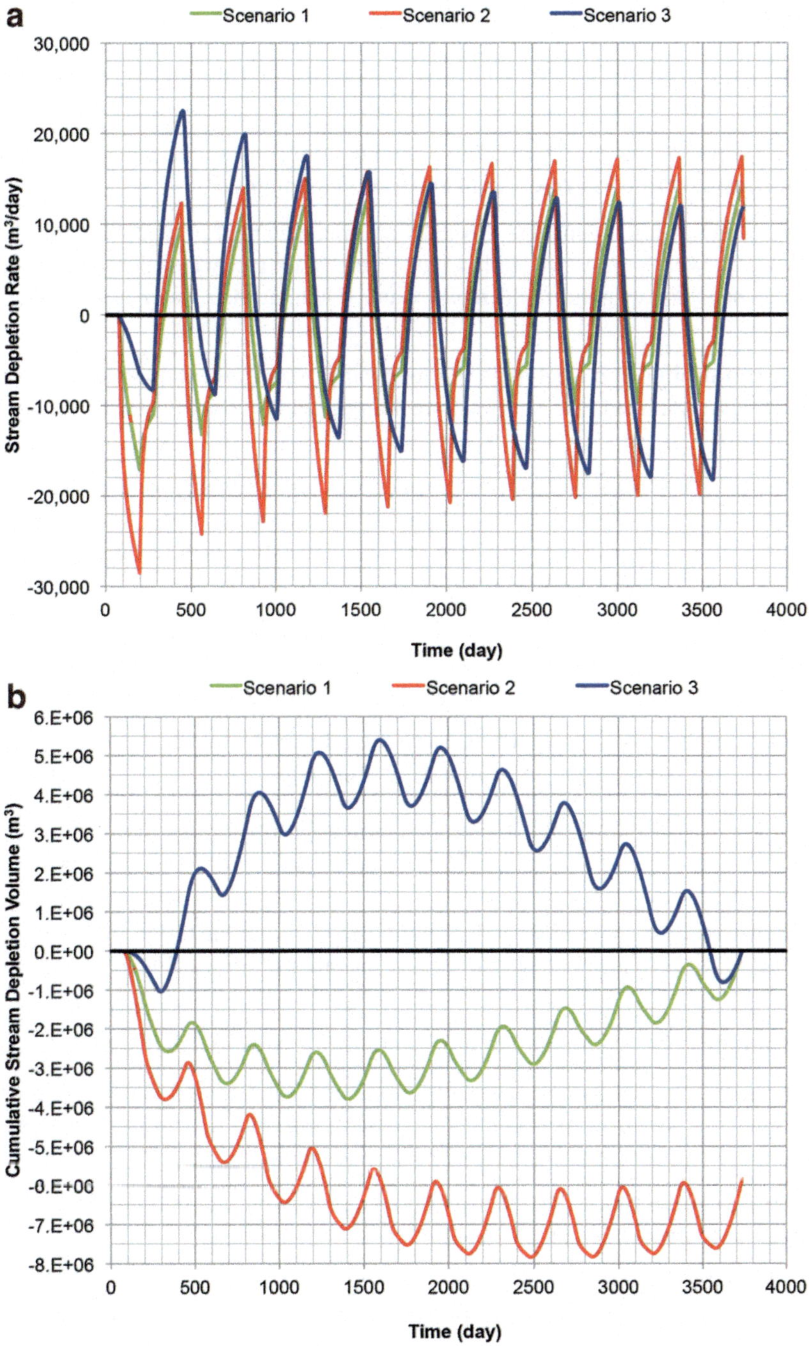

Fig. 2.37 Time series for (**a**) total stream depletion rate (m^3/d) and (**b**) total cumulative stream depletion volume (m^3) in Scenarios 1, 2, and 3, respectively

Glossary

Aquifer Underground water-saturated unconsolidated formation from which groundwater can be extracted from a well.

Alluvial aquifer Aquifer formed by sedimentation of gravel, sand, silt or clay materials deposited in river channels or on floodplains.

Aquifer storage and recovery Groundwater management practice in which each operating well is provided with a pump able to extract water during periods of water need for irrigation and inject water when surface water is available for storage.

Aquifer pumping and recharge Groundwater management practice in which, after being pumped, the aquifer is recharged with surface water at prescribed locations.

Conjunctive use Combined management of surface water and groundwater resources that optimizes common benefits by trading off water demand against water supply, while complying with a series of constraints of technical, physical, environmental and economical nature.

Convolution Integral The integral of the product of two functions, with one function reversed and shifted with respect to another. The integral is often used to model conditions of continuous, as opposed to discrete, superposition of effect.

Confined aquifer Aquifer delimited above and below by low permeable or impermeable formations, such as aquitards or aquicludes.

Cyclic well operation Groundwater management practice in which each well is operated at fixed rate for a prescribed period of each year and turned off afterwards.

Decision variables The independent variables of an optimization problem that a decision maker can control and choose in order to best achieve management goals.

Finite aquifer Constant-width aquifer, delimited by two parallel boundaries.

Glover solution Analytical model developed by Glover and Balmer (1954) to estimate the stream depletion rate due to pumping from a point source in a semi-infinite, homogeneous, and connected alluvial aquifer.

Hydraulic conductivity Physical property representing the ability of the porous medium to conduct water.

Head drawdown State variable consisting of the reduction of hydraulic head in an aquifer due to groundwater pumping.

Hydraulic head State variable consisting of the mechanical energy per unit weight of water at any point in the aquifer and at any time.

Infinite aquifer Ideal aquifer characterized by an infinite lateral extension.

Objective function The mathematical function of decision variables and state variables, whose minimization or maximization constitutes the objective of the optimization problem.

Optimization The selection of a best element, with regard to some specific criterion, expressed as an objective, from a finite or infinite set of feasible alternatives.

Optimization constraints Set of equations and inequalities defining the set of feasible alternatives within which the solution to the optimization problem is sought.

Semi-infinite aquifer Ideal aquifer characterized by an infinite lateral extension on one side and a rectilinear boundary on another.

Specific elastic storage Physical property of a porous medium, representing the volume of water released by per unit bulk volume and per unit decline in pressure head.

Specific yield Physical property given by the unitless ratio of the volume of water added or removed directly from the saturated zone of the aquifer to the resulting change in the volume of aquifer below water.

Streambed The channel bottom of a stream, representing the lower boundary of the stream flow as well as the interface between surface and subsurface flow.

Stream depletion rate Instantaneous flow rate with which a stream, idealized as a constant-head boundary, recharges a hydraulically connected aquifer in which groundwater pumping is occurring.

Stream depletion volume Cumulative volume with which a stream, idealized as a constant-head boundary, recharges a hydraulically connected aquifer in which groundwater pumping is occurring.

State variables The variables of an optimization problem that characterize the mathematical state of a dynamic system and depend upon the selection of the decision variables.

Superposition principle This principle applies to linear systems and states that the response caused by two or more forcing terms at a generic location and time is equal to the sum of the responses associated with each forcing terms as this was acting individually.

Theis equation Analytical model developed by Charles Vernon Theis for aquifer drawdown associated with two-dimensional radial flow to a point source in an infinite, homogeneous, confined aquifer.

Unconfined aquifer Aquifer whose upper boundary is the water table or phreatic surface.

Water table Surface of an unconfined aquifer at which water pressure equals the atmospheric pressure.

Water well Underground structure constructed by digging, driving, boring, or drilling to access groundwater. Water is typically lifted to the surface through a pump.

Appendix 1

Matlab code: **Drawdown2D**

This code calculates and plots the spatial distribution of head at a specified time t_{fin} for a number of wells operating cyclically or continuously in a laterally infinite aquifer, in a semi-infinite aquifer limited by a stream boundary, and in a semi-infinite aquifer limited by a no-flow boundary. If present, the boundary is represented by the $y=y_s$ straight line.

The input file **Aquifer.Parms** includes simulation time parameters and aquifer parameters:

a. time parameters

 - final simulation time t_{fin};
 - cyclical time step of simulation $\Delta t = \Delta t_{on} + \Delta t_{off}$ (e.g., 1 year)

b. aquifer parameters

 - initial hydraulic head (aquifer's saturated thickness) h_0
 - hydraulic conductivity K;
 - storativity S;
 - stream location $y_s(>0)$.

The input file **Grid.parms** includes data regarding the plotting grid:

- grid lower left corner abscissa (x_{min})
- grid lower left corner ordinate (y_{min})
- grid upper right corner abscissa (x_{max})
- grid upper right corner ordinate (y_{max})
- n. of gridblocks along x (n_x)
- n. of gridblocks along y (n_y)

The input file **Wells.dat** includes wells data:

a. total number of operating wells n_w
b. for each operating well, each of the following lines provide:

 - the pumping rate (Q_w);
 - the time at which well operation starts (t_{st}),
 - the total operation period (Δt_{on})[1]
 - well location x_w and y_w.

Examples:

Aquifer.Parms

`575.` or `730.`	`365.`	`30.`	`86.4`	`0.2`
t_{fin}(d)	$\Delta t=\Delta t_{on}+\Delta t_{off}$ (d)	b(m)	K (m/d)	S (/)

[1] If this time is set equal to Δt then the well operates continuously ($\Delta t_{off} = 0$). Otherwise the well operates cyclically.

Grid.Parms

0.	0.	4000.	4000.	40	40
x_{min}(m)	y_{min}(m)	x_{min}(m)	y_{min}(m)	n_x	n_y

Wells.dat

Description:	2 wells operating cyclically			
2 (n_w)				
+3000.	90.	120.	1000.	1000.
+3000.	90.	120.	1000.	3000.
Q_w(m^3/d)	t_{st}(d)	Δt_{on}(d)	x_w(m)	y_w (m)

```
function Drawdown2D
   clc
   clear all
   % Reading Data
   % 1- General Data
   fid1 = fopen ('Aquifer.Parms','r');
   Temp = fscanf(fid1,'%f %f %f %f %f',[1,6]);
   tfin = Temp(1); % Final Time of Simulation (d)
   delt = Temp(2); % Cycle time step of simulation (d)
   ho  = Temp(3); % Aquifer Initial Head (saturated thickness)(m)
   K   = Temp(4); % Hydraulic Conductivity (m^2/d)
   Sy  = Temp(5); % Storativity (/)
   ys  = Temp(6); % stream location (m)
   T   = K*ho;   % Transmissivity (m^2/d)
   fclose(fid1);
   % 2- Reading Grid Data
   fid2 = fopen ('Grid.parms','r');
   Temp = fscanf(fid2,'%f %f %f %f %f %f',[1,6]);
   xmin = Temp(1); % Minimum Value of X in the Grid
   ymin = Temp(2); % Minimum Value of Y in the Grid
   xmax = Temp(3); % Maximim Value of X in the Grid
   ymax = Temp(4); % Maximum Value of Y in the Grid
   nx  = Temp(5); % Number of X Divisions
   ny  = Temp(6); % Number of Y Divisions
   fclose(fid2);
   %
   % 3- Wells Data
   fid3 = fopen ('Wells.dat','r');
   Temp = fscanf(fid3,'%f',[1,1]);
   now  = Temp(1);% Wells Number
   for m = 1:now
```

```
Temp   = fscanf(fid3,'%f %f %f %f %f',[1,5]);
Qw(m)  = Temp(1); % Well Pumping Rate
tst(m) = Temp(2); % Pumping Start Time
dton(m) = Temp(3); % Pumping Period
xw(m)  = Temp(4); % Well Location X Coordinate
yw(m)  = Temp(5); % Well Location Y Coordinate
end
fclose(fid3);
%
% Creating the Grid %
nxx = nx+1;
nyy = ny+1;
dx = (xmax-xmin)/nx;
dy = (ymax-ymin)/ny;
for j= 1:nxx
for i= 1:nyy
        x(i,j)= xmin+dx*(j-1);
        y(i,j)= ymin+dy*(i-1);
end
end
%
% Computing Drawdown Distribution for number of operating wells
ssum     = zeros(nyy,nxx);% Initial Drawdown
ssum_rech  = zeros(nyy,nxx);% Initial Drawdown
ssum_noflow = zeros(nyy,nxx);% Initial Drawdown
for j= 1:nxx;
for i= 1:nyy;
for m =1:now;
t = tfin-tst(m);
if t>0
ssum(i,j)    = ssum(i,j)    + CYC_THEIS(t,delt,Sy,T,x(i,j),y(i,j),
dton(m),xw(m),yw(m),Qw(m));
ssum_rech(i,j)  = ssum_rech(i,j)  + CYC_THEIS_RECHARGE(t,delt,Sy,
T,ys,x(i,j),y(i,j),dton(m),xw(m),yw(m),Qw(m));
ssum_noflow(i,j)= ssum_noflow(i,j)+ CYC_THEIS_NOFLOW(t,delt,Sy,T,
ys,x(i,j),y(i,j),dton(m),xw(m),yw(m),Qw(m));
end
end
end
end
fid4 = fopen('results.dat','w');
fid5 = fopen('results.recharge.dat','w');
fid6 = fopen('results.noflow.dat','w');
for j= 1:nxx;
for i= 1:nyy;
```

```
temp = [x(i,j), y(i,j), ssum(i,j)];
fprintf(fid4,'%15.6E %15.6E %15.6E\n',temp);
%
temp = [x(i,j), y(i,j), ssum_rech(i,j)];
fprintf(fid5,'%15.6E %15.6E %15.6E\n',temp);
%
temp = [x(i,j), y(i,j), ssum_noflow(i,j)];
fprintf(fid6,'%15.6E %15.6E %15.6E\n',temp);
end
end
fclose(fid4);
fclose(fid5);
fclose(fid6);
%%
str1 = num2str(Qw');
cell1 = cellstr(str1);
% Figure
figure;
contour(x,y,ssum);
[C,h] = contour(x,y,ssum);
clabel(C,h);
title(['Infinite Aquifer: Drawdown (L) at time =',num2str(tfin)]);
xlabel('x (L)');
ylabel('y (L)');
hold on
scatter(xw,yw,50,'r+')
text(xw+20,yw+50,cell1,'BackgroundColor',[1 1 0],'FontSize',10);
legend('Drawdown','Operating wells')
% Figure
figure;
contour(x,y,ssum_rech);
[C,h] = contour(x,y,ssum_rech);
clabel(C,h);
title(['Semi-infinite Aquifer: Drawdown (L) at time=',num2str(tfin)]);
xlabel('x (L)');
ylabel('y (L)');
hold on
scatter(xw,yw,50,'r+')
text(xw+20,yw+50,cell1,'BackgroundColor',[1 1 0],'FontSize',10);
legend('Drawdown','Operating wells')
% Figure
figure;
contour(x,y,ssum_noflow);
[C,h] = contour(x,y,ssum_noflow);
clabel(C,h);
```

```
title(['Semi-infinite Aquifer: Drawdown (L) at time=',num2str(tfin)]);
xlabel('x (L)');
ylabel('y (L)');
hold on
scatter(xw,yw,50,'r+')
text(xw+20,yw+50,cell1,'BackgroundColor',[1 1 0],'FontSize',10);
legend('Drawdown','Operating wells')
%%%%%%%%%%%%%%%%%%%%%%%%%%%%%%%%%%%%%%%%%%%%%%%%%%%%%%%%%%
%%%%%%%%
function [dh] = CYC_THEIS(t,delt,Sy,T,xm,ym,dton,xw,yw,Qw)
dh=0.;
frac= t/delt;
int_t=fix(frac);
rest_t=t-int_t*delt;
n = int_t+1;
% Distance to wells
r1=sqrt((xm-xw)^2+(ym-yw)^2);
for i=1:n-1
t1=t-(i-1)*delt;
t2=t1-dton;
% Operating Well
u1=(Sy/(4*T))*(r1^2/t1);
% Well Function
wu1=expint(u1);
% Imaginary Compensation Well
u2=Sy/(4*T)*(r1^2/t2);
% Well Function
wu2=expint(u2);
% Drawdown
dh=dh+(Qw/(4*pi()*T))*(wu1-wu2);
end
%
if rest_t > 0 && rest_t <= dton
% Operating Well
u1=(Sy/(4*T))*(r1^2/rest_t);
% Well Function
wu1=expint(u1);
% Drawdown
dh=dh+(Qw/(4*pi()*T))*wu1;
end
if rest_t>0 && rest_t > dton
t1=rest_t;
t2=t1-dton;
% Operating Well
u1=(Sy/(4*T))*(r1^2/t1);
```

```
% Well Function
wu1=expint(u1);
% Imaginary Compensation Well
u2=(Sy/(4*T))*(r1^2/t2);
% Well Function
wu2=expint(u2);
% Drawdown
dh=dh+(Qw/(4*pi()*T))*(wu1-wu2);
end
%%%%%%%%%%%%%%%%%%%%%%%%%%%%%%%%%%%%%%%%%%%%%%%%%%%%%%%%%%%%%%%%%%%%%%%%%%%%%%%%%%%%%%%%%%%%%%%%%%%%%%%%%%%%%%%%
%%%%%%%%%
function [dh] = CYC_THEIS_RECHARGE(t,delt,Sy,T,ys,xm,ym,dton,xw,yw,Qw)
dh=0.;
frac= t/delt;
int_t=fix(frac);
rest_t=t-int_t*delt;
n = int_t+1;
% Distance to wells
r1=sqrt((xm-xw)^2+(ym-yw)^2);
r2=sqrt((xm-xw)^2+(2*ys-ym-yw)^2);
for i=1:n-1
t1=t-(i-1)*delt;
t2=t1-dton;
% Constant Head Boundary Effects
u1_O=(Sy/(4*T))*(r1^2/t1); % Operating Well
u1_I=(Sy/(4*T))*(r2^2/t1); % Image Well
% Well Function
wu1_O=expint(u1_O);
wu1_I=expint(u1_I);
% Imaginary Compensation Wells
u2_O=Sy/(4*T)*(r1^2/t2); % Imaginary Operating Well
u2_I=Sy/(4*T)*(r2^2/t2); % Imaginary Image Well
% Well Function
wu2_O=expint(u2_O);
wu2_I=expint(u2_I);
% Total Well Function
wu1 = wu1_O-wu1_I;
wu2 = wu2_O-wu2_I;
% Drawdown
dh=dh+(Qw/(4*pi()*T))*(wu1-wu2);
end
%
if rest_t > 0 && rest_t <= dton
% Constant Head Boundary Effects
u1_O=(Sy/(4*T))*(r1^2/rest_t); % Operating Well
```

```
u1_I=(Sy/(4*T))*(r2^2/rest_t); % Image Well
% Well Function
wu1_O=expint(u1_O);
wu1_I=expint(u1_I);
% Total Well Function
wu1 = wu1_O-wu1_I;
% Drawdown
dh=dh+(Qw/(4*pi()*T))*wu1;
end
if rest_t>0 && rest_t > dton
t1=rest_t;
t2=t1-dton;
% Constant Head Boundary Effects
u1_O=(Sy/(4*T))*(r1^2/t1); % Operating Well
u1_I=(Sy/(4*T))*(r2^2/t1); % Image Well
% Well Function
wu1_O=expint(u1_O);
wu1_I=expint(u1_I);
% Imaginary Compensation Wells
u2_O=(Sy/(4*T))*(r1^2/t2); % Imaginary Well
u2_I=(Sy/(4*T))*(r2^2/t2); % Imaginary Image Well
% Well Function
wu2_O=expint(u2_O);
wu2_I=expint(u2_I);
% Total Well Function
wu1 = wu1_O-wu1_I;
wu2 = wu2_O-wu2_I;
% Drawdown
dh=dh+(Qw/(4*pi()*T))*(wu1-wu2);
end
%%%%%%%%%%%%%%%%%%%%%%%%%%%%%%%%%%%%%%%%%%%%%%%%%%%%%%%%%%%%%%%%%
%%%%%%%%%
function [dh] = CYC_THEIS_NOFLOW(t,delt,Sy,T,ys,xm,ym,dton,xw,yw,Qw)
dh=0.;
frac= t/delt;
int_t=fix(frac);
rest_t=t-int_t*delt;
n = int_t+1;
% Distance to wells
r1=sqrt((xm-xw)^2+(ym-yw)^2);
r2=sqrt((xm-xw)^2+(2*ys-ym-yw)^2);
for i=1:n-1
t1=t-(i-1)*delt;
t2=t1-dton;
% No-Flow Boundary Effect
```

```
u1_O=Sy/(4*T)*(r1^2/t1); % Operating Well
u1_I=Sy/(4*T)*(r2^2/t1); % Image Well
% Well Function
wu1_O=expint(u1_O);
wu1_I=expint(u1_I);
% Imaginary Compensation Wells
u2_O=Sy/(4*T)*(r1^2/t2); % Imaginary Operating Well
u2_I=Sy/(4*T)*(r2^2/t2); % Imaginary Image Well
% Well Function
wu2_O=expint(u2_O);
wu2_I=expint(u2_I);
% Total Well Function
wu1 = wu1_O+wu1_I;
wu2 = wu2_O+wu2_I;
% Drawdown
dh=dh+(Qw/(4*pi()*T))*(wu1-wu2);
end
%
if rest_t > 0 && rest_t <= dton
% No-Flow Boundary Effect
u1_O=Sy/(4*T)*(r1^2/rest_t); % Operating Well
u1_I=Sy/(4*T)*(r2^2/rest_t); % Image Well
% Well Function
wu1_O=expint(u1_O);
wu1_I=expint(u1_I);
% Total Well Function
wu1 = wu1_O+wu1_I;
% Drawdown
dh=dh+(Qw/(4*pi()*T))*wu1;
end
if rest_t>0 && rest_t > dton
t1=rest_t;
t2=t1-dton;
% No-Flow Boundary Effect
u1_O=Sy/(4*T)*(r1^2/t1); % Operating Well
u1_I=Sy/(4*T)*(r2^2/t1); % Image Well
% Well Function
wu1_O=expint(u1_O);
wu1_I=expint(u1_I);
% Imaginary Compensation Wells
u2_O=Sy/(4*T)*(r1^2/t2); % Imaginary Well
u2_I=Sy/(4*T)*(r2^2/t2); % Imaginary Image Well
% Well Function
wu2_O=expint(u2_O);
wu2_I=expint(u2_I);
```

```
% Total Well Function
wu1 = wu1_O+wu1_I;
wu2 = wu2_O+wu2_I;
% Drawdown
dh=dh+(Qw/(4*pi()*T))*(wu1-wu2);
end
%%%%%%%%%%%%%%%%%%%%%%%%%%%%%%%%%%%%%%%%%%%%%%%%%%%%%%%%
%%%%%%%%
```

Matlab code: **Drawdown.VS.Time**

This code calculates and plots the head time series for $t \in (0, t_{fin})$ for a number of wells operating cyclically or continuously in a laterally infinite aquifer, in a semi-infinite aquifer limited by a stream boundary, and in a semi-infinite aquifer limited by a no-flow boundary. The boundary is represented by the y axis (x=0).

The input file **Aquifer.Parms** includes simulation time parameters and aquifer parameters:

a. time parameters:

  - final simulation time t_{fin};
  - cyclical time step of simulation $\Delta t = \Delta t_{on} + \Delta t_{off}$ (e.g., 1 year)

b. aquifer parameters:

  - initial hydraulic head (aquifer's saturated thickness) h_0
  - hydraulic conductivity K;
  - storativity S;
  - stream location y_s(>0).

The input file **Monitoring.Wells.dat** includes monitoring well data.

a. total number of monitoring wells n_{mw}
b. for each monitoring well, each of the following lines provides well location coordinates x_w and y_w.

The input file **Wells.dat** includes wells data:

a. total number of operating wells n_w
b. for each operating well, each of the following lines provide:

  - the pumping rate (Q_w);
  - the time at which well operation starts (t_{st}),
  - the total operation period (Δt_{on})[2]
  - well location x_w and y_w.

[2] If this time is set equal to Δt then the well operates continuously ($\Delta t_{off} = 0$). Otherwise the well operates cyclically.

Examples:

Aquifer.Parms

730.	365.	30.	86.4	0.2
t_{fin}(d)	$\Delta t = \Delta t_{on} + \Delta t_{off}$ (d)	b (m)	K (m/d)	S (/)

Grid.Parms

0.	0.	4000.	4000.	40	40
x_{min}(m)	y_{min}(m)	x_{min}(m)	y_{min}(m)	n_x	n_y

Wells.dat

Description:	2 wells operating cyclically			
2 (n_w)				
+3000.	90.	120.	1000.	1750.
+3000.	820.	120.	1000.	2250.
Q_w(m^3/d)	t_{st}(d)	Δt_{on}(d)	x_w(m)	y_w (m)

```
function Drawdown_VS_Time
   clc
   clear all
   % Reading Data Files
   % 1- Parameters File
   fid1 = fopen ('Aquifer.Parms','r');
   Temp = fscanf(fid1,'%f %f %f %f %f',[1,6]);
   tfin = Temp(1); % Final Time of Simulation (d)
   delt = Temp(2); % Cycle time step of simulation (d)
   ho   = Temp(3); % Aquifer Initial Head (saturated thickness)(m)
   K    = Temp(4); % Hydraulic Conductivity (m/d)
   Sy   = Temp(5); % Aquifer Storativity (/)
   ys   = Temp(6); % stream location (m)
   fclose(fid1);
   %
   % 2- Monitoring Wells data file
   fid2 = fopen ('Monitoring.Wells.dat','r');
   Temp = fscanf(fid2,'%f',[1,1]);
   nmw  = Temp(1);% Monitoring Wells Number
   for m = 1:nmw
   Temp  = fscanf(fid2,'%f %f',[1,2]);
   xm(m) = Temp(1);
   ym(m) = Temp(2);
   end
```

```
fclose(fid2);
% 3- Wells Data file
fid3 = fopen ('Wells.dat','r');
Temp = fscanf(fid3,'%f',[1,1]);
nw   = Temp(1);% Wells Number
for w = 1:nw
Temp    = fscanf(fid3,'%f %f %f %f %f',[1,5]);
Qw(w)   = Temp(1); % Well Pumping Rate
tst(w)  = Temp(2); % Pumping Start Time
dton(w) = Temp(3); % Pumping Period
xw(w)   = Temp(4); % Well Location X Coordinate
yw(w)   = Temp(5); % Well Location Y Coordinate
end
fclose(fid3);
%
% Calculations
% 1- General
T = K*ho;
to = 0;
dt = 1.;
t = to:dt:tfin; % time matrix
nt = length(t); % number of time steps (/)
%
% 2- Computing Drawdown Distribution for a number of operating wells
% Initialize Drawdown Arrays
H_sum = zeros(nt,nmw);
HN_sum = zeros(nt,nmw);
HR_sum = zeros(nt,nmw);
fid4 = fopen('Drawdown.VS.Time.dat','w');
for j = 1:nmw
for i= 1:nt
H_sum(i,j) = ho;
HN_sum(i,j) = ho;
HR_sum(i,j) = ho;
for w = 1:nw
dt = t(i)-tst(w);
if (dt>=0.)
H_sum(i,j)    = H_sum(i,j)  +CYC_THEIS(dt,delt,Sy,T,xm(j),ym(j),
dton(w),xw(w),yw(w),Qw(w));
HN_sum(i,j)   =  HN_sum(i,j)+CYC_THEIS_NOFLOW(dt,delt,Sy,T,ys,xm
(j),ym(j),dton(w),xw(w),yw(w),Qw(w));
HR_sum(i,j) = HR_sum(i,j)+CYC_THEIS_RECHARGE(dt,delt,Sy,T,ys,xm
(j),ym(j),dton(w),xw(w),yw(w),Qw(w));
end
end
```

```
temp = [t(i),xm(j),ym(j),H_sum(i,j),HN_sum(i,j),HR_sum(i,j)];
fprintf(fid4,'%15.6E %15.6E %15.6E %15.6E %15.6E %15.6E\n',temp);
end
[H_MAX, iMAX] = max(H_sum(:,j));
[H_MIN, iMIN] = min(H_sum(:,j));
[HN_MAX, iMAX] = max(HN_sum(:,j));
[HN_MIN, iMIN] = min(HN_sum(:,j));
[HR_MAX, iMAX] = max(HR_sum(:,j));
[HR_MIN, iMIN] = min(HR_sum(:,j));
temp = [H_MAX,HN_MAX,HR_MAX];
[hmax, iMAX] = max(temp);
temp = [H_MIN,HN_MIN,HR_MIN];
[hmin, iMIN] = min(temp);
% Fiqure
figure
hold('all')
plot(t,H_sum(:,j),'-k','LineWidth',1.5)
plot(t,HN_sum(:,j),'-b','LineWidth',1.5)
plot(t,HR_sum(:,j),'-r','LineWidth',1.5)
plot(t,ho,'--k')
title(['Hydraulic Head Time Series - Monitoring Well: ',num2str(j)]);
xlabel('Time (T)');
ylabel('Hydraulic Head (L)');
xlim ([0 tfin+10]);
ylim ([hmin-0.1 hmax+0.1]);
legend('Infinite    Aquifer','Semi-Infinite    (N)','Semi-Infinite
(R)','Location','NorthWest')
hold off
end
fclose(fid4);
%%%%%%%%%%%%%%%%%%%%%%%%%%%%%%%%%%%%%%%%%%%%%%%%%%%%%%%%%%%%%%
%%%%%%%%%%%%%
function [dh] = CYC_THEIS(t,delt,Sy,T,xm,ym,dton,xw,yw,Qw)
dh=0.;
frac= t/delt;
int_t=fix(frac);
rest_t=t-int_t*delt;
n = int_t+1;
% Distance to wells
r1=sqrt((xm-xw)^2+(ym-yw)^2);
for i=1:n-1
t1=t-(i-1)*delt;
t2=t1-dton;
% Operating Well
u1=(Sy/(4*T))*(r1^2/t1);
```

```
% Well Function
wu1=expint(u1);
% Imaginary Compensation Well
u2=Sy/(4*T)*(r1^2/t2);
% Well Function
wu2=expint(u2);
% Drawdown
dh=dh+(Qw/(4*pi()*T))*(wu1-wu2);
end
%
if rest_t > 0 && rest_t <= dton
% Operating Well
u1=(Sy/(4*T))*(r1^2/rest_t);
% Well Function
wu1=expint(u1);
% Drawdown
dh=dh+(Qw/(4*pi()*T))*wu1;
end
if rest_t>0 && rest_t > dton
t1=rest_t;
t2=t1-dton;
% Operating Well
u1=(Sy/(4*T))*(r1^2/t1);
% Well Function
wu1=expint(u1);
% Imaginary Compensation Well
u2=(Sy/(4*T))*(r1^2/t2);
% Well Function
wu2=expint(u2);
% Drawdown
dh=dh+(Qw/(4*pi()*T))*(wu1-wu2);
end
%%%%%%%%%%%%%%%%%%%%%%%%%%%%%%%%%%%%%%%%%%%%%%%%%%%%%%%%%%%%%%%%
%%%%%%%%%%%%%
function [dh] = CYC_THEIS_RECHARGE(t,delt,Sy,T,ys,xm,ym,dton,xw,yw,Qw)
dh=0.;
frac= t/delt;
int_t=fix(frac);
rest_t=t-int_t*delt;
n = int_t+1;
% Distance to wells
r1=sqrt((xm-xw)^2+(ym-yw)^2);
r2=sqrt((xm-xw)^2+(2*ys-ym-yw)^2);
for i=1:n-1
t1=t-(i-1)*delt;
t2=t1-dton;
```

```
% Constant Head Boundary Effects
u1_O=(Sy/(4*T))*(r1^2/t1); % Operating Well
u1_I=(Sy/(4*T))*(r2^2/t1); % Image Well
% Well Function
wu1_O=expint(u1_O);
wu1_I=expint(u1_I);
% Imaginary Compensation Wells
u2_O=Sy/(4*T)*(r1^2/t2); % Imaginary Operating Well
u2_I=Sy/(4*T)*(r2^2/t2); % Imaginary Image Well
% Well Function
wu2_O=expint(u2_O);
wu2_I=expint(u2_I);
% Total Well Function
wu1 = wu1_O-wu1_I;
wu2 = wu2_O-wu2_I;
% Drawdown
dh=dh+(Qw/(4*pi()*T))*(wu1-wu2);
end
%
if rest_t > 0 && rest_t <= dton
% Constant Head Boundary Effects
u1_O=(Sy/(4*T))*(r1^2/rest_t); % Operating Well
u1_I=(Sy/(4*T))*(r2^2/rest_t); % Image Well
% Well Function
wu1_O=expint(u1_O);
wu1_I=expint(u1_I);
% Total Well Function
wu1 = wu1_O-wu1_I;
% Drawdown
dh=dh+(Qw/(4*pi()*T))*wu1;
end
if rest_t>0 && rest_t > dton
t1=rest_t;
t2=t1-dton;
% Constant Head Boundary Effects
u1_O=(Sy/(4*T))*(r1^2/t1); % Operating Well
u1_I=(Sy/(4*T))*(r2^2/t1); % Image Well
% Well Function
wu1_O=expint(u1_O);
wu1_I=expint(u1_I);
% Imaginary Compensation Wells
u2_O=(Sy/(4*T))*(r1^2/t2); % Imaginary Well
u2_I=(Sy/(4*T))*(r2^2/t2); % Imaginary Image Well
% Well Function
wu2_O=expint(u2_O);
wu2_I=expint(u2_I);
```

```
% Total Well Function
wu1 = wu1_O-wu1_I;
wu2 = wu2_O-wu2_I;
% Drawdown
dh=dh+(Qw/(4*pi()*T))*(wu1-wu2);
end
%%%%%%%%%%%%%%%%%%%%%%%%%%%%%%%%%%%%%%%%%%%%%%%%%%%%%%%%%%%%%%%%%%%%%%%%%%%%%%%%%%%%%%%%%%%%%%%%%%%%%%%%%%%%%%%%%%%%%%%
%%%%%%%%%%%%%%%%%%%%%%%%
function [dh] = CYC_THEIS_NOFLOW(t,delt,Sy,T,ys,xm,ym,dton,xw,yw,Qw)
dh=0.;
frac= t/delt;
int_t=fix(frac);
rest_t=t-int_t*delt;
n = int_t+1;
% Distance to wells
r1=sqrt((xm-xw)^2+(ym-yw)^2);
r2=sqrt((xm-xw)^2+(2*ys-ym-yw)^2);
for i=1:n-1
t1=t-(i-1)*delt;
t2=t1-dton;
% No-Flow Boundary Effect
u1_O=Sy/(4*T)*(r1^2/t1); % Operating Well
u1_I=Sy/(4*T)*(r2^2/t1); % Image Well
% Well Function
wu1_O=expint(u1_O);
wu1_I=expint(u1_I);
% Imaginary Compensation Wells
u2_O=Sy/(4*T)*(r1^2/t2); % Imaginary Operating Well
u2_I=Sy/(4*T)*(r2^2/t2); % Imaginary Image Well
% Well Function
wu2_O=expint(u2_O);
wu2_I=expint(u2_I);
% Total Well Function
wu1 = wu1_O+wu1_I;
wu2 = wu2_O+wu2_I;
% Drawdown
dh=dh+(Qw/(4*pi()*T))*(wu1-wu2);
end
%
if rest_t > 0 && rest_t <= dton
% No-Flow Boundary Effect
u1_O=Sy/(4*T)*(r1^2/rest_t); % Operating Well
u1_I=Sy/(4*T)*(r2^2/rest_t); % Image Well
% Well Function
wu1_O=expint(u1_O);
```

```
wu1_I=expint(u1_I);
% Total Well Function
wu1 = wu1_O+wu1_I;
% Drawdown
dh=dh+(Qw/(4*pi()*T))*wu1;
end
if rest_t>0 && rest_t > dton
t1=rest_t;
t2=t1-dton;
% No-Flow Boundary Effect
u1_O=Sy/(4*T)*(r1^2/t1); % Operating Well
u1_I=Sy/(4*T)*(r2^2/t1); % Image Well
% Well Function
wu1_O=expint(u1_O);
wu1_I=expint(u1_I);
% Imaginary Compensation Wells
u2_O=Sy/(4*T)*(r1^2/t2); % Imaginary Well
u2_I=Sy/(4*T)*(r2^2/t2); % Imaginary Image Well
% Well Function
wu2_O=expint(u2_O);
wu2_I=expint(u2_I);
% Total Well Function
wu1 = wu1_O+wu1_I;
wu2 = wu2_O+wu2_I;
% Drawdown
dh=dh+(Qw/(4*pi()*T))*(wu1-wu2);
end
%%%%%%%%%%%%%%%%%%%%%%%%%%%%%%%%%%%%%%%%%%%%%%%%%%%%%%%%%%
%%%%%%%%
```

Appendix 2

Matlab Code: **Glover**

This code calculates and plots the stream depletion rate, Q_r, and stream depletion volume, V_r, vs Time, t, produced by a generic number of pumping wells operating either continuously or cyclically in a semi-infinite aquifer bounded by a recharge (stream) boundary. The stream is located on the y axis of the reference system (x=0).

The input file **Aquifer.dat** includes simulation time parameters and aquifer parameters:

a. time parameters

- final simulation time t_{fin};

- cyclical time step of simulation $\Delta t = \Delta t_{on} + \Delta t_{off}$ (e.g., 1 year)

b. the aquifer parameters

- aquifer's saturated thickness b;
- hydraulic conductivity K;
- storativity S.

The input file **Wells.dat** includes wells data:

a. total number of operating wells n_w
b. for each operating well, each of the following lines provide:

- the pumping rate (Q_w);
- the time at which well operation starts (t_{st}),
- the total operation period (Δt_{on})[3]
- welllocation x_w then y_w.

Examples:

Aquifer.dat

21900.	365.	30.	86.4	0.2
t_{fin}(d)	$\Delta t = \Delta t_{on} + \Delta t_{off}$ (d)	b(m)	K (m/d)	S (/)

Wells.dat

Description:	1 well operating cyclically for 5 years and then shut off			
2 (n_w)				
+3000.	90.	120.	1000.	500.
−3000.	1915.	120.	1000.	500.
Q_w(m^3/d)	t_{st}(d)	Δt_{on}(d)	x_w(m)	y_w (m)

Description:	2 wells operating cyclically (well 2 starts at time = 5 years)			
2 (n_w)				
+3000.	90.	120.	1000.	500.
+3000.	1915.	120.	500.	500.
Q_w(m^3/d)	t_{st}(d)	Δt_{on}(d)	x_w(m)	y_w (m)

Description:	1 well operating continuously			
1 (n_w)				
+3000.	0.	365.	1000.	500.
Q_w(m^3/d)	t_{st}(d)	Δt_{on}(d)	x_w(m)	y_w (m)

[3] If this time is set equal to Δt then the well operates continuously ($\Delta t_{off} = 0$). Otherwise the well operates cyclically.

Description:	1 well operating cyclically			
2 (n_w)				
+3000.	90.	120.	1000.	500.
Q_w(m^3/d)	t_{st}(d)	Δt_{on}(d)	x_w(m)	y_w (m)

Description:	1 well injecting and extracting cyclically (net pumped volume is zero)			
2 (n_w)				
+3000.	90.	120.	1000.	500.
−1479.3877	210.	245.	1000.	500.
Q_w(m^3/d)	t_{st}(d)	Δt_{on}(d)	x_w(m)	y_w (m)

```
function Glover
   clear all
   clc
   % Reading Data Files
   % 1 - General Data
   fid1 = fopen ('Aquifer.dat','r');
   Temp = fscanf(fid1,'%f %f %f %f %f',[1,5]);
   tfin = Temp(1);
   % tfin = Final Time of Simulation (d)
   delt = Temp(2);
   % delt = dton+dtoff (e.g., 365 d)
   h = Temp(3);
   % h = Thickness of the Aquifer (m)
   K = Temp(4);
   % K = Hydraulic Conductivity (m/d)
   Sy = Temp(5);
   % Sy = Storativity (/)
   fclose(fid1);
   %
   % 2 - Operating Wells Data
   fid2 = fopen ('Wells.dat','r');
   Temp = fscanf( fid2,'%f',[1,1]);
   now = Temp(1);
   for iw = 1:now
       Temp = fscanf(fid2,'%f %f %f %f %f',[1,5]);
       Qw(iw)   = Temp(1);
       tst(iw)  = Temp(2);
       dton(iw) = Temp(3);
       xw(iw)   = Temp(4);
       yw(iw)   = Temp(5);
```

```
end
fclose(fid2);
%
% Calculations
% 1 - General calculations
T = K*h;
to = 0.;
dt = 5.;
t = to:dt:tfin; % time matrix
nt = length(t); % number of time steps (/)
%
% 2 - Operation off period for operating wells
for iw = 1:now
    dtoff(iw)=delt-dton(iw);
end
%
% 3 - Calculating Qr (stream depletion rate) and Vr (stream depletion
volume)
Qr_sum = zeros(nt,1);
Qc_sum= zeros(nt,1);
Vr_sum = zeros(nt,1);
Vc_sum = zeros(nt,1);
Vrs_sum = zeros(nt,1);
for i= 1:nt
for iw= 1:now
    tt = t(i)-tst(iw);
if tt>0.
Qr_sum(i)=Qr_sum(i)+Qw(iw)*rate_sol(tt,dton(iw),delt,T,Sy,xw(iw));
Vr_sum(i)=Vr_sum(i)+Qw(iw)*vol_sol(tt,delt,dton(iw),T,Sy,xw(iw));
Vc_sum(i)=Vc_sum(i)+Qw(iw)*vol_cum(tt,delt,dton(iw));
end
end
end
for i= 1:nt
    Vrs_sum(i)= Vr_sum(i)/Vc_sum(i);
end
%
Qr_MAX = max(Qr_sum);
if Qr_MAX>=0.
    Qr_MAX=Qr_MAX+100.;
else
    Qr_MAX=0;
end
Qr_MIN = min(Qr_sum);
if Qr_MIN<=0.
```

```
    Qr_MIN=Qr_MIN-100.;
else
    Qr_MIN=0;
end
Vr_MAX = max(Vc_sum);
Vr_MIN = min(Vc_sum);
% Fiqures
figure
plot(t,Qr_sum,'-r')
%title('Total Stream Depletion Rate Vs Time');
xlabel('Time (d)');
ylabel('Total Qr (m^3/d)');
xlim ([0 tfin+1]);
ylim ([Qr_MIN Qr_MAX]);
%legend('Qr')
%
figure
plot(t,Vr_sum,'-r')
%title('Total Stream Depletion Volume Vs Time');
xlabel('Time (d)');
ylabel('Total Vr (m^3)');
xlim ([0 tfin+1]);
ylim ([0 Vr_MAX+1000]);
hold on
plot(t,Vc_sum,'-b')
%legend('Vr')
%
% Fiqures
figure
plot(t,Vrs_sum,'-r')
%title('Unitless Total Stream Depletion Volume Vs Time');
xlabel('Time (d)');
ylabel('Qr/Qc or Vr/Vc (/)');
xlim ([0 tfin+1]);
ylim ([0 1]);
%legend('Vr')
%
% Output Results
fid3 = fopen('Time.Qr.Vr.dat','w');
for i= 1:nt
temp = [t(i),Qr_sum(i),Vr_sum(i)];
fprintf(fid3,'%15.6E %15.6E %15.6E\n',temp);
end
%%%%%%%%%%%%%%%%%%%%%%%%%%%%%%%%%%%%%%%%%%%%%%%%%%%%%%%%%%%%%%%
%%%%%%%%%
```

```
function [Qr] = rate_sol(t,dton,delt,T,Sy,xw)
% Calculating Qr (stream depletion rate)
Qr = 0.;
frac=t/delt;
int_t=fix(frac);
rest_t=t-int_t*delt;
n = int_t+1;
for i=1:n-1
    t1=t-(i-1)*delt;
    t2=t1-dton;
    Qr =Qr+Qratio(T,Sy,xw,t1)-Qratio(T,Sy,xw,t2);
end
if rest_t >0 && rest_t <= dton
    Qr =Qr+Qratio(T,Sy,xw,rest_t);
end
if rest_t > dton
    Qr =Qr+Qratio(T,Sy,xw,rest_t)-Qratio(T,Sy,xw,rest_t-dton);
end
%%%%%%%%%%%%%%%%%%%%%%%%%%%%%%%%%%%%%%%%%%%%%%%%%%%%%%%%%%%%%%%%%%%%%%%%%%%%
%%%%%%%%%
function [Vr] = vol_sol(t,delt,dton,T,Sy,xw)
% Calculating Vr (depletion volume)
Vr = 0.;
frac=(t)/delt;
int_t=fix(frac);
rest_t=t-int_t*delt;
n = int_t+1;
for i=1:n-1
    t1=t-(i-1)*delt;
    t2=t1-dton;
    Vr =Vr+t1*Vratio(T,Sy,xw,t1)-t2*Vratio(T,Sy,xw,t2);
end
if rest_t >0 && rest_t <= dton
    Vr=Vr+rest_t*Vratio(T,Sy,xw,rest_t);
end
if rest_t > dton
    t1=rest_t;
    t2=t1-dton;
    Vr=Vr+t1*Vratio(T,Sy,xw,t1)-t2*Vratio(T,Sy,xw,t2);
end
%%%%%%%%%%%%%%%%%%%%%%%%%%%%%%%%%%%%%%%%%%%%%%%%%%%%%%%%%%%%%%%%%%%%%%%%%%%%
%%%%%%%%%
function [V] = vol_cum(t,delt,dton)
% Calculating cumulative pumped volume
V = 0.;
```

```
frac=t/delt;
n=fix(frac);
rest_t=t-n*delt;
V = n*dton;
if rest_t >0 && rest_t <= dton
   V=V+rest_t;
end
if rest_t > dton
   V=V+dton;
end
%%%%%%%%%%%%%%%%%%%%%%%%%%%%%%%%%%%%%%%%%%%%%%%%%%%%%%%%%%%%%%%%%%
%%%%%%%%%
function QR = Qratio (T,Sy,y,t)
u = (Sy*y^2)/(4*T*t);
QR = erfc(sqrt(u));% Stream Depletion Volume
%%%%%%%%%%%%%%%%%%%%%%%%%%%%%%%%%%%%%%%%%%%%%%%%%%%%%%%%%%%%%%%%%%
%%%%%%%%%
function VR = Vratio(T,S,y,t)
u = (S*y^2)/(4*T*t);
p = (S*y^2)/(2*T*t);
VR = (1+p)*erfc(sqrt(u))- (sqrt(u)*2/sqrt(pi()))*(exp(-u));
```

Appendix 3

MATLAB code: **BCYC_Drawdown2D**

This code calculates and plots the spatial distribution of head at a specified time t_{fin} for a number of wells operating cyclically or continuously in an aquifer bounded laterally by a either a constant-head recharge boundary and a no-flow boundary, or two constant-head recharge boundaries. One of the recharge boundaries represents a stream located on the $x=x_s$ straight line. The second boundary, whether recharge or no-flow, is located on the y-axis (the $x=0$ straight line).

The input file **Aquifer.Parms** includes simulation time parameters and aquifer parameters:

a. time parameters
   - print-out time t_{fin};
   - cyclical time step of simulation $\Delta t=\Delta t_{on}+\Delta t_{off}$ (e.g., 1 year)

b. aquifer parameters
   - initial hydraulic head (aquifer's saturated thickness) h_0
   - hydraulic conductivity K;
   - storativity S;
   - stream boundary location $x=x_s$;

– secondary boundary location $x=x_b$ ($x_s<x_b$)

The input file **Grid.dat** includes data regarding the plotting grid:

- grid lower left corner abscissa (x_{min})
- grid lower left corner ordinate (y_{min})
- grid upper right corner abscissa (x_{max})
- grid upper right corner ordinate (y_{max})
- n. of gridblocks along x (n_x)
- n. of gridblocks along y (n_y)

The input file **Wells.dat** includes wells data:

a. total number of operating wells n_w
b. for each operating well, each of the following lines provide:

   - the pumping rate (Q_w);
   - the time at which well operation starts (t_{st}),
   - the total operation period (Δt_{on})[4]
   - well location x_w and y_w.

Examples:

Aquifer.Parms

1460.	365.	30.	86.4	0.2	0.	2000.
t_{fin} (d)	$\Delta t=\Delta t_{on}+\Delta t_{off}$ (d)	b(m)	K (m/d)	S (/)	x_s (m)	x_b (m)

Grid.Parms

0.	0.	2000.	2000.	40	40
x_{min} (m)	y_{min} (m)	x_{min} (m)	y_{min} (m)	n_x	n_y

Wells.dat

Description:	2 wells operating cyclically			
2 (n_w)				
+2500.	75.	120.	1250.	1250.
+2500.	75.	120.	750.	750.
Q_w(m³/d)	t_{st}(d)	Δt_{on}(d)	x_w (m)	y_w (m)

```
function BCYC_Drawdown2D
   % Calculate the Drawdown for a Cyclically operating wells in a
   % Bounded Aquifer
```

[4] If this time is set equal to Δt then the well operates continuously ($\Delta t_{off} = 0$). Otherwise the well operates cyclically.

```
clc
clear all
% Reading Data File
% 1-Aquifer Parameters
fid1 = fopen ('Aquifer.parms','r');
Temp = fscanf( fid1,'%f',[1,2]);
tfin = Temp(1); % Print Out Time (T)
delt = Temp(2); % Cycle Time = One Year (T)
%
Temp = fscanf(fid1,'%f %f %f %f %f',[1,5]);
b      = Temp(1); % Aquifer Thickness (L)
K      = Temp(2); % Hydraulic Conductivity (L^2/T)
Sy      = Temp(3); % Aquifer Storativity (/)
x_boundary = Temp(4); % Boundary Location (L)
x_stream  = Temp(5); % Stream Location (L)
fclose(fid1);
%
% 2-Wells Data file
fid2 = fopen ('Well.dat','r');
now = fscanf( fid2,'%f',[1,1]); % Wells Number
for iw = 1:now
    Temp = fscanf( fid2,'%f %f %f %f %f',[1,5]);
    Qw(iw)  = Temp(1); % Well Pumping Rate (L^3/T)
    tst(iw)  = Temp(2); % Well Operation Starting Time (T)
    dton(iw) = Temp(3); % Pumping Period (T)
    xw(iw)  = Temp(4); % Well Location X Coordinate (L)
    yw(iw)  = Temp(5); % Well Location Y Coordinate (L)
end
fclose(fid2);
%1
% 3-Grid Data file
fid3 = fopen ('Grid.dat','r');
Temp = fscanf(fid3,'%f %f %f %f %f %f',[1,6]);
xmin = Temp(1); % Minimum Value of X in the Grid (L)
ymin = Temp(2); % Minimum Value of Y in the Grid (L)
xmax = Temp(3); % Maximim Value of X in the Grid (L)
ymax = Temp(4); % Maximum Value of Y in the Grid (L)
nx = Temp(5);  % Gridblocks Along X
ny = Temp(6);  % Gridblocks Along Y
fclose(fid3);
%
% Grid Generation
nxx = nx+1;
```

```
nyy = ny+1;
dx = (xmax-xmin)/nx;
dy = (ymax-ymin)/ny;
%
for i= 1:nyy
for j= 1:nxx
    x(i,j)= xmin+ dx*(j-1);
end
end
%
for j= 1:nxx
for i= 1:nyy
    y(i,j)= ymin+dy*(i-1);
end
end
% Calculations
% 1-General
w = x_stream - x_boundary; % Aquifer Width (L)
T = K*b; % Transmissivity
% 2-Computing Drawdown Distributions
%
fid4 = fopen('results.NR.dat','w');
ssum_noflow = zeros(nyy,nxx);% Initial Drawdown
%
fid5 = fopen('results.RR.dat','w');
ssum_rech = zeros(nyy,nxx);% Initial Drawdown
%
for i= 1:nyy;
for j= 1:nxx;
for iw =1:now
    xmw = x(i,j)-x_boundary;
    ymw = y(i,j);
    xow = xw(iw)-x_boundary;
    yow = yw(iw);
    Qow = Qw(iw);
    dtonw = dton(iw);
    t = tfin-tst(iw);
if t>0
ssum_noflow(i,j)= ssum_noflow(i,j)+BNOFLOW_CYC_THEIS(Sy,T,w,xow,
yow,Qow,delt,dtonw,t,xmw,ymw);
ssum_rech(i,j) =ssum_rech(i,j)+BRECH_CYC_THEIS(Sy,T,w,xow,yow,
Qow,delt,dtonw,t,xmw,ymw);
end
end
```

```
temp = [x(i,j), y(i,j), ssum_rech(i,j)];
fprintf(fid4,'%15.6E %15.6E %15.6E\n',temp);
fwrite(fid4,temp);
%
temp = [x(i,j), y(i,j), ssum_noflow(i,j)];
fprintf(fid5,'%15.6E %15.6E %15.6E\n',temp);
fwrite(fid5,temp);
end
end
fclose(fid4);
fclose(fid5);
%
str1 = num2str(Qw');
cell1 = cellstr(str1);
% Figure
figure;
contour(x,y,ssum_noflow);
[C,h] = contour(x,y,ssum_noflow);
clabel(C,h);
title('Drawdown in Aquifer Bounded by No-Flow and Recharge
Boundaries');
xlabel('x (L)');
ylabel('x (L)');
hold on
scatter(xw,yw,50,'r+')
text(xw+20,yw+50,cell1,'BackgroundColor',[1 1 0],'FontSize',10);
legend('Drawdown','Operating Well')
%
figure;
contour(x,y,ssum_rech);
[C,h] = contour(x,y,ssum_rech);
clabel(C,h);
title('Drawdown in Aquifer Bounded by Two Recharge Boundaries');
xlabel('x (L)');
ylabel('y (L)');
hold on
scatter(xw,yw,50,'r+')
text(xw+20,yw+50,cell1,'BackgroundColor',[1 1 0],'FontSize',10);
legend('Drawdown','Operating Well')
%%%%%%%%%%%%%%%%%%%%%%%%%%%%%%%%%%%%%%%%%%%%%%%%%%%%%%%%%%%%%%%%
%%%%%%%%%
function [BRNF] = BNOFLOW_CYC_THEIS(Sy,T,w,xw,yw,Qw,delt,dton,t,
x,y)
BRNF=0.; % Initial Drawdown in a Bounded Aquifer
a=w-xw; % Operating-well/Stream Distance
```

```
TOL=1.e-9; % Tolerance value at which the loop stops
DELTA=1.0; % Starting value of DELTA
% Loop over the number of Wells Groups
j=1; % (While) Loop Counter
while abs(DELTA)>= TOL && j<=50
    sign = (-1)^(j+1);
    Q_term = Qw/(4*pi()*T);
    DELTA = sign*Q_term*Four_Wells_BRNF(Sy,T,w,a,yw,delt,dton,t,x,
    y,j);
BRNF = BRNF+DELTA;
    j=j+1;
end
if (j>50)
    disp('convergence not reached')
end
%
function [W_Fun]= Four_Wells_BRNF(Sy,T,w,a,yw,delt,dton,t,x,y,j)
W_Fun= 0.;
% Wells x Coordinate
xw1 = (2*j-1)*w-a;
xw2 = (2*j-1)*w+a;
% xw1 = 2*(j-1)*w+a;
% xw2 = 2*j*w  -a;
%
% Wells Distance From Observation Well
r1= sqrt((x-xw1)^2+(y-yw)^2);
r2= sqrt((x-xw2)^2+(y-yw)^2);
% Image Wells
r3= sqrt((x+xw1)^2+(y-yw)^2);
r4= sqrt((x+xw2)^2+(y-yw)^2);
%
% Cycle Effect
frac =t/delt;
int_t =fix(frac);
rest_t =t-int_t*delt;
n =int_t+1; % Number of complete Operation Cycles
% Loop over the number of Operation Cycles
for i=1:n-1
% For a Number of Complete Cycles
    t1=t-(i-1)*delt;
    t2=t1-dton;
% Stream Constant Head and infinite Aquifer Effects
    u1=(Sy/(4*T))*(r1^2/t1);
    u2=(Sy/(4*T))*(r2^2/t1);
% Image Wells
```

```
    u3=(Sy/(4*T))*(r3^2/t1);
    u4=(Sy/(4*T))*(r4^2/t1);
% Well Function
    wu1=expint(u1);
    wu2=expint(u2);
% Image Well Function
    wu3=expint(u3);
    wu4=expint(u4);
% Operating And Image Wells
    wo = wu1-wu2+wu3-wu4;
% wo = wu1+wu2-wu3-wu4;
% Imaginary Compensation Wells (for the continuous pumping)
u1_I=(Sy/(4*T))*(r1^2/t2);
u2_I=(Sy/(4*T))*(r2^2/t2);
% Image Wells
u3_I=(Sy/(4*T))*(r3^2/t2);
u4_I=(Sy/(4*T))*(r4^2/t2);
% Well Function
wu1_I=expint(u1_I);
wu2_I=expint(u2_I);
% Image Well Function
wu3_I=expint(u3_I);
wu4_I=expint(u4_I);
wI = wu1_I-wu2_I+wu3_I-wu4_I;
% wI = wu1_I+wu2_I-wu3_I-wu4_I;
% Total Well Function
W_Fun =W_Fun+wo-wI;
end
%
if rest_t>0 && rest_t <= dton % During Operation Time
% Stream Constant Head and Infinite Aquifer Effects
u1=(Sy/(4*T))*(r1^2/rest_t);
u2=(Sy/(4*T))*(r2^2/rest_t);
% Image Wells
u3=(Sy/(4*T))*(r3^2/rest_t);
u4=(Sy/(4*T))*(r4^2/rest_t);
% Well Function
wu1=expint(u1);
wu2=expint(u2);
% Image Well Function
wu3=expint(u3);
wu4=expint(u4);
% Operating And Image Wells
W_Fun = W_Fun+wu1-wu2+wu3-wu4;
% W_Fun = W_Fun+wu1+wu2-wu3-wu4;
```

```
end
if rest_t>0 && rest_t > dton % During time when operation stops
t1=rest_t;
t2=t1-dton;
% Stream Constant Head and infinite Aquifer Effects
u1=(Sy/(4*T))*(r1^2/t1);
u2=(Sy/(4*T))*(r2^2/t1);
% Image Wells
u3=(Sy/(4*T))*(r3^2/t1);
u4=(Sy/(4*T))*(r4^2/t1);
% Well Function
wu1=expint(u1);
wu2=expint(u2);
% Image Well Function
wu3=expint(u3);
wu4=expint(u4);
% Operating And Image Wells
wo = wu1-wu2+wu3-wu4;
% wo = wu1+wu2-wu3-wu4;
% Imaginary Compensation Wells (for continous pumping)
u1_I=(Sy/(4*T))*(r1^2/t2);
u2_I=(Sy/(4*T))*(r2^2/t2);
% Image Wells
u3_I=(Sy/(4*T))*(r3^2/t2);
u4_I=(Sy/(4*T))*(r4^2/t2);
% Well Function
wu1_I=expint(u1_I);
wu2_I=expint(u2_I);
% Image Well Function
wu3_I=expint(u3_I);
wu4_I=expint(u4_I);
% Imaginary And Image Wells
wI = wu1_I-wu2_I+wu3_I-wu4_I;
% wI = wu1_I+wu2_I-wu3_I-wu4_I;
% Total Well Function
W_Fun =W_Fun+wo-wI;
end
%%%%%%%%%%%%%%%%%%%%%%%%%%%%%%%%%%%%%%%%%%%%%%%%%%%%%%%%%%%%%%%%%%%%%%%%%%%%%%%%
%%%%%%%%%%
function [BRR] = BRECH_CYC_THEIS(Sy,T,w,xw,yw,Qw,delt,dton,t,x,y)
a=w-xw; % Operating-well/Stream Distance
BRR=0.; % Initial Drawdown in a Bounded Aquifer
TOL=1.e-9; % Tolerance value at which the loop stops
DELTA=1.0; % Starting value of DELTA
% Loop over the number of Wells Groups
```

```
j=1; % (While) Loop Counter
while abs(DELTA)>= TOL && j<=50
Q_term = Qw/(4*pi()*T);
DELTA = Q_term*Four_Wells_BRR(Sy,T,w,a,yw,delt,dton,t,x,y,j);
BRR = BRR+DELTA;
j=j+1;
end
if (j>50)
disp('convergence not reached')
end
%
function [W_Fun]=Four_Wells_BRR(Sy,T,w,a,yw,delt,dton,t,x,y,j)
W_Fun= 0.;
% Wells x Coordinate
xw1 = (2*j-1)*w-a;
xw2 = (2*j-1)*w+a;
% xw1 = 2*(j-1)*w+a;
% xw2 = 2*j*w   -a;
%
% Wells Distance From Observation Well
r1= sqrt((x-xw1)^2+(y-yw)^2);
r2= sqrt((x-xw2)^2+(y-yw)^2);
% Image Wells
r3= sqrt((x+xw1)^2+(y-yw)^2);
r4= sqrt((x+xw2)^2+(y-yw)^2);
%
% Cycle Effect
frac =t/delt;
int_t =fix(frac);
rest_t =t-int_t*delt;
n =int_t+1; % Number of complete Operation Cycles
% Loop over the number of Operation Cycles
for i=1:n-1
% For a Number of Complete Cycles
t1=t-(i-1)*delt;
t2=t1-dton;
% Stream Constant Head and infinite Aquifer Effects
u1=(Sy/(4*T))*(r1^2/t1);
u2=(Sy/(4*T))*(r2^2/t1);
% Image Wells
u3=(Sy/(4*T))*(r3^2/t1);
u4=(Sy/(4*T))*(r4^2/t1);
% Well Function
wu1=expint(u1);
wu2=expint(u2);
```

```
% Image Well Function
wu3=expint(u3);
wu4=expint(u4);
% Operating And Image Wells
wo = wu1-wu2-wu3+wu4;
% Imaginary Compensation Wells (for the continuous pumping)
u1_I=(Sy/(4*T))*(r1^2/t2);
u2_I=(Sy/(4*T))*(r2^2/t2);
% Image Wells
u3_I=(Sy/(4*T))*(r3^2/t2);
u4_I=(Sy/(4*T))*(r4^2/t2);
% Well Function
wu1_I=expint(u1_I);
wu2_I=expint(u2_I);
% Image Well Function
wu3_I=expint(u3_I);
wu4_I=expint(u4_I);
% Imaginary And Image Wells Well Function:
wI = wu1_I-wu2_I-wu3_I+wu4_I;
% Total Well Function
W_Fun =W_Fun+wo-wI;
end
%
if rest_t>0 && rest_t <= dton % During Operation Time
% Stream Constant Head and Infinite Aquifer Effects
u1=(Sy/(4*T))*(r1^2/rest_t);
u2=(Sy/(4*T))*(r2^2/rest_t);
% Image Wells
u3=(Sy/(4*T))*(r3^2/rest_t);
u4=(Sy/(4*T))*(r4^2/rest_t);
% Well Function
wu1=expint(u1);
wu2=expint(u2);
% Image Well Function
wu3=expint(u3);
wu4=expint(u4);
% Operating And Image Wells
W_Fun = W_Fun+wu1-wu2-wu3+wu4;
end
if rest_t>0 && rest_t > dton % During time when operation stops
t1=rest_t;
t2=t1-dton;
% Stream Constant Head and infinite Aquifer Effects
u1=(Sy/(4*T))*(r1^2/t1);
u2=(Sy/(4*T))*(r2^2/t1);
```

```
% Image Wells
u3=(Sy/(4*T))*(r3^2/t1);
u4=(Sy/(4*T))*(r4^2/t1);
% Well Function
wu1=expint(u1);
wu2=expint(u2);
% Image Well Function
wu3=expint(u3);
wu4=expint(u4);
% Operating And Image Wells
wo = wu1-wu2-wu3+wu4;
% Imaginary Compensation Wells (for continous pumping)
u1_I=(Sy/(4*T))*(r1^2/t2);
u2_I=(Sy/(4*T))*(r2^2/t2);
% Image Wells
u3_I=(Sy/(4*T))*(r3^2/t2);
u4_I=(Sy/(4*T))*(r4^2/t2);
% Well Function
wu1_I=expint(u1_I);
wu2_I=expint(u2_I);
% Image Well Function
wu3_I=expint(u3_I);
wu4_I=expint(u4_I);
% Imaginary And Image Wells
wI = wu1_I-wu2_I-wu3_I+wu4_I;
% Total Well Function
W_Fun =W_Fun+wo-wI;
end
%%%%%%%%%%%%%%%%%%%%%%%%%%%%%%%%%%%%%%%%%%%%%%%%%%%%%%%%%%%%%
%%%%%%%%%
```

Matlab code: **BCYC_Drawdown_Time**

This code calculates and plots the head time series for t ∈ $(0, t_{fin})$ for a number of wells operating cyclically or continuously in an aquifer bounded laterally by a either a constant-head recharge boundary and a no-flow boundary, or two constant-head recharge boundaries. One of the recharge boundaries represents a stream located on the $x=x_s$ straight line. The second boundary, whether recharge or no-flow, is located on the *y*-axis (the $x=0$ straight line).

The input file **Aquifer.Parms** includes simulation time parameters and aquifer parameters:

a. time parameters

   - print-out time t_{fin};
   - cyclical time step of simulation $\Delta t=\Delta t_{on}+\Delta t_{off}$ (e.g., 1 year)

b. aquifer parameters

- initial hydraulic head (aquifer's saturated thickness) h_0
- hydraulic conductivity K;
- storativity S;
- stream boundary location $x=x_s$;
- secondary boundary location $x=x_b$ ($x_s<x_b$)

The input file **Monitoring.Wells.dat** includes monitoring well data.

a. total number of monitoring wells n_{mw}
b. for each monitoring well, each of the following lines provides well location coordinates x_w and y_w.

The input file **Wells.dat** includes wells data:

a. total number of operating wells n_w
b. for each operating well, each of the following lines provide:

- the pumping rate (Q_w);
- the time at which well operation starts (t_{st}),
- the total operation period (Δt_{on})[5]
- well location x_w and y_w.

Examples:
Aquifer.Parms

0.2	365.	30.	86.4	0.2	0.	2000.
t_{fin} (d)	$\Delta t=\Delta t_{on}+\Delta t_{off}$ (d)	b(m)	K (m/d)	S (/)	x_s (m)	x_b (m)

Monitoring.Wells.dat

2 (n_{mw})	
1000.	1000.
750.	1250.
x_{mw}(m)	Y_{mw}(m)

Wells.dat

Description:	2 wells operating cyclically			
2 (n_w)				
+2500.	+2500.	+2500.	+2500.	+2500.
+2500.	+2500.	+2500.	+2500.	+2500.
Q_w(m^3/d)	t_{st}(d)	Δt_{on}(d)	x_w (m)	y_w (m)

[5] If this time is set equal to Δt then the well operates continuously ($\Delta t_{off} = 0$). Otherwise the well operates cyclically.

```
function BCYC_Drawdown_Time
   clc
   clear all
   % Reading Data Files
   % 1-Aquifer Parameters
   fid1 = fopen ('Aquifer.parms','r');
   Temp = fscanf(fid1,'%f',[1,2]);
   tfin = Temp(1); % Final Time of Simulation (D)
   delt = Temp(2); % Cycle Time = One Year (d)
   %
   Temp = fscanf(fid1,'%f %f %f %f %f',[1,5]);
   ho      = Temp(1); % Aquifer Thickness (Initial Head) (L)
   K       = Temp(2); % Hydraulic Conductivity (L/T)
   Sy      = Temp(3); % Aquifer Storativity (/)
   x_boundary = Temp(4); % Boundary Location (L)
   x_stream   = Temp(5); % Stream Location (L)
   fclose(fid1);
   %
   % 2-Wells Data file
   fid2 = fopen ('Well.dat','r');
   now = fscanf( fid2,'%f',[1,1]); % Well Number
   for iw = 1:now
       Temp = fscanf( fid2,'%f %f %f %f %f',[1,5]);
       Qw(iw)   = Temp(1);  % Well Pumping Rate (L^3/T)
       tst(iw)  = Temp(2);  % Well Operation Starting Time (T)
       dton(iw) = Temp(3);  % Pumping Period (T)
       xw(iw)   = Temp(4);  % Well Location X Coordinate (L)
       yw(iw)   = Temp(5);  % Well Location Y Coordinate (L)
   end
   fclose(fid2);
   %
   % 3- Monitoring Wells data file
   fid3 = fopen ('Monitoring.Wells.dat','r');
   Temp = fscanf(fid3,'%f',[1,1]);
   nmw  = Temp(1);% Monitoring Wells Number
   for m = 1:nmw
   Temp  = fscanf(fid3,'%f %f',[1,2]);
   xm(m) = Temp(1);
   ym(m) = Temp(2);
   end
   fclose(fid3);
   %%
   % Calculations
   % 1- General
```

```
w = x_stream - x_boundary; % Aquifer Width (L)
T = K*ho; % Aquifer Transmissivity (L^2/T)
to = 0;
dt = 1.;
t = to:dt:tfin; % time matrix
nt = length(t); % number of time steps (/)
% 2- Computing Drawdown Distribution for number of operating wells
BR_sum = zeros(nt,nmw);% Initial Drawdown
BN_sum = zeros(nt,nmw);% Initial Drawdown
fid4 = fopen('Drawdown_VS_Time.dat','w');
for j = 1:nmw
for i= 1:nt
BR_sum(i,j) = ho;
BN_sum(i,j) = ho;
for iw = 1:now
xmw = xm(j)-x_boundary;
ymw = ym(j);
xow = xw(iw)-x_boundary;
yow = yw(iw);
Qow = Qw(iw);
dtonw = dton(iw);
dt = t(i)-tst(iw);
if (dt>0.)
BR_sum(i,j)  =  BR_sum(i,j)+BRECH_CYC_THEIS(Sy,T,w,xow,yow,Qow,
delt,dtonw,dt,xmw,ymw);
BN_sum(i,j) = BN_sum(i,j)+BNOFLOW_CYC_THEIS(Sy,T,w,xow,yow,Qow,
delt,dtonw,dt,xmw,ymw);
end
end
temp = [t(i), xm(j),ym(j), BR_sum(i,j),BN_sum(i,j)];
fprintf(fid4,'%15.6E %15.6E %15.6E %15.6E %15.6E\n',temp);
end
[HR_MAX, iMAX] = max(BR_sum(:,j));
[HR_MIN, iMIN] = min(BR_sum(:,j));
[HN_MAX, iMAX] = max(BN_sum(:,j));
[HN_MIN, iMIN] = min(BN_sum(:,j));
temp = [HR_MAX,HN_MAX];
[H_MAX, iMAX] = max(temp);
temp = [HR_MIN,HN_MIN];
[H_MIN, iMIN] = min(temp);
% Fiqure
figure
%  plot(t,BR_sum(:,j),'-r',t,ho,'-.k')
plot(t,BR_sum(:,j),'-r')
hold;
```

```
plot(t,BN_sum(:,j),'-b')
hold;
% plot(t,ho,'-.k')
% hold
title(['Hydraulic Head Time Series - Monitoring Well: ',num2str(j)]);
xlabel('Time (T)');
ylabel('Hydraulic Head (L)');
xlim ([0 tfin+10]);
ylim ([H_MIN-0.1 H_MAX+0.1]);
legend('Rech./Rech. Bs','No-Flow/Rech. Bs')
hold off;
end
fclose(fid4);
%
%%%%%%%%%%%%%%%%%%%%%%%%%%%%%%%%%%%%%%%%%%%%%%%%%%%%%%%
%%%%%%%%
function [BRNF] = BNOFLOW_CYC_THEIS(Sy,T,w,xw,yw,Qw,delt,dton,t,x,y)
BRNF=0.; % Initial Drawdown in a Bounded Aquifer
a=w-xw; % Operating-well/Stream Distance
TOL=1.e-9; % Tolerance value at which the loop stops
DELTA=1.0; % Starting value of DELTA
% Loop over the number of Wells Groups
j=1; % (While) Loop Counter
while abs(DELTA)>= TOL && j<=50
    sign = (-1)^(j+1);
    Q_term = Qw/(4*pi()*T);
DELTA = sign*Q_term*Four_Wells_BRNF(Sy,T,w,a,yw,delt,dton,t,x,
y,j);
BRNF = BRNF+DELTA;
    j=j+1;
end
if (j>50)
    disp('convergence not reached')
end
%
function [W_Fun]= Four_Wells_BRNF(Sy,T,w,a,yw,delt,dton,t,x,y,j)
W_Fun= 0.;
% Wells x Coordinate
xw1 = (2*j-1)*w-a;
xw2 = (2*j-1)*w+a;
% xw1 = 2*(j-1)*w+a;
% xw2 = 2*j*w  -a;
%
% Wells Distance From Observation Well
r1= sqrt((x-xw1)^2+(y-yw)^2);
```

```
r2= sqrt((x-xw2)^2+(y-yw)^2);
% Image Wells
r3= sqrt((x+xw1)^2+(y-yw)^2);
r4= sqrt((x+xw2)^2+(y-yw)^2);
%
% Cycle Effect
frac =t/delt;
int_t =fix(frac);
rest_t =t-int_t*delt;
n =int_t+1; % Number of complete Operation Cycles
% Loop over the number of Operation Cycles
for i=1:n-1
% For a Number of Complete Cycles
t1=t-(i-1)*delt;
t2=t1-dton;
% Stream Constant Head and infinite Aquifer Effects
u1=(Sy/(4*T))*(r1^2/t1);
u2=(Sy/(4*T))*(r2^2/t1);
% Image Wells
u3=(Sy/(4*T))*(r3^2/t1);
u4=(Sy/(4*T))*(r4^2/t1);
% Well Function
wu1=expint(u1);
wu2=expint(u2);
% Image Well Function
wu3=expint(u3);
wu4=expint(u4);
% Operating And Image Wells
wo = wu1-wu2+wu3-wu4;
% wo = wu1+wu2-wu3-wu4;
% Imaginary Compensation Wells (for the continuous pumping)
u1_I=(Sy/(4*T))*(r1^2/t2);
u2_I=(Sy/(4*T))*(r2^2/t2);
% Image Wells
u3_I=(Sy/(4*T))*(r3^2/t2);
u4_I=(Sy/(4*T))*(r4^2/t2);
% Well Function
wu1_I=expint(u1_I);
wu2_I=expint(u2_I);
% Image Well Function
wu3_I=expint(u3_I);
wu4_I=expint(u4_I);
wI = wu1_I-wu2_I+wu3_I-wu4_I;
% wI = wu1_I+wu2_I-wu3_I-wu4_I;
% Total Well Function
```

```
W_Fun =W_Fun+wo-wI;
end
%
if rest_t>0 && rest_t <= dton % During Operation Time
% Stream Constant Head and Infinite Aquifer Effects
u1=(Sy/(4*T))*(r1^2/rest_t);
u2=(Sy/(4*T))*(r2^2/rest_t);
% Image Wells
u3=(Sy/(4*T))*(r3^2/rest_t);
u4=(Sy/(4*T))*(r4^2/rest_t);
% Well Function
wu1=expint(u1);
wu2=expint(u2);
% Image Well Function
wu3=expint(u3);
wu4=expint(u4);
% Operating And Image Wells
W_Fun = W_Fun+wu1-wu2+wu3-wu4;
% W_Fun = W_Fun+wu1+wu2-wu3-wu4;
end
if rest_t>0 && rest_t > dton % During time when operation stops
t1=rest_t;
t2=t1-dton;
% Stream Constant Head and infinite Aquifer Effects
u1=(Sy/(4*T))*(r1^2/t1);
u2=(Sy/(4*T))*(r2^2/t1);
% Image Wells
u3=(Sy/(4*T))*(r3^2/t1);
u4=(Sy/(4*T))*(r4^2/t1);
% Well Function
wu1=expint(u1);
wu2=expint(u2);
% Image Well Function
wu3=expint(u3);
wu4=expint(u4);
% Operating And Image Wells
wo = wu1-wu2+wu3-wu4;
% wo = wu1+wu2-wu3-wu4;
% Imaginary Compensation Wells (for continous pumping)
u1_I=(Sy/(4*T))*(r1^2/t2);
u2_I=(Sy/(4*T))*(r2^2/t2);
% Image Wells
u3_I=(Sy/(4*T))*(r3^2/t2);
u4_I=(Sy/(4*T))*(r4^2/t2);
% Well Function
```

```
wu1_I=expint(u1_I);
wu2_I=expint(u2_I);
% Image Well Function
wu3_I=expint(u3_I);
wu4_I=expint(u4_I);
% Imaginary And Image Wells
wI = wu1_I-wu2_I+wu3_I-wu4_I;
% wI = wu1_I+wu2_I-wu3_I-wu4_I;
% Total Well Function
W_Fun =W_Fun+wo-wI;
end
%%%%%%%%%%%%%%%%%%%%%%%%%%%%%%%%%%%%%%%%%%%%%%%%%%%%%%%%%%%%%%
%%%%%%%%%
function [BRR] = BRECH_CYC_THEIS(Sy,T,w,xw,yw,Qw,delt,dton,t,x,y)
a=w-xw; % Operating-well/Stream Distance
BRR=0.; % Initial Drawdown in a Bounded Aquifer
TOL=1.e-9; % Tolerance value at which the loop stops
DELTA=1.0; % Starting value of DELTA
% Loop over the number of Wells Groups
j=1; % (While) Loop Counter
while abs(DELTA)>= TOL && j<=50
Q_term = Qw/(4*pi()*T);
DELTA = Q_term*Four_Wells_BRR(Sy,T,w,a,yw,delt,dton,t,x,y,j);
BRR = BRR+DELTA;
j=j+1;
end
if (j>50)
disp('convergence not reached')
end
%
function [W_Fun]=Four_Wells_BRR(Sy,T,w,a,yw,delt,dton,t,x,y,j)
W_Fun= 0.;
% Wells x Coordinate
xw1 = (2*j-1)*w-a;
xw2 = (2*j-1)*w+a;
% xw1 = 2*(j-1)*w+a;
% xw2 = 2*j*w   -a;
%
% Wells Distance From Observation Well
r1= sqrt((x-xw1)^2+(y-yw)^2);
r2= sqrt((x-xw2)^2+(y-yw)^2);
% Image Wells
r3= sqrt((x+xw1)^2+(y-yw)^2);
r4= sqrt((x+xw2)^2+(y-yw)^2);
%
```

```
% Cycle Effect
frac =t/delt;
int_t =fix(frac);
rest_t =t-int_t*delt;
n =int_t+1; % Number of complete Operation Cycles
% Loop over the number of Operation Cycles
for i=1:n-1
% For a Number of Complete Cycles
t1=t-(i-1)*delt;
t2=t1-dton;
% Stream Constant Head and infinite Aquifer Effects
u1=(Sy/(4*T))*(r1^2/t1);
u2=(Sy/(4*T))*(r2^2/t1);
% Image Wells
u3=(Sy/(4*T))*(r3^2/t1);
u4=(Sy/(4*T))*(r4^2/t1);
% Well Function
wu1=expint(u1);
wu2=expint(u2);
% Image Well Function
wu3=expint(u3);
wu4=expint(u4);
% Operating And Image Wells
wo = wu1-wu2-wu3+wu4;
% Imaginary Compensation Wells (for the continuous pumping)
u1_I=(Sy/(4*T))*(r1^2/t2);
u2_I=(Sy/(4*T))*(r2^2/t2);
% Image Wells
u3_I=(Sy/(4*T))*(r3^2/t2);
u4_I=(Sy/(4*T))*(r4^2/t2);
% Well Function
wu1_I=expint(u1_I);
wu2_I=expint(u2_I);
% Image Well Function
wu3_I=expint(u3_I);
wu4_I=expint(u4_I);
% Imaginary And Image Wells Well Function:
wI = wu1_I-wu2_I-wu3_I+wu4_I;
% Total Well Function
W_Fun =W_Fun+wo-wI;
end
%
if rest_t>0 && rest_t <= dton % During Operation Time
% Stream Constant Head and Infinite Aquifer Effects
u1=(Sy/(4*T))*(r1^2/rest_t);
```

```
u2=(Sy/(4*T))*(r2^2/rest_t);
% Image Wells
u3=(Sy/(4*T))*(r3^2/rest_t);
u4=(Sy/(4*T))*(r4^2/rest_t);
% Well Function
wu1=expint(u1);
wu2=expint(u2);
% Image Well Function
wu3=expint(u3);
wu4=expint(u4);
% Operating And Image Wells
W_Fun = W_Fun+wu1-wu2-wu3+wu4;
end
if rest_t>0 && rest_t > dton % During time when operation stops
t1=rest_t;
t2=t1-dton;
% Stream Constant Head and infinite Aquifer Effects
u1=(Sy/(4*T))*(r1^2/t1);
u2=(Sy/(4*T))*(r2^2/t1);
% Image Wells
u3=(Sy/(4*T))*(r3^2/t1);
u4=(Sy/(4*T))*(r4^2/t1);
% Well Function
wu1=expint(u1);
wu2=expint(u2);
% Image Well Function
wu3=expint(u3);
wu4=expint(u4);
% Operating And Image Wells
wo = wu1-wu2-wu3+wu4;
% Imaginary Compensation Wells (for continous pumping)
u1_I=(Sy/(4*T))*(r1^2/t2);
u2_I=(Sy/(4*T))*(r2^2/t2);
% Image Wells
u3_I=(Sy/(4*T))*(r3^2/t2);
u4_I=(Sy/(4*T))*(r4^2/t2);
% Well Function
wu1_I=expint(u1_I);
wu2_I=expint(u2_I);
% Image Well Function
wu3_I=expint(u3_I);
wu4_I=expint(u4_I);
% Imaginary And Image Wells
wI = wu1_I-wu2_I-wu3_I+wu4_I;
% Total Well Function
```

```
W_Fun =W_Fun+wo-wI;
end
%%%%%%%%%%%%%%%%%%%%%%%%%%%%%%%%%%%%%%%%%%%%%%%%%%%%%%%%%%%%
%%%%%%%%%
```

Appendix 4

Matlab Code: **BGlover**

This code calculates and plots the stream depletion rate, Q_r, and stream depletion volume, V_r, vs. Time t for a number of operating wells operating either cyclically or continuously in a bounded aquifer comprised between a recharge stream and another boundary, which can be either constant-head or impermeable. The stream is parallel to the x axis at $y=y_{stream}$. The second boundary is located at $y=y_{boundary}< y_{stream}$.

The input file **Aquifer.dat** includes simulation time parameters and aquifer parameters:

a. time parameters
   - final simulation time t_{fin};
   - cyclical time step of simulation $\Delta t=\Delta t_{on}+\Delta t_{off}$ (e.g., 1 year)
b. the aquifer parameters
   - aquifer's saturated thickness b;
   - hydraulic conductivity K;
   - storativity S;
   - aquifer boundary coordinates $y_{boundary}$ and y_{stream}.

The input file **Wells.txt** includes wells data:

a. total number of operating wells n_w
b. for each operating well, each of the following lines provide:
   - the pumping rate (Q_w);
   - the time at which well operation starts (t_{st}),
   - the total operation period (Δt_{on})[6]
   - well location x_w and y_w.

[6] If this time is set equal to Δt then the well operates continuously ($\Delta t_{off} = 0$). Otherwise the well operates cyclically.

Examples:

Aquifer.dat

3650.	365.	30.	86.4	0.2	0.	4000.
t_{fin}(d)	Δt=Δt_{on}+Δt_{off} (d)	b(m)	K (m/d)	S (/)	$y_{boundary}$	$y_{boundary}$

Wells.dat

Description:	1 well operating cyclically for 5 years and then shut off			
2 (n_w)				
+3000.	90.	120.	500.	2000.
−3000.	1915.	120.	500.	2000.
Q_w(m^3/d)	t_{st}(d)	Δt_{on}(d)	x_w(m)	y_w (m)
Description:	1 well injecting and extracting cyclically (net pumped volume is zero)			
2 (n_w)				
+3000.	90.	120.	1000.	500.
−1479.3877	210.	245.	1000.	500.
Q_w(m^3/d)	t_{st}(d)	Δt_{on}(d)	x_w(m)	y_w (m)

```
function BGLOVER
   clear all
   clc
   %
   % Reading data
   % 1 - General data
   fid1 = fopen ('Aquifer.dat','r');
   Temp = fscanf(fid1,'%f %f %f %f %f %f %f',[1,7]);
   % tfin = Final Time of Simulation (d)
   tfin = Temp(1);
   % delt = dton+dtoff [e.g. 365] (d)
   delt = Temp(2);
   % b = Thickness of the Aquifer (m)
   b = Temp(3);
   % K = Hydraulic Conductivity (m/D)
   K = Temp(4);
   % Sy = storativity (/)
   Sy = Temp(5);
   % Boundary Location (m)
   y_boundary = Temp(6);
   % Stream Location (m)
   y_stream = Temp(7);
   %
```

```
fclose(fid1);
%
% 2 - Well Field data
fid2 = fopen ('Wells.dat','r');
Temp = fscanf( fid2,'%f',[1,1]);
now = Temp(1);
for ow = 1:now
    Temp = fscanf(fid2,'%f %f %f %f %f',[1,5]);
    Qw(ow)  = Temp(1);
    tst(ow) = Temp(2);
    dton(ow) = Temp(3);
    xw(ow)  = Temp(4);
    yw(ow)  = Temp(5) ;
end
fclose(fid2);
%
% General calculations
%
T = K*b;
to = 0;
w = y_stream-y_boundary;
index = (tfin/delt)-1;
dt = 1;
t = to:dt:tfin; % time matrix
nt = length(t); % number of time steps(/)
%
% Off time for operating wells
for ow = 1:now
    dtoff(ow) = delt-dton(ow);
end
%
% Calculating stream depletion rate (Qr) and stream depletion volume (Vr)
Qr_Nsum = zeros(nt,1);
Vr_Nsum = zeros(nt,1);
Qr_Rsum = zeros(nt,1);
Vr_Rsum = zeros(nt,1);
%
for i= 1:nt
for iw= 1:now
a = y_stream -yw(iw);
if (t(i)-tst(iw)>=0.)
%
Qr_Nsum(i)=Qr_Nsum(i)+     BNrate_sol_CYC(t(i)-tst(iw),dton(iw),
delt,Qw(iw),T,Sy,a,w);
```

```
Vr_Nsum(i)=Vr_Nsum(i)+    BNvol_sol_CYC(t(i)-tst(iw),delt,dton
(iw),Qw(iw),T,Sy,a,w);
%
Qr_Rsum(i,1)=Qr_Rsum(i)+ BRrate_sol_CYC(t(i)-tst(iw),dton(iw),
delt,Qw(iw),T,Sy,a,w);
Vr_Rsum(i,1)=Vr_Rsum(i)+  BRvol_sol_CYC(t(i)-tst(iw),delt,dton
(iw),Qw(iw),T,Sy,a,w);
%
end
end
end
%
Qr_NMAX = max(Qr_Nsum);
Qr_NMIN = min(Qr_Nsum);
if Qr_NMAX>0.
Qr_NMAX=Qr_NMAX+100.;
else
Qr_NMAX=0;
end
if Qr_NMIN<0.
Qr_NMIN=Qr_NMIN-100.;
else
Qr_NMIN=0;
end
Vr_NMAX = max(Vr_Nsum);
Vr_NMIN = min(Vr_Nsum);
%
% Fiqures
figure
plot(t,Qr_Nsum,'-r')
title('Aquifer delimited by a no-flow and a recharge boundary');
xlabel('Time (d)');
ylabel('Total Q_r (m^3/d)');
xlim ([0 tfin]);
ylim ([Qr_NMIN Qr_NMAX]);
%legend('Qr')
%
figure
plot(t,Vr_Nsum)
title('Aquifer delimited by a no-flow and a recharge boundary');
xlabel('Time (d)');
ylabel('Total V_r (m^3)');
xlim ([0 tfin]);
ylim ([Vr_NMIN Vr_NMAX]);
%legend('Vr')
```

```
%
Qr_RMAX = max(Qr_Rsum);
Qr_RMIN = min(Qr_Rsum);
if Qr_RMAX>0.
Qr_RMAX=Qr_RMAX+100.;
else
Qr_RMAX=0;
end
if Qr_RMIN<0.
Qr_RMIN=Qr_RMIN-100.;
else
Qr_RMIN=0.;
end
Vr_RMAX = max(Vr_Rsum);
Vr_RMIN = min(Vr_Rsum);
% Fiqures
figure
plot(t,Qr_Rsum,'-r')
title('Aquifer delimited by two recharge boundaries');
xlabel('Time (d)');
ylabel('Total Q_r (m^3/d)');
xlim ([0 tfin]);
ylim ([Qr_RMIN Qr_RMAX]);
%legend('Qr')
%
figure
plot(t,Vr_Rsum)
title('Aquifer delimited by two recharge boundaries');
xlabel('Time (d)');
ylabel('Total V_r (m^3)');
xlim ([0 tfin]);
ylim ([Vr_RMIN Vr_RMAX]);
%legend('Vr')
%
% Output Results
fid3 = fopen('Time.Qr.Vr.dat','w');
for i= 1:nt
temp = [t(i),Qr_Nsum(i),Vr_Nsum(i),Qr_Rsum(i),Vr_Rsum(i)];
fprintf(fid3,'%15.6E %15.6E %15.6E %15.6E %15.6E\n',temp);
end
%%%%%%%%%%%%%%%%%%%%%%%%%%%%%%%%%%%%%%%%%%%%%%%%%%%%%%%%%%%%%%%%%%%%%
%%%%%%%%%
function [Qr] = BNrate_sol_CYC(t,dton,delt,Q,T,S,a,w)
% Calculating Qr (stream depletion rate)
Qr = 0.;
```

```
frac= t/delt;
int_t=fix(frac);
rest_t=t-int_t*delt;
n = int_t+1;
for i=1:n-1
t1=t-(i-1)*delt;
t2=t1-dton;
Qr =Qr+Q*(BNQRatio(T,S,a,w,t1)-BNQRatio(T,S,a,w,t2));
end
if rest_t>0 && rest_t <= dton
Qr =Qr+Q*BNQRatio(T,S,a,w,rest_t) ;
end
if rest_t>0 && rest_t > dton
t1=rest_t;
t2=t1-dton;
Qr =Qr+Q*(BNQRatio(T,S,a,w,t1)-BNQRatio(T,S,a,w,t2));
end
%%%%%%%%%%%%%%%%%%%%%%%%%%%%%%%%%%%%%%%%%%%%%%%%%%%%%%%%%%%%%%%%%%%%%%%%%%%%%%
%%%%%%%%%
function [Vr] = BNvol_sol_CYC(t,delt,dton,Q,T,S,a,w)
% Calculating Vr (depletion volume)
Vr = 0.;
frac=t/delt;
int_t=fix(frac);
rest_t=t-int_t*delt;
n = int_t+1;
for i=1:n-1
t1=t-(i-1)*delt;
t2=t1-dton;
Vr =Vr+Q*(t1*BNVRatio(T,S,a,w,t1)-t2*BNVRatio(T,S,a,w,t2));
end
if rest_t>0 && rest_t <= dton
Vr=Vr+Q*rest_t*BNVRatio(T,S,a,w,rest_t);
end
if rest_t>0 && rest_t > dton
t1=rest_t;
t2=t1-dton;
Vr=Vr+Q*(t1*BNVRatio(T,S,a,w,t1)-t2*BNVRatio(T,S,a,w,t2));
end
%%%%%%%%%%%%%%%%%%%%%%%%%%%%%%%%%%%%%%%%%%%%%%%%%%%%%%%%%%%%%%%%%%%%%%%%%%%%%%
%%%%%%%%%
function [Qr] = BRrate_sol_CYC(t,dton,delt,Q,T,S,a,w)
% Calculating Qr (stream depletion rate)
Qr = 0.;
frac= t/delt;
```

```
int_t=fix(frac);
rest_t=t-int_t*delt;
n = int_t+1;
for i=1:n-1
t1=t-(i-1)*delt;
t2=t1-dton;
Qr =Qr+Q*(BRQRatio(T,S,a,w,t1)-BRQRatio(T,S,a,w,t2));
end
if rest_t>0 && rest_t <= dton
Qr =Qr+Q*BRQRatio(T,S,a,w,rest_t) ;
end
if rest_t>0 && rest_t > dton
t1=rest_t;
t2=t1-dton;
Qr =Qr+Q*(BRQRatio(T,S,a,w,t1)-BRQRatio(T,S,a,w,t2));
end
%%%%%%%%%%%%%%%%%%%%%%%%%%%%%%%%%%%%%%%%%%%%%%%%%%%%%%%
%%%%%%%%%
function [Vr] = BRvol_sol_CYC(t,delt,dton,Q,T,S,a,w)
% Calculating Vr (depletion volume)
Vr = 0.;
frac=t/delt;
int_t=fix(frac);
rest_t=t-int_t*delt;
n = int_t+1;
for i=1:n-1
t1=t-(i-1)*delt;
t2=t1-dton;
Vr =Vr+Q*(t1*BRVRatio(T,S,a,w,t1)-t2*BRVRatio(T,S,a,w,t2));
end
if rest_t>0 && rest_t <= dton
Vr=Vr+Q*rest_t*BRVRatio(T,S,a,w,rest_t);
end
if rest_t>0 && rest_t > dton
t1=rest_t;
t2=t1-dton;
Vr=Vr+Q*(t1*BRVRatio(T,S,a,w,t1)-t2*BRVRatio(T,S,a,w,t2));
end
%%%%%%%%%%%%%%%%%%%%%%%%%%%%%%%%%%%%%%%%%%%%%%%%%%%%%%%
%%%%%%%%%
function BNRQ=BNQRatio(T,S,a,w,t)
BNRQ = erfc(sqrt(S/4/T*a^2/t))+FBNQRatio(T,S,a,w,t);
%%%%%%%%%%%%%%%%%%%%%%%%%%%%%%%%%%%%%%%%%%%%%%%%%%%%%%%
%%%%%%%%%
function FQR=FBNQRatio(T,S,a,w,t)
```

```
TOL=1.e-9; i=1; DELTA=1.0; FQR=0;
while abs(DELTA) >=TOL && i<=30
DELTA=(-1)^(i+1)*A(T,S,a,w,t,i);
FQR=FQR+DELTA;
i=i+1;
end
%%%%%%%%%%%%%%%%%%%%%%%%%%%%%%%%%%%%%%%%%%%%%%%%%%%%%%%%%%%%%%%
%%%%%%%%%
function BGVR=BNVRatio(T,S,a,w,t)
TOL=1.e-9; i=0; DELTA=1.0; BGVR=0;
while abs(DELTA) >=TOL && i<=30
DELTA=(-1)^i*(C(T,S,a,w,t,i)+D(T,S,a,w,t,i));
BGVR=BGVR+DELTA;
i=i+1;
end
%%%%%%%%%%%%%%%%%%%%%%%%%%%%%%%%%%%%%%%%%%%%%%%%%%%%%%%%%%%%%%%
%%%%%%%%%
function BRRQ=BRQRatio(T,S,a,w,t)
BRRQ = erfc(sqrt(S/4/T*a^2/t))+FBRQRatio(T,S,a,w,t);
%%%%%%%%%%%%%%%%%%%%%%%%%%%%%%%%%%%%%%%%%%%%%%%%%%%%%%%%%%%%%%%
%%%%%%%%%
function FQR=FBRQRatio(T,S,a,w,t)
TOL=1.e-9; i=1; DELTA=1.0; FQR=0;
while abs(DELTA) >=TOL && i<=50
DELTA=-A(T,S,a,w,t,i);
FQR=FQR+DELTA;
i=i+1;
end
%%%%%%%%%%%%%%%%%%%%%%%%%%%%%%%%%%%%%%%%%%%%%%%%%%%%%%%%%%%%%%%
%%%%%%%%%
function BGVR=BRVRatio(T,S,a,w,t)
TOL=1.e-9; i=0; DELTA=1.0; BGVR=0;
while abs(DELTA) >=TOL && i<=30
DELTA=C(T,S,a,w,t,i)-D(T,S,a,w,t,i);
BGVR=BGVR+DELTA;
i=i+1;
end
%%%%%%%%%%%%%%%%%%%%%%%%%%%%%%%%%%%%%%%%%%%%%%%%%%%%%%%%%%%%%%%
%%%%%%%%%
function Ai=A(T,S,a,w,t,i)
Ai=erfc(sqrt(S/4/T*(2*w*i-a)^2/t))-erfc(sqrt(S/4/T*(2*w*i+a)^2/t));
%%%%%%%%%%%%%%%%%%%%%%%%%%%%%%%%%%%%%%%%%%%%%%%%%%%%%%%%%%%%%%%
%%%%%%%%%
function Ci=C(T,S,a,w,t,i)
```

```
Ci=(1+S/2/T*(2*w*i+a)^2/t)*erfc(sqrt(S/4/T*(2*w*i+a)^2/t))-2/
sqrt(pi())*sqrt(S/4/T*(2*w*i+a)^2/t)*exp(-S/4/T*(2*w*i+a)^2/t);
%%%%%%%%%%%%%%%%%%%%%%%%%%%%%%%%%%%%%%%%%%%%%%%%%%%%%%%%%
%%%%%%%%%
function Di=D(T,S,a,w,t,i)
Di=(1+S/2/T*(2*w*(i+1)-a)^2/t)*erfc(sqrt(S/4/T*(2*w*(i+1)-a)
^2/t))-2/sqrt(pi())*sqrt(S/4/T*(2*w*(i+1)-a)^2/t)*exp(-S/4/T*
(2*w*(i+1)-a)^2/t);
%%%%%%%%%%%%%%%%%%%%%%%%%%%%%%%%%%%%%%%%%%%%%%%%%%%%%%%%%
%%%%%%%%%
```

Appendix 5

Matlab code: **LP_APR_BNR**

This code uses the linear programming (LP) algorithm available in MATLAB to minimize the absolute value of the stream depletion volume over a given time window (0, t_{fin}) due to a number of operating wells at given locations in an aquifer delimited between the stream and a no-flow boundary. Wells can be activated either cyclically or continuously. The pumping rates at candidate well locations constitute the decision variables of the optimization problem. The LP algorithm chooses whether a well is used or not, and at which rate it is operated. The algorithm also selects if a well is to be activated in injection ($Q>0$) or extraction ($Q<0$) mode.

Constraints are imposed on:

- Minimum ($Q_{min} \leq 0$) and a maximum pumping rates ($Q_{max} \geq 0$);
- Minimum (h_{min}) and a maximum (h_{max}) water levels at a number n_{mw} of monitoring wells and check times;
- available injection rate (≥ 0) Q_{rech}
- irrigation demand rate (≤ 0) Q_{dem}

Following is a description of the required input files.

The input file **Aquifer.Parms.dat** includes simulation time parameters and aquifer parameters:

a. the aquifer parameters

   – hydraulic conductivity K (L/T);
   – storativity $S(/)$;

b. time parameters

   – simulation time horizon t_{fin} (T);
   – cyclical time step of simulation $\Delta t=\Delta t_{on}+\Delta t_{off}$ (T);

c. hydraulic parameters and constraints

   – initial hydraulic head (saturated thickness) h_0(L);

- maximum hydraulic head (saturated thickness) h_{max} (L);
- minimum hydraulic head (saturated thickness) h_{min} (L);
- available injection rate (≥ 0) Q_{rech} (L^3/T);
- irrigation demand rate (≤ 0) Q_{dem} (L^3/T);
- no-flow Boundary Location $y_{no\text{-}flow}$ (L);
- stream boundary location y_{stream} ($>y_{no\text{-}flow}$) (L);

The input file **Operation.Wells.dat** includes pumping welldata:

a. Numbers of extraction and injection wells, n_{ew} and n_{iw}
b. For each extraction well provide:

- the well index (i=1,..,n_{ew});
- the well location coordinates, x_w and y_w (L);
- the time at which well operation starts t_{st} (T);
- operation period Δt_{on}[7](T);
- maximum extraction rate $Q_{w,min}$ (≤ 0) (L^3/T);
- maximum injection rate $Q_{w,max}$ (≥ 0) (L^3/T).

c. For each injection well provide:

- the well index (i=n_{ew}+1,..,n_{ew} +n_{iw});
- the well location coordinates, x_w and y_w (L);
- the time at which well operation starts t_{st} (T);
- operation period Δt_{on}[7] (T);
- maximum extraction rate $Q_{w,min}$ (≤ 0) (L^3/T);
- maximum injection rate $Q_{w,max}$ (≥ 0) (L^3/T).

The input file **Monitoring.Wells.dat** includes monitoring welldata:

a. Total number of monitoring wells n_{mw}.
b. For each operating well provide:

- well location x_{mw} and y_{mw} (L);
- number of hydraulic head check times $n_{t,ch}$;
- times at which heads checks are performed t_{ch}(T);

Following is a description of the output files produced by the code.
The output file **Results.dat** includes[8]:

a. Total numbers of activated (both extraction and injection) wells for the optimal well layout.
b. For each extraction well provide:

- The design pumping rate Q (L^3/T);
- The time at which well operation starts t_{st} (T);

[7] If this time is set equal to Δt then the well operates continuously ($\Delta t_{off} = 0$). Otherwise the well operates cyclically.

[8] This file is formatted for direct reading from the MATLAB code BCYC_Drawdown2D.

- Operation period Δt_{on}(T);
- The well location coordinates, x_w and y_w (L);

c. Absolute Value of the Net Cumulative Stream Recharge Volume (Vr') (L^3);
d. Net Cumulative Stream Recharge Volume (Vr) (L^3);
e. Cumulative Extraction Rate (L^3/T);
f. Cumulative Recharge (L^3/T).

The output file **Optimal.Scheme.dat** includes[9]:

a. The well location coordinates, x_w and y_w (L);
b. The design pumping rate Q (L^3/T).

for each well activated in the optimal layout.

Examples:

Aquifer.Parms.dat

86.4	0.2	1460.	365.	30.	30.5	29.5	70000.	−100000.	0.	6000.
K (m/d)	S (/)	t_{fin} (d)	$\Delta t=\Delta t_{on}+\Delta t_{off}$ (d)	h_0 (m)	h_{max} (m)	h_{min} (m)	Q_{ava} (m^3/d)	Q_{rech} (m^3/d)	$y_{no\text{-}flow}$ (m)	y_{stream} (m)

Operation.Wells.dat

42	48	n_{ew}	n_{iw}			
1	500.00	500.00	75.00	120.00	−5000.00	0.00
...	...	...	...	...	...	...
42	6500.00	5500.00	75.00	120.00	−5000.00	0.00
43	0.00	0.00	270.00	180.00	0.00	+5000.00
...	...	...	...	...	...	...
100	7000.00	5000.00	270.00	180.00	0.00	+5000.00
well index	x_w (m)	y_w (m)	t_{st} (d)	Δt_{on} (d)	Q_{min} (m^3/d)	Q_{max} (m^3/d)

Monitoring.Wells.dat

12	n_{mw}	
1	500.00	1000.00
	2	
		2020.
		2275.
...	...	...
	...	
		...
		...
12	6500.00	5000.00
	2	

(continued)

[9] This file is formatted for direct reading from SURFER in order to plot the optimal well layout.

		2020.
		2275.
well index	x_{mw} (m)	x_{mw} (m)
	$n_{t,ch}$	
		$t_{ch}(1)$ (d)
		...
		$t_{ch}(n_{t,ch})$ (d)

```
function LP_APR_BNR
  %
  % Optimization of Well Locations in an aquifer delimited between a
  no-flow
  % boundary and a stream. Wells operate cyclically. The objective is to
  % minimize the absolute value of the volume of water drawn from the
stream over a
  % prescribed time interval.
  % Constraints are imposed on:
  % - groundwater demand rates;
  % - recharge availability rates;
  % - maximum and minimum pumping rates;
  % - maximum and minimum water levels at given monitoring wells and
  times.
  %
  clear
  clc
  % Reading Data
  % 1- Simulation and Aquifer Parameters
  fid1 = fopen ('Aquifer.Parms.dat','r');
  Temp = fscanf(fid1,'%f %f %f %f %f %f %f %f %f %f %f',[1,11]);
  % Hydraulic Conductivity (m/d)
  K =Temp(1);
  % Storativity (/)
  S = Temp(2);
  % Simulation Time Horizon (T)
  tfin = Temp(3);
  % Cycle simulation time (T) (e.g. 365 d)
  delt = Temp(4);
  % Aquifer Initial Head (saturated thickness) (L)
  ho = Temp(5);
  % Aquifer Maximum Head (L) - constraint
```

```
hmax = Temp(6);
% Aquifer Minimum Head (L) - constraint
hmin = Temp(7);
% Available water for Injection (>0) (L^3/T)
Q_ava = Temp(8);
% Irrigation Extraction Demand (<0) (L^3/T)
Q_dem = Temp(9);
% No-Flow Boundary Location (L)
y_boundary = Temp(10);
% Stream Boundary Location (L)
y_stream = Temp(11);
fclose(fid1);
%
% 2- Operating Wells
fid2 = fopen ('Operation.Wells.dat','r');
Temp = fscanf (fid2 ,'%f',[1,2]);
new = Temp(1); % Number of Extraction Wells
niw = Temp(2); % Number of Injection Wells
now=new+niw;
sprintf('Total N. of Candidate Wells: %d',now)
%
xw = zeros(now);
yw = zeros(now);
tst = zeros(now);
dt = zeros(now);
Qw_min = zeros(now);
Qw_max = zeros(now);
%
for j = 1:new
    Temp = fscanf( fid2,'%f %f %f %f %f %f %f',[1,7]);
    xw(j)    = Temp(2);
    yw(j)    = Temp(3);
    tst(j)   = Temp(4); % Operation Starting Time (T)
    dt(j)    = Temp(5); % Operation Period (T)
    Qw_min(j) = Temp(6); % Minimum extraction rate (L^3/T)
    Qw_max(j) = Temp(7); % Maximum extraction rate (L^3/T)
if (Qw_max(j)>0.)
    disp('Warning: extraction rate may be positive.')
end
end
for j = new+1:now
    Temp = fscanf( fid2,'%f %f %f %f %f %f %f',[1,7]);
    xw(j)    = Temp(2);
    yw(j)    = Temp(3);
    tst(j)   = Temp(4); % Operation starting time (T)
```

```
    dt(j)    = Temp(5); % Operation Period (T)
    Qw_min(j) = Temp(6); % Minimum injection rate (L^3/T)
    Qw_max(j) = Temp(7); % Maximum injection rate (L^3/T)
if (Qw_min(j)<0.)
    disp('Warning: injection rate may be negative.')
end
end
fclose(fid2);
%
% 3- Monitoring Wells
fid3 = fopen ('Monitoring.Wells.dat','r');
Temp = fscanf( fid3,'%f',[1,1]);
nmw = Temp(1); % Number of Monitoring Wells
%
xm = zeros(nmw);
ym = zeros(nmw);
ntch = zeros(nmw);
tch= zeros(nmw,10);
%
for mw = 1:nmw % mw = monitoring wells
    Temp    = fscanf( fid3,'%f %f %f',[1,3]);
    xm(mw)   = Temp(2); % Monitoring Well x-coordinate (L)
    ym(mw)   = Temp(3); % Monitoring Well y-coordinate (L)
    Temp    = fscanf( fid3,'%f',[1,1]);
% Number of Head Check Times
ntch(mw)   = Temp(1);
% Read in Head Check Times
for ich=1:ntch(mw)
tch(mw,ich)= fscanf( fid3,'%f',[1,1]); % Head-Check Time (T)
end
end
fclose(fid3);
%
% Start Calculations
%
T = K*ho;
h1 = hmax-ho; % Maximum Allowed Head Increase (L)
h2 = ho-hmin; % Maximum Allowed Head Decrease (L)
w = y_stream-y_boundary; % Aquifer Width (L)
%
% Calculate Cumulative number of head check times
cntch=0;
for i= 1:nmw
    cntch = cntch + ntch(i);
end
```

```
%
% Assembling Linear Optimization Problem Coefficients
%
% Allocation of Inequality Matrix A
% The total number of rows of Matrix A is given by the sum of:
% -> 2*now (at each candidate well, the pumping rate must be below the
%      maximum and above the minimum, which requires two inequalities
%       per well)
% -> 2*cntch (at each monitoring well, the head must be below the
maximum
%       and above the minimum at each prescribed check time,
%       which requires two inequalities per well per check time)
% -> 2 (at any time, the sum of extraction rates (<0) must be below
Q_dem (<0),
%      and the sum of injection rates (>0) must be below Q_ava, which
requires
%     two inequalities overall)
% -> 2 (extra inequalities for objective function modification
accounting
%     for the module of the volume of stream depletion, Vr',which
%     requires two inequalities overall)
%
nrow = 2*now + 2*cntch + 2 + 2;
sprintf('N. of Constraint Inequalities: %d',nrow)
%
% The total number of columns of Matrix A is given by the total number of
% candidate operating wells plus one (for Vr'):
%
ncol = now + 1;
sprintf('N. of Decision Variables: %d',ncol)
%
% Allocate matrix A
A = zeros(nrow,ncol);
%
% Allocate Inequality RHS Vector: the total number of rows of this
vector
% is the same as in Matrix A:nrow = 2*now + 2*cntch + 2 + 2;
b = zeros(nrow,1);
%
% Calculate and assemble Matrix A and RHS Vector b coefficients
%
irow=0;
for j = 1:now
% set coefficients for two pumping rate constraints at each candidate
well
```

```
irow=irow+2;
A(irow-1,j) = +1.;
A(irow,j)  = -1.;
% pumping rate constraints at each candidate well
b(irow-1)  = +Qw_max(j);
b(irow)    = -Qw_min(j);
end
%
for i=1:nmw
% set coefficients for two hydraulic head constraints at each
monitoring
% well at each check time
for ich = 1:ntch(i)
    irow=irow+2;
for j = 1:now
% calculate unit response coefficient for operating well j
% obtained using BRC_CYC(a Theis derived solution for a well
% operating cyclically in a bounded aquifer)
if (tch(i,ich)>tst(j))
    xow = xw(j);
    yow = yw(j)-y_boundary;
    Qow = 1.0;
    dtonw = dt(j);
    xmw = xm(i);
    ymw = ym(i)-y_boundary;
% aw=y_stream-yw(j);
% am=y_stream-ym(i);
% A(irow-1,j) = BRNF_CYC(S,T,w,xw(j),aw,delt,dt(j),tch(i,ich)-
tst(j),xm(i),am);
A(irow-1,j) = BNOFLOW_CYC_THEIS(S,T,w,yow,xow,Qow,delt,dtonw,
tch(i,ich)-tst(j),ymw,xmw);
A(irow ,j) = -A(irow-1,j);
end
% hydraulic head constraints at each monitoring well at each check
time
b(irow-1) = h1;
b(irow)  = h2;
end
end
end
%
% set coefficients for sum of extraction rates to be below Q_dem <0;
irow=irow+1;
```

```
   for j = 1:new
      A(irow,j) = +1.;
   end
   b(irow) = Q_dem;
   % set coefficients for sum of injection rates to be below Q_ava >0;
   irow=irow+1;
   for j = new+1:now
      A(irow,j) = +1.;
   end
   b(irow) = Q_ava;
   %
   % Set up inequality coefficients for objective function modification
   % accounting for the module of the volume of stream depletion, Vr')
   %
   irow=irow+2;
   for j = 1:now
   if ((tfin-tst(j))>0.)
   A(irow-1,j) = BNvol_sol_CYC(tfin-tst(j),delt,dt(j),T,S,y_stream
-yw(j),w);
   A(irow ,j) = -A(irow-1,j);
   end
   end
   % set up Vr' cofficients
   A(irow-1,ncol) = -1;
   A(irow,ncol)  = -1;
   % Constraints for objective function modification accounting for the
module
   % of the volume of stream depletion, Vr', are equal to zero.
   b(irow-1)    = 0.;
   b(irow)     = 0.;
   %
   % Matrix of Linear Objective Function Cofficients
   f = zeros(ncol,1);
   f(ncol) = 1; % for Vr' column
   %
   %%%%%%%%%%%%%%%%%%%%%%%%%%%%%%%%%%%%%%%%%%%%%%%%%%%%%%%%%%
   %%%%%%%%% Solution of the formulated Linear Optimization Problem
   %
   options = optimset('LargeScale','off','Simplex','on');
   [Qw,fval,exitflag,output,lambda] = linprog(f,A,b,[],[]);
   %
   disp('Exit Condition:')
   if (exitflag==1)
   exitflag
   disp('optimum is found')
```

```
else
exitflag
disp('optimum is NOT found: check linprog help')
end
%disp('Optimal Pumping Rate Set:')
%Qw
fprintf('Objective Function Value at Optimum: Minimum Stream Volume
Depletion')
sprintf('fval= %f',fval)
%
%output
%lambda
% Calculated Aquifer Recharge and Extraction Cumulative Rates,
% Stream Recharge Volume over the simulated period and
% Reduce Optimal Solution by eliminating non active wells
naow=0;
Q_extract=0.;
Q_recharge=0.;
Vr=0.;
Qlim=101.;
for i=1:now
Vr=Vr+A(nrow-1,i)*Qw(i);
if (abs(Qw(i))>Qlim)
naow=naow+1;
if (Qw(i)<0.)
Q_extract=Q_extract+Qw(i);
end
if (Qw(i)>0.)
Q_recharge=Q_recharge+Qw(i);
end
end
end
% Output File
fid4 = fopen('Results.dat','w');
fprintf(fid4,'%d \n',naow);
fid5 = fopen('Optimal.Scheme.dat','w');
for i=1: now
%   temp = [Qw(i),xw(i),yw(i),tst(i),dt(i),Qw_min(i),Qw_max(i)];
%    fprintf(fid4,'%15.6E %15.6E %15.6E %15.6E %15.6E %15.6E %15.6E
\n',temp);
if (abs(Qw(i))>Qlim)
temp = [Qw(i),tst(i),dt(i),xw(i),yw(i)];
fprintf(fid4,'%15.6E %15.6E %15.6E %15.6E %15.6E\n',temp);
temp = [xw(i),yw(i),Qw(i)];
fprintf(fid5,'%15.6E %15.6E %15.6E\n',temp);
```

```
end
end
%
fprintf(fid4,'Net Cumulative Stream Recharge Volume (Vr_prime)
(m^3/d)= %15.3E \n',fval);
fprintf(fid4,'Cumulative Stream Recharge Volume (Vr) (m^3)= %15.3E
\n',Vr);
fprintf(fid4,'Cumulative Extraction Rate (m^3/d) = %15.3E \n',
Q_extract);
fprintf(fid4,'Cumulative Recharge Rate (m^3/d) = %15.3E \n',
Q_recharge);
fclose(fid4);
fclose(fid5);
%
%%%%%%%%%%%%%%%%%%%%%%%%%%%%%%%%%%%%%%%%%%%%%%%%%%%%%%%%%%%%%%
%%%%%%%%
function [BRNF] = BNOFLOW_CYC_THEIS(Sy,T,w,xw,yw,Qw,delt,dton,t,
x,y)
BRNF=0.; % Initial Drawdown in a Bounded Aquifer
a=w-xw; % Operating-well/Stream Distance
TOL=1.e-9; % Tolerance value at which the loop stops
DELTA=1.0; % Starting value of DELTA
% Loop over the number of Wells Groups
j=1; % (While) Loop Counter
while abs(DELTA)>= TOL && j<=50
sign = (-1)^(j+1);
Q_term = Qw/(4*pi()*T);
DELTA = sign*Q_term*Four_Wells_BRNF(Sy,T,w,a,yw,delt,dton,t,x,
y,j);
BRNF = BRNF+DELTA;
j=j+1;
end
if (j>50)
disp('convergence not reached')
end
%
function [W_Fun]= Four_Wells_BRNF(Sy,T,w,a,yw,delt,dton,t,x,y,j)
W_Fun= 0.;
% Wells x Coordinate
xw1 = (2*j-1)*w-a;
xw2 = (2*j-1)*w+a;
% xw1 = 2*(j-1)*w+a;
% xw2 = 2*j*w  -a;
%
% Wells Distance From Observation Well
```

```
r1= sqrt((x-xw1)^2+(y-yw)^2);
r2= sqrt((x-xw2)^2+(y-yw)^2);
% Image Wells
r3= sqrt((x+xw1)^2+(y-yw)^2);
r4= sqrt((x+xw2)^2+(y-yw)^2);
%
% Cycle Effect
frac =t/delt;
int_t =fix(frac);
rest_t =t-int_t*delt;
n =int_t+1; % Number of complete Operation Cycles
% Loop over the number of Operation Cycles
for i=1:n-1
% For a Number of Complete Cycles
t1=t-(i-1)*delt;
t2=t1-dton;
% Stream Constant Head and infinite Aquifer Effects
u1=(Sy/(4*T))*(r1^2/t1);
u2=(Sy/(4*T))*(r2^2/t1);
% Image Wells
u3=(Sy/(4*T))*(r3^2/t1);
u4=(Sy/(4*T))*(r4^2/t1);
% Well Function
wu1=expint(u1);
wu2=expint(u2);
% Image Well Function
wu3=expint(u3);
wu4=expint(u4);
% Operating And Image Wells
wo = wu1-wu2+wu3-wu4;
% wo = wu1+wu2-wu3-wu4;
% Imaginary Compensation Wells (for the continuous pumping)
u1_I=(Sy/(4*T))*(r1^2/t2);
u2_I=(Sy/(4*T))*(r2^2/t2);
% Image Wells
u3_I=(Sy/(4*T))*(r3^2/t2);
u4_I=(Sy/(4*T))*(r4^2/t2);
% Well Function
wu1_I=expint(u1_I);
wu2_I=expint(u2_I);
% Image Well Function
wu3_I=expint(u3_I);
wu4_I=expint(u4_I);
wI = wu1_I-wu2_I+wu3_I-wu4_I;
% wI = wu1_I+wu2_I-wu3_I-wu4_I;
```

```
% Total Well Function
W_Fun =W_Fun+wo-wI;
end
%
if rest_t>0 && rest_t <= dton % During Operation Time
% Stream Constant Head and Infinite Aquifer Effects
u1=(Sy/(4*T))*(r1^2/rest_t);
u2=(Sy/(4*T))*(r2^2/rest_t);
% Image Wells
u3=(Sy/(4*T))*(r3^2/rest_t);
u4=(Sy/(4*T))*(r4^2/rest_t);
% Well Function
wu1=expint(u1);
wu2=expint(u2);
% Image Well Function
wu3=expint(u3);
wu4=expint(u4);
% Operating And Image Wells
W_Fun = W_Fun+wu1-wu2+wu3-wu4;
% W_Fun = W_Fun+wu1+wu2-wu3-wu4;
end
if rest_t>0 && rest_t > dton % During time when operation stops
t1=rest_t;
t2=t1-dton;
% Stream Constant Head and infinite Aquifer Effects
u1=(Sy/(4*T))*(r1^2/t1);
u2=(Sy/(4*T))*(r2^2/t1);
% Image Wells
u3=(Sy/(4*T))*(r3^2/t1);
u4=(Sy/(4*T))*(r4^2/t1);
% Well Function
wu1=expint(u1);
wu2=expint(u2);
% Image Well Function
wu3=expint(u3);
wu4=expint(u4);
% Operating And Image Wells
wo = wu1-wu2+wu3-wu4;
% wo = wu1+wu2-wu3-wu4;
% Imaginary Compensation Wells (for continous pumping)
u1_I=(Sy/(4*T))*(r1^2/t2);
u2_I=(Sy/(4*T))*(r2^2/t2);
% Image Wells
u3_I=(Sy/(4*T))*(r3^2/t2);
u4_I=(Sy/(4*T))*(r4^2/t2);
```

```
% Well Function
wu1_I=expint(u1_I);
wu2_I=expint(u2_I);
% Image Well Function
wu3_I=expint(u3_I);
wu4_I=expint(u4_I);
% Imaginary And Image Wells
wI = wu1_I-wu2_I+wu3_I-wu4_I;
% wI = wu1_I+wu2_I-wu3_I-wu4_I;
% Total Well Function
W_Fun =W_Fun+wo-wI;
end
%%%%%%%%%%%%%%%%%%%%%%%%%%%%%%%%%%%%%%%%%%%%%%%%%%%%%%%%%%%%%%%%%%%
%%%%%%%%%
function [Vol_coeff] = BNvol_sol_CYC(t,delt,dton,T,S,a,w)
% Calculating Vr (depletion volume)
Vol_coeff = 0.;
frac=(t)/delt;
int=fix(frac);
rest_t=t-int*delt;
n = int+1;
for i=1:n-1
t1=t-(i-1)*delt;
t2=t1-dton;
Vol_coeff=Vol_coeff+(t1*BNVRatio(T,S,a,w,t1)-t2*BNVRatio(T,S,
a,w,t2));
end
if rest_t>0 && rest_t <= dton
Vol_coeff=Vol_coeff+rest_t*BNVRatio(T,S,a,w,rest_t);
end
if rest_t > dton
t1=rest_t;
t2=t1-dton;
Vol_coeff=Vol_coeff+t1*BNVRatio(T,S,a,w,t1)-t2*BNVRatio(T,S,a,
w,t2);
end
%
%%%%%%%%%%%%%%%%%%%%%%%%%%%%%%%%%%%%%%%%%%%%%%%%%%%%%%%%%%%%%%%%%%%
%%%%%%%%%
function BGVR=BNVRatio(T,S,a,w,t)
TOL=1.e-9; i=0; DELTA=1.0; BGVR=0;
while abs(DELTA) >=TOL && i<=50
DELTA=(-1)^i*(C(T,S,a,w,t,i)+D(T,S,a,w,t,i));
BGVR=BGVR+DELTA;
i=i+1;
end
```

```
function Ci=C(T,S,a,w,t,i)
Ci=(1+S/2/T*(2*w*i+a)^2/t)*erfc(sqrt(S/4/T*(2*w*i+a)^2/t))-2/
sqrt(pi())*sqrt(S/4/T*(2*w*i+a)^2/t)*exp(-S/4/T*(2*w*i+a)^2/t);
function Di=D(T,S,a,w,t,i)
Di=(1+S/2/T*(2*w*(i+1)-a)^2/t)*erfc(sqrt(S/4/T*(2*w*(i+1)-a)
^2/t))-2/sqrt(pi())*sqrt(S/4/T*(2*w*(i+1)-a)^2/t)*exp(-S/4/T*
(2*w*(i+1)-a)^2/t);
%%%%%%%%%%%%%%%%%%%%%%%%%%%%%%%%%%%%%%%%%%%%%%%%%%%%%
%%%%%%%%%
```

References

1. Theis, C. V. (1935). The relation between the lowering of the piezometric surface and the rate and duration of discharge of well using groundwater storage. *AGU Transactions, 16*(2), 519–524.
2. Glover, R. E., & Balmer, C. G. (1954). River depletion resulting from pumping a well near a river. *AGU Transactions, 35*(3), 468–470.
3. Jenkins, C. T. (1968). Techniques for computing rate and volume of stream depletion by wells. *Ground Water, 6*(2), 37–46.
4. Harbaugh, A. W., & McDonald, M. G. (1996). *User's documentation for MODFLOW-96, an update to the U.S. Geological Survey modular finite-difference groundwater flow model* (U.S. Geological Survey Open-File Report 96-485, 56 pp.).
5. Harbaugh, A. W. (2005). MODFLOW-2005, the U.S. Geological Survey modular ground-water model -- The Ground-Water Flow Process: U.S. Geological Survey Techniques and Methods 6-A16, variously p.
6. Reilly, T. E., Franke, O. L., & Bennett, G. D. (1987). *The principle of superposition and its application in ground-water hydraulics* (USGS Open File Report 84-459). Washington, DC.
7. Bear, J. (1972). *Dynamics of fluids in porous media*. New York: Elsevier.
8. McWhorter, D. B., & Sunada, D. K. (1995). *Ground water hydrology and hydraulics*. Highlands Ranch, CO: Water Resources Publication, LLC.
9. Todd, D. K., & Mays, L. W. (2005). *Groundwater hydrology*. Hoboken, NJ: Wiley.
10. Willis, R., & Yeh, W. W. G. (1987). *Groundwater system planning and management* (416 pp.). Prentice Hall.
11. Hantush, M. S., & Jacob, C. E. (1955). Non-steady radial flow in an infinite leaky aquifer. *American Geophysical Union Transactions, 36*(1), 95–100.
12. Hantush, M. S. (1960). Modification of the theory of leaky aquifers. *Journal of Geophysical Research, 65*(11), 3713–3725.
13. Voss, C. I., & Provost, A. M. (2002). *SUTRA, A model for saturated-unsaturated variable-density ground-water flow with solute or energy transport* (U.S. Geological Survey Water-Resources Investigations Report 02-4231, 291 pp. (9/22/2010 version))
14. Miller, C. D., Durnford, D., Halstead, M. R., Altenhofen, J., & Flory, V. (2007). Stream depletion in alluvial valleys using the SDF semi-analytical models. *National Groundwater Association, 45*(4), 506–514.
15. Polubarinova-Kochina, P. Y. (1962). *Theory of groundwater movement*. Princeton, NJ: Princeton University Press.
16. Theis, C. V. (1941). The effect of a well on the flow of a nearby stream. *AGU Transactions, 22* (3), 734–738.
17. Grigg, N. S. (2005). *Water manager's handbook: A guide to the water industry*. Fort Collins, CO: Aquamedia Publishing.

18. Colorado Foundation For Water Education. (2003). Colorado water law. Denver, CO: Author.
19. Bouwer, H. (2002). Artificial recharge of groundwater: Hydrogeology and engineering. *Hydrogeology Journal, 10*, 121–142.
20. Pyne, R., & David, G. (1994). *Groundwater recharge and wells: A guide to aquifer storage recovery*. Boca Raton, FL: CRC Press.
21. O'Hare, M. P. (1986). *Artificial recharge of ground water: Status and potential in the contiguous United States*. Chelsea, MI: Lewis.
22. Ruffino, L. (2009). Rainwater harvesting and artificial recharge to groundwater, Water Conservation Technical Briefs, SAI Platform.
23. Central Groundwater Board. (2000). *Guide on artificial recharge to groundwater*. New Delhi, India: Ministry of Water Resources.
24. Dantzig, G. B., Orden, A., & Wolfe, P. (1955). Generalized simplex method for minimizing a linear form under linear inequality restraints. *Pacific Journal Mathematics, 5*, 183–195.
25. Mehrotra, S. (1992). On the implementation of a primal-dual interior point method. *SIAM Journal on Optimization, 2*, 575–601.
26. Zhang, Y. (1995). *Solving large-scale linear programs by interior-point methods under the MATLAB environment* (Technical Report TR96-01). Baltimore, MD: Department of Mathematics and Statistics, University of Maryland, Baltimore County.
27. Molden, D., Sunada, D. K., & Warner, J. W. (1984). Microcomputer model of artificial recharge using Glover solution. *Groundwater, 22*(1), 73–79.

Chapter 3
One-Dimensional Model of Water Quality and Aquatic Ecosystem/Ecotoxicology in River Systems

Podjanee Inthasaro and Weiming Wu

Contents

Abstract A one-dimensional water quality and aquatic ecology/ecotoxicology model has been incorporated into a package for the modeling of hydrodynamic, sediment transport, contaminant transport, water quality, aquatic ecosystem, and ecotoxicology in river systems. The water quality model alone can be used to determine water temperature, dissolved oxygen, biological oxygen demand, nitrogen, phosphorus, and conservative chemical such as chloride. The aquatic ecosystem model considers a basic food web structure consisting of four trophic

---

P. Inthasaro, Ph.D.
Previously at, 1157 Crane Crest Way, Orlando, FL 32825, USA
e-mail: pinthasaro@gmail.com

W. Wu, Ph.D. (✉)
Department of Civil and Environmental Engineering,
Wallance H. Coulter School of Engineering, Clarkson University, Box 5710,
8 Clarkson Avenue, Potsdam, NY 13699, USA
e-mail: wwu@clarkson.edu

L.K. Wang, C.T. Yang, and M.-H.S. Wang (eds.), *Advances in Water Resources Management*, Handbook of Environmental Engineering, Volume 16,
DOI 10.1007/978-3-319-22924-9_3

levels: phytoplankton, zooplankton, forage fish, and predatory fish, undergoing various biological processes such as photosynthesis, grazing, respiration, excretion, defecation, mortality, gamete, and reproduction. The model simulates the bio-accumulation of toxic chemicals in organisms by uptake, depuration and dietary, and takes into account the effects of toxicity on organisms through modification factors of photosynthesis, grazing, and gamete mortality. The modeling package has been tested by simulating the water quality parameters in the Tualatin River, Oregon and the water quality, aquatic ecosystem, polychlorinated biphenyl (PCB) transport and bioaccumulation in the Upper Hudson River, New York. The simulated water quality parameters, phytoplankton and zooplankton biomass, fish populations, and PCB concentrations in fish are in generally good agreement with the measurement data.

Keywords Water quality model • Aquatic ecosystem model • Ecotoxicology model • Freshwater riverine system • Contaminant transport • Food web

Nomenclature

$[H]^+$	Molar concentration of hydrogen ion, mol/m^3
$[OH]^-$	Molar concentration of hydroxide ion, mol/m^3
A	Cross-sectional flow area, m^2
C_a	Biomass concentration of phytoplankton, g/m^3, or μg/L
c_b	Bowen coefficient
C_{CBOD}	Concentration of carbonaceous biological oxygen demand (CBOD), g/m^3
C_{DO}	Dissolved oxygen (DO) concentration, g/m^3
C'_{DO}	Saturation DO concentration, g/m^3
C_g	Gas-phase concentration of the contaminant, g/m^3
C_L	Fraction of cloud cover
C_m	Concentration of suspended solid, g/m^3
C_{NH3}	Concentration of ammonia nitrogen, g/m^3
C_{NO3}	Concentration of nitrate nitrogen, g/m^3
C_{ON}	Concentration of organic nitrogen, g/m^3
C_{OP}	Concentration of organic phosphorus, g/m^3
c_p	Specific heat capacity
C_{PO4}	Concentration of orthophosphate, g/m^3
$C_{tb,i}$	Total concentration of contaminant in bed layer i
C_{ti}	Contaminant concentration associated with organism i in unit volume of water column, g/m^3
C_{tw}	Total concentration of contaminant in the water column, g/m^3
D_b	Sediment deposition rate, m/d
D_x	Longitudinal dispersion coefficient, m^2/s
e_{air}	Air vapor pressure, mb

ε_{air}	Emissivity value of air
E_b	Sediment erosion rate, m/d
e_s	Saturation vapor pressure, mb
ε_{water}	Emissivity value of water
f_{act}	Factor for respiratory rate associated with swimming or active respiratory fraction
$f_{db,1}$	Fraction of dissolved contaminant in bed surface layer
f_{den}	Density-dependent respiration factor
f_{dw}	Fraction of dissolved concentration to the total concentration of contaminant in water column
f_{dyn}	Proportion of assimilated energy lost to specific dynamic action
f'_{ib}	Increase factor in the gamete due to toxic chemicals
f'_{ig}	Reduction factor in animal growth due to toxic chemicals
f_{ij}	Relative preference factor of predator j feeding on organism i as food
f_L	Light limitation factor
f_N	Nutrient limitation factor
f_{NH3}	Fraction of ammonia in dead organic material
f_{pb}	Fraction of particulate contaminant in the bed sediment
f_{PBOD}	Fraction of particulate CBOD in total CBOD
f_{PO4}	Fraction of phosphate in dead organic material
f_{PON}	Fraction of particulate organic nitrogen to organic nitrogen
f_{POP}	Fraction of particulate organic phosphorus to organic phosphorus
f_{pw}	Fraction of particulate contaminant in the water column
f_{shade}	Shading factor defined as the fraction of potential solar radiation that is blocked due to riparian vegetation and landscape
f_T	Temperature limitation factor
f_{TOX}	Reduction factor due to toxic chemicals
H	Henry's law constant, atm m^3/mol
h_{CBOD}	Half-saturation DO concentration for CBOD decay, g/m^3
h_L	Half-saturation light intensity for phytoplankton growth
h_N	Half-saturation concentration for nitrogen, g/m^3
h_N	Michaelis–Menten constant for nitrogen uptake, mgN/L
h_{NH3}	Half-saturation DO concentration for nitrification, g/m^3
h_{NO3}	Half-saturation DO concentration for denitrification, g/m^3
h_{OP}	Half-saturation phytoplankton conc. for mineralization of phosphorus, g/m^3
h_P	Half-saturation concentration for phosphorus, g/m^3
I_0	Light intensity at the water surface
IC_{i50}	Internal concentration of the contaminant in the biotic organism
J_{dbw}	Vertical diffusion fluxes between water column and bed surface layer, g/m^2d
K_b	Biodegradation rate, 1/d
K_{CBOD}	CBOD decay rate, 1/d
K_d	Sorption–desorption coefficient, m^3/g

$k_{dbi,i+1}$ Diffusional transfer coefficient of dissolved contaminant between layers i and i + 1
k_{dbw} Diffusional transfer coefficient of dissolved contaminant across the bed surface
K_H Acid-catalyzed hydrolysis rate, m^3/mold
K_i Carrying capacity of fish i, g/m^3
K_{i1} Uptake rate of contaminant of organism i, 1/d
K_{i2} Depuration rate of contaminant of organism i, 1/d
K_{ib} Gamete loss rate of organism i, 1/d
K_{ib0} Intrinsic gamete mortality rate, 1/d
K_{id} Defecation rate of biotic organism i, 1/d
K_{ie} Excretion rate of biotic organism i, 1/d
$K_{ie,max}$ Maximum rate of excretion of organism i, 1/d
K_{ig} Grazing rate of organism i, 1/d
$K_{ig,max}$ Maximum grazing rate of organism i, 1/d
K_{im} Nonpredatory mortality rate of organism i, 1/d
$K_{im,max}$ Maximum rate of nonpredatory mortality of organism i, 1/d
K_{ir} Respiration rate of biotic organism i, 1/d
$K_{ir,max}$ Maximum respiration rate of organism i, 1/d
K_{ir0} Basal or standard respiratory rate, 1/d
K_{ire} Reproduction rate of organism i, 1/d
K_N Neutral hydrolysis rate, 1/d
K_{NH3} Nitrification rate, 1/d
K_{NO3} Denitrification rate, 1/d
K_{OH} Base-catalyzed hydrolysis rate, m^3/mol d
K_{ON} Mineralization rate of organic nitrogen, 1/d
K_{OP} Mineralization rate of organic phosphorus, 1/d
K_p Photolysis rate, 1/d
K_{RE} Depth-averaged reaeration rate, 1/d
$k_{tb,i}$ Decay coefficient of contaminant at layer i
KT_{g1} Coefficient representing the relationships of growth on temperature below the optimal temperature
KT_{g2} Coefficient representing the relationships of growth on temperature above the optimal temperature
K_v Volatilization rate, m/d
LC_{i50} Internal concentration (the concentration of contaminant in water that causes 50 % mortality for a given period of exposure)
m Suspended sediment concentration by volume
p_{ji} Preference of predator i feeding on organism j as food
p_{NH3} Ammonia preference factor
Q Flow discharge, m^3/s
q_l Latent heat flux
q_{lw} Long-wave atmospheric radiation
q_s Convective heat flux
q_{sw} Solar radiation

$q_{sw,clear}$	Short-wave radiation reaching the water surface on a clear day after atmospheric attenuation
$q_{t,ex}$	Total exchange rate of contaminant due to sediment erosion and deposition, g/m^2d
$q_{tbi,i+1}$	Total exchange rate of contaminant between layers i and i + 1 due to lowering and rising of the interface
q_{tw}	Total loading rate of contaminant per unit volume, g/m^3d
R	Universal gas constant, atm m^3/mol °K
R_{lw}	Reflectivity of water surface for long-wave radiation
R_{sw}	Albedo or reflection coefficient
S_{SOD}	Sediment oxygen demand flux, g/m^2s
t	Time, s
T	Water temperature, °C
t_1	Exposure time in toxicity test
t_2	Period of exposure
T_{air}	Air temperature, °K
T_K	Water temperature in °K
T_{opt}	Optimal temperature for biological growth
T_{water}	Water temperature, °K or °C
U	Flow velocity, m/s
α	Velocity correction coefficient
α_{NC}	Stoichiometric ratio of nitrogen to carbon, gN/gC
α_{PC}	Stoichiometric ratio of phosphorus to carbon, gP/gC
γ	Light extinction, 1/m
γ_0	Background light extinction, 1/m
δ_i	Thickness of layer i
θ	Temperature coefficient
λ_i	Grazing limitation factor
ν_i	Concentration of contaminant in biotic organism i, g/g
ρ	Water density, kg/m^3
ρ_d	Dry density of the bed sediment, g/m^3
σ	Stefan–Boltzmann constant, W/m^2 $°K^4$
ϕ	Porosity
ω_a	Settling velocity of phytoplankton, m/d
ω_{CBOD}	Settling velocity of CBOD, m/s
ω_{ON}	Settling velocity of organic nitrogen, m/s
ω_{OP}	Settling velocity of organic phosphorus, m/s

1 Introduction

During the past decades, many streams and rivers all over the world have been impacted by point and nonpoint source pollutants from residential area, industry, agriculture and so on. Human or animals can be exposed to the toxic pollutants through food chains in the ecosystems and experience health problems. Because ecosystems are highly dependent on the hydrodynamic, morphodynamic and water quality factors and a large number of physical, chemical, biological, and ecological processes are involved, it is desired to study the integrated dynamics of flow, sediment transport, water quality, aquatic ecosystem, and ecotoxicology in river systems. Simulation of them is quite challenging but important.

Most of the early water quality models focused on dissolved oxygen (DO) and biological oxygen demand (BOD) and launched with the Streeter–Phelps simple BOD–DO model [1]. Then, the models evolved to investigate eutrophication for environmental management by incorporating more processes and components that influence water quality and cope with complex hydrodynamics. In recent years, because of advanced computer technology and increased public health and environmental awareness, several well-established water quality models have become available, such as the Water Quality Analysis Simulation Program (WASP) [2, 3], the river and stream water quality model QUAL2K [4], the multi-dimensional water quality model CE-QUAL-ICM [5], and CCHE_WQ [6]. Further developments have led to a number of aquatic ecological models which represent biotic and abiotic structures in combination with physical, chemical, biological, and ecological processes. Examples of well-established aquatic ecosystem models are Ecopath with Ecosim (EwE) [7], the Computational Aquatic Ecosystem Dynamics Model (CAEDYM) [8], EcoNetwrk [9], and AQUATOX [10, 11]. A recent review of the state of the art for water quality modeling can be found at [12, 13].

Among the above-mentioned water quality and ecosystem models, WASP and AQUATOX are two versatile programs and have been applied widely. Recent studies for model applications include [14–19]. WASP can simulate both phytoplankton and benthic algae in an eutrophication system, but it does not include higher trophic compartments such as zooplankton and fish. In contrast, AQUATOX has a robust aquatic ecosystem model that can simulate a complex aquatic food web with age-structure and trophic interactions. WASP can perform channel flow calculations itself or be linked to external hydrodynamic models such as the Environmental Fluid Dynamics Code (EFDC), the hydrodynamic program DYNHYD, RIVMOD, and the one-dimensional dynamic flow and water quality model CE-QUAL-RIV1 [20]. Similar to WASP, AQUATOX is linked to the Hydrological Simulation Program—Fortran (HSPF) for external hydrodynamic computation. Recently, AQUATOX introduced the multi-segment version which includes linkage of individual AQUATOX segments into a single simulation [11]. However, both models do not have robust sediment transport models. Although AQUATOX is a powerful tool for the simulation of aquatic ecosystems, it involves a large number of variables and parameters that can possibly be used only by experienced users [21]. To reduce the complexity of AQUATOX and incorporate water quality computations from WASP, a new water quality and aquatic ecosystem/ecotoxicology model has been developed in this study.

The present water quality and ecological model is intended for prediction of riverine ecosystems with the effects of toxic chemicals. The model schemes are developed by adopting the merits of the water quality model WASP and the aquatic ecosystem model AQUATOX. The developed water quality model simulates the temporal and spatial variations of water temperature, conservative substances such as chloride, and non-conservative substances such as dissolved oxygen, biological oxygen demand, nitrogen, and phosphorus. The aquatic food web model simulates dynamic interactions of phytoplankton, zooplankton, forage fish, and predatory fish. The model can compute the fate and transport of a contaminant in water column and sediment bed. The bioaccumulation model involves the direct transfer of the contaminant from water through surface sorption or gill uptake and the accumulation throughout the trophic levels of the food web. The governing equations, kinetic relations, numerical solution algorithms and tests of the developed model are presented in the following sections.

2 Governing Equations

The model simulates the fate and transport of constituents carried by water and/or sediment in channel networks. The present model is developed as an add-on to the existing 1-D flow and sediment transport model, CCHE1D [22]. CCHE1D is a one-dimensional channel-network hydrodynamic and sediment transport model, which was developed and maintained by the National Center for Computational Hydroscience and Engineering (NCCHE) of the University of Mississippi. The CCHE1D flow model simulates unsteady flow in channels of compound cross-sections, accounting for in-stream hydraulic structures. The sediment transport model computes the non-equilibrium transport of non-uniform sediment mixtures. The flow is governed by the 1-D St. Venant equations and the multiple-sized sediment transport is described by a non-equilibrium total-load transport equation. The flow and sediment transport equations and the corresponding numerical solution procedures refer to [22]. The transport of a constituent in the water quality and ecosystem is described by the following advection–dispersion equation:

$$A\frac{DC}{Dt} = \frac{\partial (AC)}{\partial t} + \frac{\partial (\alpha QC)}{\partial x} - \frac{\partial}{\partial x}\left(D_x A \frac{\partial C}{\partial x}\right) = AS \tag{3.1}$$

where

t = time, s
x = coordinate along the channel, m
A = cross-sectional flow area, m^2
Q = flow discharge, m^3/s
C = concentration of the constituent in water column, g/m^3
D_x = longitudinal dispersion coefficient, m^2/s
S = net source/sink term due to biochemical and physical processes and/or due to lateral input to the channel by runoff, g/m^3s
α = velocity correction coefficient.

Equation (3.1) is a general transport equation. The constituent can be any substance transporting in the water column, including sediment, heat (water temperature), pollutants, dissolved oxygen, nitrogen, phosphorus, phytoplankton, zooplankton, and fish. The coefficient α is given a value of 1 for the water quality constituents, phytoplankton, zooplankton and small fish which are assumed to move with the flow, whereas α is set as 0 for large fish which is assumed to move randomly in the domain in the present study. The value of α for migration fish needs to be investigated further. In addition, large fish does not experience the turbulent diffusion or mechanical dispersion as the water quality constituents do, but this difference is ignored for simplicity because the dispersion term is usually much smaller than the advection term in 1-D river systems.

Note that Eq. (3.1) also defines the operator (DC/Dt), which represents the storage, convection and dispersion terms divided by A.

3 Numerical Procedures

Equation (3.1) is discretized using a finite-volume scheme. The control volume for point i is embraced by faces i − 1/2 and i + 1/2 as shown in Fig. 3.1. For α = 1.0, Eq. (3.1) is integrated over the control volume as [23]

$$\frac{(A_iC_i)^{n+1}-(A_iC_i)^n}{\Delta t}\Delta x+\left(QC-AD_x\frac{\partial C}{\partial x}\right)_{i+1/2}^{n+1}-\left(QC-AD_x\frac{\partial C}{\partial x}\right)_{i-1/2}^{n+1}$$
$$=A_i^{n+1}S_i^{n+1}\Delta x \tag{3.2}$$

where

Δx = length of the control volume
Δt = computational time step
n = superscript which denotes time level
i = subscript which denotes grid point.

Using the analytical solution expressions of the steady, homogeneous, linearized form of Eq. (3.1) in the control volume, Eq. (3.2) is reformulated as

$$a_iC_i=a_{i+1}C_{i+1}+a_{i-1}C_{i-1}+b \tag{3.3}$$

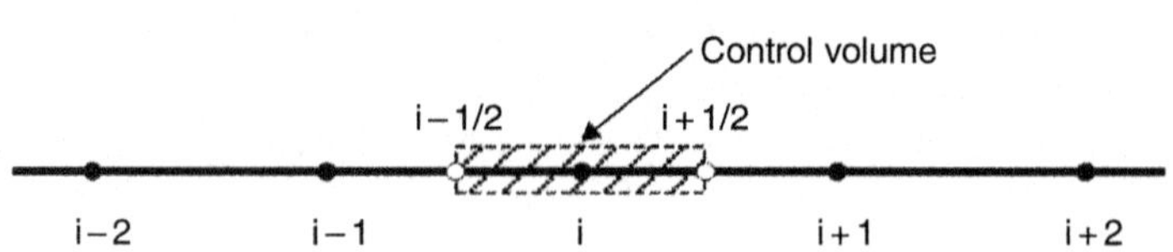

Fig. 3.1 1-D finite volume mesh

$$a_{i+1} = \left(\frac{Q_{i+1/2}}{\exp\left[(U/D_x)_{i+1/2}(x_{i+1} - x_i) \right] - 1} \right)^{n+1} \tag{3.4}$$

$$a_{i-1} = \left(\frac{Q_{i-1/2}}{\exp\left[(U/D_x)_{i-1/2}(x_i - x_{i-1}) \right] - 1} + Q_{i-1/2} \right)^{n+1} \tag{3.5}$$

$$a_i = a_{i+1} + a_{i-1} + Q_{i+1/2}^{n+1} - Q_{i-1/2}^{n+1} + A_i^{n+1} \frac{\Delta x}{\Delta t} \tag{3.6}$$

$$b = S_i^{n+1} A_i^{n+1} \Delta x + A_i^n \frac{\Delta x}{\Delta t} C_i^n \tag{3.7}$$

where

U = flow velocity.

When $\alpha = 0$, Eq. (3.1) becomes a diffusion-type equation. The dispersion term can be discretized using the central difference scheme. The final discretized equation can be written as Eq. (3.3) with different coefficients.

The discretized equations at the internal control volumes and boundary conditions at the inlet and outlet form a system of algebraic equations with a tridiagonal coefficient matrix, which can be solved using the Thomas algorithm, also called TDMA (TriDiagonal Matrix Algorithm). The details can be found in many text books and thus are not introduced here.

4 Water Temperature

Water temperature is a key factor for water quality and ecological studies. It affects water chemistry such as gas solubility, chemical reactions, contaminant toxicity, and biological activities. It is influenced by heat fluxes across the water and bed surfaces, the temperature of upstream and lateral inflows, water depth, shading from river's bank landscape and vegetation, time of year, and latitude of the river.

The water temperature model describes heat transfer in the water column based on the first law of thermodynamics. The 1-D transport equation (3.1) can be applied here, with $S = q_T/(h\rho c_p)$, in which q_T is the net heat flux, ρ is the water density, h is the water depth, and c_p is the specific heat capacity. The net heat flux is considered as the exchange of heat across the air–water interface and the subsequent distribution of heat source throughout the water column. The surface heat flux consists of four major components: solar radiation (q_{sw}), long-wave atmospheric radiation (q_{lw}), latent heat flux (q_l), and convective heat flux (q_s), shown in Fig. 3.2. The

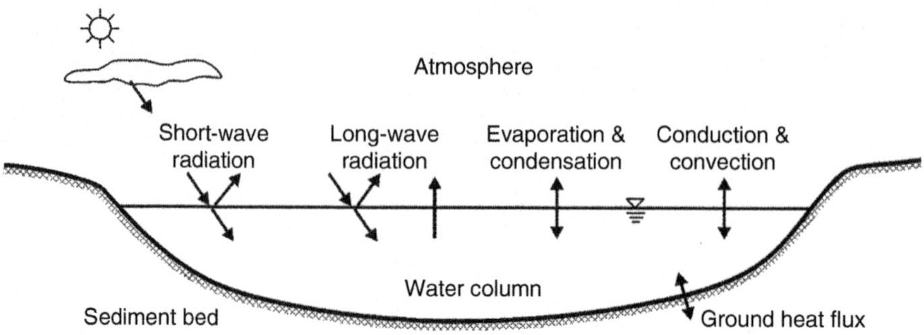

Fig. 3.2 Heat budget in water column

Table 3.1 Cloud cover and reflection coefficients [18]

Cloud cover	Overcast	Broken	Scattered	Clear
C_L	1.0	0.5–0.9	0.1–0.5	0
A	0.33	0.95	2.20	1.18
B	−0.45	−0.75	−0.97	−0.77

heat exchange with the underlying sediment is not considered in the present model, but can be readily added in the source/sink term q_T using methods presented in cited references such as [24]. The net heat flux is calculated as

$$q_T = q_{sw} + q_{lw} + q_l + q_s \tag{3.8}$$

The solar radiation is either measured directly or computed from a number of available formulas. It is a function of geographical location, time of year, hour of day, and cloudiness. The net solar radiation can be determined by [25, 26]

$$q_{sw} = q_{sw,clear}\left(1 - 0.65C_L^2\right)(1 - R_{sw})(1 - f_{shade}) \tag{3.9}$$

where

$q_{sw,clear}$ = short-wave radiation reaching the water surface on a clear day after atmospheric attenuation
C_L = fraction of cloud cover as given in Table 3.1
R_{sw} = albedo or reflection coefficient
f_{shade} = shading factor defined as the fraction of potential solar radiation that is blocked due to riparian vegetation and landscape.

Clouds have the greatest effect in reducing the amount of radiation energy received on the Earth [27]. The attenuation of solar radiation by clouds is difficult to predict due to a variety of types, distributions, and albedos of clouds. The reflection coefficient is calculated by [25, 26]

$$R_{sw} = A\left(\frac{180}{\pi}\alpha\right)^B \tag{3.10}$$

where

A and B = coefficients depending on the cloud cover as given in Table 3.1
α = altitude of the Sun in radians.

The long-wave or thermal radiation is radiation emitted by terrestrial object and the earth's atmosphere. The long-wave radiation depends on the surface temperature of the emitting object, air temperature, and water temperature. It is computed from the empirical formula of an overall atmospheric emissivity and the Stefan–Boltzmann law. The net long-wave radiation is determined by

$$q_{lw} = \varepsilon_{air}\sigma T_{air}^4\left(1 + 0.17C_L^2\right)(1 - R_{lw}) - \varepsilon_{water}\sigma T_{water}^4 \tag{3.11}$$

where

ε_{air} = emissivity value of air (= 0.96 for an approximation for normal and hemispherical emissivity)
ε_{water} = emissivity value of water (= $0.938 \times 10^{-5}\ T^2_{air}$)
σ = Stefan–Boltzmann constant (= 5.669×10^{-8} W/m^2 °K^4)
T_{air} = air temperature, °K
T_{water} = surface water temperature, °K
R_{lw} = reflectivity of water surface for long-wave radiation (= 0.03).

The latent heat flux is a gain or loss of energy during a change in the state of water between liquid and vapor. The latent heat flux in natural water depends on vapor pressure, air temperature, wind speed, and dew point temperature. It is calculated as

$$q_l = f(U_w)(e_{air} - e_s) \tag{3.12}$$

where

$f(U_w)$ = function of wind speed, W/m^2 mb, as given in Table 3.2
e_{air} = air vapor pressure, mb
e_s = saturation vapor pressure, mb.

Table 3.2 Wind speed functions in W/m^2 mb [18]

Wind speed function formula[a]	$f(U_w)$
Meyer (1928)	$4.18 \times 10^{-9} + 0.95 \times 10^{-9}\ U_w$
Marciano and Harbeck (1952)	$1.02 \times 10^{-9}\ U_w$
Harbeck et al. (1959)	$1.51 \times 10^{-9}\ U_w$
Morton (1965)	$3.59 \times 10^{-9} + 1.26 \times 10^{-9}\ U_w$
Ryan and Harleman (1973)	$2.83 \times 10^{-9} + 1.26 \times 10^{-9}\ U_w$

[a] U_w is the wind speeds (m/s), typically specified as measured at a height of 2 m over the water surface

The saturation vapor pressure is the highest pressure of water vapor that can exist in equilibrium with a plane, free water surface at a given temperature. It is approximated by the Tetens formula as [26]

$$e_s = 6.108\exp\left(\frac{17.27T_{water}}{T_{water} + 273.3}\right) \tag{3.13}$$

where

T_{water} = water temperature, °C.

The air vapor pressure, e_{air}, is calculated in a similar way by substituting T_{water} in Eq. (3.13) with the dew point temperature.

The convective or sensible heat is described as the heat flux transferred between air and water by conduction and transported away from/or toward the air–water interface. The amount of heat gained or lost through the sensible heat depends on the gradient of temperature in the vertical direction. The Bowen ratio describes the relationship between heat and vapor transport. The surface heat conduction is related to the evaporative heat flux and the Bowen ratio. It is estimated by [26]

$$q_s = c_b f(U_w)(T_{air} - T_{water}) \tag{3.14}$$

where
c_b = Bowen coefficient (=0.62 mb).

5 Kinetic Relations of Water Quality

The relationships of constituents in the developed water quality model are illustrated in Fig. 3.3. Nutrients and other constituents move in circular paths through biotic and abiotic components, which are known as biogeochemical cycles [28]. In the water column, four biogeochemical cycles are considered: oxygen, carbon, nitrogen, and phosphorus cycles.

Dissolved oxygen (DO) plays an important role in aquatic ecosystems. It is essential for living organisms and controls many chemical and biological reactions. Oxygen can be removed from or added to water by various physical, chemical, and biological processes. It is governed by

$$\begin{aligned}\frac{DC_{DO}}{Dt} = &\left(\frac{32}{12} + \frac{48}{14}\alpha_{NC}\left(1 - p_{NH_3}\right)\right)K_{ag}C_a - \frac{32}{12}\sum_{i\in\{a,z,f,p\}} K_{ir}C_i + K_{RE}\theta_{RE}^{T-20}\left(C'_{DO} - C_{DO}\right)\\ &-K_{CBOD}\frac{C_{DO}}{h_{CBOD} + C_{DO}}\theta_{CBOD}^{T-20}C_{CBOD} - \frac{64}{14}K_{NH_3}\frac{C_{DO}}{h_{NH_3} + C_{DO}}\theta_{NH_3}^{T-20}C_{NH_3} + \frac{S_{SOD}}{h}\end{aligned} \tag{3.15}$$

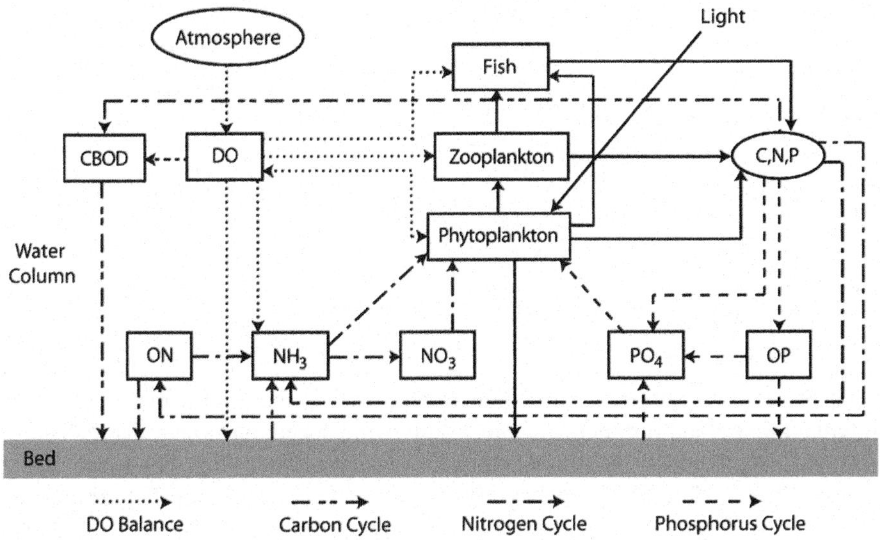

Fig. 3.3 Kinetic processes in water column

where

C_{DO} = DO concentration, g/m^3
C'_{DO} = saturation DO concentration, g/m^3
C_{NH3} = concentration of ammonia nitrogen, g/m^3
C_{CBOD} = concentration of carbonaceous biological oxygen demand (CBOD), g/m^3
C_a = biomass concentration of phytoplankton, g/m^3
C_i = biomass concentration of biotic organism i, g/m^3
i = organism index which specifies a as phytoplankton, z as zooplankton, f as forage fish, and p as predatory fish
K_{ag} = photosynthesis rate of phytoplankton, 1/d
K_{ir} = respiration rate of biotic organism i, 1/d
K_{RE} = depth-averaged reaeration rate, 1/d
K_{CBOD} = CBOD decay rate, 1/d
K_{NH3} = nitrification rate, 1/d
α_{NC} = stoichiometric ratio of nitrogen to carbon, gN/gC
p_{NH3} = ammonia preference factor, from Eq. (3.21)
h_{CBOD} = half-saturation DO concentration for CBOD decay, g/m^3
h_{NH3} = half-saturation DO concentration for nitrification, g/m^3
S_{SOD} = sediment oxygen demand flux, g/m^2 s
T = water temperature, °C
θ = temperature coefficient
h = water depth, m.

The saturation DO concentration is calculated as

$$C_{DO}^{'} = \exp\left[c_0 + c_1/T_K + c_2/T_K^2 + c_3/T_K^3 + c_4/T_K^4\right] \tag{3.16}$$

where

$c_0 = -139.3441$
$c_1 = 1.5757 \times 10^5$
$c_2 = -6.6423 \times 10^7$
$c_3 = 1.2438 \times 10^{10}$
$c_4 = -8.6219 \times 10^{11}$
T_K = water temperature in °K ($T_K = T + 273.15$).

The reaeration rate K_{RE} in natural rivers depends on several factors, such as internal mixing and turbulence, temperature, wind speed, and water depth. Therefore, K_{RE} is given as a temporally and spatially varying rate which can be calculated from several existing formulas [29, 30], such as O'Connor–Dobbins (1958), Churchill (1962), Owen and Gibbs (1964), and Langbein and Durum (1967). These formulas adopt different functions of flow velocity and water depth for K_{RE}. Each formula was developed for certain hydrodynamic and topographic conditions and thus is adequate in different types of rivers. The Churchill formula is used in the test cases of the present study.

Sediment oxygen demand (SOD) is the rate at which dissolved oxygen is removed from the overlying water column by biological processes in the river bed sediments. SOD rate is mainly affected by biological factors such as organic content of the benthic sediment and microbial concentrations [31]. The sediment oxygen demand flux (S_{SOD}) is treated in the present model as input from measurement data. The flux can be given as a constant but mostly a spatially varying or spatially and temporally varying flux depending upon the availability of the data.

Biological oxygen demand (BOD) is one of the common water quality indicators. It is a measure of the amount of oxygen required to stabilize organic matter in the water. BOD_5 is determined from a standardized test, which measures the amount of oxygen available after incubation of the sample at 20 °C for a specific length of time, usually 5 days. The BOD kinetic processes are represented by the BOD formation, carbonaceous deoxygenation, nitrogenous deoxygenation, and BOD settling. Carbonaceous biological oxygen demand (CBOD) testing is similar to BOD testing with the exception that a nitrification inhibitor is added at the start of the process to eliminate nitrifying bacteria from water sample. Therefore only the carbonaceous demand is measured. The rate of change in CBOD concentration is determined as

$$\begin{aligned}\frac{DC_{CBOD}}{Dt} = {} & \frac{32}{12}K_{am}C_a + \frac{32}{12}\sum_{i\in\{z,f,p\}}(K_{im} + K_{ie} + K_{id})C_i - K_{CBOD}\frac{C_{DO}}{h_{CBOD} + C_{DO}}\theta_{CBOD}^{T-20}C_{CBOD} \\ & - \frac{5}{4}\frac{32}{14}K_{NO_3}\frac{h_{NO_3}}{h_{NO_3} + C_{DO}}\theta_{NO_3}^{T-20}C_{NO_3} + \frac{\omega_{CBOD}}{h}f_{PBOD}C_{CBOD}\end{aligned} \tag{3.17}$$

where

C_{NO3} = concentration of nitrate nitrogen, g/m^3
K_{im} = mortality rate of biotic organism i, 1/d
K_{ie} = excretion rate of biotic organism i, 1/d
K_{id} = defecation rate of biotic organism i, 1/d
K_{NO3} = denitrification rate, 1/d
h_{NO3} = half-saturation DO concentration for denitrification, g/m^3
ω_{CBOD} = settling velocity of CBOD, m/s
f_{PBOD} = fraction of particulate CBOD in total CBOD.

The major components of nitrogen in aquatic ecosystems are organic nitrogen (ON), ammonia (NH_3), and nitrate (NO_3). Mortality of biotic organisms produces organic nitrogen, which is converted into ammonia through bacteria decomposition. In the presence of nitrifying bacteria and oxygen, ammonia is oxidized to nitrite (NO_2) and nitrate via nitrification. The uptake of ammonia and nitrate by plants is through the assimilation process. In natural water, the presence of nitrogen gives rise to nitrification problem, eutrophication, and ammonia toxicity [29]. Nitrification reduces the oxygen level. One of the byproducts of nitrification is nitrate, which is a pollutant. Depending on the pH and temperature, ammonia can manifest itself into an un-ionized form, which is toxic to aquatic organisms. Both ammonia and nitrate are essential nutrients for photosynthesis, but high levels of ammonia and nitrate can result in excessive phytoplankton growth and in turn water quality problems. The kinetic processes of nitrogen are described by the following equations:

$$\frac{DC_{ON}}{Dt} = \sum_{i\in\{a,z,f,p\}} (1 - f_{NH_3})K_{im}\alpha_{NC}C_i - K_{ON}\frac{C_a}{h_{ON} + C_a}\theta_{ON}^{T-20}C_{ON} - \frac{\omega_{ON}}{h}f_{PON}C_{ON} + \frac{S_{ON}}{h} \tag{3.18}$$

$$\frac{DC_{NH_3}}{Dt} = \sum_{i\in\{a,z,f,p\}} f_{NH_3}K_{im}\alpha_{NC}C_i + \sum_{i\in\{z,f,p\}} K_{ie}\alpha_{NC}C_i + K_{ON}\frac{C_a}{h_{ON} + C_a}\theta_{ON}^{T-20}C_{ON} - K_{ag}p_{NH_3}\alpha_{NC}C_a - K_{NH_3}\frac{C_{DO}}{h_{NH_3} + C_{DO}}\theta_{NH_3}^{T-20}C_{NH_3} + \frac{S_{NH_3}}{h} \tag{3.19}$$

$$\frac{DC_{NO_3}}{Dt} = K_{NH_3}\frac{C_{DO}}{h_{NH_3} + C_{DO}}\theta_{NH_3}^{T-20}C_{NH_3} - K_{ag}(1 - p_{NH_3})\alpha_{NC}C_a - K_{NO_3}\frac{h_{NO_3}}{h_{NO_3} + C_{DO}}\theta_{NO_3}^{T-20}C_{NO_3} + \frac{S_{NO_3}}{h} \tag{3.20}$$

where

C_{ON} = concentration of organic nitrogen, g/m^3
K_{ON} = mineralization rate of organic nitrogen, 1/d
f_{NH3} = fraction of ammonia in dead organic material
ω_{ON} = settling velocity of organic nitrogen, m/s
f_{PON} = fraction of particulate organic nitrogen to organic nitrogen
S = concentration flux from the sediment bed, g/m^{2}s.

The ammonia preference factor is introduced to take into account the preference of ammonia over nitrate when both are available for phytoplankton to uptake, and calculated by

$$p_{NH_3} = \frac{C_{NH_3} C_{NO_3}}{(h_N + C_{NH_3})(h_N + C_{NO_3})} + \frac{C_{NH_3} h_N}{(C_{NH_3} + C_{NO_3})(h_N + C_{NO_3})} \quad (3.21)$$

where

h_N = Michaelis–Menten constant for nitrogen uptake, mgN/L.

Phosphorus in natural water exists in several states. The soluble reactive phosphorus (SRP), also called orthophosphate or soluble inorganic phosphorus, is the form that is readily available to phytoplankton. Particulate organic phosphorus is the form that mainly stays with living plants, animals, bacteria, and organic detritus. Nonparticulate organic phosphorus can be dissolved or colloidal organic compounds containing phosphorus. They are usually from the decomposition of particulate organic phosphorus. Particulate inorganic phosphorus consists of phosphate mineral, sorbed orthophosphate, and phosphate complex with solid matter. Nonparticulate inorganic phosphorus includes condensed phosphate such as those found in detergents. In the present model, phosphorus is divided into two main groups: organic phosphorus (OP) and inorganic phosphorus (orthophosphate, PO_4). The kinetic processes of phosphorus are governed by

$$\frac{DC_{OP}}{Dt} = \sum_{i \in \{a,z,f,p\}} (1 - f_{PO_4}) K_{im} \alpha_{PC} C_i - K_{OP} \frac{C_a}{h_{OP} + C_a} \theta_{OP}^{T-20} C_{OP} - \frac{\omega_{OP}}{h} f_{POP} C_{OP} + \frac{S_{OP}}{h} \quad (3.22)$$

$$\frac{DC_{PO_4}}{Dt} = \sum_{i \in \{a,z,f,p\}} f_{PO_4} K_{im} \alpha_{PC} C_i + \sum_{i \in \{z,f,p\}} K_{ie} \alpha_{PC} C_i + K_{OP} \frac{C_a}{h_{OP} + C_a} \theta_{OP}^{T-20} C_{OP} - K_{ag} \alpha_{PC} C_a + \frac{S_{PO_4}}{h} \quad (3.23)$$

where

C_{OP} = concentration of organic phosphorus, g/m^3
C_{PO4} = concentration of orthophosphate, g/m^3

α_{PC} = stoichiometric ratio of phosphorus to carbon, gP/gC
K_{OP} = mineralization rate of organic phosphorus, 1/d
h_{OP} = half-saturation phytoplankton conc. for mineralization of phosphorus, mg/L
ω_{OP} = settling velocity of organic phosphorus, m/s
f_{POP} = fraction of particulate organic phosphorus to organic phosphorus
f_{PO4} = fraction of phosphate in dead organic material.

6 Food Web Relations

The food webs in river systems are quite complex, and modeling of food web dynamics coupled with the water quality model is usually case-dependent. Therefore, several assumptions are made in order to simplify prey–predator relationships. Firstly, biotic organisms are considered separately as groups according to trophic levels. The upper trophic level can feed on the lower level as its food. Secondly, the feeding preference of predator on a particular group of prey is the same regardless of size, density, and distribution of prey. However, the feeding preference can be different when predator feeds on different groups of prey. Finally, age-structure is not considered in the model, and consequently the kinetic rates such as grazing and respiration rates of each group are given as constants.

In this study, the food web model consists of four trophic levels: phytoplankton, zooplankton, forage fish, and predatory fish, and phytoplankton is assumed to be the lowest trophic level or the main food source of the upper trophic levels. Certainly this assumption has limitation because some stream ecosystems are also based on insects, benthic fauna, and benthic algae, which are not included here.

The dynamic processes of phytoplankton are described by

$$\frac{DC_a}{Dt} = (K_{ag} - K_{ar} - K_{ae} - K_{am})C_a - \sum_{i\in\{z,f,p\}} K_{ig} f_{ai} C_i - \frac{\omega_a}{h} C_a \tag{3.24}$$

where

K_{ag} = rate of photosynthesis, 1/d
K_{ar} = rate of respiration of phytoplankton, 1/d
K_{ae} = rate of excretion of phytoplankton, 1/d
K_{am} = rate of nonpredatory mortality of phytoplankton, 1/d
K_{ig} − grazing rate of predator i (− z, f, p), 1/d
f_{ai} = relative preference of predator i on phytoplankton as food
ω_a = settling velocity of phytoplankton, m/d.

The photosynthesis rate is modeled as the maximum photosynthesis rate ($K_{ag,max}$) multiplied by environmental factors as

$$K_{ag} = K_{ag,\max} f_N f_L f_T f_{TOX} \tag{3.25}$$

where

f_N = nutrient limitation factor
f_L = light limitation factor
f_T = temperature limitation factor
f_{TOX} = reduction factor due to toxic chemical from Eq. (3.57).

The nutrient limitation factor for photosynthesis process is computed using a Michaelis–Menten equation as follows [32]

$$f_N = \min\left(\frac{C_{NH_3} + C_{NO_3}}{h_N + C_{NH_3} + C_{NO_3}}, \frac{C_{PO_4}}{h_P + C_{PO_4}}\right) \tag{3.26}$$

where

h_N = half-saturation concentration for nitrogen, g/m^3
h_P = half-saturation concentration for phosphorus, g/m^3.

The effect of light on phytoplankton growth is complex. The interception and utilization of light by phytoplankton determine net productivity, species succession, and abundance of higher trophic organisms [33]. Several factors, such as the light attenuation through water depth and the dependence of growth on light, can be integrated to come up with the total effect. The depth-averaged light limitation factor is modeled as [34]

$$f_L = \frac{1}{\gamma h} \ln\left(\frac{h_L + I_0}{h_L + I_0 e^{-\gamma h}}\right) \tag{3.27}$$

where

I_0 = light intensity at the water surface
h_L = half-saturation light intensity for phytoplankton growth
γ = light extinction, 1/m.

The light extinction is calculated by the modified equation from the WASP6 model as [6]

$$\gamma = \gamma_0 + 0.0088C_a + 0.054C_a^{0.67} + 0.0458C_m \tag{3.28}$$

where

γ_0 = background light extinction, 1/m
C_a = phytoplankton concentration as total Chlorophyll-a, μg/L
C_m = concentration of suspended solid, g/m^3.

Aquatic organisms have preferred temperature ranges. Biological production increases as a function of temperature until an optimum temperature. The temperature limitation factor for biological growth is calculated using the Cerco and Cole's formula as [5]

$$f_T = \begin{cases} \exp\left[-KT_{g1}\left(T - T_{opt}\right)^2\right] & T \le T_{opt} \\ \exp\left[-KT_{g2}\left(T_{opt} - T\right)^2\right] & T > T_{opt} \end{cases} \tag{3.29}$$

where

T_{opt} = optimal temperature for biological growth
KT_{g1} = coefficient representing the relationships of growth on temperature below the optimal temperature
KT_{g2} = coefficient representing the relationships of growth on temperature above the optimal temperature.

Compared to the theta model described below, the Cerco and Cole's formulation may become necessary when several individual species or groups of algae are modeled because each group of algae is sensitive to temperature differently [29].

For respiration, excretion, and non-predatory mortality rates of phytoplankton, the temperature limitations are computed by the theta model, $f_T = \theta^{T-20}$, as follows:

$$K_{ar} = K_{ar,\max}\theta_a^{T-20} \tag{3.30}$$

$$K_{ae} = K_{ae,\max}\theta_a^{T-20} \tag{3.31}$$

$$K_{am} = \left(1 + f'_{am}\right)K_{am,\max}\theta_a^{T-20} \tag{3.32}$$

where

$K_{ar,max}$ = maximum rate of respiration of phytoplankton, 1/d
$K_{ae,max}$ = maximum rate of excretion of phytoplankton, 1/d
$K_{am,max}$ = maximum rate of nonpredatory mortality of phytoplankton, 1/d
f'_{am} = increase factor in the mortality due to toxic chemicals which can be calculated from the general form shown in Eq. (3.55).

For a higher trophic level (i = z, f, p), the dynamic process is modeled as

$$\frac{DC_i}{Dt} = \left(K_{ig} - K_{ir} - K_{ie} - K_{im} - K_{id} - K_{ib} + K_{ire}\right)C_i - \sum_j K_{jg} f_{ij} C_j \tag{3.33}$$

where

K_{ig} = grazing rate of organism i, 1/d
K_{ir} = respiration rate of organism i, 1/d
K_{ie} = excretion rate of organism i, 1/d
K_{im} = nonpredatory mortality rate of organism i, 1/d
K_{id} = defecation rate of organism i, 1/d
K_{ib} = gamete loss rate of organism i, 1/d

K_{ire} = reproduction rate of organism i, 1/d
f_{ij} = relative preference factor of predator j feeding on organism i as food.

The general formulation of the grazing rate is computed by

$$K_{ig} = K_{ig,\max}\lambda_i f_T f'_{ig} \tag{3.34}$$

where

$K_{ig,max}$ = maximum grazing rate of predator i, 1/d
λ_i = grazing limitation factor
f_T = temperature limitation factor from Eq. (3.29)
f'_{ig} = reduction factor in animal growth due to toxic chemical from Eq. (3.58).

The maximum grazing rate can be given as a constant, but it is generally calculated from the body weight with $K_{ig,max} = a_i W_i^{bi}$, where a_i is the weight specific consumption, b_i is the slope of the allometric function, and W_i is the average body weight of organism i.

The grazing limitation factor λ_i reduces the grazing rate of predator when food concentrations are low. λ_i is calculated by using a modified Michaelis–Menten factor and a threshold food concentration μ_j, which is revised from the equation of Rounds et al. [35]:

$$\lambda_i = \frac{\sum_j (p_{ji}C_j - \mu_j)}{h_i + \sum_j (p_{ji}C_j)} \tag{3.35}$$

where

h_i = half-saturation food concentration for grazing, g/m^3
p_{ji} = preference of predator i feeding on organism j as food.

The general form of relative preference of predator i on organism j as food, used in the grazing term of each prey–predator relationship, is calculated as $f_{ji} = p_{ji}C_i / \sum_j (p_{ji}C_i)$ [10].

It should be noted that detritus, which is derived from the mortality, excretion, and defecation of living organisms, can be a food source for zooplankton and fish and thus is considered in the present food web model. However, the concentration of detritus, C_d, is not computed directly, but related to the CBOD by a simple relation: $C_d = 12/32 \times C_{CBOD}$.

The excretion and non-predatory mortality rates of zooplankton and fish are modeled as single first-order kinetics similar to the phytoplankton model:

$$K_{ie} = K_{ie,\max}\theta_i^{T-20} \tag{3.36}$$

$$K_{im} = \left(1 + f'_{im}\right) K_{im,\max}\theta_i^{T-20} \tag{3.37}$$

where

$K_{ie,max}$ = maximum rate of excretion of organism i, 1/d
$K_{im,max}$ = maximum rate of nonpredatory mortality of organism i, 1/d
f'_{im} = increase factor in the mortality due to toxic chemicals from Eq. (3.55).

A fraction of ingested food can be egested as feces or discarded as organic material. The defecation rate of unassimilated food depends on the assimilating efficiency e_{ig} and is determined as

$$K_{id} = \left(1 - e_{ig}\right) f'_{id} K_{ig} \tag{3.38}$$

where

f'_{id} = increase factor in the defecation due to toxic chemicals from Eq. (3.59).

The respiration of zooplankton and fish can be modeled as single first-order kinetics with a rate as

$$K_{ir} = K_{ir,\max}\theta_i^{T-20} \tag{3.39}$$

where

$K_{ir,max}$ = maximum respiration rate of organism i, 1/d.

The respiration in fish is comprised of three components: standard, active, and dynamic respirations. Standard respiration is a rate at resting in which the organism is expending energy without consumption. Active respiration depends on swimming speed and temperature. The dynamic action is the metabolic action due to digesting and assimilating prey. The maximum respiratory rate of fish (i = f, p) can be calculated as [10]

$$K_{ir,\max} = K_{ir0} f_{den} f_{act} + f_{dyn}\left(K_{ig} e_{ig} - K_{id}\right) \tag{3.40}$$

where

K_{ir0} = basal or standard respiratory rate, 1/d
f_{act} = factor for respiratory rate associated with swimming or active respiratory fraction
f_{dyn} = proportion of assimilated energy lost to specific dynamic action
f_{den} = density-dependent respiration factor and can be computed by [10, 36]

$$f_{den} = 1 + 0.25 C_i / K_i \tag{3.41}$$

where
C_i = biomass concentration of fish i, g/m^3
K_i = carrying capacity of fish i, g/m^3 which depends on species and location. In this study, the carrying capacity value is taken from [10, 37].

The gamete and reproduction rates are only used in fish dynamic models. Eggs and sperm (gametes) in adult fish are a significant fraction of fish biomass. The production of gametes is influenced by environmental factors, such as temperature, genetic factors, hormones, and nutrition. Gametogenesis and spawning occurs during a defined period when the environmental conditions are optimal in terms of survival. It is assumed that spawning occurs when the water temperature enters an appropriate range of optimal temperature which is between $0.6T_{opt}$ and $T_{opt} - 1$, where T_{opt} is the optimal water temperature for fish spawning [10]. This gamete loss can be determined by

$$K_{ib} = K_{ib0}\left(1 + f'_{ib}\right)\min(1, C_i/K_i) \tag{3.42}$$

where

K_{ib0} = intrinsic gamete mortality rate, 1/d
f'_{ib} = increase factor in the gamete due to toxic chemicals from Eq. (3.60).

In general, only a fraction of the gametes results in the biomass of young fish and subsequently adult fish. Some of them unsuccessfully reproduce and become organic materials. The increase in the biomass of small fish due to spawning in large fish when both are in the same species is referred to as reproduction. There are many environmental factors resulting in reproduction failure such as predator and toxic chemical. Due to the uncertainty of the reproduction process, these factors are neglected in this study. For the simplified single-age structure of fish dynamics, the reproduction rate K_{ire} depends on the gamete loss in adult fish, the percentage of success in reproduction, r_{ire}, and the biomass ratio between young-of-the-year (YOY) fish and adult fish, r_{iYA}, for a given fish species, as expressed with $K_{ire} = K_{ib0}\min(1, C_i/K_i)r_{ire}r_{iYA}$.

7 Fate and Transport of Contaminants

When a contaminant is discharged into a river, it is subject to fate and transport processes as shown in Fig. 3.4. It is usually dissolved in the water or absorbed by the moving sediments. Changes in concentration of the contaminant in the water column are caused by advection, diffusion (mixing), external loading, sorption, desorption, volatilization, photolysis, microbial decay, settling with sediments, exchange with the bed, uptake and depuration by the aquatic organisms, etc. One may determine the dissolved and absorbed contaminants separately using the non-equilibrium partition model or the total concentration by assuming the dissolved and absorbed phases in the equilibrium state [24]. The latter approach is used here, so that the contaminant transport in the water column is governed by Eq. (3.1) with the source/sink terms:

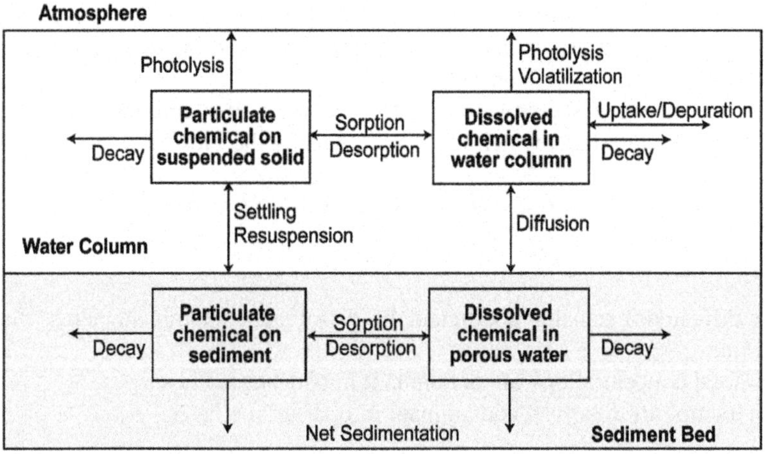

Fig. 3.4 Fate and transport of contaminants in water column and sediment bed

$$\frac{DC_{tw}}{Dt} = q_{tw} + \frac{J_{dbw}}{h} + \frac{q_{t,ex}}{h} - \left(K_N + K_H[H]^+ + K_{OH}[OH]^-\right) f_{dw} C_{tw} - K_p C_{tw} - K_b C_{tw}$$
$$-\frac{K_v}{h}\left(\frac{C_g}{H/RT_K} - f_{dw}C_{tw}\right) - K_{a1} f_{dw} C_{tw} C_a + K_{a2} C_{ta} - \sum_{j=\{z,f,p\}} \left(K_{j1} f_{dw} C_{tw} - K_{j2} C_{tj}\right) \tag{3.43}$$

where

C_{tw} = total concentration of contaminant in the water column, g/m^3
f_{dw} = fraction of dissolved concentration to the total concentration of contaminant in water column
q_{tw} = total loading rate of contaminant per unit volume, g/m^{3}d
J_{dbw} = vertical diffusion fluxes between water column and bed surface layer, g/m^{2}d
$q_{t,ex}$ = total exchange rate of contaminant due to sediment erosion and deposition, g/m^{2}d
K_N = neutral hydrolysis rate, 1/d
K_H = acid-catalyzed hydrolysis rate, m^3/mold
K_{OH} = base-catalyzed hydrolysis rate, m^3/mol d
$[H]^+$ = molar concentration of hydrogen ion, mol/m^3
$[OH]^-$ = molar concentration of hydroxide ion, mol/m^3
K_p = photolysis rate, 1/d
K_b = biodegradation rate, 1/d
K_v = volatilization rate, m/d
H = Henry's law constant, atm m^3/mol
R = universal gas constant, atm m^3/mol °K
C_g = gas-phase concentration of the contaminant, g/m^3.

Note that the last three terms in Eq. (3.43) are related to the changes in contaminant concentrations due to biotic organisms, and the related variables and explanations are given in Sect. 8.

The flux between the bed pore water and the water column occurs through diffusion at the interface, which is calculated using the Fick's law as

$$J_{dbw} = k_{dbw}\left(f_{db,1}C_{tb,1} - f_{dw}C_{tw}\right) \tag{3.44}$$

where

k_{dbw} = diffusional transfer coefficient of dissolved contaminant across the bed surface
$C_{tb,1}$ = total concentrations of contaminant in bed surface layer
$f_{db,1}$ = fraction of dissolved contaminant in bed surface layer.

By using the linear isotherm of sorption–desorption process, the fractions of dissolved and particulate contaminants in the water column and sediment bed are computed by [29]

$$f_{dw} = 1/(1 + K_d C_m), \quad f_{pw} = 1 - f_{dw} \tag{3.45}$$

$$f_{db} = \phi/(\phi + K_d \rho_d), \quad f_{pb} = 1 - f_{db} \tag{3.46}$$

where

f_{pw} = fraction of particulate contaminant in the water column
f_{pb} = fraction of particulate contaminant in the bed sediment
K_d = sorption–desorption coefficient, m^3/g
C_m = suspended sediment concentration, g/m^3
ϕ = porosity
ρ_d = dry density of the bed sediment, g/m^3.

The exchange rate of contaminant due to deposited and eroded sediments is calculated by [24]

$$q_{t,ex} = \max(E_b - D_b, 0)\frac{C_{tb,1}}{1-\phi} + \min(E_b - D_b, 0)\left(\frac{\phi}{1-\phi}\frac{f_{dw}C_{tw}}{1-m} + \frac{f_{pw}C_{tw}}{m}\right) \tag{3.47}$$

where

D_b = sediment deposition rate, m/d
E_b = sediment erosion rate, m/d
m = suspended sediment concentration by volume.

These sediment quantities in the above equation are computed in the sediment model. The first term on the right-hand side accounts for the net erosion case, with $1/(1 - \phi)$ converting the net eroded bed sediment rate $\max(E_b - D_b, 0)$ to the net erosion rate of the bed sediment and pore water mixture in which $C_{tb,1}$ is defined (contaminant mass over the total volume of the bed sediment and pore water mixture). Note that the pore water between the eroded sediment particles is also entrained into the water column. The second term on the right-hand side accounts for the net deposition case, in which the net deposited sediment rate is $\min(E_b - D_b, 0)$. This sediment is equivalent to a volume of $\min(E_b - D_b, 0)/m$ in the water column in which C_{tw} is defined. In the meantime, this net deposited sediment accompanies with water from the water column to bed pores, and the volume rate of this water is $\min(E_b - D_b, 0)\,\phi/(1-\phi)$ in the bed surface layer, which is equivalent to a volume of $\min(E_b - D_b, 0)\phi/[(1-\phi)(1-m)]$ in the water column in which C_{tw} is defined. Therefore, the second term on the right-hand side includes the contaminants dissolved in the water column and absorbed with moving sediment that both deposit onto the bed. Detailed derivation of Eq. (3.47) can be found in [24].

Contaminant in the sediment bed is usually transported by the pore water flow, and the thin surface layer in the bed may be mixed by bioturbation. For simplicity, a vertical diffusion model is used in this study, which divides the sediment bed into a suitable number of layers (three layers are used here as example) and determines the fate and transport of contaminant in the bed layers as [24, 38]

$$\frac{\partial(\delta_1 C_{tb,1})}{\partial t} = Q_{tb,1} - k_{tb,1}\delta_1 C_{tb,1} - k_{dbw}\left(f_{db,1}C_{tb,1} - f_{dw}C_{tw}\right) - q_{t,ex} + k_{db12}\left(f_{db,2}C_{tb,2} - f_{db,1}C_{tb,1}\right) + q_{tb12} \tag{3.48}$$

$$\frac{\partial(\delta_2 C_{tb,2})}{\partial t} = Q_{tb,2} - k_{tb,2}\delta_2 C_{tb,2} - k_{db12}\left(f_{db,2}C_{tb,2} - f_{db,1}C_{tb,1}\right) - q_{tb12} + k_{db23}\left(f_{db,3}C_{tb,3} - f_{db,2}C_{tb,2}\right) \tag{3.49}$$

$$\frac{\partial(\delta_3 C_{tb,3})}{\partial t} = Q_{tb,3} - k_{tb,3}\delta_3 C_{tb,3} - k_{db23}\left(f_{db,3}C_{tb,3} - f_{db,2}C_{tb,2}\right) \tag{3.50}$$

where

$C_{tb,i}$ = total concentration of contaminant in bed layer i
δ_i = thickness of layer i
$Q_{tb,i}$ = total contaminant loading rate in layer i
$k_{tb,i}$ = decay coefficient of contaminant at layer i
$k_{dbi,i+1}$ = diffusional transfer coefficient of dissolved contaminant between layers i and i + 1
$q_{tbi,i+1}$ = total exchange rate of contaminant between layers i and i + 1 due to lowering and rising of the interface.

Note that it is assumed in Eqs. (3.48)–(3.50) that the interface between bed layers 1 and 2 may lower or rise due to bed change, while the interface between bed layers 2 and 3 does not change, as explained in [22].

8 Bioaccumulation Processes

The transfers of contaminant in water through surface sorption of phytoplankton and gill and dietary uptakes in fish are important routes of contaminant uptake in aquatic ecosystems. The contaminant concentration in the aquatic organisms is governed by the advection–dispersion equation as shown in Eq. (3.1). The dynamic processes for contaminant concentration in phytoplankton can be described by a simple linear reversible sorption–desorption equation suggested in [39]. The resulting net source term in Eq. (3.1) is [10]

$$\frac{DC_{ta}}{Dt} = K_{a1} f_{dw} C_{tw} C_a - K_{a2} C_{ta} - (K_{ae} + K_{am}) C_{ta} - \sum_{i \in \{z,f,p\}} \left(K_{ig} f_{ai} C_i\right) \nu_a \quad (3.51)$$

where

C_{ta} = concentration of contaminant associated with phytoplankton in unit volume of water column, g/m^3
K_{a1} = uptake rate of contaminant, m^3/g d
K_{a2} = depuration rate of contaminant, 1/d
ν_a = concentration of contaminant in phytoplankton, g/g.

The concentration of contaminant in organism i is calculated by

$$\nu_i = C_{ti}/C_i \quad (3.52)$$

where

C_{ti} = contaminant concentration associated with organism i in unit volume of water column, g/m^3
C_i = biomass concentration of organism i, g/m^3.

For higher trophic levels, the input of contaminant due to ingestion of contaminated food plays an important role. The rate of change in contaminant concentration in biotic organism i (=z, f, p) is determined by

$$\frac{DC_{ti}}{Dt} = K_{i1} f_{dw} C_{tw} + \sum\nolimits_k K_{ig} e_{ig} C_i f_{ki} \nu_k - (K_{i2} + K_{im} + K_{id} + K_{ib}) C_{ti} - \sum\nolimits_j \left(f_{ij} K_{jg} C_j\right) \nu_i \quad (3.53)$$

where

K_{i1} = uptake rate of contaminant of organism i, 1/d
K_{i2} = depuration rate of contaminant of organism i, 1/d
j = organism index representing predator of organism i
k = organism index representing prey of organism i.

Connolly et al. [40] proposed a formula to determine the uptake rate of contaminant by an aquatic organism. The uptake rate depends on the respiration rate and transfer efficiency across the organism's membrane. The depuration of contaminant is related to the characteristics of organism such as body weight and lipid content and the chemical properties of the contaminant, i.e., the octanol–water partition coefficient, K_{ow} [10]. Due to a single-age class of the current fish dynamic model, body weight and lipid content are kept as constants. Therefore, the uptake and depuration rates of contaminant cannot be determined from the existing formula. They are treated as calibrated constant parameters, similar to the other kinetics rates in the fish model, for instance, grazing and respiration rates.

9 Effects of Toxic Chemicals

Biomass loss due to acute toxicity is estimated based on the internal concentration of the contaminant in the biotic organism, IC_{i50}. The internal concentration depends on the concentration of contaminant in water that causes 50 % mortality for a given period of exposure, LC_{i50}, and the bioconcentration factor, BCF, where $IC_{i50} = LC_{i50} \times BCF$. The constant uptake and depuration rates of contaminant are used, thus the constant BCF is applied for all aquatic organisms in the current model. The internal concentrations of contaminant vary due to the depuration process of organisms. The time-varying concentration of the contaminant C_{i50} is calculated as [10]

$$C_{i50} = IC_{i50}\left(1 - e^{-K_{i2}\,t_1}\right) / \left(1 - e^{-K_{i2}\,t_2}\right) \tag{3.54}$$

where

t_1 = exposure time in toxicity test
t_2 = period of exposure.

The fraction killed by a given internal concentration of toxicant is estimated using the time-dependent C_{i50} in the cumulative form that is determined by [10]

$$f'_{im} = 1 - \exp(-v_i / C_{i50})^{1/K_s} \tag{3.55}$$

where

f'_{im} = fraction of organism i killed for a given period of exposure
K_s = parameter representing toxic response (= 0.33).

When the concentration level of the contaminant is less than the level causing death, organisms still experience some adverse effects. The ratio of chronic to acute concentration is used to predict the chronic effect, and is calculated by

$$r_{i50} = EC_{i50}/LC_{i50} \tag{3.56}$$

where

r_{i50} = chronic to acute ratio
EC_{i50} = contaminant concentration in water that causes 50 % reduction in photosynthesis, growth, or reproduction.

The effects of contaminant on phytoplankton photosynthesis and animal growth and reproduction are considered through the reduction factor f_{TOX}. The general form is expressed as [10]:

$$f_{TOX} = \exp(-v_i/C_{i50}r_{i50})^{1/K_S} \tag{3.57}$$

For phytoplankton, the effect of chemical from Eq. (3.57) is directly applied into the photosynthesis rate of phytoplankton in Eq. (3.25). However, in animals, the reduction factor for growth is related to assimilation and defecation processes. It is assumed that 20 % of the assimilation is reduced while the amount of food that is not assimilated increases by 80 % [10]

$$f'_{ig} = 1 - 0.2f_{TOX} \tag{3.58}$$

$$f'_{id} = 1 + 0.8f_{TOX} \tag{3.59}$$

where

f'_{ig} = reduction factor for animal assimilation
f'_{id} = increase factor for the amount of unassimilated food.

The effect of contaminant on the reproduction of animals is complex since several factors are involved in the reproductive failure. For simplification, the reduction factor for reproduction is applied only for the increase of gamete mortality, which is written as

$$f'_{ib} = 1 - f_{TOX} \tag{3.60}$$

where

f'_{ib} = increase factor in gamete loss due to contaminant.

Note that f_{TOX} in Eq. (3.60) is calculated from Eqs. (3.56) and (3.57) by using the contaminant concentration in water that causes 50 % reduction in reproduction EC_{i50}.

10 Model Test

10.1 Model Test in the Tualatin River, Oregon

10.1.1 Study Area

The Tualatin River is located in the west side of the Portland metropolitan area, northwestern Oregon, USA. Its watershed drains 1844 km^2. The main stem of the river is approximately 128 km, originates in the Coast Range, and flows eastward before joining the Willamette River near West Linn, as shown in Fig. 3.5. The average slope and width of the river ranges 0.01524–14.1 m/km and 4.6–46 m, respectively [35]. Historically, the wastewater treatment plants in the urban area of the Tualatin River watershed discharged high concentrations of ammonia and phosphorus into the river [35]. The river at the lower reaches encountered algal blooms and consequently faced water quality problem. The water quality violations in the river included the minimum DO level, the maximum pH standard, and the exceedance of phytoplankton concentration.

Water quality and ecological properties of the Tualatin River in the period from May 1, 1991 to October 31, 1993 are simulated in this study. The simulation domain is approximately 50 km long, from the Rood Bridge at Hillsboro or at river mile (RM, 1 mile = 1.61 km) 38.4 to the Stafford Road near Lake Oswego (RM5.5). It is represented by 132 cross-sections, and each cross-section is divided into 11 vertical panels. The time step is 15 min. Several tributaries, irrigation withdrawals, and two wastewater treatment plants (WWTP) at Rock Creek (RM38.1) and Durham (RM9.3) are included as side discharges. The measurement data for hydrodynamic, water quality, and ecological properties are reported in [41], and the estimated irrigation withdrawals are published in [35]. Daily air temperature, dew point temperature, wind speed, wind direction, and precipitation (rainfall) measured at the Tualatin Valley Irrigation District (TVID) Agrimet Weather Station located in Verboort, Oregon [41] are used as inputs for water temperature simulation. Details on data interpretation and assumption are summarized in [42].

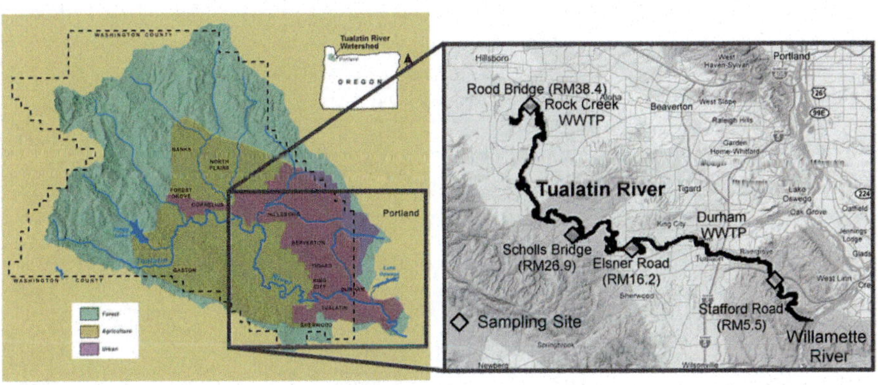

Fig. 3.5 Study domain: Tualatin River, Oregon (http://www.trwc.org and www.maps.google.com)

10.1.2 Estimation and Calibration of Model Parameters

Model parameters are important for determining the transport and transformation of each constituent in the model. Some parameters are taken from technical reports such as [35, 43, 44], and some are calibrated. For example, the Manning's roughness coefficient n is calibrated as a constant value of 0.025 $s/m^{1/3}$ in the entire simulation domain. The longitudinal dispersion coefficient D_x is calibrated as 5.0 m^2/s by simulating a conservative tracer, chloride. Selected parameters related to water quality and ecological simulations are given in Tables 3.3 and 3.4, respectively.

The zooplankton abundance may be affected by planktivory fish [35]. However, because there is no fish data available, the fish is not considered in this case. Therefore, the predatory and non-predatory mortality rates of zooplankton are combined as a single mortality rate, which is calibrated as 0.005 1/d. The zooplankton grazing rate coefficient is allowed to vary during the simulation period. It is calibrated as a constant value of 0.9 1/d in the reach upstream of RM12.25, and varies seasonally between 0.6 and 1.2 1/d in the reach downstream of RM12.25 as presented in Fig. 3.6. Such difference between the upstream and downstream reaches may reflect different biotic processes due to a large amount of organic materials accumulated at the river bed in downstream reaches [44].

Table 3.3 Summary of model parameters for water quality simulations

Symbol	Unit	Value	Symbol	Unit	Value
K_{ON}	1/d	0.20	f_{PO4}	–	0.75
K_{NH3}	1/d	0.05	f_{PBOD}	–	0.5
K_{NO3}	1/d	0.10	h_{NH3}	mg O_2/L	2.5
K_{OP}	1/d	0.25	h_{NO3}	mg O_2/L	2.0
K_{CBOD}	1/d	0.25	h_{CBOD}	mg O_2/L	0.5
α_{NC}	gN/gC	0.16	ω_{CBOD}	m/d	0.01
α_{PC}	gP/gC	0.022	θ_{RE}	–	1.0241
f_{NH3}	–	0.50	θ	–	1.047
			θ_{SOD}	–	1.065

Table 3.4 Summary of model parameters for phytoplankton and zooplankton dynamics

Symbol	Unit	Value	Symbol	Unit	Value
p_{dz}	–	0.15	$K_{zm,max}$	1/d	0.005
p_{az}	–	0.85	h_L	W/m^2	177
$K_{ag,max}$	1/d	2.0	h_N	mg/L	0.01
$K_{zg,max}$	1/d	0.6–1.2	h_P	mg/L	0.005
$K_{ar,max}$	1/d	0.35	h_{gz}	mg/L	0.08
$K_{ae,max}$	1/d	0.0025	γ_0	m^{-1}	1.002
$K_{am,max}$	1/d	0.20	r_{CChla}	gC/gChl a	25.0
$K_{zr,max}$	1/d	0.005	θ	–	1.072
$K_{ze,max}$	1/d	0.0002	ω_a	m/d	0.05

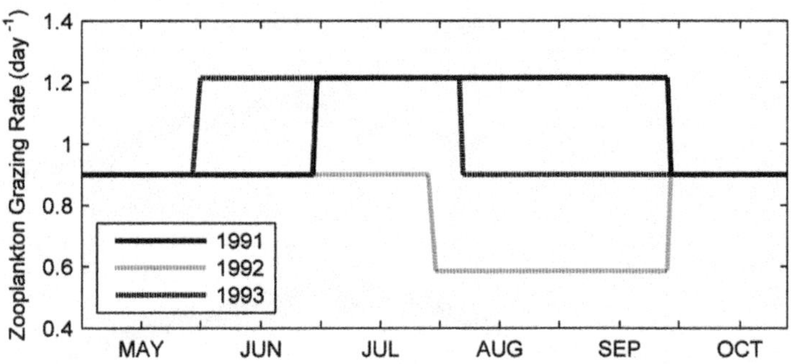

Fig. 3.6 Zooplankton grazing rates in the downstream reaches of the Tualatin River

10.1.3 Simulation Results

Figure 3.7 compares the measured and calculated water temperatures at Station RM16.2. Due to the use of 1-D heat transport equation, the simulated water temperature is cross-sectionally averaged. The field data obtained from [41] was collected at depths of 3, 6, 9, 12, 15, and 18 ft (1 ft = 0.3048 m) below the water surface. The cross-sectionally averaged water temperature is comparable to the measured water temperatures averaged over the depths, with 0.952 for R^2.

The depth-averaged reaeration rate K_{RE} and the sediment oxygen demand (SOD) are two key factors affecting DO concentration. The SOD in the main stem and several tributaries of the Tualatin River was measured in 1992–1994 during the summer period from May through October each year [44]. The temporally and spatially varied measurement SOD rates are used in the present simulation in 1992–1994, whereas the SOD rates in 1991 are unavailable and are approximated using the 1992 data. Because no measurement data for the reaeration rate, we tested the reaeration rate formulas of O'Connor and Dobbins (1958), Churchill et al. (1962), and Cadwallader and McDonnell (1969) by matching the simulated DO results with the measurement data. The simulated flow depths of the study reach in 1991–1993 range between 1.6 and 3.1 m and the flow velocities are approximately 0.02–1.2 m/s. According to the hydrodynamic properties of the river, the Churchill formula, $K_{RE} = 5.02Uh^{-1.67}$ (1/d), is suitable in this river reach and thus provides the simulated DO concentrations most comparable to the measurements. Here, U is the flow velocity in m/s and h is the water depth in m. The general trend of DO seasonal variation is reproduced by the model, as shown in Fig. 3.7, with R^2 of 0.606 when comparing the simulated and measured DO concentrations.

Figure 3.8 shows the temporal variations of simulated and measured ammonia and nitrate concentrations at RM5.5, and Fig. 3.9 presents the longitudinal profiles of mean concentrations of ammonia and nitrate averaged during the summer months May–October. The simulated and measured nitrogen concentrations agree well, with R^2 of 0.890 and 0.792 for ammonia and nitrate respectively. Measurement and simulation show low concentrations of ammonia in the upper reach from the upstream end to RM11.6. The ammonia and nitrate concentrations increase significantly at RM38.1 and RM9.3 due to the lateral discharges from the Rock

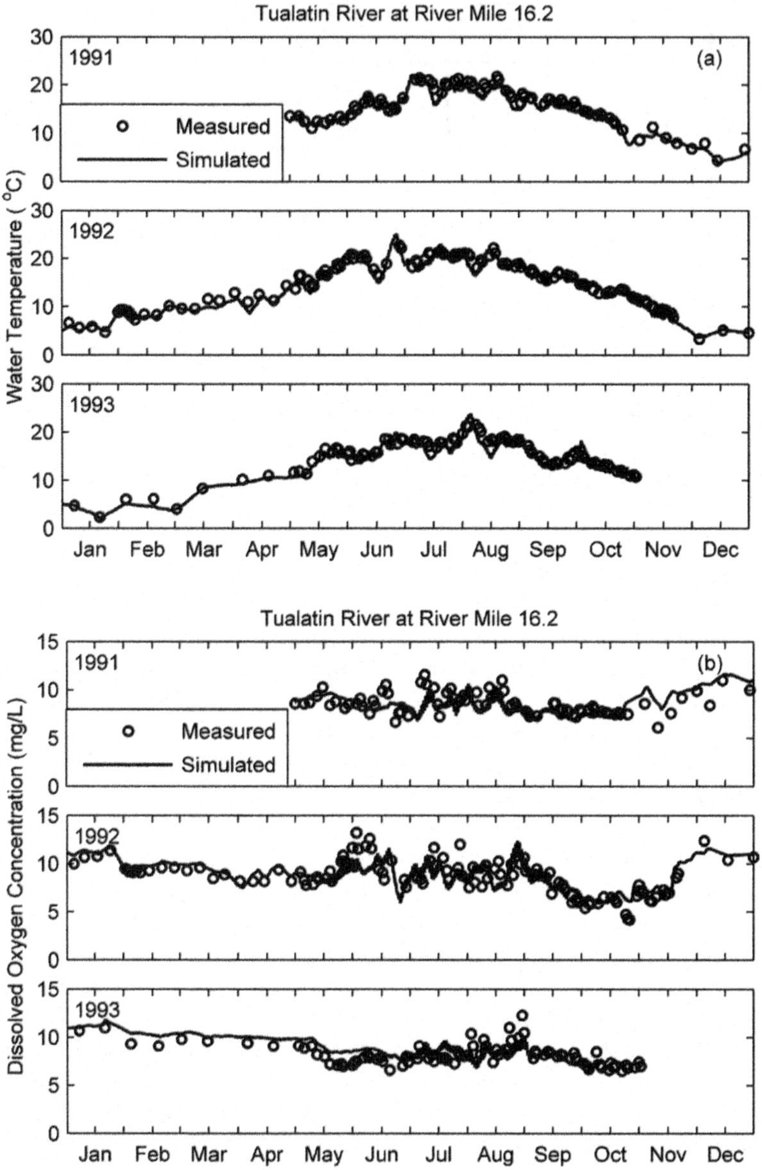

Fig. 3.7 Simulated vs. measured water temperature (**a**) and DO (**b**) concentrations at RM16.2 of the Tualatin River (measurements from Doyle and Caldwell [41])

Creek and Durham WWTPs. The figures show that the model can predict the variations of instream concentrations due to lateral inputs.

Figure 3.10 compares the simulated and measured temporal variations of phosphate concentrations at RM26.9 and longitudinal profiles of phosphate

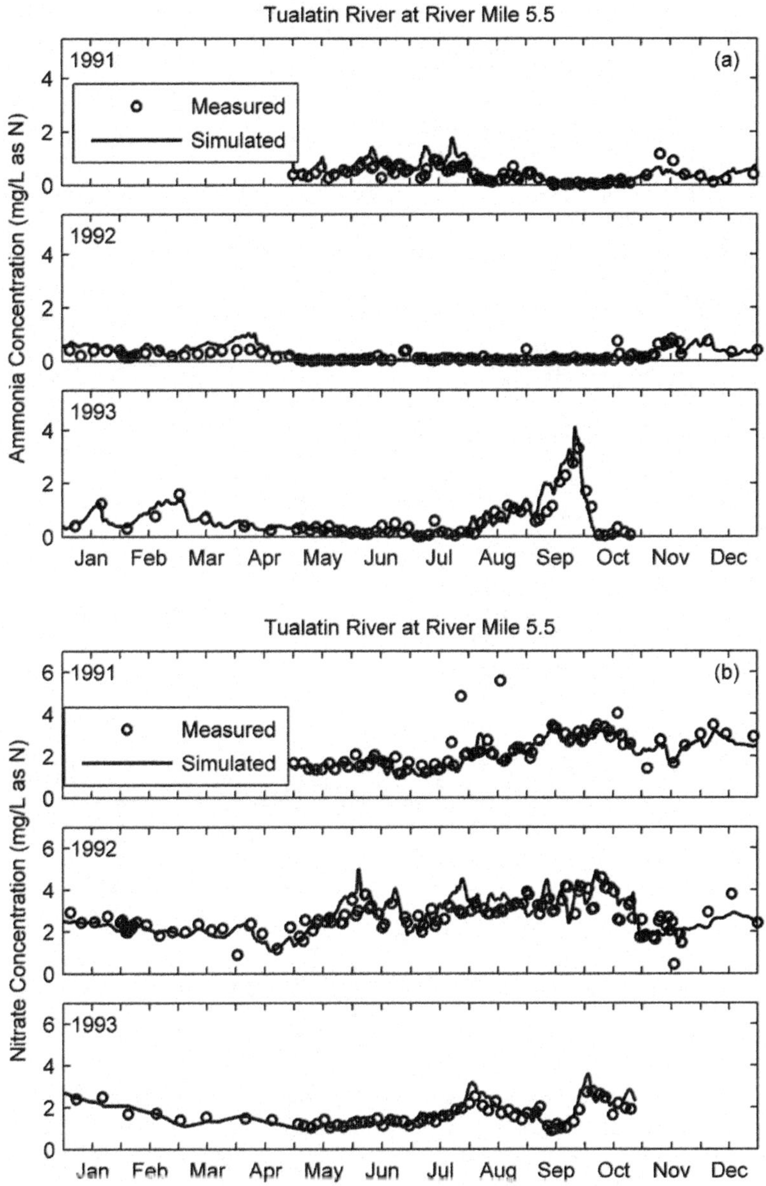

Fig. 3.8 Simulated vs. measured ammonia (**a**) and nitrate (**b**) concentrations at RM5.5

concentrations averaged over the summer months. The overall trend of simulated phosphate concentrations is comparable to the measurements, with R^2 of 0.922. In particular, reduction in the seasonally averaged concentration of phosphorus due to operation of the WWTPs at peak phosphorus-removal efficiency in 1992 [24] is presented in both measurement and simulation.

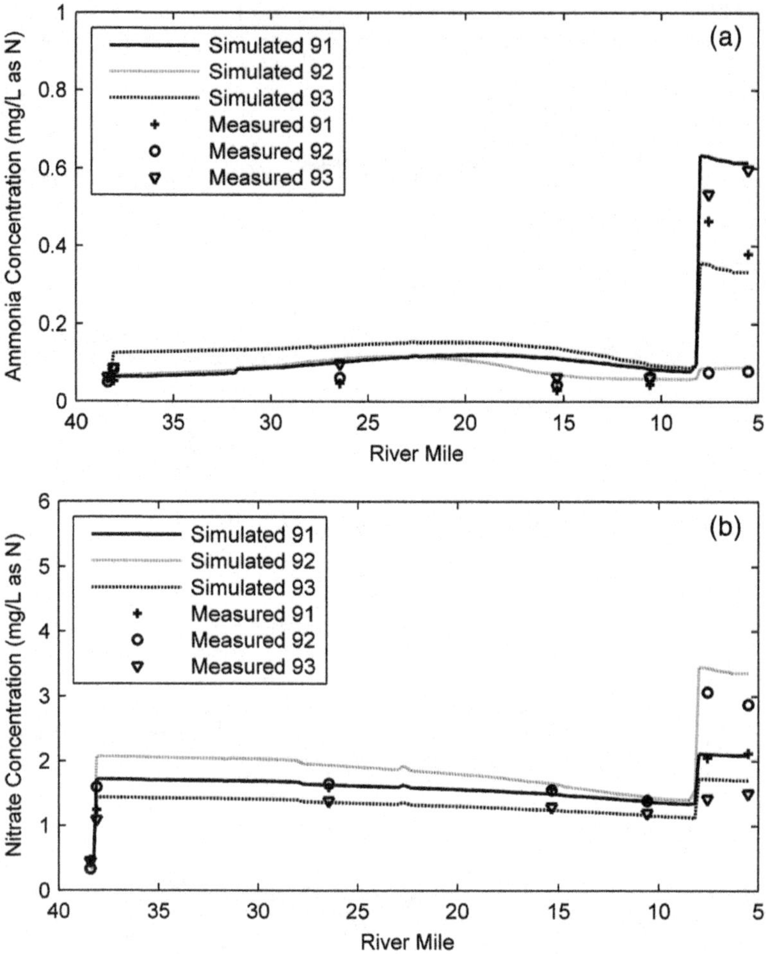

Fig. 3.9 Measured vs. simulated mean ammonia (**a**) and nitrate (**b**) concentrations during May–October

Figure 3.11 presents the comparison of the simulated and measured phytoplankton concentrations at RM5.5 and longitudinal profiles of phytoplankton biomasses averaged during the summer months. The simulation results and the field data are in a good agreement in the upstream locations. However, the simulated biomass in 1992 is much lower than the measurement in the downstream reaches. This may be due to the phytoplankton growth is limited by low concentration of phosphorus during this period. In addition, one can see that the model can predict the daily fluctuations due to the daily growth cycle.

The comparisons of measured and simulated zooplankton biomass at RM5.5 and other sections are presented in Fig. 3.12. The measured zooplankton concentrations have strong seasonal and interannual variability [35], especially at downstream

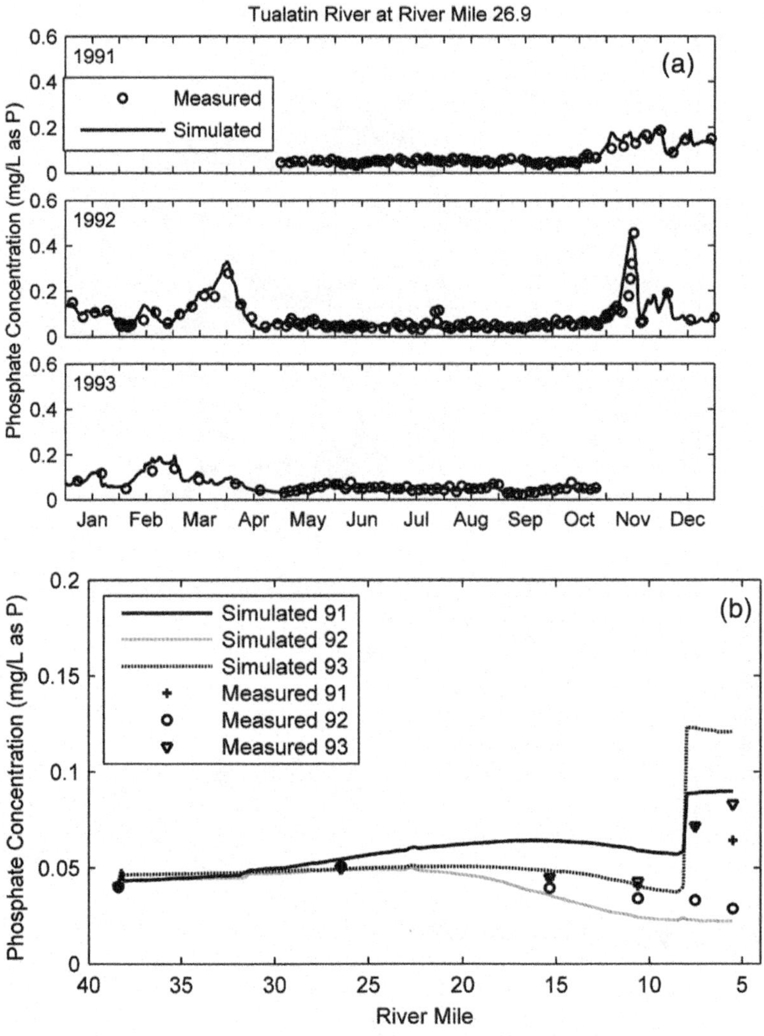

Fig. 3.10 Simulated vs. measured phosphate concentrations: (**a**) temporal variations at RM26.9 and (**b**) longitudinal profiles of mean values during May–October

locations. Using these calibrated grazing rates shown in Fig. 3.5, the model can reasonably reproduce the temporally and spatially varying zooplankton biomass, although the biomass peak at the RM5.5 during 1991 could not be observed. The calculated and measured zooplankton concentrations agree generally well, with $R^2 = 0.747$.

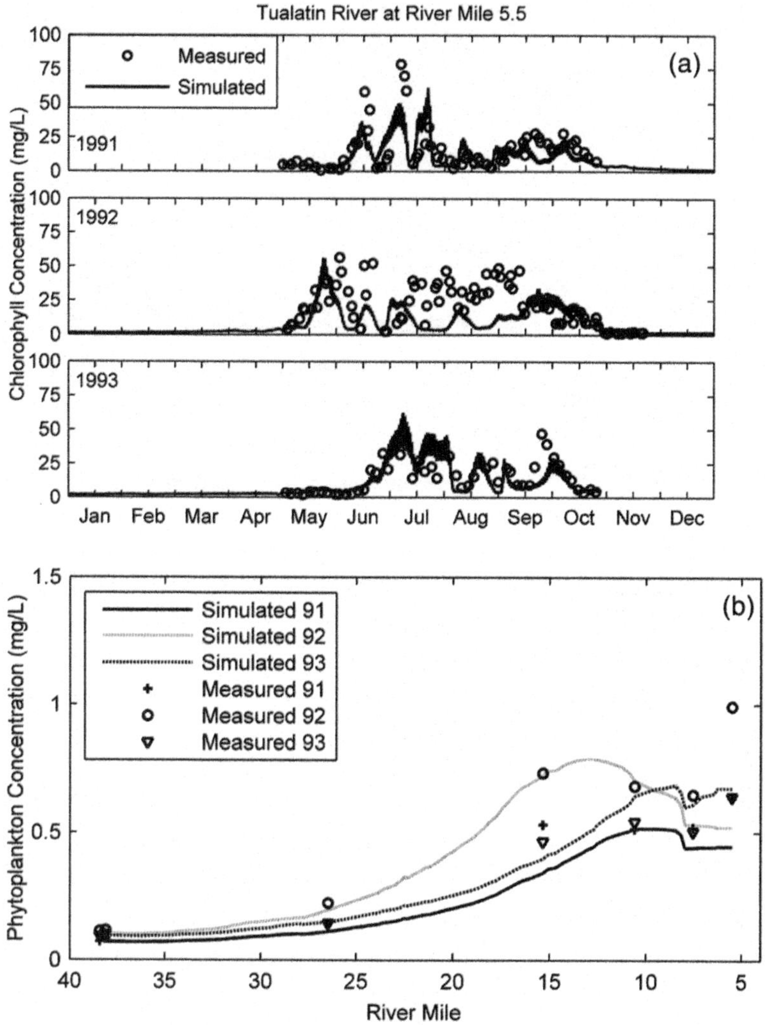

Fig. 3.11 Simulated vs. measured phytoplankton concentrations: (**a**) temporal variations at RM5.5 and (**b**) longitudinal profiles of mean values during May–October

10.2 Model Test in the Upper Hudson River, New York

10.2.1 Study Area

The Hudson River watershed encompasses 13,400 square miles in New York, Massachusetts, and Vermont, USA. The primary health risk of the river is the accumulation of PCBs discharged from plants of the General Electric Company that were located at Hudson Falls and Fort Edward. The reach is divided into the Upper and Lower Hudson River (UHR, LHR) by the Federal Dam at Troy. The UHR is a

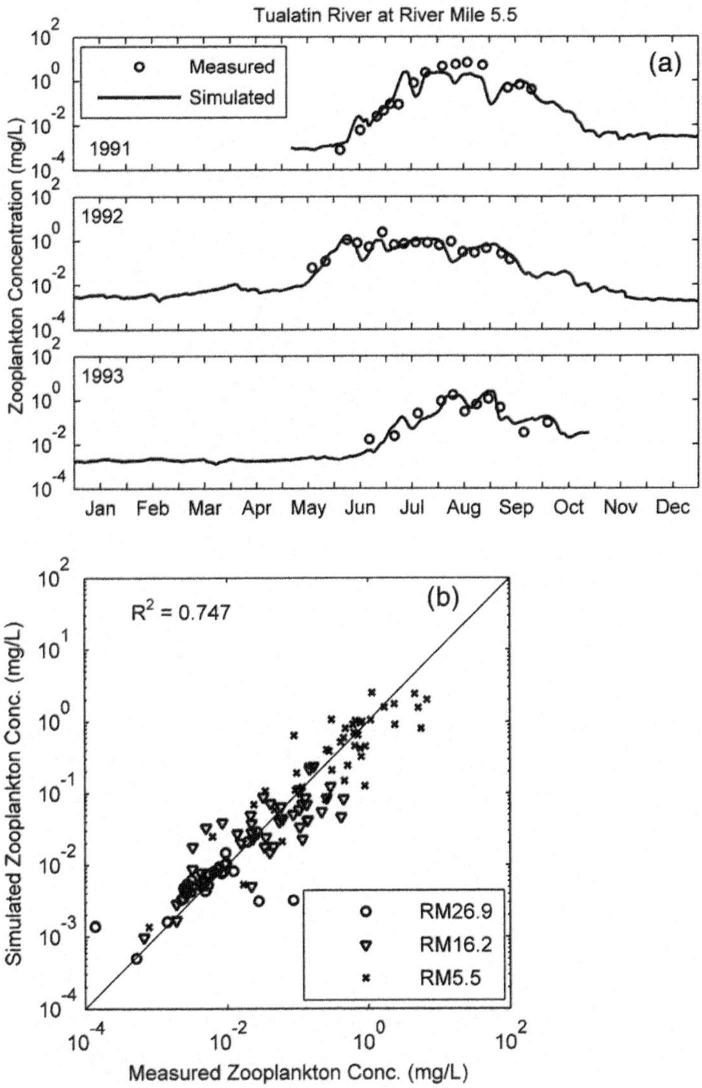

Fig. 3.12 Measured vs. simulated zooplankton concentrations: (**a**) temporal variations at RM5.5 and (**b**) comparison at three cross-sections

river-reservoir system comprised of a series of eight dams and associated backwater that extends from Fort Edward to Troy [45]. Due to the discontinuity along the river, each of the river reaches between the dams can be studied separately. The model developed in this study is applied to simulate the fate and transport of PCBs in a 13.3-mile reach of the UHR extending from Schuylerville (RM181.3) to Stillwater Dam (RM168.0), as shown in Fig. 3.13. The simulation domain is divided into 163 cross-sections. Each section is divided into 25 vertical panels.

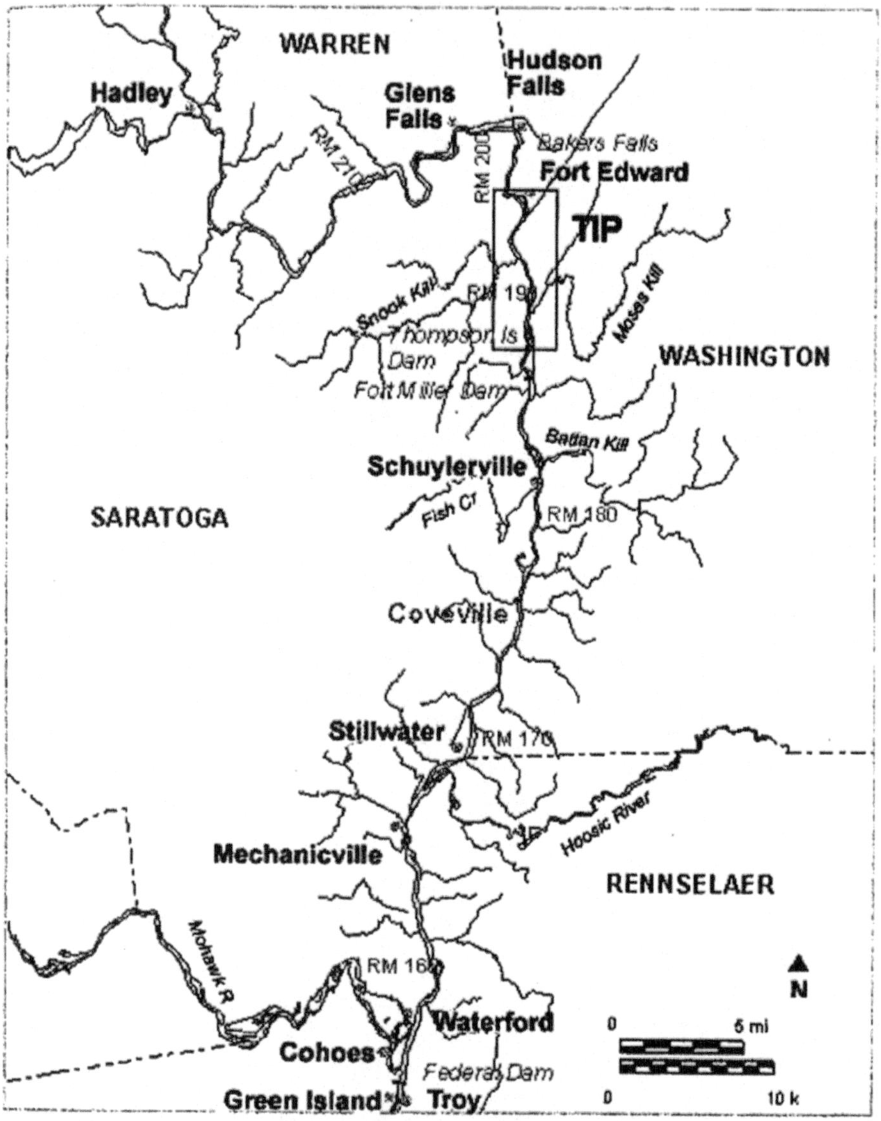

Fig. 3.13 Upper Hudson River (http://www.epa.gov/hudson/slide6.gif)

The simulation period is from March 1977 to December 1986. The time step is 15 min. The water velocities along the simulation domain range between 0.003 and 1.4 m/s with the average velocity of 0.37 m/s over the simulation period. The travel time is approximately 16 h. The study reach perhaps is too short to demonstrate the capability of modeling transport of water quality constituents, but it is a good case to test the model components of aquatic ecosystem and chemical bioaccumulation because of the controlled boundary conditions and abundant measurement data.

10.2.2 Model Inputs and Parameters

Discharge hydrograph from the USGS gauging station at Schuylerville and the staff gauge readings from the New York State Department of Transportation (NYSDOT) at the downstream location are used as boundary conditions for the hydrodynamic simulation. The upstream inflow is the main source for the flow discharge in the study reach. Small tributaries between Schuylerville and Stillwater are not considered in this study. Water quality of the UHR at Schuylerville and Stillwater were surveyed by New York State Department of Environmental Conservation (NYSDEC), and the data was reposited in STORET (STOrage and RETrieval) which is accessible through http://www.epa.gov/storet/. The times series of water quality constituents except phytoplankton measured at Schuylerville are used as the upstream inputs for water quality simulation. Because phytoplankton data is not available at Schuylerville, the phytoplankton biomass of the UHR at Waterford, which is located downstream of Stillwater, is used as the upstream phytoplankton loading. This substitution may contribute to errors in the model results.

The food web structure in the study reach can be divided to four trophic levels: phytoplankton, zooplankton, forage fish, and predatory fish. From 1976 to 1985, the New York State Department of Health (NYSDOH) conducted long-term biomonitoring studies using caddisfly and chironomid larvae as part of the Hudson River PCB Reclamation Demonstration Project [46]. The samplings were made in June through September of each year. The study showed that the most abundant taxa were chironomids and oligochaetes [45], which are used as the second trophic level representative in this study. Fish data were surveyed and collected using electrofishing by NYSDEC between 1970 and 1993, and weight, length, and the number of radius of annual growth rings on scales were measured. The number of fish caught in the Stillwater pool and the general detail of fish characteristics and behavior are published in [45]. Fish community of the UHR is composed of more than 30 species, which are classified into two groups: forage and predatory fish according to diet nature. The most common forage fish species are yellow perch, pumpkinseed fish, white sucker, golden fish, and brown bullhead. Large number of pumpkinseed fish is annually found in the study site starting from 1980, so that it is used as a representative for the forage fish in the model. The predatory fish species include largemouth bass and American eel. Largemouth bass is used to represent the predatory fish due to its general abundance.

The pathways of PCBs in aquatic organisms are the direct uptake from the water column and the transfer through food web via predation. A challenge to developing a modeling framework for PCB bioaccumulation is that PCBs consists of 209 individual congeners, which exhibit varying degrees of bioaccumulation potential [46]. The total PCB concentrations in fish were collected as part of NYSDEC monitoring program, and measured in fish on an Aroclor basis [45]. Therefore, the total PCBs is considered in the present model for simplicity.

The water quality model parameters in the UHR case are similar to those in the Tualatin River case shown in Tables 3.3 and 3.4 [42]. The feeding preference of the

Table 3.5 Preference consumption of the UHR food web model

Species	Detritus	Phytoplankton	Zooplankton	Forage fish	Predatory fish
Zooplankton	0.15	0.85	–	–	–
Forage fish	0.15	0.25	0.5	0.1	–
Predatory fish	0.025	0.175	0.25	0.85	0.15

Table 3.6 Summary of some model parameters used for the UHR ecological model

Symbol	Unit	Value	Symbol	Unit	Value
$K_{ag,max}$	1/d	1.0	e_{fg}	–	0.75
$K_{ar,max}$	1/d	0.25	a_f	–	0.45
$K_{am,max}$	1/d	0.35	b_f	–	−0.36
ω_a	m/d	0.25	W_p	g	525
$K_{zg,max}$	1/d	0.75	$K_{pb,max}$	1/d	0.99
$K_{zr,max}$	1/d	0.015	$K_{pm,max}$	1/d	0.002
$K_{zm,max}$	1/d	0.035	e_{pg}	–	0.70
e_{zg}	–	0.7	a_p	–	0.33
W_f	g	380	b_p	–	−0.325
$K_{fb,max}$	1/d	0.9	T_{opt}	°C	22.5
$K_{fm,max}$	1/d	0.0015	θ	–	1.072

Table 3.7 Summary of some model parameters used for the UHR ecotoxicological model

Species	K_1 (1/d)	K_2 (1/d)	LC50 (μg/L)	t_1 (h)
Phytoplankton	1.0E-7	5.0E-4	1E-8	24
Zooplankton	1.0E-7	5.0E-5	31	96
Forage fish	5.0E-4	2.5E-3	2740	96
Predatory fish	7.5E-5	5.0E-3	236.4	96

UHR food web model is presented in Table 3.5. The selected model parameters used in ecological and ecotoxicological models are shown in Tables 3.6 and 3.7. These model parameters obtained from several literature sources, such as [2, 10, 34, 35].

10.2.3 Simulation Results

The comparisons of simulated results and field data of ammonia, nitrate, phosphate, DO and CBOD at Stillwater are shown in Figs. 3.14 and 3.15. Because lateral inflows from tributaries and fields are not taken into account in this study, the simulation results largely depend on the nutrient concentrations of the upstream input. From Schuylerville to Stillwater, the river flows mostly through suburban and agricultural areas, and the usage of fertilizer might contribute to the measured instream nutrient of the river. Nevertheless, the simulation results and field data of water quality are in generally good agreement, with R^2 of 0.856, 0.665, 0.495, 0.865 and 0.514 for ammonia, nitrate, phosphate, DO and CBOD, respectively.

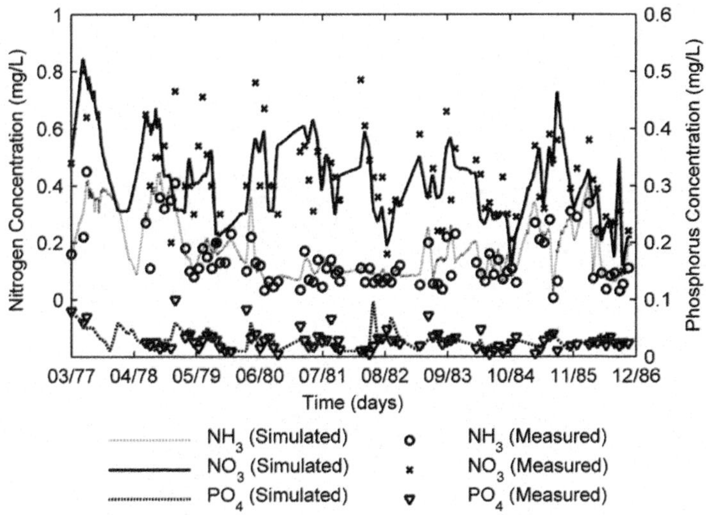

Fig. 3.14 Simulated and measured ammonia, nitrate, and phosphate concentrations at Stillwater (measurements from NYSDEC)

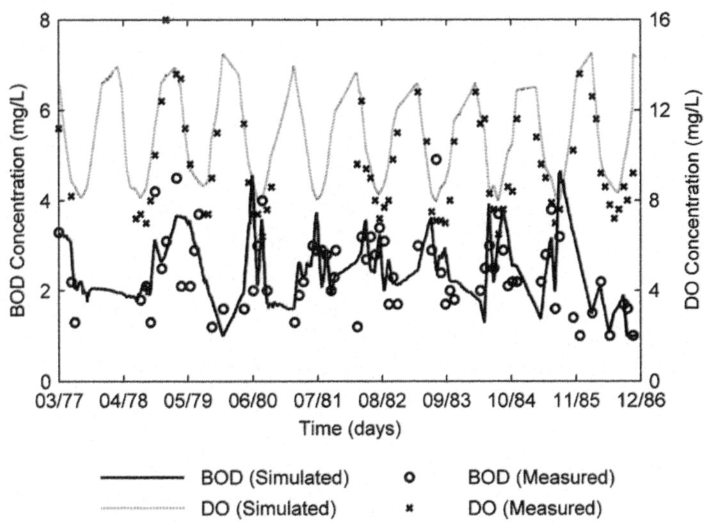

Fig. 3.15 Simulated and measured DO and BOD concentrations at Stillwater (measurements from NYSDEC)

The simulated and measured biomass concentrations of zooplankton, forage fish and predatory fish at Stillwater are shown in Fig. 3.16. The simulation shows annual zooplankton biomass peaks, which may be caused by the seasonal growth of phytoplankton. Although zooplankton in August 1983 is under-estimated, its biomass concentrations estimated over 11-year simulation period are consistent with

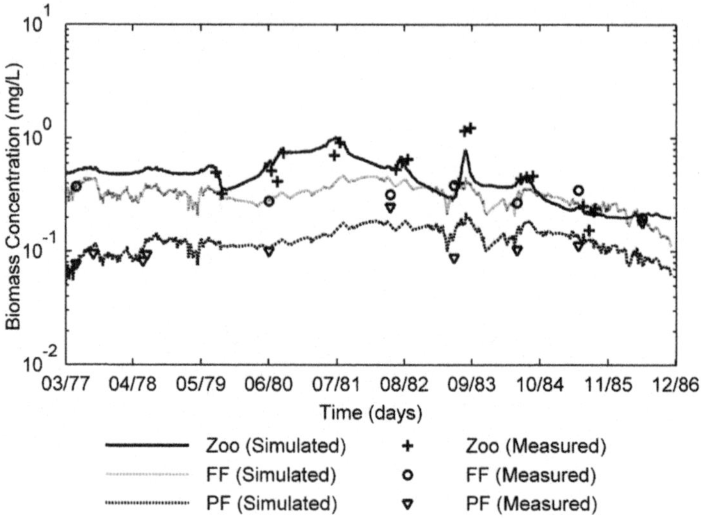

Fig. 3.16 Simulated and measured zooplankton (Zoo), forage fish (FF), and predatory fish (PF) biomass concentrations at Stillwater (measurements from NYSDEC)

the field measurements. The simulated biomasses of forage fish are comparable to the survey data. The numerical results of the predatory fish in 1982 and 1986 are lower than the measurements. One of the reasons is that from the field record, American eel was caught only in 1982 and 1986, and there is no evidence to show the cause of its nonexistence for other years. Therefore, the measured total biomass of the top predators including largemouth bass and American eel suddenly increases in 1982 and 1986. In addition, the initial fish biomass in 1977, which is used for modeling setup, is only the biomass of largemouth bass. It means that there is no American eel at the beginning of the simulation, and the model basically simulates the biomass of largemouth bass for the entire simulation period. Without considering American eel, the simulated results are more comparable to the measurement data. Moreover, the model assumes that fish did not leave the system because the downstream end of the study reach is the Stillwater Dam. Since the predatory fish is considered as the top trophic level, the biomass loss depends solely on non-predatory and gamete mortalities as well as defecation of unassimilated food. In reality, they can be consumed by other animals, caught by humans, or leave the system domain. These unconsidered factors may contribute to the difference between the simulation results and measurements. Other factors include the lack of data and uncertainties in the real nature.

Figure 3.17 compares the simulated and measured total PCB concentrations in the water column at Stillwater, and the general trend is reproduced well by the model. Figure 3.18 compares the simulated and measured PCB concentrations in pumpkinseed and largemouth bass, which are used as representatives for forage and predatory fish, respectively. The adult largemouth bass samples were collected in Spring, while small pumpkinseed were collected in late Summer or early Fall. The

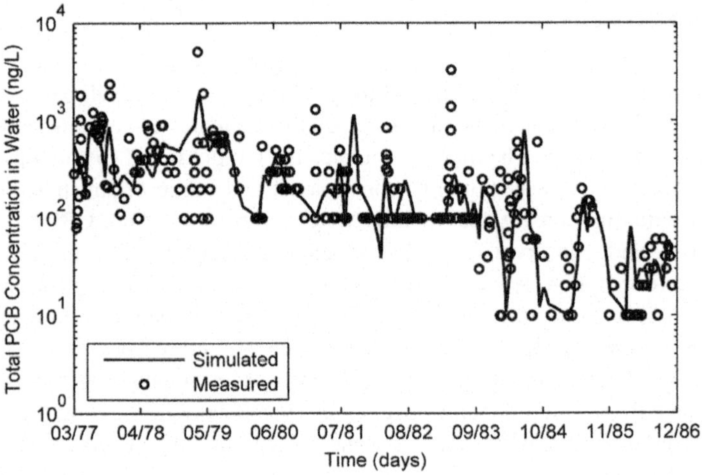

Fig. 3.17 Simulated and measured the total PCB concentrations in water column at Stillwater (measurements from USGS)

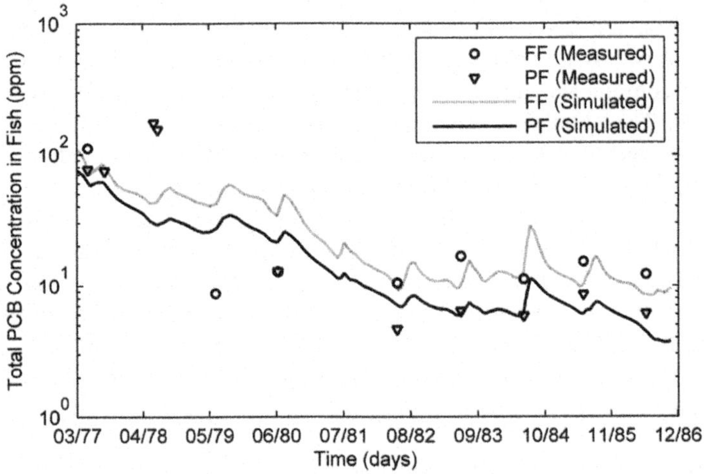

Fig. 3.18 Simulated and measured total PCB concentrations in fish at Stillwater (measurements from NYSDEC)

PCB levels in pumpkinseed slowly decline from the late 1970s to 1982. Slight fluctuation of PCB is observed after 1982. Similar to pumpkinseed, the PCB concentrations in largemouth bass gradually decrease until 1982 and are subsequently steady. The results show that the model is able to predict reasonably well the bioaccumulation of PCB in both fish species, with R^2 of 0.425 and 0.373.

11 Conclusions

An integrated one-dimensional modeling package has been developed to simulate hydrodynamic, sediment transport, water quality, aquatic ecosystem and ecotoxicology in river systems. The model simulates the temporal and spatial variations of the concentrations of water quality constituents and biotic organisms. The simulated water quality constituents include water temperature, DO, CBOD, nitrogen, phosphorus, and conservative chemical such as chloride. The used food web consists of four trophic levels: phytoplankton, zooplankton, forage fish, and predatory fish, which undergo the biological processes of photosynthesis, grazing, respiration, excretion, defecation, mortality, gamete, and reproduction. The model simulates the bioaccumulation of toxic chemicals in aquatic organisms through direct uptake from water, depuration and dietary, and takes into account the toxicity effects through modification factors for the growth, grazing, and gamete mortality of the organisms.

The developed model is applied to simulate water quality in the Tualatin River, Oregon, which has high influences of lateral inputs from wastewater treatment plants and tributary discharges. The model reproduces well the time-series concentrations of water quality constituents and biomass of phytoplankton and zooplankton in the Tualatin River. The model is also applied to simulate the water quality, aquatic ecosystems, as well as fate and transport of the total PCB concentrations in the water column and aquatic organisms in the Upper Hudson River (UHR), New York. The simulated water quality parameters, zooplankton biomass, fish populations, and total PCBs concentrations in both forage and predatory fish are in generally good agreement with the measurement data.

The developed model is comparable to the WASP model in terms of water quality modeling, and has the basic features but is much simpler than the AQUATOX model in terms of aquatic ecology and ecotoxicology modeling. It is integrated with a well-developed model of flow and sediment transport in channel networks. It is relatively more convenient to use in assessment of the impacts of flood, river restoration, dam construction, morphological change, chemical spill and etc. on river ecosystems.

References

1. Jørgensen, S. E. (1980). *Lake management*. New York: Pergamon Press.
2. Di Toro, D. M., Fitzpatrick, J. J., & Thomann, R. V. (1981). *Water Quality Analysis Simulation Program (WASP) and Model Verification Program (MVP) Documentation* (EPA/600/3-81/044). Washington, DC: U.S. Environmental Protection Agency.
3. Wool, T. A., Ambrose, R. B., Martin, J. L., & Comer, E. A. (2001). *The Water Quality Analysis Simulation Program, WASP6. Part A: Model documentation*. Athens, GA: U.S. Environmental Protection Agency, Center for Exposure Assessment Modeling.

4. Chapra, S. C., Pelletier, G. J., & Tao, H. (2008). *QUAL2K: A modeling framework for simulating river and stream water quality, version 2.11: Documentation and users manual.* Medford, MA: Civil and Environmental Engineering Department, Tufts University.
5. Cerco, C., & Cole, T. (1994). *Three-dimensional eutrophication model of Chesapeake Bay* (Technical Report EL-94-4). Vicksburg, MS: U.S. Army Engineer Waterways Experiment Station.
6. Chao, X., Jia, Y., & Zhu, T. T. (2006). *CCHE_WQ water quality module* (Technical Report No. NCCHE-TR-2006-01). Oxford, MS.
7. Christensen, V., & Walters, C. J. (2004). Ecopath with ecosim: Methods, capabilities, and limitations. *Ecological Modelling, 172*, 109–139.
8. Hipsey, M. R., Romero, J. R., Antenucci, J. P., & Hamilton, D. (2006). *Computational aquatic ecosystem dynamics model: CAEDYM v2* (WP 1387.1 MH). Australia.
9. Ulanowicz, R. E. (2006). *EcoNetwrk – A user's guide.* http://www.glerl.noaa.gov/EcoNetwrk (accessed in 2009).
10. Park, R. A., & Clough, J. S. (2004). *AQUATOX (release 2) modeling environmental fate and ecological effects in aquatic ecosystems. Volume 2: Technical documentation* (EPA-823-R-04-002). Washington, DC: Office of Water, U.S. Environmental Protection Agency.
11. Park, R. A., & Clough, J. S. (2009). *AQUATOX (release 3) modeling environmental fate and ecological effects in aquatic ecosystems. Volume 2: Technical documentation* (EPA-823-R-09-004). Washington, DC: Office of Water, U.S. Environmental Protection Agency.
12. Bahadur, R., Amstutz, D. E., & Samuels, W. B. (2013). Water contamination modeling – A review of the state of the science. *Journal of Water Resource and Protection, 5*, 142–155.
13. Sharma, D., & Kansal, A. (2013). Assessment of river quality models: A review. *Reviews in Environmental Science and Bio/Technology, 12*(3), 285–311.
14. Zhang, W., & Rao, Y. R. (2012). Application of a eutrophication model for assessing water quality in Lake Winnipeg. *Journal of Great Lakes Research, 38*(3), 158–173.
15. Zhang, L., Liu, J., Li, Y., & Zhao, Y. (2013). Applying AQUATOX in determining the ecological risk assessment of polychlorinated biphenyl contamination in Baiyangdian Lake, North China. *Ecological Modelling, 265*, 239–249.
16. Scholz-Starke, B., Ottermanns, R., Rings, U., Floehr, T., Hollert, H., Hou, J., et al. (2013). An integrated approach to model the biomagnification of organic pollutants in aquatic food webs of the Yangtze Three Gorges Reservoir ecosystem using adapted pollution scenarios. *Environmental Science and Pollution Research, 20*(10), 7009–7026.
17. Joyner, T. A., & Rohli, R. V. (2013). Atmospheric influences on water quality: A simulation of nutrient loading for the Pearl River Basin, USA. *Environmental Monitoring and Assessment, 185*(4), 3467–3476.
18. Karaaslan, Y., Akkoyunlu, A., Erturk, F., & Citil, E. (2013). The effect of toxic organic chemicals on Mogan lake. *International Journal of Environmental Research, 7*(3), 595–604.
19. Chen, Y., Niu, Z., & Zhang, H. (2013). Eutrophication assessment and management methodology of multiple pollution sources of a landscape lake in North China. *Environmental Science and Pollution Research, 20*(6), 3877–3889.
20. Ambrose, R. B., & Wool, T. A. (2009). *WASP7 stream transport – Model theory and user's guide, Supplement to Water Quality Analysis Simulation Program (WASP), user documentation* (EPA/600/R-09/100). Washington, DC: U.S. Environmental Protection Agency.
21. Tsegaye, T., & Wagaw, M. (2007). *Stream water quality modeling using AQUATOX.* National Water Quality Meeting, Savannah, GA.
22. Wu, W., & Vieira, D. A. (2002). *One-dimensional channel network model CCHE1D version 3.0 – Technical manual* (Technical Report No. NCCHE-TR-2002-1). Oxford, MS.
23. Vieira, D. (2004). *Integrated modeling of watershed and channel processes.* Oxford, MS: National Center for Computational Hydroscience and Engineering, The University of Mississippi.
24. Wu, W. (2007). *Computational river dynamics.* New York: Taylor & Francis.
25. Tennessee Valley Authority. (1972). *Heat and mass transfer between a water surface and the atmosphere.* Division of Water Control Planning.

26. Deas, M. L., & Lowney, C. L. (2000). *Water temperature modeling review*. Central Valley, CA.
27. Genetti, A. J. J. (1998). *Engineering and design – Runoff from Snowmelt*. Washington, DC: U.S. Army Corps of Engineers.
28. Kaushik, C. P., Bhavikatti, S. S., & Kaushik, A. (2010). *Basic civil and environmental engineering*. New Delhi, India: New Age International.
29. Chapra, S. (1997). *Surface water-quality modeling*. New York: McGraw-Hill.
30. Haider, H., Ali, W., & Haydar, S. (2012). Evaluation of various relationships of reaeration rate coefficient for modeling dissolved oxygen in a river with extreme flow variations in Pakistan. *Hydrological Processes*. doi:10.1002/hyp.9528.
31. Liu, W.-C., & Chen, W.-B. (2012). Monitoring sediment oxygen demand for assessment of dissolved oxygen distribution in river. *Environmental Monitoring and Assessment, 184*(9), 5589–5599.
32. Brown, L. C., & Barnwell, T. O. (1987). *The enhanced stream water quality model QUAL2E and QUAL2E-UNCAS: Documentation and user manual*. Athens, GA.
33. Xu, K., Jiang, H., Juneau, P., & Qiu, B. (2012). Comparative studies on the photosynthetic responses of three freshwater phytoplankton species to temperature and light regimes. *Journal of Applied Phycology, 24*(5), 1113–1122.
34. Bowie, G. L., Mills, W. B., Porcella, D. B., Campbell, C. L., Pagenkopf, J. R., Rupp, G. L., et al. (1985). *Rates, constants, and kinetics formulations in surface water quality modeling* (EPA/600/3-85/040). Athens, GA: U.S. Environmental Protection Agency.
35. Rounds, S. A., Wood, T. M., & Lynch, D. D. (1999). *Modeling discharge, temperature, and water quality in the Tualatin River, Oregon*. USGS Water Supply Paper; 2465-B, U.S. Geological Survey, Oregon.
36. Kitchell, J. F., Koonce, J. F., O'Neill, R. V., Shugart, H. H., Jr., Magnuson, J. J., & Booth, R. S. (1974). Model of fish biomass dynamics. *Transactions of the American Fisheries Society, 103*, 786–798.
37. Leidy, G. R., & Robert, M. J. (1977). *The development of fishery compartments and population rate coefficients for use in reservoir ecosystem modeling* (Contract Rept. CR-Y-77-1, 134 pp.). Vicksburg, MS: U.S. Army Engineer Waterways Experiment Station.
38. Di Toro, D. M. (2001). *Sediment flux modeling*. New York: Wiley.
39. Thomann, R. V., & Connolly, J. P. (1984). *An age dependent model of PCB in a Lake Michigan food chain*. New York: Manhattan College.
40. Connolly, J. P., Zahakos, H. A., Benaman, J., Ziegler, K. C., Rhea, J. R., & Russell, K. (2000). A model of PCB fate in the Upper Hudson River. *Environmental Science & Technology, 34*, 4076–4087.
41. Doyle, M. C., & Caldwell, J. M. (1996). *Water-quality, streamflow, and meteorological data for the Tualatin River Basin, Oregon, 1991–1993* (Open-File Report 96-173). Portland, OR: U.S. Geological Survey.
42. Inthasaro, P. (2010). *A one-dimensional aquatic ecology and ecotoxicology model in river systems*. PhD dissertation, National Center for Computational Hydroscience and Engineering, The University of Mississippi, Mississippi.
43. Thomann, R. V., & Mueller, J. A. (1988). *Principles of surface water quality modeling and control*. New York: Harper & Row.
44. Rounds, S. A., & Doyle, M. C. (1997). *Sediment oxygen demand in the Tualatin River basin, Oregon, 1992–96* (Water-Resources Investigations Report 97-4103). Portland, OR: U.S. Geological Survey.
45. QEA. (1999). *PCBs in the Upper Hudson River Volume 2: A model of PCB fate, transport, and bioaccumulation*.
46. TAMS Consultant, Limno-Tech, Menzie-Cura & Associates, & Tetra Tech. (2000). *Volume 2D – Revised baseline modeling report: Hudson River PCBs reassessment RI/FS*, Book 3 of 4 – Bioaccumulation models. New York.

Chapter 4
Hydraulic Fracturing and Shale Gas: Environmental and Health Impacts

Hsue-Peng Loh and Nancy Loh

Contents

H.-P. Loh, M.L.S., Ph.D. (✉) • N. Loh, M.A. (✉)
Wenko Systems Analysis, 230 Beverly Road, Pittsburgh, PA 15216, USA
e-mail: hp_loh@yahoo.com; hp.loh2@gmail.com; nancyloh1@gmail.com

L.K. Wang, C.T. Yang, and M.-H.S. Wang (eds.), *Advances in Water Resources Management*, Handbook of Environmental Engineering, Volume 16,
DOI 10.1007/978-3-319-22924-9_4

Abstract Shale gas development means the nation's energy independence and economic benefits in terms of employment, manufacturing, services, tax revenues, local economies as well as lease and royalty payments to state and local governments and land owners. However, there are great concerns about its potential impacts and risks on environment, human health, and ecosystems as evidenced by numerous reported incidents and litigation. Among the major issues are: use of fracturing fluids that contain carcinogens and toxins; migration of hydrocarbons, heavy metals, and radioactive substances from shale and nearby formations that causes groundwater and surface water contamination; substantial consumption of water for fracturing; air emissions; sedimentation; earthquakes; explosion; and human health effects. These issues are discussed in this chapter.

Keywords Shale gas • Hydraulic fracturing • Fracking • Hydrofracking • Water contamination • Carcinogen • Sedimentation • Air emissions • Earthquake • Explosion • Blowout • Migration • Fracking chemicals • Environmental impact • Health impact

Acronyms

Bcf	Billion cubic feet
CAA	Clean Air Act
CERCLA	Comprehensive Environmental Response, Compensation, and Liability Act (Superfund)
CWA	Clean Water Act
EIA	Energy Information Administration
ELG	Effluent Limit Guide
EMC	(Turbidity) Event Mean Concentration
EPACT	Energy Policy Act
GAO	General Accounting Office
GHG	Greenhouse gas
GWPC	Ground Water Protection Council
HAP	Hazardous air pollutants

IOGCC	Interstate Oil and Gas Compact Commission
MMcf	Million cubic feet
NETL	(DOE) National Energy Technology Laboratory
NPDES	National Pollutant Discharge Elimination System
NSPS	New Source Performance Standards
NTU	Nephelometric Turbidity Unit
NYDEC	New York State Department of Environmental Conservation
OGAP	Oil and Gas Accountability Project
PADEP	Pennsylvania Department of Environmental Protection
PAEQB	Pennsylvania Energy Quality Board
POTW	Publicly Owned Treatment Works
RCRA	Resource Conservation and Recovery Act
Scf	Standard cubic feet
Scfd	Standard cubic feet per day
SPE	Society of Petroleum Engineers
STRONGER	State Review of Oil and Natural Gas Environmental Regulations
SWDA	Safe Water Drinking Act
Tcf	Trillion cubic feet
TDS	Total dissolved solids
tpd	Tons per day
TSS	Total suspended solids
USDOE	US Department of Energy
USEPA	US Environmental Protection Agency
USGS	US Geological Survey
VOC	Volatile organic compounds

1 Introduction

1.1 Shale Gas Basins

The lower 48 states have a wide distribution of highly organic shale, but they cannot be economically recovered because of very tight shale formations. Hydraulic fracturing (also known as "fracking" or "hydrofracking") is a well stimulation technology that creates fractures in targeted zones that improves the shale formation's permeability and allows oil and gas to be recovered economically. Hydraulic fracturing was developed in the late 1940s to maximize production of crude oil and natural gas from unconventional reservoirs. But large scale, economically viable horizontal drilling along with hydraulic fracturing began only in the 1990s.

This technology has generated the sudden appearance of a new energy source, responsible for the decrease of U.S. petroleum imports from 60 % of domestic consumption in 2005 to today's 46 %. And natural gas produced from shale formation using fracturing has become an increasingly important source in the United States over the past decade. In 2000 shale gas accounted for only 1 % of

U.S. natural gas production; it now accounts for 34 %. And it is projected by EIA to be 49 % by 2035 [1].

The shift in oil and gas sources means that our energy supplies will become more secure, and the nation will have more flexibility in dealing with crisis. It also means that economic benefits—in terms of jobs, manufacturing and services—will register on the ground in North America. As the use of fracturing has increased, so have the concerns about its potential environmental and health impacts, which have generated spirited and heated debates between proponents and opponents of the use of this technology.

There are 26 shale gas basins in the United States containing vast resources of natural gas (Fig. 4.1), but only seven have significant shale gas production. Table 4.1

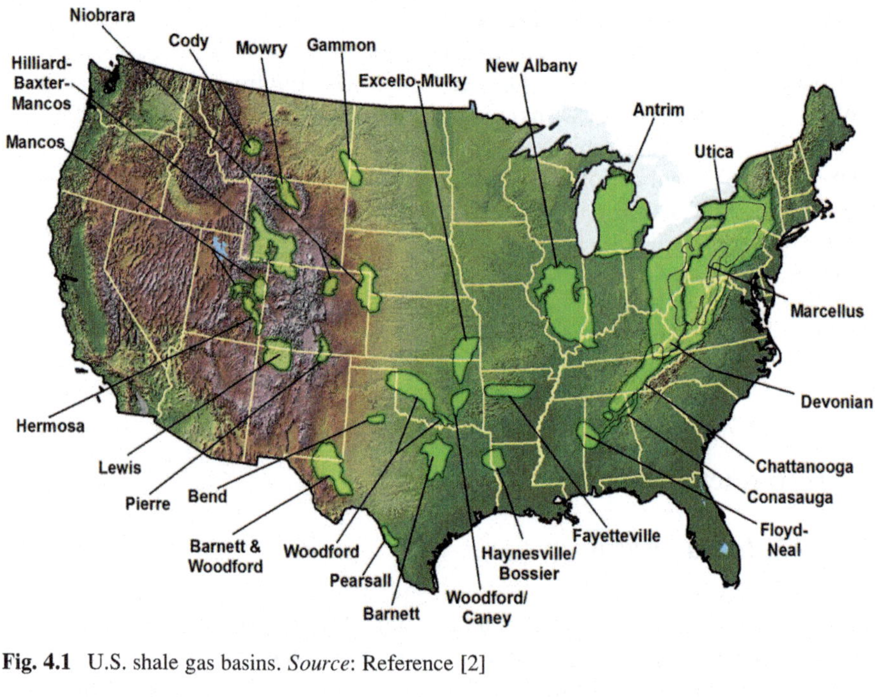

Fig. 4.1 U.S. shale gas basins. *Source*: Reference [2]

Table 4.1 Shale gas location

Shale gas basin	Site
Marcellus	Pennsylvania, New York, Ohio, and West Virginia
Haynesville	Northern Louisiana and eastern Texas
Barnett	Central Texas
Fayetteville	Northern Arkansas and eastern Oklahoma
Antrim	Upper portion of lower peninsula of Michigan
New Albany	Southwest Illinois, Indiana, Kentucky
Woodford	South-central Oklahoma

shows the sites of seven shale gas basins with largest technically recoverable resources.

Table 4.2 provides a summary of the seven major shale gas basins [2].

- Marcellus has the largest technically recoverable resources—262 trillion cubic feet (Tcf), followed by Haynesville's 251 Tcf.
- Marcellus Shale covers the largest area (95,000 square miles), followed by New Albany, which is 43,500 square miles.
- The gas content varies between 40 and 350 standard cubic feet (Scf) per ton.
- The thickness of shale formation varies between 50 and 600 feet.
- The depth of base of treatable water formation varies between 300 and 1200 feet.
- The depth of the shale formation varies between 500 and 13,500 feet. These depths are typically much deeper than the depth of treatable water formation.

The Barnett Shale Basin is the first in the United States to explore shale gas and has the greatest accumulated shale gas production since the onset of hydraulic fracturing, followed by Haynesville Basin and Fayetteville Basin. However, the Marcellus Shale, which has the largest reserves, ranks only fourth in accumulated production because Marcellus' drilling only began in 2004. But the Marcellus shale now accounts for 26 % of the total shale gas production in the United States and ranks first in the average daily production (7.4 billion cubic feet per day as of July 2012).

1.2 Horizontal Wells and Hydraulic Fracturing

The casing of the gas wells, whether it is vertical or horizontal, is accomplished in multiple phases from the largest diameter casing to the smallest as shown in Fig. 4.2 [3].

There is no one-size-fits-all technological routine for drilling and fracturing because of the wide variation in the different basin characteristics. For instance, different service providers use different wellbore diameters and lengths tailored to different depths and thicknesses of shale formation. The drilling and fracturing details described below by Fracfocus.org [4] is an illustrative, design example.

The first phase involves the setting of the conductor casing (26 inch diameter, 40 feet deep). The purpose of this casing is to prevent the sides of the hole from caving into the wellbore. After the conductor casing string is set in place, drilling continues inside the conductor casing string to a depth below the lowest groundwater zone.

Next, a wellbore ($17^{1/2}$ inch diameter, 2,000 feet deep) is drilled. Surface casing ($13^{3/8}$ inch diameter) is then run from the surface to just above the bottom of the hole. Cement is pumped down the inside of the casing, forcing it up from the bottom of the surface casing into the space between the outside of the casing and the wellbore. This space is called the annulus.

Once a volume of cement sufficient to fill the annulus is pumped into the casing, fresh water is then pumped into the casing until the cement begins to return to the

Table 4.2 Major shale gas basins in the United States

Shale basin	Marcellus	Haynesville	Barnett	Fayetteville	Antrim	New Albany	Woodford
Technically recoverable resources, Tcf	262	251	44	41.6	20	19.2	11.4
Estimated basin area, square mile	95,000	9,000	5,000	9,000	12,000	43,500	11,000
Shale zone depth, ft	4,000–8,500	10,500–13,500	6,500–8,500	1,000–7,000	600–2,200	500–2,000	6,000–11,000
Shale zone thickness, ft	50–200	200–300	100–600	20–200	70–120	50–100	120–220
Depth to base of treatable water, ft	~850	~400	~1,200	~500	~300	~400	~400
Gas content, Scf/ton	60–100	100–330	300–350	60–220	40–100	40–80	200–300

Source: Reference [2]

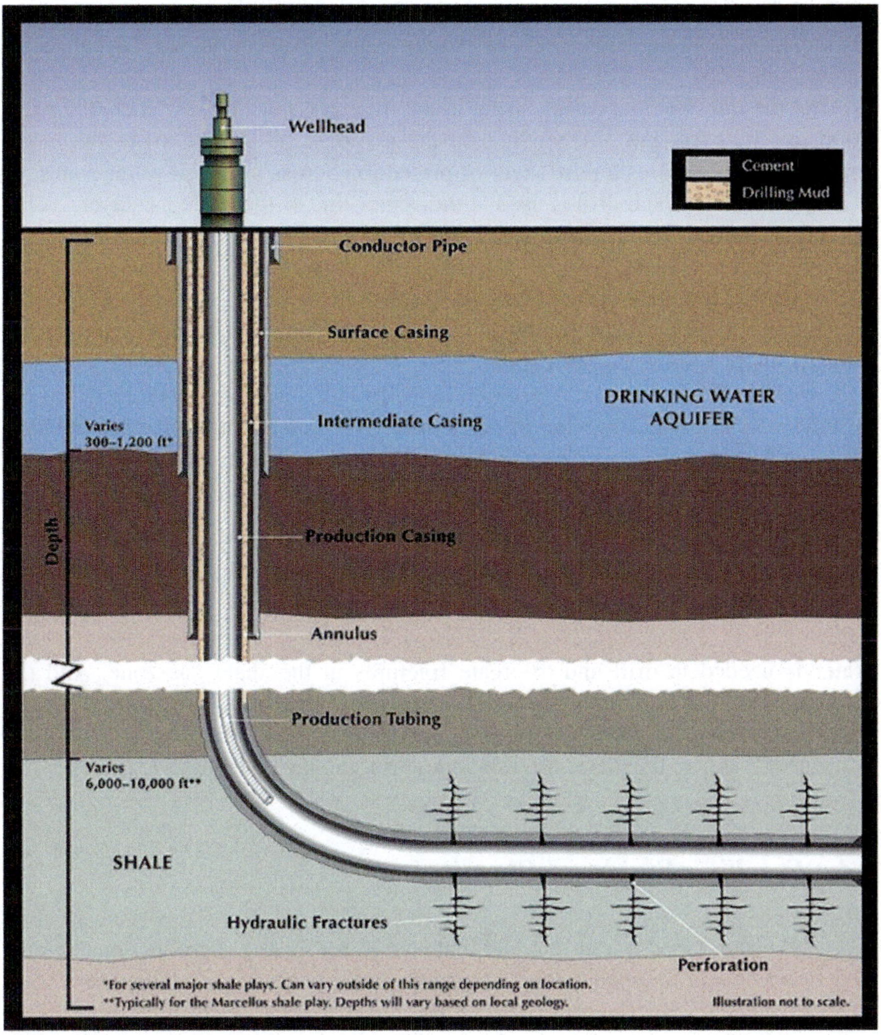

Fig. 4.2 Drilling and fracturing. *Source*: Reference [3]

surface in the annular space. The cementing of casing from bottom to top using this method is called circulation. The circulation of cement behind the surface casing ensures that the entire annular space fills with cement from below the deepest groundwater zone to surface. The surface casing depth wholly covers the fresh water zone, at least 50 feet below the deepest fresh water zone.

Once the surface casing is set and the cement has time to cure, the wellbore (9 inch diameter, at least 200 feet of cement above the shallowest formation capable of contaminating fresh water) is drilled down to next zone where casing will be set. This casing is the intermediate casing.

Combined, the surfacing casing and intermediate casing should have at least 250 feet of water-tight cement between water formation and anything that can cause contamination.

After the intermediate casing string is set, the well is drilled ($8^{3/4}$ inch diameter, 5,000 to 7,000 feet long, depending on the geology the formation) to the target formation. This ensures a third layer of protection across the fresh water zones.

A special directional drill is used at the center line of the shale gas layer, called the kickoff point, while the wellbore turns 90° and begins the horizontal drilling. It takes 900 feet to achieve a complete transition from vertical well to horizontal well at 10° per 100 feet of horizontal length. Another string of pipe (usually $5^{1/2}$ inches in diameter called production tubing is run through the vertical production casing and extends in a horizontal direction.

A special perforation gun is inserted in the production tube and pierces through the tube wall to create a number of holes through which the high-pressure fracking fluids will ultimately pass through the wells and penetrate the shale formation around the tubing to create fractures in the shale layer.

1.3 Water Use

Water is needed to drill and to create fractures in the shale gas zone. And the amounts of water vary from well to well, and from play to play. The estimated water usage estimations are shown in Table 4.3 [3] and Table 4.4 [5].

Table 4.5 shows the water use efficiency in four shale gas plays [5].

1.4 Fracking Fluid and Chemicals

A typical water-based fracturing fluid that is used for shale gas production contains 99.5 % water and proppants, with proppants making up anywhere between 0 and 10 % of the 99.5 %. The rest (0.5 %) are additives that perform essential tasks during different stages of fracturing such as cleanup, formation stabilization, surface tension reduction, corrosion prevention, etc.

Table 4.3 Estimated water needs for drilling and fracturing

Shale gas play	Volume of drilling water per well (gallons)	Volume of fracturing water per well (gallons)	Total volume of water per well (gallons)
Barnett	400,000	2,300,000	2,700,000
Fayetteville	60,000	2,900,000	3,060,000
Haynesville	1,000,000	2,700,000	3,700,000
Marcellus	80,000	3,800,000	3,880,000

Note: These volumes are approximate and may vary substantially between wells
Source: Reference [2]

Table 4.4 Alternate estimated water needs for drilling and fracturing

Shale play	Average drilling water use per well (gallons)	Average fracking water use per well (gallons)	Total average water use (million gallons per well)
Gas shale play (dry gas)			
Barnett	250,000	3,800,000	~4.0
Fayetteville	65,000	4,900,000	~4.9
Haynesville	600,000	5,000,000	~5.6
Marcellus	85,000	5,500,000	~5.6
Liquid shale play (gas, oil, condensate)			
Eagle Ford	125,000	6,000,000	~6.1

Source: Reference [5]

Table 4.5 Water use efficiency in natural gas plays

	Avg. water use per well (million gallons)	Est. avg. natural gas over lifetime (billion cubic feet)	Energy from production per well (trillion Btu)	Water use efficiency (gallons per million Btu)
Marcellus	5.6	5.2	5.35	1.05
Haynesville	5.6	6.5	6.68	0.84
Barnett	4.0	3.0	3.08	1.30
Fayetteville	4.9	2.6	2.67	1.84

Source: Reference [5]

Table 4.6 shows the function of each additive [2].

1.4.1 Proppants

The purpose of a proppant is to prop open a hydraulic fracture. An ideal proppant should produce maximum permeability in a fracture. Fracture permeability is a function of proppant grain roundness, proppant purity, and crush strength.

1.4.2 Gelled Fluids

Water alone is not always adequate for fracturing certain formations because its low viscosity limits its ability to transport proppant. In response to this problem, the industry developed gel fluids, which have higher viscosity. Gellant selection is based on formation characteristics such as pressure, temperature, permeability, porosity, and zone thickness.

Table 4.6 Fracturing fluid additives

Additive type	Main component	Function
Diluted acid (15 %)	Hydrochloric acid or muriatic acid	Help dissolve minerals and initiate cracks in the rock
Biocide	Glutaraldehyde	Eliminates bacteria in the water that produce corrosive byproducts
Breaker	Ammonium persulfate	Allows a delayed breakdown of the gel polymer chains
Corrosion inhibitor	*N*-Dimethyl formamide	Prevent the corrosion of the pipe
Crosslinker	Borate salts	Maintains fluid viscosity as temperature increases
Friction reducer	Polyacrylamide, mineral oil	Minimizes friction between the fluid and the pipe
Gel	Guar gum or hydroxyethyl cellulose	Thickens the water in order to suspend the sand
Iron control	Citric acid	Prevents precipitation of metal oxides
KCl	Potassium chloride	Creates a brine carrier fluid
Oxygen scavenger	Ammonium bisulfite	Removes oxygen from the water to protect the pipe from corrosion
pH adjusting agent	Sodium or potassium carbonate	Maintains the effectiveness of other components, such as crosslinkers
Proppant	Silica, quartz sand	Allows the fractures to remain open so the gas can escape
Scale inhibitor	Ethylene glycol	Prevents scale deposits in the pipe
Surfactant	Isopropanol	Used to increase the viscosity of the fracture fluid

Note: The specific compounds used in a given fracturing operation will vary depending on company preference, source water quality, and site-specific characteristics of the target formation. The components shown above are representative compounds used in hydraulic fracturing
Source: Reference [2]

1.4.3 Breakers

Breaker fluids are used to degrade the fracturing fluid viscosity, which helps to enhance post-fracturing fluid recovery, or flowback. Breakers can be mixed with the fracturing fluid during pumping, or they can be introduced later as an independent fluid.

1.4.4 Acids

Acids are used in limestone formations that overlay or are inter-bedded within formation to dissolve the rock and to create a conduit through which formation water and methane can travel. In addition, acid can be used to clean up perforations of the cement surrounding the well casing prior to fracturing fluid injection. Acids may also be used as a component of breaker fluids. Acids are corrosive, and can be extremely hazardous in concentrated form. Acids are substantially diluted with water-based fluids prior to injection into the subsurface.

1.4.5 Biocides

One hydraulic fracturing design problem that arises when using organic polymers in fracturing fluids is the incidence of bacterial growth within the fluids. Due to the presence of organic constituents, fracturing fluids provide a medium for bacterial growth. As the bacteria grow, they secrete enzymes that break down the gelling agent, which reduces the viscosity of the fracturing fluid. Reduced viscosity translates into poor proppant placement and poor fracturing performance. To alleviate this degradation in performance, biocides, bactericides, or microbiocides are added to the mixing tanks with the polymeric gelling agents to kill any existing microorganisms.

1.4.6 Friction Reducers

To optimize the fracturing process, water-based fluids must be pumped at maximum rates and fluids must be injected at maximum pressures. Increasing flow velocities and pressures in this manner can lead to undesirable levels of friction within the injection well and the fracture itself. In order to minimize friction, friction reducers are added to water-based fracturing fluids.

1.4.7 Acid Corrosion Inhibitors

Corrosion inhibitors are required in acid fluid mixtures because acids will corrode steel tubing, well casings, tools, and tanks. These products can negatively affect the liver, kidney, heart, central nervous system, and lungs.

Service companies have developed a number of different fluids and treatment methods to more efficiently induce and maintain permeable and productive fractures. The required characteristics are:

- Be viscous enough to create a fracture of adequate depth
- Maximize fluid travel distance to extend fractured length
- Be able to transport large amounts of proppants into fracture
- Require minimal gelling agent to allow for easier degradation to reduce cost

The compositions of these fluids, tailored to the specific properties of each shale formation and each well, vary significantly. Each type of fracturing fluid has unique characteristics, and each possesses its own positive and negative performance traits.

2 Water Contamination

Shale gas development poses risks to the water due to contamination of surface water and groundwater. The number one issue with regard to water contamination is flowback water and produced water. About 15–40 % of the water that is used to fracture the shale formation returns to the surface which is called flowback water. Produced water is the water produced during the shale gas production phase. It is difficult to differentiate between flowback water and produced water.

The contaminant may include drill muds and cutting, chemicals, salts, metals, hydrocarbons, fracturing fluids, and dissolved hydrocarbons.

Contamination can occur through spills, pipeline breaks, leaks from storage ponds, leaks or overflow from pits or tanks that store wastewater leading to soil contamination, or leaks of diesel or other fuel used to power the compressors. This can result from inadequate drilling practice (casing, cementing, and completion); inadequate transport or treatment of waste waters; equipment failure; spills of chemicals and flowback water; human errors; or migration of methane from lower rock formation.

2.1 Contaminants

The contaminant compositions vary, depending on formation, fracking fluids, and drilling/fracturing routines, and are site specific. Table 4.7 shows typical flowback water analysis for Marcellus shale in northeastern Pennsylvania [6]. The chloride, sodium, total dissolved solids, and total suspended solids are very high.

Table 4.8 shows how the flowback water compositions vary with the total dissolved solids concentration [6]. The composition of each component increases with increasing dissolved solids concentration.

2.2 Chemicals of Concern

One concern about fracking is that the fracking fluids used to fracture rock formations contain numerous chemicals that could harm human health and the environment, especially if they enter drinking water supplies.

In the last session of Congress, the Committee on Energy and Commerce launched an investigation to examine the practice of fracking in the United States. The Committee asked 14 leading oil and gas service companies to disclose the types and volumes of the fracking fluids they used between 2005 and 2009 [7]. Between 2005 and 2009, the 14 companies used more than 2,500 fracking products

Table 4.7 Typical analysis of flowback water for northeastern Pennsylvania

Parameter	Typical analysis
pH	6.0
Aluminum, mg/L	3.0
Barium, mg/L	6,500
Calcium, mg/L	18,000
Chloride, mg/L	116,900
Iron, mg/L	60
Lithium, mg/L	150
Magnesium, mg/L	1,300
Manganese, mg/L	5.0
Sodium, mg/L	48,000
Strontium, mg/L	4,000
Sulfate, mg/L	130
Total dissolved solids, mg/L	195,000
Total suspended solids	1,200
Total hardness (as $CaCO_3$), mg/L	54,500

Source: Reference [6]

Table 4.8 Contaminant composition for samples with different dissolved solids

Component	Low dissolved solids	Moderate dissolved solids	High dissolved solids
Barium, mg/L	2,300	3,310	4,300
Calcium, mg/L	5,140	14,100	31,300
Iron, mg/L	11.2	52.5	134.1
Magnesium, mg/L	438	938	1,630
Manganese, mg/L	1.9	5.17	7.0
Strontium, mg/L	1,390	6,830	2,000
Total dissolved solids, mg/L	17,941	49,416	90,633
Total hardness, mg/L as $CaCO_3$	69,640	175,268	248,428

Source: Reference [6]

containing 750 chemicals and other components, including 17 that are carcinogens or Safe Water Drinking ACT (SWDA) regulated chemicals (Table 4.9).

The impact of these chemicals on human health will be discussed in a later section.

2.3 *Treatment of Flowback Water*

There is a high incentive to reuse these waters for the next fracturing job because of reduced transportation, procurement, water demand, and disposal cost for the operators. However, the flowback water has presence of:

Table 4.9 Chemicals of concern: carcinogens; SDWA-regulated chemicals

Chemical component	Chemical category
Di (2-ethylhexyl) phthalate	SDWA, carcinogen
Benzene	SDWA, carcinogen
Diesel	SDWA, carcinogen
Lead	SDWA, carcinogen
Acrylamide	SDWA, carcinogen
Nitrilotriacetic acid	Carcinogen
Acetaldehyde	Carcinogen
Ethylene oxide	Carcinogen
Propylene oxide	Carcinogen
Formaldehyde	Carcinogen
Sulfuric acid	Carcinogen
Thiourea	Carcinogen
Benzyl chloride	Carcinogen
Naphthalene	Carcinogen
Ethylbenzene	SDWA
Copper	SDWA
Toluene	SDWA

Source: Reference [7]

Table 4.10 Options for produced water management

Shale gas play	Technology	Availability
Barnett Shale	Injection wells	Commercial and non-commercial
	Recycling	On-site treatment and recycling
Fayetteville Shale	Injection wells	Non-commercial
	Recycling	On-site recycling
Haynesville Shale	Injection wells	Commercial and non-commercial
Marcellus Shale	Injection wells	Commercial and non-commercial
	Treatment and discharge	Municipal wastewater treatment facilities
	Recycling	On-site recycling

Source: Reference [2]

- Salt that causes fouling of heat exchangers and membranes
- Total dissolved solids (TDS) that interfere with the functioning of friction reducers
- Total suspended solids (TSS) and bacteria that cause down hole plugging
- Metal (Barium, Strontium), sulfates, and carbonates that form precipitates

Flowback water management options include injection wells, treatment (onsite or offsite), recycling for reuse, or disposal. Table 4.10 shows the current produced water management options for four shale gas plays.

Produced water is typically produced for the lifespan of a well, although quantities vary significantly by well and by play. For the purpose of this treatment technology discussion, produced water includes all water that is returned to the surface (flowback water) through a well borehole and is made up of water injected during the fracture stimulation process, as well as migrated natural formation water.

The feasibility of produced water reuse is dependent on:

- Quantity of the produced water generated during the first few weeks after stimulation
- Duration in time of produced water generation and how it declines over time
- Quality of the produced water

The TDS concentrations vary tremendously depending on the type of water: brackish water (5,000–35,000 mg/L TDS), saline water (35,000–50,000 mg/L TDS), brine water (50,000–150,000+ mg/L TDS).

TDS, TSS, the larger suspended particulate in water, scale-causing compounds (calcium, magnesium, barium, sulfate), and bacteria growth all have a major effect on the feasibility of reusing produced water.

The objectives of produced water treatment are oil/grease removal, scale control, suspended solids removal, and brine volume reduction. To achieve beneficial use of produced water it is often necessary to use more than one unit process for each objective. The required unit processes for beneficial use vary from well to well and play to play.

Table 4.11 shows a brief description of well-established unit processes for produced water treatment. Other less well-established technologies are available [8–11].

2.4 Water Management Technologies

The management and technology options for produced water treatment vary depending on geological characteristics, shale layer formation, fracturing fluids used, site specific considerations, and preferences of the service company involved. What follows is how Marcellus Shale in Pennsylvania, the basin with the largest reserves and the highest average shale gas daily production, now accounting for 26 % of total shale gas production in the United States, manages the technology options [13].

2.4.1 Underground Injection

All of the shale gas plays employ injection wells as a primary means of disposal. But few, if any, onsite injection wells are used in Pennsylvania or New York, nor are there any commercial disposals wells used for Marcellus Shale flowback and produced water located in these states because of the lack of geological formation suitable for injection.

Table 4.11 Established produced water treatment processes

Unit process	Process description
Oil removal	
API separator	The separator is designed to separate oil from water. It is useful as a first-line treatment process. Oil is mechanically collected as a floated material; a variant of the process uses corrugated plates to collect oil.
Sand bed filter	A bed of sand or walnut shell granular media that is at least four feet deep in a vertical tank.
Hydrocyclone	A cylindrical construction with tangential inlet(s) causes the entering fluid to follow a circular path around the wall of cyclone. Rotation of fluid generates a centrifugal force that causes heavier solids to move toward the wall; lighter material to move toward the center. And the light oil is rejected from the process.
Induced gas flotation	Fine gas bubbles are generated and dispersed in a chamber to suspend particles which ultimately rise to the surface forming a froth layer. The foam contaminating the oil is skimmed from the surface.
Ultrafiltration and microfiltration	Ultrafiltration is a membrane process that is capable of retaining solutes as small as 3.6×10^{-24} lb
Primary treatment	
Sedimentation	A long retention time tank or pond designed to establish quiescent conditions for settling particulates.
Multimedia sand filter	Consist of a bed of stratified granular materials designed to achieve removal of particle matter.
Cartridge filter	Process is comprised of a tube support system that holds filter cartridges. Often used as a pretreatment device. Water is pumped through the filter.
Secondary treatment	
Biological treatment	A number of biological processes have the capability to degrade dissolved oils, volatile acids, and other soluble organics to carbon dioxide.
Activated carbon	A fixed bed column that promotes the adsorption of organic compounds on the surface carbon media as the water passes through the column.
Iron removal	
Aeration and sedimentation	Water is aerated, settled in a sedimentation tank and filtered. Soluble iron Fe^{+++} is oxidized to Fe^{++} which forms an insoluble iron hydroxide precipitate. The oxidized iron floc is then removed by sedimentation and/or filtration.
Lime soda ash softening	Hydrated lime or caustic soda is added to produce water to adjust the pH to above 10.
Ion exchange	A process based on the exchange of ions between water and resin. For example, zeolite resins exchange sodium ions for calcium and magnesium ions that cause hardness in the water.
Desalination	
Reverse osmosis	A membrane process capable of separating chemicals (solute) from an aqueous solution by forcing the water through a semipermeable membrane by applying pressure greater than the osmotic pressure of the solute.

(continued)

Table 4.11 (continued)

Unit process	Process description
Vapor compression distillation	The process includes a multi-effect evaporator that uses a compressor to pull a vacuum on the vessel that induces the boiling of water at lower temperatures.
Freeze thaw evaporation	A process that combines the natural processes of freezing and evaporation to provide driving forces for the demineralization of produced water.
Electrodialysis	An electrically-driven membrane separation process that is capable of separating, concentrating, and purifying selected ions from an aqueous solution.

Source: Reference [12]

Where injection is available (e.g., at other shale gas plays and in portions of Ohio or West Virginia), the injection wells can be either onsite wells operated by the gas producer or offsite third-party commercial disposal wells. Flowback and produced water are delivered by tank truck and are transferred into storage tanks. The flowback and produced water are then injected into a deep formation that has sufficient porosity and injectivity to accept the water.

2.4.2 Discharge to Surface Water Body

Many types of industrial wastewater are discharged to streams, rivers, and other surface water bodies. Permission to discharge wastewater is granted through National Pollutant Discharge Elimination System (NPDES) permits issued by state agencies. However, discharging flowback or produced water directly from a well site presents various challenges.

First, the water typically contains high levels of TDS and other constituents that would require treatment. In response to concerns over flowback and produced water discharge, the Pennsylvania Department of Environmental Protection (PADEP) in April 2009 proposed a new strategy that would add effluent standards for oil and gas wastewaters of 500 mg/L for TDS, 250 mg/L for sulfates, 250 mg/L for chlorides, and 10 mg/L for total barium and total strontium. On May 17, 2010, the Pennsylvania Environmental Quality Board (PAEQB) approved amendments—the new discharge requirements—to the Pennsylvania regulations [14].

The amendments state that no discharge of oil and gas wastewater can be made directly from an oil and gas sites to surface waters. Oil and gas wastewater can be sent to either a commercial industrial wastewater treatment plant or to a publicly owned treatment works (POTW).

Second, the USEPA has adopted national discharge standards for many industries, known as effluent limitations guidelines (ELG). The ELG for the oil and gas industry are promulgated at 40 CFR Part 435 (Title 40, Part 435 of the Code of Federal Regulations).

2.4.3 Haul to POTWs

Prior to the recent rapid development in the Marcellus Shale region, oil and gas development activities in the region generated relatively small volumes of produced water. Some POTWs accepted limited quantities of produced water from oil and gas operators. The produced water was trucked from tanks or pits at the well site and discharged into the treatment facility. The treatment processes found at most POTWs are designed to remove suspended solids and biodegradable materials, but not salinity or TDS.

As the Marcellus Shale development grew in popularity, operators sought permission to bring more truckloads of salty flowback and produced water to the treatment plants. The increased input of TDS resulted in increased levels of TDS in the discharge. The amendments to the PADEP discharge regulations place restrictions on the volume of flowback and produced water that POTWs can accept.

2.4.4 Haul to Commercial Industrial Wastewater Treatment Plant

Several Pennsylvania companies have provided wastewater disposal services to the oil and gas community for many years. As the volume of flowback and produced water has increased rapidly over the past few years, new commercial disposal companies are opening their doors.

The amendments include an important provision relating to existing commercial industrial disposal companies. Any commercial industrial disposal company with a valid NPDES permit is allowed to continue discharging at the permitted levels until such time as the facility seeks an increase in discharge allowance.

2.4.5 Reuse for a Future Fracking Job

Gas companies are interested in finding water to use in fracking jobs and in managing the subsequent flowback and produced water from those wells in ways that minimize costs and environmental impacts. One way to accomplish this goal is to collect the flowback water and reuse it for fracking fluids in other wells. Several gas companies are currently using this approach.

The amendments to the PADEP discharge regulations also include a requirement that any oil and gas wastewater having TDS of less than 30,000 mg/L cannot be discharged; it must be recycled and reused.

The chemical composition of fracking fluids is designed to optimize the performance of the fracking. Generally, the TDS concentration in flowback and produced water is higher than the desired TDS range for new fracking fluids. Several Marcellus Shale operators start with flowback and produced water and blend it with enough freshwater from some other sources to reduce TDS and other constituents so they fall within an acceptable concentration range.

2.5 *Water Management Options Used by Selected Operators*

Many options are currently being employed to manage flowback and produced water by service providers. Most of the operators are recycling some or all of their flowback and produced water. The flowback and produced water that is not being recycled is hauled offsite to POTWs, commercial wastewater disposal facilities, or commercial injection wells.

2.5.1 Chesapeake Energy

Chesapeake Energy conducts ongoing research to identify environmentally safer methods of byproduct management. At various locations, Chesapeake has transported flowback and produced water offsite to a commercial wastewater disposal company, transported it offsite to a sewage treatment plant, and has treated the water for reuse. The company reuses 100 % of its flowback water in some wells.

2.5.2 Range Resources

Range Resources is trying to reuse 100 % of its flowback, production brine, and drill pit water. The only processes involved are settling and dilution. Working backwards from the well performance it sees that it gets just as good of a result with diluted reuse water.

2.5.3 EQT

EQT reuses all of its flowback water without treating it. Flowback water is trucked to the next well location where it is blended with freshwater. Some of the ongoing produced water is hauled to a commercial disposal facility in West Virginia, while other produced water is hauled to a commercial disposal well in Ohio.

2.5.4 East Resources

East Resources recycles all of its produced water and drilling pit fluids into fracking fluids used in other wells. The produced water is not treated for TDS but is blended with freshwater. East Resources generally gets 18–20 % of the water back to the surface. The company is looking at alternate sources of freshwater such as mine water, produced water from shallow formations, and treated POTW effluent.

2.5.5 BLX

BLX is a small producer and does not drill as many wells as some larger companies. Therefore, the treatment of water for reuse does not work for them because of the length of time between fracking jobs. BLX has hauled water to another site, if available. The rest of the water goes to a disposal well, a sewage plant in New Castle, Pennsylvania, or one of the offsite commercial disposal facilities.

2.5.6 Norse Energy

Norse Energy operates in New York and currently has only one operating Marcellus well. For its existing Marcellus well, Norse has disposed of fluid at two facilities located in Warren, Pennsylvania and Franklin, Pennsylvania. Norse is looking at additional disposal sites in the Williamsport, Pennsylvania, area for future wells. Trucking costs are a major portion of the total disposal cost. Norse would also consider transporting the flowback and produced water to a sewage treatment plant, if available, in order to reduce transportation costs.

3 Migration of Water and Gas

There have been heated debates regarding fluid migration from shale formations to water aquifers.

As shown in Table 4.12 the depth of the aquifer (i.e., the depth to the base of treatable water) for four major shale gas basins are between 400 and 1,200 feet, and the depth of the shale formations are between 4,000 and 13,500 feet.

The estimated distances between shale formation and base of treatable water (between 2125 and 13,100 feet) are considered by many as safe as far as aquifer contamination is concerned. It is safe until some unexpected events emerged, however.

Rex Tillerson, the Chairman and CEO of Exxon/Mobil, declared [16] at a congressional hearing in 2010:

Table 4.12 Depth of treatable water and shale formation

Shale play	Depth to base of treatable water (feet)	Depth to shale (feet)	Distance between shale and base of treatable water (feet)
Marcellus	850	4,000–8,500	2,125–7,650
Haynesville	400	10,500–13,500	10,100–13,100
Barnett	1,200	6,500–8,500	5,300–7,300
Fayetteville	500	1,000–7,000	5,600–6,500

Source: Reference [15]

"There have been over a million wells hydraulically fractured in the history of the industry, and there is not one – not one – reported case of a freshwater aquifer having ever been contaminated."

Lisa Jackson testified [17] under oath before the House Committee on Oversight and Government Reform, answering questions before Senator James Inhofe, Ranking Member on the Senate Committee on Environment and Public Health:

"I am not aware of any proven case where the fracking process had affected water, although there are investigations ongoing."

For decades, oil and gas industry executives as well as regulators have maintained that hydraulic fracturing has never contaminated underground drinking water.

As a matter of fact, there is a cause-effect link between fracturing and contamination. USEPA published a report in 1987 [18] concluding that fracking fluids migrated from drilled wells into aquifers. The USEPA Administrator Lisa Jackson was either unaware of the conclusion of USEPA's own report 25 years ago (see below, *EPA 1987 Report*) or trying to avoid the fact when she testified before the House Committee.

3.1 Three USEPA Reports

3.1.1 USEPA 1987 Report

In 1982, Kaiser Gas Co. drilled and hydraulically fractured a natural gas well on the property of Mr. James Parson in Jackson County, West Virginia. A year and a half after Kaiser fractured the gas well Parson's well water became polluted.

The well was fractured using a typical fracturing fluid—gel. The residual fracturing fluid migrated into Mr. Parson's water well, according to an analysis of well samples taken from the property by the West Virginia Environmental Health Services Lab. Dark and light gelatinous fracturing fluid was found. The gel found in the water is consistent with contamination from hydraulic fracturing fluid [19].

USEPA investigated the case and published a report [18]. The report concluded that the fracking fluids did indeed migrate from drilled wells into aquifers. The USEPA stated in the report that fractures can be produced, allowing migration of native brine, fracturing fluid, and hydrocarbon from the oil and gas well to nearby water wells. This finding contradicts what the industry has been saying for years. It also showed that fractures from one well could spread unpredictably and were known to have caused fracturing fluid to migrate into other nearby natural gas and oil wells [19].

The USEPA pointed out at the time that one of the biggest problems for researchers has been that the oil and gas industry often reaches sealed settlements with the people whose drinking water was rendered undrinkable, so documentation is under wraps. The result is that the public, including scientists and regulators, is prevented from learning about cases where groundwater has been contaminated.

These findings also contradict another USEPA report (see below, *USEPA 2004 Report*) with the conclusion that hydraulic fracturing in coal bed methane natural gas wells posed no risks to underground water supplies.

3.1.2 USEPA 2004 Report

Citizens living near coal bed methane production using hydraulic fracturing expressed concern about contaminated drinking water wells. The USEPA has contacted and been contacted by citizens who believed their water wells were affected by coalbed methane production in San Juan Basin (Colorado and New Mexico), Black Warrior Basin (Wyoming and Montana), Central Appalachian (Alabama), and Powder River Basin (Virginia and West Virginia).

The drinking water contamination incidents and complaints of citizens are many, including:

- Methane gas seeped into the river
- Pump house door was blown off because of a methane explosion
- In addition to methane there are hydrogen sulfide, anaerobic bacteria, grass and trees turning brown
- Water flow decreased
- Water wells producing more and more gas with milky white water
- Water with odor, globs of black, jelly-like grease, and smelling of petroleum
- Soap bubbles (used in fracturing) flowing from residential household fixtures
- Fluids drained from fracturing sites with soap bubbles in the water—killing all animal and plant life in its path
- Water with dark black suspended sediments
- Water supplies diminishing or drying up entirely
- Etc.

Most of the residents said that their complaints to the state usually resulted in investigation without resolution. Some residents mentioned that the gas companies were providing them with water to compensate for the contamination or loss of their drinking water wells. However, the residents said that this was not adequate compensation for the impacts, or loss of, their private drinking water supplies.

Responding to the complaints, USEPA undertook a study and issued a report [20].

The report concluded that the injection of hydraulic fracturing fluids into coalbed methane wells *posed little or no threat to drinking water and did not justify additional study at that time.* Although potentially hazardous chemicals could be introduced into drinking water when fracturing fluids are injected into coal seams that lie within the drinking water, the risk posed to drinking water by introduction of these chemicals was reduced significantly by groundwater production and injected fluid recovery, combined with the mitigating effects of dilution and dispersion, adsorption, and potentially biodegradation.

Inexplicably, the USEPA failed to mention in this 2004 report the contradictory conclusions of the USEPA 1987 report. *In the debate over these risks, USEPA and Congress have never cited the USEPA 1987 report.*

The report's conclusion—no confirmed case of linking hydraulic fracturing to contaminating drinking water wells—was used by the Bush Administration to justify exemption of hydraulic fracturing from the SDWA oversight as part of Energy Policy Act 2005.

Criticism for the 2004 Report

The 2004 report whitewashed industry and was dismissed by experts as superficial and politically motivated [21].

The review board for the USEPA 2004 study consisted of seven independently appointed professionals, and five of the seven "deciders" came directly from the ranks of the oil and gas industry itself. No qualified, experienced professionals employed by the USEPA were included on the peer review team [22].

The 2004 USEPA study has been labeled "scientifically unsound" by USEPA whistleblower Weston Wilson. In an October 2004 letter to Colorado's Congressional delegation, Wilson recommended that the USEPA continue investigating hydraulic fracturing and form a new peer review panel that would be less heavily weighted with members of the regulated industry.

In March of 2005, USEPA Inspector General Nikki Tinsley found enough evidence of potential mishandling of the USEPA hydraulic fracturing study to justify a review of Wilson's complaints [23].

There were two versions of the study, a draft version and a final version. The differences between the two versions were controversial and at times the two versions contradicted each other. For instance, the draft version included water samples from Fruitland aquifer in San Juan Basin of Colorado and New Mexico, which showed evidence of residual contamination from previous fracturing treatments. However, this information was deleted from the final study.

The Oil and Gas Accountability Project conducted a review [24] of the USEPA study and found that the USEPA removed information from earlier drafts that suggested unregulated fracturing poses a threat to human health, and that the Agency did not include information suggesting that fracturing fluids may pose a threat to drinking water long after drilling operations are completed.

New Study

On June 9, 2009 Senators Bob Casey (D-PA) and Charles Schumer (D-NY), and Representatives Dianna DeGette (D-CO), Jared Polis (D-CO), and Maurice Hinchey (D-NY), introduced bills in the Senate and House to close the

"**Halliburton Loophole**" (see Box 4.1) in the SWDA. Local governments expressed support for ending the loophole. This time Congress demanded a transparent peer-reviewed process.

Box 4.1 Halliburton Loophole

Halliburton developed hydraulic fracturing in the 1940s and remains one of three largest manufacturers of fracturing fluids. Dick Cheney served as CEO (1995–2000) of Halliburton just before joining the Bush Administration in 2001. Cheney convened a secret energy task force in his second week in office as Vice President.

The task force recommended, among others, exemption of hydraulic fracturing from the SWDA oversight. The Energy Policy Act 2005, crafted by Cheney, amended SWDA so that it no longer applied to hydraulic fracturing. Congress passed an amendment of SWDA in 2005 and officially exempted hydraulic fracturing.

The exemption from SWDA oversight has become known as "Halliburton Loophole" because it is widely perceived to have come about as a result of the efforts of then Vice President Dick Cheney.

The Halliburton loophole authorized oil and gas drillers, exclusively, to inject known hazardous materials—unchecked—directly into or adjacent to underground drinking water supplies. It passed as part of Bush Administration's Energy Policy Act 2005.

The study came at the behest of Democratic lawmakers who inserted language into the agency's fiscal year 2010 spending bill that directed the USEPA to conduct the study.

3.1.3 USEPA 2011 Report

Domestic well owners near the Town of Pavillion, Colorado, complained about smells, tastes, and adverse changes in the quality of the water quality coming from their domestic wells and petitioned the USEPA in 2008, asking the agency to investigate whether groundwater contamination existed, its extent, and possible sources. In response to the petition, USEPA sampled 39 individual wells, collecting data to assess groundwater conditions and to evaluate potential threats to human health and environment.

On December 8, 2011, the USEPA issued a draft report on its investigation [25]. The USEPA identified certain constituents in groundwater above the production zone of the Pavillion natural gas well that were consistent with some of the constituents used in natural gas operation, including the process of hydraulic fracturing. In its report, the USEPA claimed that its approach to the investigation best supported the explanation that inorganic and organic compounds associated with hydraulic fracturing had contaminated the aquifer at or below the depths used for domestic water supply in the Pavillion area. The USEPA also stated that gas production activities had likely enhanced the migration of natural gas into the aquifer and migration of gas to domestic wells in the area. Because the draft report linked groundwater contamination to activities related to hydraulic fracturing during natural gas production in the area, it raised concerns about hydraulic fracturing in general. Organizations representing portions of the natural gas industry and other stakeholders took issue with some of the findings in the draft report, and questioned the scientific validity of the USEPA's contention. On December 14, 2011 the USEPA began a 45-day public comment period (which was later extended until January 15, 2013) for the draft report. Additionally, the report will be peer-reviewed by a panel of independent scientists.

3.2 *Methane Contents in Water*

Methane concentrations in groundwater for the 170 wells in West Virginia ranged from not detected to 68.5 mg/L. Methane was detected in 131 of 170 of wells of these wells and was present in concentrations greater than 28 mg/L (a dangerous level) in 13 of these wells. That is, 77 % of the wells were found to be detectable, but dangerous concentrations only occurred in about 8 % of these wells. Another 13 wells had methane concentrations ranging from 11.9 to 14.3 mg/L; 32 wells had concentrations ranging from 1.00 to 10.00 mg/L; and 73 wells had detectable methane concentration less than 1.00 mg/L. Methane was detected in 43 of 47 counties sampled, but methane concentrations exceeding 10 mg/L were found in only 11 counties [26].

3.3 *Three Controversial Research Papers*

Three research papers on migration generated heated debates.

3.3.1 Osborn et al. 2011

A paper was published in the Proceedings National Academy of Science (PNAS) by Stephen Osborn [27] on methane contamination of drinking water. It was observed that dissolved methane concentrations in water from the 34 wells located more than 1 km from fracking operations averaged about 1.1 mg/L. But in water taken from 26 wells within 1 km, methane concentrations averaged 19.2 mg/L. It strongly suggested that drilling could lead to elevated methane concentration to at least some nearby drinking water supplies. But the study did not find any evidence of impacts to drinking water caused by fracking itself.

The paper was contested by Richard Davies [28]:

> "Their data showed that contaminations may have occurred, but the association with hydraulic fractures remains unproven. To test whether hydraulic fracturing could cause aquifer contamination requires baseline measurements of levels of methane in aquifers before and after hydraulic fracturing, preferably elsewhere in the world where there has been less historical drilling and natural seepage."

3.3.2 Warner et al. 2012

A new research study [29] found that salty, mineral-rich fluids deep beneath Pennsylvania's Marcellus natural gas fields are likely seeping upward thousands of feet into drinking water supplies. The occurrences of saline water did not correlate with the location of shale gas wells. However, the study concluded that the presence of these fluids suggests conducive pathways and specific geostructural and hydrodynamic regimes in northeastern Pennsylvania that are at increased risk for contamination of shallow drinking water resources.

This research was contested by Chris Tucker [30]: (1) No discussion of time scale of migration, (2) No discussion of migration pathway, (3) No discussion of whether Marcellus even contains enough brine water to leak, (4) No discussion of transport mechanism.

The paper was also criticized by Penn State professor Terry Engelder [31]. Engelder was one of the peer reviewers of this paper who wrote a letter to PNAS objecting to publication of the paper.

3.3.3 Myers 2012

A study by Tom Myers [32] using computer modeling concluded that natural faults and fractures in the Marcellus, exacerbated by the effects of fracking itself, could allow chemicals to reach the surface in as little as "just a few years." Their study challenged conventional wisdom that drinking water sources are insulated from mile-deep fracking. Simply put, the rock layers are not impermeable. This is the first peer-reviewed research evaluating this possibility.

While the study was covered heavily by groups opposing shale gas development, several scientists, including Don Siegel [33] criticized that mistaken assumptions were used for the rocks above the Marcellus shale and groundwater movement as well as wildly exaggerated fracture-length assumptions.

3.4 Migration Pathway Discussion

Fracturing may create new fractures that intersect natural geologic vertical faults that communicate with the surface or with upper geological zones. The formation pressure would force newly liberated oil and gas, as well as the residual toxic fracking fluids, through these new fractures and into a freshwater aquifer or to the surface at a distance as great as a mile from the well [34].

While oil and gas companies have geological data, they cannot identify every natural fault or irregularity near the wellbore.

As the borehole extends deeper into the earth, previously isolated layers of formation are exposed to one another, with the hole as the conductive path. Isolating these layers, or establishing zonal isolation, is a key to minimizing the migration of formation fluids between zones or to the surface. Since formations are generally far from being homogeneous, the horizontal laterals can intersect with existing natural fractions or faults in the formation, or an aggressive hydraulic fracking could induce fractures into existing natural fractures or faults which can penetrate a lower water sand layer [35].

The relative risk of hydraulic fracturing varies substantially by local geological context, including the nature and depth of the source rock, lithology of overlying rocks, and the nature of existing fractures and fault networks.

The Marcellus shale represents a dynamic system that has been changing for almost 400 million years, and the shale is not homogeneous; different regions will react differently to the same hydraulic fracturing stimulation. There are environmental concerns about hydraulic fracturing in the Marcellus shale, in part because the shale surrounding and including the Marcellus has already experienced natural fracturing, or jointing, as a part of its geologic history. It has been suggested that stimulating the Marcellus with hydraulic fracturing may cause the pre-existing fractures to connect and create a pathway that leads drilling mud, hydraulic fracture fluid, formation water, and methane gas to contaminate drinking water aquifers or water sources [36].

The propagation of fractures caused by fracking and thus the full effects of the fracking process itself are not fully known, since the underlying geology thousands of feet below the surface and up to a thousand feet away from the wellbore cannot be identified with exactitude [37].

In the scenario that the fractures intersect, the formation pressure would force the newly liberated oil and gas and toxic fracking fluids through these new fractures into the natural fault. These toxic fluids could then travel upward past the reservoir cap into a freshwater aquifer near the surface.

Possible mechanisms for leakage of gas to water resources:

- Leakage of pressurized gas through uncompleted casing to a shallow fracture system
- Migration from target formation via a fracturing system (could be enhanced by fracturing)

Leakage pathways:

- Poorly cemented casing/hole annulus
- Casing failure
- Well abandonment failure

Except for the cases of preexisting faults or fractures—either naturally formed or created by hydraulic fracturing—the probability of migration from shale layers or geological layers above shale layers to water layers is very low. The probability is from 1:200,000 to 1:200,000,000 [38].

The industry correctly claims that fracking is 60 years old. But using this convenient shorthand cloaks and diverts the much more complex issue on hand. What's at issue here isn't really just fracking; it's the entire process of producing gas from shale using high-volume, fracking fluid with long laterals from clustered, multi-well pads. The debate about whether leaking shale methane comes from migration of heavily fracked zones creating faults into groundwater, or along poorly cemented wellbores is immaterial to landowners or home owners. A leak is a leak. They are concerned about whether it leaks or not; how it leaks is secondary.

4 Sedimentation

4.1 Runoff Water and Sedimentation

Water runoff is unfiltered water that reaches streams, lakes, sounds, and oceans by means of flowing across impervious earth surfaces. Storm water runoff is generated when precipitation from rain and snowmelt events flow over land or impervious surfaces and does not percolate into the ground. As stormwater flows over a construction site, it can pick up pollutants like debris, and chemicals from paint, concrete, washout, etc. and transport them to nearby sewer systems, or rivers. The uncontrolled runoff can greatly affect water bodies by decreasing water clarity, conveying toxic chemicals into waterways, increasing pathogen concentrations, increasing the need for dredging, and raising the cost of drinking water treatment.

Consequences of stormwater pollution include:

- Flooding and property damage
- Stream bank and streambed erosion
- Siltation and sedimentation
- Increased water temperature, impacting aquatic ecosystems
- Harm to aquatic life
- Harm to sport fishing and coastal shellfisheries
- Impact on drinking water supply
- Human illness
- Aesthetic losses

Erosion is a natural process that wears away rocks and soil deposits on the earth's surface through the action of water, ice, or wind. Erosion can remove soil from the land and carry it away by water action. When water eventually drops the soil, it settles into the bottom of the waterway. If sufficient erosion control is not established, then surfaces are exposed to precipitation.

The soil accumulating at the bottom of the waterway (rivers, lakes, and streams) is known as ***sediment***. Much of the impact of this sediment is aesthetic—muddy water is unsightly and can decrease property values. However, turbid water can also decrease the quantity of sunlight that penetrates to submerged plants in lakes and oceans, thereby potentially harming entire ecosystems. The process of depositing sediment is known as ***sedimentation***. Soil sedimentation is the result of water erosion.

Sedimentation deposition can result in:

- Increased dissolved solids
- Increased storm water runoff
- More long-term infrastructure items: removal of forest cover, change in land use
- Loss of critical habitat
- Changes in ecosystem diversity, vegetation type
- Soil compaction

The resulting siltation can cause physical, chemical, and biological harm to our nation's waters. For example, excess sediment can quickly fill rivers and lakes, requiring dredging and destroying aquatic habitats [39].

The erosion process adversely affects the land, while sedimentation harms waterways. Sources of sedimentation include agriculture, urban runoff, forestry, and ***construction sites***.

Land left exposed or undergoing construction is vulnerable to an increased rate of erosion. Runoff from construction sites can deposit sediment and other harmful pollutants into rivers, lakes, and streams. The primary environmental concerns of sediment with construction activities are that it clouds water, decreases photosynthetic activity, reduces the viability of aquatic plants and animals, and ultimately destroys organisms and their habitats [40].

The construction and operation of shale gas extraction facilities can have significant and adverse environmental impacts on water quality. Specifically, the impacts associated with erosion and sediment discharge and storm water discharge during construction, operation, and after well closure can negatively and significantly impact water quality.

Construction sites can contribute more sediment to streams in a short period of time than can be deposited naturally during several decades [39]. The sediment runoff rates from cleared and graded construction sites are typically 10–20 times greater than those from agricultural lands and forest lands [40].

4.2 Runoff Water Strategy

There is presently no regulatory oversight of oil and gas-related construction or operations under the National Pollutant Discharge Elimination System (NPDES) permit program, except in very limited circumstances. While NPDES stormwater regulations cover a large amount of the construction and industrial activity, the Congress mandated that oil and gas construction is specifically exempt from stormwater regulations in the Energy Policy Act of 2005. To help local governments decide whether drilling activities do, in fact, impact their water resources, and how to minimize those impacts, the USEPA awarded a grant in 2005 to the City of Denton, Texas, to monitor and assess the impact of gas well drilling on stormwater runoff, and to provide, if necessary, management strategies for these activities. Runoff, primarily from the sites' well pad areas, was monitored and analyzed, as were the contents of on-site drilling mud pits. The findings [41]:

Gas well sites have the potential to produce sediment loads comparable to traditional construction sites.

- Total suspended solids (TSS) and turbidity event mean concentrations (EMC)—pollutant mass/runoff volume—at gas sites were significantly greater than at reference sites (the median TSS EMC at gas sites was 136 times greater than reference sites).
- Compared to the median EMCs of storms sampled by Denton near one of their outfalls, the gas well site median EMC was 36 times greater.

Other pollutants in gas well runoff were found in high concentrations.

- The EMCs of total dissolved solids, conductivity, calcium, chlorides, hardness, alkalinity and pH were higher at gas well sites compared to reference sites, and differences were statistically significant for all parameters except conductivity.
- Generally, the presence of metals was higher at gas well sites compared to reference sites and EMCs were statistically significantly greater for Fe, Mn, and Ni.
- Overall, the concentrations of metals tend to be higher at gas well sites compared to both nearby reference sites and as measured in runoff from local mixed-use watersheds.

Conclusions based on runoff sampling results:

- Gas well sites have the potential to negatively impact surface waters due to increased sedimentation rates and an increase in the presence of metals in storm water runoff.
- Pad sites also have the potential to produce other contaminants associated with equipment and general site operations.
- Gas wells do not appear to result in high concentrations of petroleum hydrocarbons in runoff, but accidental spills and leaks are still a potential source of impact.

4.3 New USEPA Regulation

On December 1, 2009 the USEPA published in the Federal Register (74 FR 62995) effluent limitation guidelines (ELG) and new source performance standards (NSPS) for the construction and Development Point Source category, governing discharge of pollutants from construction sites. The new numeric limit, 280 nephelometric turbidity units (NTU), applies regardless of the conditions at the project, including the soil, the site, the water or the natural levels of sediment present in the waters into which the construction site runoff may drain [42].

This was the first time that the USEPA adopted a numeric limit for storm water flow from construction sites. This new limit on turbidity is being phased in over a 4 year period that began February 1, 2012. This requirement applies to construction sites disturbing 20 acres or more starting August 2, 2011 and sites disturbing 10 acres or more no later than February 2, 2014 [43].

The construction rule requires construction site owners and operators that disturb 1 or more acres to use best management practices to ensure that soil disturbed during construction does not pollute nearby water bodies.

More studies and investigations are needed to determine the full scope of this issue.

4.4 Soil Contamination

Soil contamination can be caused by:

- Inadequate casing and cementing practice
- Surface spills
- Human errors
- Inadequate liners or leaks from storage tanks, ponds, surface impoundment
- Migration of chemicals in fracking fluids from formation

Wastewaters returning to the surface can contain radioactive materials including strontium, uranium, radon, and heavy metals, all of which contaminate the soil through spills and leaks during venting or flaring. Heavy metals found in soils near gas sites have included lead, mercury, cadmium, chromium, barium, and arsenic. Though radioactive materials are not necessarily present in the soil at every well or drilling site, soils in the upper Midwest or Gulf Coast states are more likely to contain radioactive materials. Contamination also comes from produced water, drilling mud, sludge, slimes, evaporation ponds, pits, and from mineral scales that form in pipes, storage tanks or other extraction equipment. Radon gas, radium-226, and radium-228 can be released into the atmosphere from produced water and mud, which is placed in ponds, or pits to evaporate, re-use, or recover [44].

Because flowback water is exempt from Resource Conservation and Recovery Act (RCRA) regulations, it is important to test and inspect lines, vessels, dump valves, hoses, and other pollution prevention equipment where failures or malfunctions could result in a potential spill incident. Installation of contaminant, Best Management Practices (BMP), barriers or response equipment on site is deemed necessary [44].

5 Air Emissions

5.1 *Air Emissions from Shale Gas Production*

Shale gas development poses risks to air quality. These risks are generally results of the engine exhaust from increased traffic, emissions from diesel-powered pumps used to power equipment, intentional flaring or venting of gas for operational reasons, and unintentional emissions of pollutants from faulty equipment or impoundments, plus emissions resulting from hydraulic fracturing [15].

According to the USEPA analysis, natural gas well completion involving horizontal wells and fracturing vents approximately 230 times more natural gas and volatile organic compounds than natural gas well completions that do not involve hydraulic fracturing [15]. Construction of well pads, access roads, and drilling facilities requires substantial truck traffic.

There is a scarcity of pollutants emission data for horizontal drilling and hydraulic fracturing in any shale gas basins.

In 2009, a report [45] on pollutant emissions in the Barnett shale area was prepared for the Environmental Defense Fund. Table 4.13 provides the estimated pollutant emissions for the Barnett shale area in 2009. The estimations are

Table 4.13 Estimated pollutant emission for Barnett shale area in 2009

Emission source	VOC (tpd)	HAP (tpd)	CH_4 (tpd)	CO_2e (tpd)	NOx (tpd)
Well drilling, fracturing, and well completion	21	0.49	183	4,061	5.5
Production fugitives	26	0.62	232	4,884	
Natural gas processing	15	0.37	50	1,056	
Natural gas transmission fugitives	28	0.67	411	8,643	
Condensates and oil tanks, annual average	30	0.60	7.0	149	
Compressor engines	19	3.6	61	13,877	46
Total daily emissions, annual average	139	6.4	945	32,670	51
Total daily emissions, peak summer	255	17	961	33,004	

Source: Reference [45]

Table 4.14 Emission factors used for estimating Barnett shale area pollutants in 2009

	VOC (lbs/MMcf)	HAP (lbs/MMcf)	CH_4 (lbs/MMcf)	CO_2 (lbs/MMcf)
Production fugitives	11	0.26	99	1.9
Processing fugitives	1.4	0.3	45	1.0
Transmission fugitives	12	0.28	175	3.3
Condensate	10	0.20	1.7	0.23
Oil tank	1.3	0.013	0.26	0.70

Source: Reference [45]

calculations based on emission factors listed in Table 4.14. Barnett Shale Basin covers 5000 square miles and extends into at least 21 Texas counties.

The emission factors can be used to estimate pollutant emissions in other shale gas basins.

5.2 *DISH, Texas Study*

The Town of DISH, Texas—where the Barnett Shale Basin lies—released a study [46] undertaken in response to residents' complaints about human health effects. Thirty-five chemicals were detected in the air in association with shale gas drilling, production, and distribution (Table 4.15).

These chemicals released into the environment by shale gas activities are toxic and pose threats to human health.

5.3 *Greenhouse Gas Discussion*

A number of studies for greenhouse gas emissions have been published [47–52].

Compared to conventional natural gas extraction, horizontal drilling with hydraulic fracturing has higher emissions because the wellbore is longer, extra energy is needed for puncturing casing, high pressure fracking fluid, and transportation of water and chemicals to the well site, as well as removal of the waste water/chemicals after fracturing. A paper by Cornell University's Robert Howarth [47] argues that the natural gas from fracking operations can be worse for the atmosphere than coal because of methane seepage into the atmosphere. The Cornell study suggests that life cycle greenhouse gas (GHG) emissions from shale gas are 20–100 % higher than coal on a 20-year timeframe basis; the paper has not been well received, and has been criticized for incomplete data. A study by NETL [48] shows that natural gas base load has 50 % lower GHG emissions than coal on a 20-year timeframe basis. Two other studies [49, 51] also have different conclusions

Table 4.15 Human health impact associated with the development of shale gas play

Benzene	Dimethyl disulfide	Toluene	Isobutane
1,3-Butadiene	Ethyl benzene	1,1,2-Trichloroethane	Methane
n-Butyl alcohol	Ethylene	Trichloroethylene	Propane
Carbon disulfide	Ethylene oxide	Trimethyl benzene	Propylene
Carbonyl sulfide	Methyl-ethyl disulfide	1,2,4-Trimethyl benzene	Nitrogen oxide
Chlorobenzene	Formaldehyde	Xylene	Carbon monoxide
Chloromethane	Methyl ethyl benzene	Ethane	Sulfur dioxide
1,2-Dichloroethane	Methyl pyridine	1,1,1,2-Tetrachloroethane	
Diethyl benzene	Naphthalene		
Dimethyl pyridine	Tetramethyl benzene		

Source: Reference [46]

Table 4.16 Greenhouse gas study conclusion

Investigator	Conclusion
Robert Howarth 2011	Natural gas from shale fracking operations can be worse for the atmosphere than coal because of methane seepage into atmosphere. The study suggests that life cycle greenhouse gas emissions from shale gas are 20 %-1005 higher than coal on a 20-year timeframe basis.
Skone 2011	On electricity-generation comparison basis the natural gas base load has 50 % lower GHG emissions than coal on a 20-year timeframe basis.
Worldwatch 2011	On average U.S. natural gas-fired electricity generation emits 47 % less GHG than coal from source to use using 100-year timeframe basis.
Jiang 2011	The life cycle GHG footprint for shale gas is 20–50 % lower than that for coal.
Cathles III 2012	Concludes that the Howarth study was "seriously flawed," and that the shale gas has a GHG footprint that is only one-third to one-half that of coal.

Source: Reference [52]

than Howarth's. All of these studies have different design basis, system boundaries, and assumed data. This controversy has not settled yet, and a study based on the same design basis and system boundary is needed for prudent policy consideration.

Conclusions from five investigators [52] are provided in Table 4.16.

6 Earthquake, Explosion and Blowout

6.1 Earthquake

While earthquakes are a result of movements in the earth's crust, they can also be caused by humans. It was pretty much established in the 1960s that injecting fluids into the ground sometimes causes earthquakes [53]. Numerous earthquakes between the 1960s and 1990 that were clearly associated with fluid injection to deep geological formations were studied and well documented [54].

Table 4.17 Seismic activity in Arkansas

Year	Number of events
2001	17
2002	19
2003	10
2004	16
2005	28
2006	9
2007	22
2008	31
2009[a]	37
2010	772
2011	788

Source: Reference [55]
[a]Year fracking started

The earth's crust is pervasively fractured at depth by faults. These faults can sustain high stresses without slipping because natural tectonic stress and the weight of the overlying rock push the opposing fault blocks together, increasing the frictional resistance to fault slip. The injected wastewater counteracts the frictional forces on faults and, in effect, pries them apart, thereby facilitating earthquake slip.

Fracking requires up to 4 million gallons of fluid per horizontal well. It has been reported that anywhere between 25 and 100 % of the chemicals-laced hydraulic fracturing fluid (known as flowback) returns to the surface at the Marcellus Shale operation while the rest remains in the formations. The flowback fluid can be treated for reuse for other fracking operations or can be injected to dedicated disposal wells (different from the fracking and gas production wells) nearby. It is the flowback fluid injected into the waste disposal well, rather than the original fracture fluid used to fracture shale rock and extract gas, that triggers earthquakes.

The flowback fluid, when injected to a disposal well, can infiltrate a nearby, pre-existing fault and act as a lubricant making it easier for the sides of the fault to slip past each other and relieve the build-up pressure that triggers earthquakes.

In Arkansas, the seismic events from 2001 to 2008 were 32 or fewer (Table 4.17). When large-scale fracking started in 2009 the number of events increase to 37 (2009), 772 (2010), and 778 (2011). Obviously a causative relationship between injection and seismic events exists. The number of quakes in the state of Arkansas in 2010 nearly equals all of Arkansas' quakes for the entire twentieth century. Arkansas' history of earthquakes and drilling reveals a shocking surge in quake frequency following advanced drilling and fracking [55].

On February 27, 2011, a magnitude 4.7 earthquake occurred near the Town of Greenbrier in central Arkansas, which was the largest quake to hit the state in 35 years. The service operators (Chesapeake Energy and Clarita) shut down two injection wells on March 4, 2011. The week after the two wells went offline, earthquake frequency dropped by half. Once the wells were shut down, only two quakes had a magnitude 3.0 or greater. The majority were between magnitudes 1.2

and 2.8. That is, both the number and strength of earthquakes have noticeably dropped since the shutdown [56, 57].

In Ohio there were 12 earthquakes near Youngstown in 2011, and the state shut down injection wells after several minor earthquakes occurred on December 24 (2.7-magnitude) and December 25 (4.0-magnitude) within a 5-mile radius. Another earthquake (4.0-magnitude) that was felt across the Youngstown area occurred December 31, just one day after regulators shut down a suspected disposal well [58, 59].

Ohio doesn't have much history with earthquakes, and the frequency of the 11 earthquakes in an 8-month period within two miles of the wells is definitely unusual [60].

Between November 2009 and September 2011, there were 67 earthquakes in the Barnett Basin with magnitude 1.5 and larger. All of the epicenters occurred within 5 miles of one or more injection wells. These included wells near Dallas-Fort Worth and Cleburne, Texas, where earthquakes near injection wells were reported in 2008 and 2009, as well as in other locations where no earthquakes had previously been reported. Cliff Frohlich concluded in a paper published in the Proceedings of the National Academy of Sciences [61] that there is strong correlation between fracking and earthquakes.

In late May, England stopped operation of its only hydrofracking project after two earthquakes occurred near the site within an 8-week period. Dr. Brian Baptie, the seismology project leader for the British Geological Survey says, "It seems quite likely that they are related." [57].

6.2 Explosion

Unplanned release of natural gas in the subsurface during drilling may release gas from the well (called blowout) or migration of gas from deeper formation to the surface of nearby houses. The blowout gas may cause an explosion when the methane concentration exceeds the explosion limit.

On December 15th, 2007, an explosion was reported in a home in Bainbridge Township (Geauga County, Ohio). Early investigation determined that methane was entering homes in the vicinity of the explosion through domestic water wells.

The Ohio Department of Natural Resources, Division of Mineral Resources Management (DMRM) inspected local gas wells and identified that the source of gas (English No. 1 well) was owned by Ohio Valley Energy Systems Corp (OVESC). The primary causes of gas invasion into drinking water aquifers are: (a) poor primary cement job; (b) decision to hydraulically fracture the well despite the poor cement job; (c) the decision to shut in the surface-production casing annulus for 31 days allowing the annulus to become over-pressured and gas to migrate from the high-pressure annulus, through fractures, to the groundwater aquifer and eventually into domestic water wells [62].

6.3 Blowout

In 2004, citizens notified the Colorado Oil and Gas Conservation Commission (COGCC) of the presence of gas bubbling in the West Divide Creek, Garfield County, Colorado, near the Mamm Creek Gas Field. Subsequent investigations identified the gas as thermogenic gas from the William Fork Formation, which is the primary gas-bearing target in the Mamm Creek Field. Some key observations and conclusions emerging from various studies are [62]:

- Some domestic water samples contain methane and deep formation water which may have migrated to water wells through natural pathways or gas wellbores, or both.
- The study area is naturally faulted and fractured. Fault and fracture density increases near structural features.
- Domestic water wells with elevated methane and chloride concentrations are often coincident with structural fractures.
- Natural fractures and faults may provide migration pathways for gas and fluids, both to groundwater and to the uncemented annular space of wellbores. Fractures and faults may also cause complications in well drilling, construction, and completion, resulting in well integrity problems.

On August 17, 2012, a spark from a Marcellus gas drilling operation in Harrison County, West Virginia ignited methane gas several hundred feet underground, sending up a fireball and triggering a blaze that burned for about an hour on the floor of the rig [63]. It was a spark that occurred when they began to withdraw the drill that ignited methane.

On June 3, 2010, shale gas driller EOG Resources in Clearfield County, Pennsylvania lost control of a wellbore during post-fracturing cleanout activities, releasing natural gas, flowback water, and brine as high as 75 feet into the air. It was determined that blowout prevention equipment was inadequate and that certified well-control personnel were not on-site [64].

7 Health Impacts

7.1 General Health Impacts

According to the New York State Department of Environmental Conservation the human health concerns related to hydraulic fracturing of shale formations include, but are not limited to [65]:

- *Neurological systems:* Developmental disorders involving cognitive, behavioral and psychosocial disorders among children.
- *Endocrine disruptors:* Affecting hormonal and metabolic processes, leading to infertility, early puberty and other reproductive issues affecting both men and women.

- *Immune-suppressants:* Decreasing the immunological defenses of the general population, leading to greater vulnerability to existing and emerging infectious agents.
- *Mutagens and carcinogens:* leading to a greater incidence of all cancers especially among children, adolescents and young adults.
- *Other chemicals* which do damage to the renal system, gastrointestinal system and cardiac and respiratory systems, as well as skin, eyes, ears, and nasopharyngeal tissues.

Lisa McKenzie's testimony on potential human health impacts [66] includes:

- Based on the ambient air studies in Colorado, Texas, and Wyoming, natural gas development processes can result in direct and fugitive air emissions of a complex mixture of petroleum hydrocarbons and other pollutants.
- The natural gas resource itself contains petroleum hydrocarbons, including alkanes, benzene, and other aromatic hydrocarbons. Petroleum hydrocarbons and other pollutants also may originate from diesel engines, tanks containing produced water, and on site materials used in production, such as drilling muds and fracking fluids. This complex mixture of chemicals can result in the formation of secondary air pollutants, such as ozone. The public health concern is the transport of these air pollutants to nearby residences and population centers.
- Multiple studies on inhalation exposure to petroleum hydrocarbons in occupational settings as well as homes near refineries, oil spills and gas stations indicate an increased risk of eye irritation and headaches, asthma symptoms, leukemia, and myeloma. Many of the petroleum hydrocarbons, such as benzene, observed in these studies are the same as those that have been observed in and around natural gas development sites.
- The ambient benzene levels observed in the natural gas development area of Garfield County, Colorado demonstrate an increased potential risk of developing cancer as well as chronic and acute non-cancer health effects.
- Health effects associated with benzene include leukemia, anemia, and other blood disorders and immunological effects. In addition, there is a link between maternal exposure to ambient levels of benzene and an increase in prevalence of neural tube birth defects.
- Inhalation of trimethylbenzenes and xylenes can irritate the respiratory system and mucous membranes with effects ranging from eye, nose, and throat irritation to difficulty in breathing and impaired lung function.
- Inhalation of trimethylbenzenes, xylenes, benzene, and alkanes can adversely affect the nervous system with effects ranging from dizziness, headaches, and fatigue at lower exposures to numbness in the limbs, incoordination, tremors, temporary limb paralysis, and unconsciousness at higher exposures.
- Subchronic health effects, such as headaches and throat and eye irritation reported by residents during well completion activities occurring in Garfield County, are consistent with some of these health effects.

- In the 2007 Garfield County emission inventory, the Colorado Department of Public Health and Environment attributed the bulk of benzene, xylene, toluene, and ethyl benzene emissions in the county to natural gas development.
- The non-cancer health risks from air emissions due to natural gas development is greatest for residents living near wells during the relatively short-term, but high emission, well completion period. Furthermore, the risk is driven principally by exposure to trimethylbenzenes, alkanes, and xylenes, all of which have neurological and/or respiratory effects.
- Residents living nearer to wells have higher cancer health risks as compared to residents residing further from wells. Benzene is the major contributor to lifetime excess cancer risk.

7.2 Chronic and Acute Health Effects of Residents at DISH, Texas

Wilma Subra conducted a health survey on residents of DISH, Texas [67] requested by the Town of DISH. DISH is located over the Barnett Shale, and the survey was ordered because of the residents' complaints about health issues.

Subra identified the acute and chronic health effects based on the survey (Table 4.18).

Table 4.18 Acute and chronic health effects of residents of DISH, Texas

Acute health effects	Chronic health effects
Irritates skin, eyes, nose, throat and lungs	Damage to liver and kidneys
Headaches	Damage to lungs
Dizziness, light headed	Damage to developing fetus
Nausea, vomiting	Causes reproductive damage
Skin rashes	Damages nerves causing weakness
Fatigue	Poor coordination
Tense and nervous	Affects nervous system
Personality changes	Affects the brain
Depression, anxiety, irritability	Leukemia
Confusion	Aplastic anemia
Drowsiness	Changes in blood cells
Weakness	Affects blood clotting ability
Muscle cramps	Carcinogen
Irregular heartbeat (arrhythmia)	Mutagen
Teratogen—developmental malformations	

Source: Reference [67]

7.3 Carcinogens

One concern about fracking is that the fracking fluids used to fracture rock formations contain numerous chemicals that could harm health and the environment, especially if they enter drinking water supplies. The opposition of many oil and gas companies to public disclosure of the chemicals they use has compounded this concern.

The U.S. House of Representatives, the Committee on Energy and Commerce issued a report [7] on the fracturing fluid chemicals. The report identified chemicals which are carcinogens:

Benzene	Diesel	Di (2-ethylhexyl) phthalate
Acrylamide	Nitrilotriacetic acid	Acetaldehyde
Ethylene oxide	Propylene oxide	Formaldehyde
Sulfuric acid	Thiourea	Benzyl chloride
Naphthalene	Lead	

7.4 Deadly Chemicals

The most lethal chemicals known to be used are [68]:

Methanol

- Methanol appeared most often in hydraulic fracturing products (in terms of the number of compounds containing the chemical).
- Vapors can cause eye irritation, headache, and fatigue, and can be fatal in high enough doses. Swallowing may cause eye damage or death.

BTEX compounds

- The BTEX compounds—benzene, toluene, xylene, and ethylbenzene—are listed as hazardous air pollutants in the Clean Air Act and contaminants in the Safe Drinking Water Act.
- Benzene, commonly found in gasoline, is also a known human carcinogen. Longtime exposure can cause cancer, bone marrow failure, or leukemia. Short term effects include dizziness, weakness, headache, breathlessness, chest constriction, nausea, and vomiting.
- Toluene, ethyl benzene, and xylenes have harmful effects on the central nervous system.

Diesel fuel

- A carcinogen listed as a hazardous air pollutant under the Clean Air Act and a contaminant in the Safe Drinking Water Act.

- Hydraulic fracturing companies injected more than 30 million gallons of diesel fuel or hydraulic fracturing fluids containing diesel fuel in wells in 19 states.
- Diesel fuel contains toxic constituents, including BTEX compounds.
- Contact with skin may cause redness, itching, burning, severe skin damage and cancer.

Lead

- A carcinogen found in paint, building construction materials and roofing joints.
- It is listed as a hazardous air pollutant in the Clean Air Act and a contaminant in the Safe Drinking Water Act.
- Lead is particularly harmful to children's neurological development. It also can cause reproductive problems, high blood pressure, and nerve disorders in adults.

Hydrogen fluoride

- Found in rust removers, aluminum brighteners and heavy duty cleaners.
- Listed as a hazardous air pollutant in the Clean Air Act.
- Fumes are highly irritating, corrosive, and poisonous. Repeated ingestion over time can lead to hardening of the bones, and contact with liquid can produce severe burns. A lethal dose is 1.5 g.
- Absorption of substantial amounts of hydrogen fluoride by any route may be fatal.

Naphthalene

- A carcinogen found in mothballs.
- Listed as a hazardous air pollutant in the Clean Air Act.
- Inhalation can cause respiratory tract irritation, nausea, vomiting, abdominal pain, fever or death.

Sulfuric acid

- A carcinogen found in lead-acid batteries for cars.
- Corrosive to all body tissues. Inhalation may cause serious lung damage and contact with eyes can lead to a total loss of vision. The lethal dose is between 1 teaspoonful and one-half ounce.

Crystalline silica

- A carcinogen found in concrete, brick mortar and construction sands.
- Dust is harmful if inhaled repeatedly over a long period of time and can lead to silicosis or cancer.

Formaldehyde

- A carcinogen found in embalming agents for human or animal remains.
- Ingestion of even one ounce of liquid can cause death. Exposure over a long period of time can cause lung damage and reproductive problems in women.

7.5 *Human and Animal Health*

Michelle Bamberger and Robert Oswald published a report [69] titled "The Impacts of Gas Drilling on Human and Animal Health". The findings illustrate which aspects of the drilling process may lead to health problems. The research provides evidence of several possible links between gas drilling and negative health effects such as upper-respiratory symptoms and burning of the eyes, vomiting and diarrhea, rashes, nosebleeds, headaches and neurological problems.

Complete evidence regarding health impacts of gas drilling cannot be obtained due to incomplete testing and non-disclosure agreements. Without rigorous scientific studies, the gas drilling boom sweeping the world will remain an uncontrolled health experiment on an enormous scale.

More questions regarding hydraulic fracturing are asked partially answered by governments and news media worldwide [70, 71].

References

1. EIA. (2012). *EIA annual energy outlook 2012*. Energy Information Administration, July 2012. DOE/EIA-0380.
2. NETL. (2009). *Modern shale gas development in the United States, a primer*. April 2009. Prepared by GWPC and ALL Consulting.
3. Clark, C., Burnham, A., Harto, C., & Horner, R. (2012). *Hydraulic fracturing and shale gas production: Technology, impacts, and policy*. Argonne National Laboratory. September 10, 2012.
4. Fracfocus. (2012). *Well construction & groundwater protection*. http://fracfocus.org/hydraulic-fracturing-how-it-works/casing/
5. Mantell, M. (2011). *Produced water reuse and recycling challenges and opportunities across major shale plays*. Chesapeake Energy Corporation. EPA Hydraulic Fracturing Study Technical Workshop #4 Water Resources Management. March 29–30, 2011.
6. Keister, T. (2010). Marcellus *Hydrofracture flowback and produced wastewater treatment, recycle, and disposal technology*. The Science of Marcellus Shale, hosted by Lycoming College in Williamsport, PA. January 29, 2010.
7. U.S. House. (2011). *Chemicals used in hydraulic fracturing*. U.S. House of Representatives Committee on Energy and Commerce Minority Staff. April 2011.
8. CSM. (2009). *An integrated framework for treatment and management of produced water*. Colorado School of Mines. November 2009.
9. Parsons. (2008). *Emerging technologies for waste-water treatment and in-plant wet weather management, EPA-832-R-06-006*. Parsons Corporation. February 2008.
10. NETL. (2004). *A white paper describing produced water from production of crude oil, natural gas, and coal bed methane*, Prepared for National Energy Technology Laboratory by John Veil, Puder M, Elock D, and Redweik R. Argonne National Laboratory. January 2004.
11. Igunnu, E., & Chen, G. (2012). Produced water treatment technologies. *International Journal of Low-Carbon Technologies, 2012*, 1–21.
12. Hayes, T., & Arthur, D. (2004). Overview of emerging produced water treatment technologies. In *The 11th Annual International Petroleum Environmental Conference*, Albuquerque, NM. October 12–15, 2004.

13. NETL. (2010). *Water management technologies used by Marcellus shale gas producers*, Prepared for National Energy Technology Laboratory by John Veil. July 2010.
14. PAEQB. (2011). *Pennsylvania environmental quality board final amendments to regulations on wastewater treatment requirements*. November 3, 2011.
15. GAO. (2012). *Information on shale resources, development, and environmental and public health risks*. Government Accounting Offices, GAO-12-732. September, 2012.
16. Tillerson, R. (2011). ExxonMobil Testimony to U.S. House of Representatives, Committee on Energy and Commerce, "*The ExxonMobil-XTO Merger: Impact on U.S. Energy Markets*" January 20, 2011.
17. Jackson, L. (2011). *EPA Administrator Lisa Jackson feels the anti-fracking rage*, May 24, 2011. http://lonelyconservative.com/2011/05/epa-administrator-lisa-jackson-feels-the-anti-fracking-rage/
18. USEPA. (1987). *Report to Congress: Management of wastes from the exploration, development, and production of crude oil, natural gas, and geothermal energy*, Vol. 1 of 3. EPA/530/SW-88/003A. Prepared by Versar Inc. December 1987.
19. Horwitt, D. (2011). *Cracks in the Façade, 25 years ago EPA linked fracking to water contamination*. Environmental Working Group, August 31, 2011.
20. USEPA. (2004). *Evaluation of impacts to underground sources of drinking water by hydraulic fracturing of coalbed methane reservoirs, final report (EPA 816-R-04-003)*. June 2004.
21. NY Times. (2009). *New York Times Editorial*, November 3, 2009.
22. Hunziker, R. (2012). *The unregulated, indefensible effects of fracking*. Fireband Magazine, January 24, 2012.
23. Earthworksaction. (2012). *Halliburton Loophole*. http://www.earthworksaction.org/issues/detail/inadequate_regulation_of_hydraulic_fracturing/. Accessed 4 Nov 2012.
24. OGAP. (2005). *Our drinking water at risk*. Oil and Gas Accountability Project, April 2005.
25. USEPA. (2011). *(Draft) investigation of groundwater contamination near Pavilion, Wyoming*. USEPA, December 2011.
26. USGS. (2006). *Methane in West Virginia Groundwater*. USGS Fact Sheet 2006-3011.
27. Osborn, S., Vengogh, A., Warner, N., & Jackson, R. (2011, May 17). Methane contamination of drinking water accompanying gas-well drilling and hydraulic fracturing. *Proceedings of the National Academy of Sciences of the United States of America, 108*(20).
28. Davies, R. (2011). Methane contamination of drinking water caused by hydraulic fracturing remains unproven. *Proceedings of the National Academy of Sciences of the United States of America, 108*(43), E871.
29. Warner, N., Jackson, R., Darrah, T., Osborn, S., Down, A., & Zhao, K. (2012). *Geochemical evidence for possible natural gas migration of Marcellus formation brine to shallow aquifers in Pennsylvania*. http://www.pnas.org/cgi/10.1073/pnas.1121181109, approved May 10, 2012.
30. Tucker, C. (2012). *Another Duke Rebuke*. http://eidmarcellus.org/Marcellus-shale/another-duke-rebuke/10832/. July 9, 2012.
31. Engelder. (2012). http://marcellusdrilling.com/2012/07/mdn-in-depth-new-duke-study-on-brine-migration-into-aquifers/
32. Myers, T. (2012). Potential contaminant pathways from hydraulically fractured shale to aquifers. *Ground Water*. National Ground Water Association, 2012.
33. Siegel, D. (2012). http://www.energyindepth.org/tag/tom-myers/
34. Fracturingofamerica. (2012). *Migration of residual fracking fluid into fresh water aquifers*. http://www.fracturingofamerica.com. October 24, 2012.
35. Heddleston, D. (2009). *Horizontal well production logging deployment and measurement techniques for U.S. land shale hydrocarbon plays*. Society of Petroleum Engineers. SPE 120591, 2009.
36. Marcellus Shale. (2011). Jointing and fracturing in the Marcellus Shale. *Paleontological Research Institution*, Issue No. 5, August 2011.
37. Frackingofamerica.com. (2012). *Fracking of America*. Accessed October 24, 2012.

38. Arthur, D., Bohm, B., Bobb, P., & Coughlin, J. (2010). Summary of environmental issues, mitigation strategies, and regulatory challenges associated with shale gas development in the United States and applicability to development and operations in Canada. In *Canadian Unconventional Resources an International Petroleum Conference*, Calgary, Alberta, Canada, October 20, 2010.
39. USEPA. (2005). *Construction site runoff control minimum control measure*. EPA-833-F-00-008. Revised December, 2005.
40. GAO. (2005). *Information needed on the implications of permitting oil and gas construction activities, GAO-05-240*. General Accounting Office. February, 2005.
41. USEPA. (2007). *Summary of the results of the investigation regarding gas well site surface water impacts*. http://www.epa.gov/npdes/pubs/oilandgas_gaswellsummary.pdf
42. USEPA. (2010). *Stay and correction of the numeric limit for the construction and development, EPA-821-F-10-003*. October 2010.
43. Meaker, M., & Tijsseling, C. (2010). *New EPA regulations significantly impact storm water flow at construction sites*. Spring 2010.
44. Herridge, A. (2012). *The consequence of hydraulic fracturing*. http://prtl.uhcl.edu/portal/page/portal/SCR/Home/Files-%20ePres/consequence_of_hydraulic_fracturing.pdf
45. Armendariz, A. (2009). *Emissions from natural gas production in the Barnett shale area and opportunities for cost-effective improvement*. January 26, 2009.
46. Subra, W. (2012). Human health impacts associated with chemicals and air pathways of exposure from the development of shale gas plays. In *Presentation to the US Environmental Protection Agency Community Air Conference*. June 27, 2012.
47. Howarth, R., Santoro, R., & Ingraffea, A. (2011). Methane and the greenhouse-gas footprint of natural gas shale formation, *Climate Change*, March 13, 2011.
48. Skone, T. (2011). *Life cycle greenhouse gas analysis of natural gas extraction & delivery in the United States*. Revised May 23, 2011.
49. Jiang, M., Griffin, W. M., Hendrickson, C., Jaramillo Van Briesen, P. J., & Venkatesh, A. (2011, August 5). Life cycle greenhouse gas emissions of Marcellus shale gas. *Environmental Research Letter, 6*(3).
50. Fulton, M., Mellquist, N., Kitasei, S., & Bluestein, J. (2011). *Comparing life-cycle greenhouse gas emissions from natural gas and coal*. Worldwatch Institute. August 25, 2011.
51. Cathles III, L. M., Brown, L., Taam, M., & Hunter, A. (2012). *A commentary on "The Greenhouse-Gas Footprint of Natural Gas in Shale Formations" by RW Howarth*. http://www.springerlink.com/content/x001g12t2332462p. January 3, 2012.
52. Thorn, T. (2012). *Environmental issues surrounding shale gas production, the US experience, a primer*. International Gas Union, April 2012.
53. Henry, T. (2012). *How fracking, drilling, and earthquake are linked*. February 8, 2012. http://stateimpact.npr.org/texas/2012/02/08/how-fracking-drilling-and-earthquakes-are-linked/
54. Nicholson, C., & Wesson, R. (1990). *Earthquake hazard associated with deep well injection – A report to the U.S. Environmental Protection Agency*. U.S. Geological Survey. 1990.
55. Arkansas Geology. (2012). *Historic earthquakes*. Arkansas Geology, accessed October 10, 2012. http://www.geology.ar.gov/geohazards/historic_earthquates.htm
56. Eddington, S. (2011). *Arkansas earthquakes decline after fracking injection well closures*. http://www.huffingtonpost.com/2011/03/15/arkansas-earthquakes-2011-fracking_n_835868.html. March 15, 2011.
57. Bradbury, M. (2011). *Fracking earthquakes*. http://www.realscience.us/2011/10/25/fracking-earthquakes. October 25, 2011.
58. Power Engineering. (2012). Ohio officials shut down gas production sites over earthquake concerns. *Power Engineering*, January 3, 2012
59. Ludwig, M. (2012). *Regulators say fracking wastewater well caused 12 earthquakes in Ohio*. Truthout, March 13, 2012.
60. Dayen, D. (2012). *Fracking linked to earthquakes in Ohio*. News.firedoglake.com, January 2, 2012.

61. Frohlich, C. (2012). Two-year survey comparing earthquake activity and injection well locations in Barnett Shale, Texas. *Proceedings of the National Academy of Sciences of the United States of America*. August 6, 2012.
62. Mordick, B. (2011). *Risks to drinking water from oil and gas wellbore construction and integrity: Case studies and lessons learned*. May, 2011.
63. Associated Press. (2012). *Explosion at Harrison County Marcellus Shale Gas Well Drilling Site*. August 17, 2012.
64. PA Environmental Digest. (2012). *DEP plans investigation into Marcellus shale well blowout in Clearfield county*. June 7, 2010.
65. NYDEC. (2012). *Summary report, human health risks and exposure pathways of proposed horizontal hydraulic fracking in New York State*. New York State Department of Environmental Conservation and the New York State Department of Health, October 9, 2012.
66. McKenzie, L. (2012). *Testimony on "Federal Regulation: Economics. Job, and Energy Implication of Federal Hydraulic Fracturing Regulation"*, May 2, 2012.
67. Subra, W. (2009). *Health survey results of current and former DISH/Clark, Texas residents*. December 2009.
68. Nelson, J. (2012). *Deadly chemicals used in hydraulic fracturing revealed*. May 31, 2012. http://www.opednews.com/articles/Deadly-chemicals-used-in-h-by-Jerry-Nelson-120531-979.html
69. Bamberger, M., & Oswald, R. (2012). Impacts of gas drilling on human and animal health. *New Solutions, 22*(1), 51–77.
70. NICNAS. (2014). *National assessment of chemicals associated with coal seam gas extraction in Australia*. National Industrial Chemicals Notification & Assessment Scheme, Australia. http://www.nicnas.gov.au/
71. Washington Examiner. (2014). Examiner editorial: EPA officially backs fracking, then uses tax dollars to oppose it. *Washington Examiner*, Washington, DC, June 7, 2014.

Chapter 5
Bioaugmentation for Water Resources Protection

Erick Butler and Yung-Tse Hung

Contents

E. Butler, Dr. Eng. (✉)
School of Engineering and Computer Science, West Texas A&M University, Canyon, TX 79016, USA
e-mail: erick.ben.butler@gmail.com

Y.-T. Hung, Ph.D., P.E., D.EE, F-ASCE.
Department of Civil and Environmental Engineering, Cleveland State University, Cleveland, OH, USA

L.K. Wang, C.T. Yang, and M.-H.S. Wang (eds.), *Advances in Water Resources Management*, Handbook of Environmental Engineering, Volume 16,
DOI 10.1007/978-3-319-22924-9_5

Abstract This chapter provides a comprehensive study of bioaugmentation in wastewater, groundwater, and soil contamination remediation applications as applied to water resources protection. It defines major key terms and concepts, including many of the contaminants present within these media. Several case studies were discussed as written concerning in situ bioremediation treatment for groundwater, soil contamination, and wastewater treatment. In addition, laboratory studies highlight the success of bioaugmentation given the proper conditions within the system.

Keywords Biotechnology • Bioremediation • Biological waste treatment • Groundwater contamination • Biodegradation • Biotransformation • Water resources protection

Nomenclature

A_m	Membrane area (ft^2)
1/n	Describes adsorption favorability, where <1 is unfavorable, greater than 1 is favorable, and 1 is linear
b	Constant of net enthalpy of adsorption
C_e	Solute concentration within the solution
C_7H_8	Toluene
cl	Variable factor concentration (mg/L)
CO_2	Carbon dioxide
$Cost_n$	Costs for new projects
$Cost_o$	Costs for old projects
C_w	Concentration of the contamination in the water phase (mg/dm^3)

$\delta C/\delta x$	Change of concentration to change in time
dH/dx	Hydraulic gradient in the x-direction
Dp	Diffusion coefficient
dρ/dx	Pressure gradient in the x-direction
Exp	Additional equipment required aboveground (typically 0.7)
Fe(III)	Ferric (Iron III) Ion
$Fe(OH)_3$	Ferric hydroxide
H_2S	Hydrogen sulfide
K_d	Adsorption coefficient (dm^3/kg); K_d is proportionate to the organic carbon present within the soil
K_k	Permeability coefficient of water (m/day)
K_m	Conversion speed $v_{max}/2$ (mg/L)
N	Mass flux of pollutant
N_2	Diatomic nitrogen
NO_3^-	Nitrate ion
O_2	Diatomic oxygen
OH^-	Hydroxide ion
q	Content level of contaminant in soil
$Size_n$	Size of new projects
$Size_o$	Size of old projects
SO_3^-	Sulfur Dioxide ion
v	Conversion speed (mg/kg ds/d)
v_{max}	Maximum degradation speed (mg/kg ds/d)
V_w	Flow velocity of pore water (m/day)

1 Introduction

Bioaugmentation, a subset of biotechnology is a major proponent for the removal of contaminants in soils and water. Generally, bioaugmentation has been extended to the remediation of substances such as hazardous and toxic wastes in various treatment plants and for the removal of polycyclic aromatic hydrocarbons (PAHs) within contaminated soils. While persistence is prevalent in practice, it is still considered an innovative method of environmental remediation despite the fact that augmentation occurs naturally and has been prevalent since the existence of microorganisms. While the technology is currently accepted in the environmental world, many question its success rate, leading many scientists and engineers to question its ability to perform in real world applications. This setback should or will not deter continual research of this technology.

Nevertheless, an issue of concern within treatment is the biological removal of xenobiotic contaminants. Xenobiotic are manmade chemicals formed from the production of everyday products. These xenobiotic chemicals commonly aggregate within contaminated soils and are a cause for great concern within our environment.

The synthetic chemicals introduced into the system are foreign to indigenous microbial populations and therefore can remain without proper biodegradation. This can lead to the transfer of chemicals into ground and surface water sources, many of which are carcinogenic and mutanogenic. Therefore, there is much concern over the proper protection of various microbial species [1].

Because of the prevalence of xenobiotic contaminants and major environmental incidents in the last 30 years, debate has occurred over the proper removal and control of contaminants entering the soil. Research compiled from these previous historical incidents concludes that contaminant reduction is related to the population of microorganisms present—a philosophy of regenerating the population of microorganisms for the purpose of contaminant reduction is born. Using bioaugmentation as the solution to contaminated soil and water is a major breakthrough in bioremediation technology.

The primary goal of this chapter is to investigate the realm of bioaugmentation—compare and contrast biotechnological terminology include bioremediation, biotechnology, and bioaugmentation; provide purpose and reason for application, its practical presence within actual systems including in situ case studies involving various forms of proper application, which will also briefly mention additional applications outside of contaminated soils. Finally, the discussion will connect this technology as appropriate within ecological risk assessment. This chapter will enlighten readers on the subject, but not to become an all-encompassing guide of all bioaugmentation practices. Research still continues at the present date improving the way to provide more feasibility within this solution to improve current practices to solve environmental problems.

2 Terminology

There are many significant definitions that describe bioaugmentation. In wastewater treatment, bioaugmentation supplies additional microorganisms in the Activated Sludge (AS) process. This specific treatment application is a subset of bioremediation. However, bioaugmentation is a subset in the category of biotechnology. In order to understand how bioaugmentation fits as a major proponent in waste treatment, one must understand the different between bioremediation and biotechnology. Figure 5.1 is a diagram indicating the relationship between the terms.

2.1 Definition of Biotechnology

Singh and Ward [3] define biotechnology as "the application of biological organisms, systems or processes manufacturing and service industries." This definition of biotechnology is very broad and can include various areas such as biochemistry,

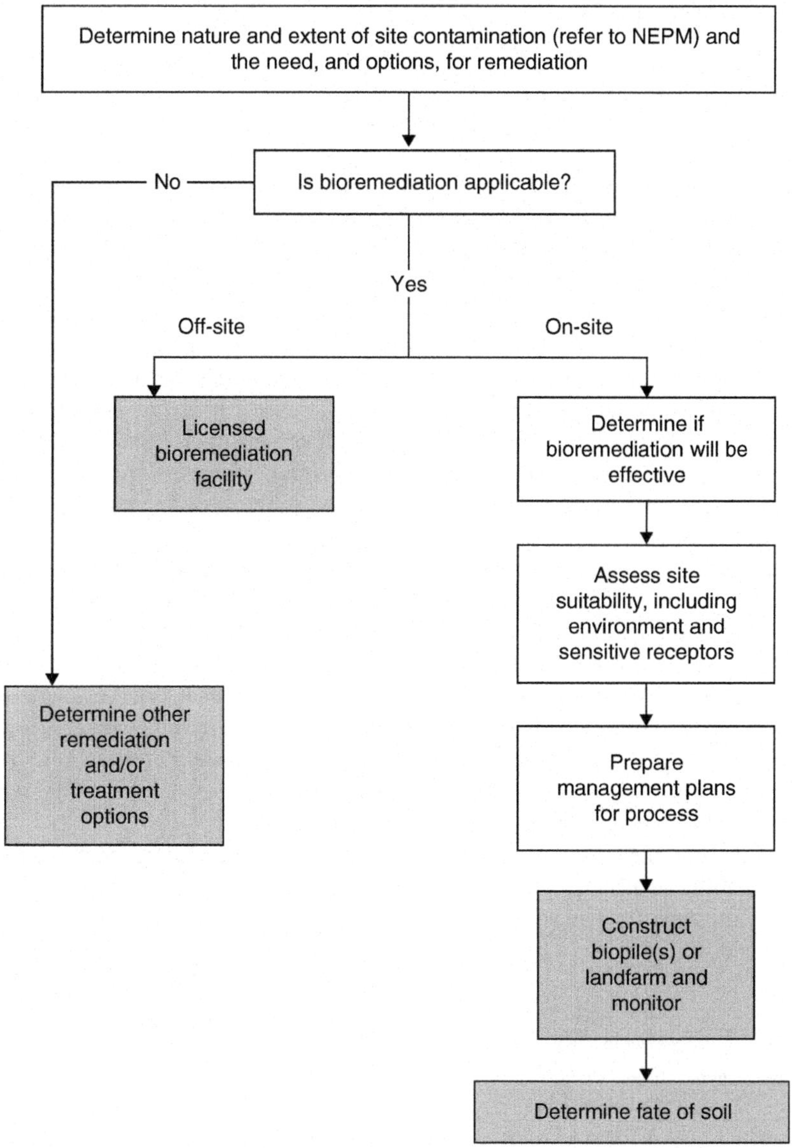

Fig. 5.1 Process flow diagram for bioremediation process [2]

microbiology, and engineering sciences. Because biotechnology has a loose definition, its applications may not be directly related to the sciences. For example, the public considers biotechnology as an opportunity to develop alternative sources to solve problems such as fossil fuels. Therefore, this chapter will refrain from using biotechnology and focus on the area of concern—bioremediation.

2.2 Definition of Bioremediation

Scragg [4] defines bioremediation "as the biological treatment and removal of pollution from the environment." Bacteria and fungi are the major microorganisms in bioaugmentation, although there others can be used. This is dependent on the microorganism tolerance of the environmental conditions and factors such as oxygen, nutrients, moisture, and temperature. Table 5.1 provides a summary of the optimum levels for each of these factors needed to provide sufficient treatment.

There are four major advantages for using bioremediation. First, bioremediation can help reduce the historical build-up of industrial wastes that causes land, water, and air pollution. Second, biological treatment can successfully remove complex organic compounds and heavy metals. Organic compounds and heavy metals are volatile, can evaporate into the atmosphere and cause further air pollution. They also can become reactive if in contact with soil or soil-microorganisms. Furthermore, wastes that are integrated in the food chain can be ingested, or remain within the soil. Clearly, bioremediation can reduce the problems of volatility and food chain integration.

Third, bioremediation reduces the infiltration of xenobiotic chemical pollution. Xenobiotic pollutants are derived from the atmospheric release of hydrocarbons from petroleum. The list of pollutants includes polycylic aromatic hydrocarbons (PAHs), dioxins, and dibenzoforms. The incomplete combustion of carbon causes cancer-causing agents. Additional chemicals such as benzene, xylene, alkanes, and alkenes provide further problems within the environment. Unfortunately, legislation for control and prevention in water and soil did not appear until the mid-1960s and early-1970s when the generation of xenobiotics peaked [5]. The use of bioaugmentation can immobilize many of these pollutants.

Finally, bioremediation allows for the restoration of the environment and reduction of public health concerns by maintaining clean water and soil. Clean water supplies must be free of all pollutant material so populations can remain viable, while clean soil provides a place for crop growth. Agricultural practices such as the

Table 5.1 Environmental factors for soil activity [6]

Environmental factor	Optimum levels
Oxygen	Aerobic: levels greater than 0.2 mg/L; 10 % air pore space minimum with air
	Anaerobic: less than 0.2 mg/L dissolved oxygen; less than 1 % minimum pore space with air
Nutrients	Nitrogen, phosphorus, other nutrients (C:N:P; 120:10:1)
Moisture	25–85 % water holding capacity
pH	5.5–8.5
Redox	Aerobic: 50 mV and higher Anaerobic: less than 50 mV
Temperature (°C)	15–45 (mesophilic)

release of pesticides for the reduction of insects effect on crop growth release contaminants into streams and associated systems during rainstorms. Additional fertilizer application increases nutrient availability end potential for eutrophication and nitrate and nitrite concentration in groundwater. Bioaugmentation reduces the impacts of chemicals by means of discharges, runoffs, and spills [5].

It is important to understand how engineers use bioremediation to solve environmental problems. Generally, there are four major methods in which microorganisms are used within bioremediation. The first method uses indigenous microbial populations to increase hydrocarbon degradation that naturally occurs within soils. The second method, known as ingenious growth encouragement involves compound organic biodegradation. The challenges with this method relate to the oxygen levels, pH, and soil moisture. These factors are important in the success rate of removing organic compounds by microorganisms. Bioaugmentation or the third method adds additional selected microbial populations. Genetic manipulation, or the final method, uses cultured organisms. This method is not popular because of questionable applications in the environment. Deciding the appropriate bioremediation method is contingent on degrading contaminants such as dehalorespiration [1].

3 Bioaugmentation

3.1 History of Bioaugmentation

Historically, microorganism presence has been noted since the existence of life on earth. In fact, scientists suggest that biodegradation is a process over 3 billion years old, and the degradation of materials is contingent on the presence of bacteria cultures. Scientific development of theories of the degradation of organics became a reality in 1729, when Pier Antonia Micheli studied micro-sized fungal species and conducted experiments on their preference of fruit. Following Micheli's studies, scientists continued to conduct experiments. Processes such as fermentation and food spoilage are still used today. One of the more renowned scientists Louis Pasteur in the 1860s developed fermentation products from the presence of microorganisms in the air [7].

Following continuous work from scientists as John Tyndall and Ferdinand Cohn, many scientists began developing pure culture of microorganisms. John Lister in 1878 developed the use of bacteria placement on gelatin by a flame-sterilized platinum wire. He also developed the use of algal polysaccharide agar-agar [7].

Another important scientific discovery is the development of enrichment culture. Wackett and Hershberger [7] write in detail about this procedure:

> "...the use of selective medium conditions that favor one or a small group of organisms which favor one or a small group of organisms that one desires to obtain from an environmental sample. This becomes very important within bioaugmentation because one is able to prepare the proper medium within a laboratory, provide continuous transfers of

the culture within the medium, for the purpose of increasing or 'augmenting' the amount of bacteria within the system [7]."

As seen throughout the history of microorganism development, bioaugmentation has been noted for its use in major discoveries in the laboratory.

In on-site treatment, bioaugmentation is notarized in the application of the removal of crude oil. The most infamous case study is the Exxon *Valdez* oil spill. Scientists applied microorganisms capable of degrading hydrocarbons. However, the first application on-site application of bioaugmentation was in the industrial wastewater treatment industry. Because of its toxic and inhibitory materials from food, chemical, and oil plants, industrial wastewater is too difficult to degrade. While bacteria from sludge are available from municipal wastewater treatment plants, an additional set of bacteria is necessary to remove materials from industrial plants. Additional considerations are also necessary to use bioaugmented microorganisms in the degradation of contaminants from soil and ground water [4].

3.2 Bioaugmentation Cultivation in Laboratory

There are four major classes of bioaugmentation cultivation techniques in the laboratory—the first two classes of bioaugmentation are known as "natural." These are a series of bacteria created from an original indigenous population by means of in vitro mutation or adaptation. The third class uses several applications of microorganisms for the purpose of degrading specific organic compounds. The fourth class of bioaugmentation applies various forms of genetic material to create the microorganisms [8].

Any discussion on the most suitable bioaugmentation cultivation technique cannot be void of discussing survival of the bioaugmented microorganisms. This is important because some species may have a hard time surviving outside a non-laboratory environment. Therefore, there are two methods applied to control the inoculum species from being harmed in the environment. First, immobilization is the process of placing inoculated species on a medium. *Flavobacterium*, a pentachlorophenol (PCP) degrading bacteria can be contained onto granular clay or a lava slag. Other immobilized microorganisms include loofah sponges. Second, encapsulation protects the strains microorganisms and also allows for the slow release of nutrients within the system. *Flavobacterium* have been encapsulated using an agar, alginate, or polyurethane in a 2–50 μm diameter bead dropped by a nozzle [8].

3.3 Bioaugmentation Techniques

Bioremediation techniques are either in situ, direct application of treatment processes onto the contaminated region, or ex situ, which removes contaminants and are treated away from the site. This chapter will focus on in situ treatment [4]. An important in situ technique includes biosparging, the process of supplying air to the soil to increase biological activity. Typically bioaugmentation experiments use air compressors and blowers to supply air at rates between 0.5 and 3 ft^3/m/point of injection [9]. Another possible technique involves the mixture of soils by ploughing or tilling. This can be an important solution for resolving problems of soil clumping which can prevent the transfer of contaminants to the microorganisms specifically in soil bioaugmentation [4].

3.4 *Constituents of Bioaugmentation*

Bioaugmentation consists of fermented single strains or multiple cultures which are then either dried on a support or organisms that are suspended in solution. These mixtures are then blended for the purposes of either single strain or multiple culture use. Before use, these mixtures first undergo rigorous analytical tests ranging from determining nutritional needs, tolerances in high and low temperatures, contaminant concentrations, and heavy metals. Once the selection has been created, strains are improved by the processes of adaptation or mutagenesis. Commercial products maybe added if the strains require additional assistance [5].

There are two types of bioaugmentation—seed inoculation anticipates an increase of microorganisms when exposed ideal environmental conditions. Therefore, few microorganisms are introduced into the contaminated area. Mass inoculation provides microorganisms to the contaminated area to prevent lag phase microbial growth [9]. An example of mass inoculation is in activated sludge systems where bacteria in the form of sludge are constantly returned to an aeration tank to reduce contaminants and avoid the loss of microbial populations [9].

3.5 *Advantages of Bioaugmentation*

Devinny and Chang [9] provide many advantages to bioaugmentation. First, bioaugmentation has been deemed pertinent to address additional contaminants outside of the reach of current microorganism populations. Second, bioaugmentation provides an additional microorganism presence when predation, environmental conditions, or products which are unfavorable for use by current microorganism within these soils. These populations can handle low water volume and poor diffusibility conditions. Third, bioaugmentation cultivates microorganisms to handle low concentrations that are not beneficial to natural species. Fourth, bioaugmentation applies genetic material to assist current microorganisms that are deficient in contamination degradation [9].

Other advantages of using bioremediation as compared with other treatment processes include the ability to complete treatment without disruption of natural processes. By direct application, bioaugmentation reduces the probability of introduced toxic constituents and keeps cost of treatment down [10].

Table 5.2 The purposes of bioaugmentation [5]

Purpose	Reason
Low microbe count	The amount of bacteria can only be sufficient if the growth rate of bacteria is high enough to be able to reduce concentration of contaminants. Therefore, if the microbe count is less than 10^5, bioaugmentation can supplement for this decreasing count.
Complex waste	Bioaugmentation can treat waste on-site eliminating the need for pretreatment by physical or chemical processes.
Expedition of waste material	When the decontamination of materials is needed quickly to lower costs of contamination removal, bioaugmentation is a prime candidate for waste material because it does not contain inhibitory materials.
Assurance	Bioaugmentation adds to a population of highly successful indigenous populations at a contaminated site. This reduces cost and need for research.

3.6 Purposes of Bioaugmentation

Forsynth et al. [5] describe the four major purposes of bioaugmentation shown in Table 5.2.

4 Legislation

Prior to legislation an incident occurred in a neighborhood in Niagara Falls, New York. The residents and school in Love Canal were unaware of the developed hazardous and toxic wastes. It was not until many began to develop cancer and birth defects that residents noticed the region was contaminated. Following relocation of all families, the United States developed its first legislature, the passing of the Reconstruction and Recovery Act of 1976 (RCRA). This law provides special consideration of hazardous waste generation and proper disposal into a landfill. RCRA coins the term 'cradle to grave to describe the monitoring of a landfill from infancy (cradle) until closure (grave) [11].'

The law first defines the parameters for identifying hazardous wastes, treatment, and disposal and storage facilities available, standards for the generation and hazardous waste transport, and permits oversight by the individual states. Next, RCRA provides conditions to induce cleanup. However, one of the important developments from RCRA legislation are the identification of sources for—unknown source, specific source, all chemical productions, species, and containers discarded. Finally, the development of off-site closure plans was instituted into the law. This means that any facility that produces hazardous waste must have proper closure of all facilities and include hazardous waste removal and decontamination of equipment. RCRA provides the provisions for the design of landfills [11, 12].

In similar fashion, the Toxic Substances Control Act of 1976 (TSCA) aims to protect public health and the environment by controlling toxic chemicals [12]. TSCA accomplishes this protection by documenting 75,000 industrial emissions as environmental or human health hazards [13]. In addition, the EPA receives proper provisions for inventory of industrial chemicals developed within the year. Section 8(e) of TSCA summarizes this provision:

> "Any person who manufactures, [(includes imports)] processes or distributes in [US] commerce a chemical substance or mixture, and who obtains information which reasonably supports the conclusion that such substance or mixture presents a substantial risk of injury to human health or the environment, shall immediately inform the [EPA] Administrator of such information unless such person has actual knowledge that the Administrator has been adequately informed of such information [13]."

However, the landmark and most significant legislature passed the Comprehensive Environmental Response, Compensation, and Liability Act (CERCLA or Superfund) on December 11, 1980. There are five important provisions within this legislation. First, CERCLA requires a tax on any release or threatened release of hazardous substances that can impact human health or the environment. This includes such substances as chemical or petroleum products [14]. Second, CERCLA provides methodology in regards for abandoned hazardous waste sites, imposes sanctions for respective parties for hazardous wastes, and sets up a trust fund for clean-up. Third, CERCLA includes the National Priorities List, a list of pollutants from uncontrolled hazardous waste sites [12].

Fourth, CERCLA holds the United States Environmental Protection Agency (USEPA) responsible to address any release or threats of hazardous substances or contaminants within the environment, and also ensure proper clean-up. These clean-up measures are covered by Section 121 of CERCLA. Section 121 develops the parameters and remedial design to confirm the effectiveness of remedial reactions. The clean-up measures begin prior to remediation where the proper precautions such as investigation, feasibility, and the assessment of long-term threat are required to certify that remediation practices can be completed. After preliminary precautions, removal actions remove contaminants that cause short term or threats to the environment. Then, the EPA examines the remedial actions are feasible as a long term solution [12].

Fifth, CERCLA contains the National Contingency Plan, a series of response procedures to any clean-up of hazardous substances. There are nine major criteria considered within the plan—human health and environmental protection, state and federal law compliance, effectiveness within the short and long term, toxicity, mobility, cost, and the perception by the state and community. To create a National Contingency Plan for cleanup, potentially responsible parties (PRPs) are selected as the financial caretakers for the contaminant cleanup. These parties develop measurements to resolve the problem, receive money for Superfund cleanup, comply with the steps outlined in the National Contingency Plan, and begin the proper procedure in cleaning up contaminants. The rights and repercussions of the responsible party will be negotiated and can be shared between multiple parties responsible for cleanup [12].

Table 5.3 Superfund groundwater treatment projects (FY 1982–1999) [19]

Technology	Number of sites
Ex situ technologies	
Pump and treat	638
In situ technologies	
Air sparging	48
Bioremediation	20
Dual-phase extraction	10
Permeable reactive barrier	8
Phytoremediation	4
Chemical treatment	2
In-well air stripping	2

However, if parties are not zealous in resolving the problem the responsibility returns to the EPA to either research all responsible parties and/or beginning cleanup while attempting to conduct investigations and receive funds for the cleanup. In addition, the EPA will conduct feasibility and remedial investigation studies and will begin to direct penalties to those that are found as PRPs by the EPA [12].

An additional amendment was passed a few years following CERCLA. Known as the Superfund Amendment and Reauthorization Act (SARA) of 1986, the Environmental Protection Agency presented several key components—the introduction of treating hazardous wastes, Superfund sites [15], and how waste impacts not only the environment but also human health [16]. SARA also describes how to handle emergencies and presents proactive measures to prevent future incidents from happening [17]. But the most important component found within SARA is the implementation of the applicable or relevant appropriate remedial actions (ARAR). This states that the stringency of cleanup must comply with both Federal and State legislations. Finally, SARA introduced the hazard ranking system (HRS) or the methodology that places a hazardous waste site on the National Priorities List (NPL). Sites are evaluated based on the amount and type of hazardous substances escaping into the environment, along with their impacts to human health and the environment [18]. Currently, there have been many efforts to use bioremediation for the purpose of treating Superfund sites. Table 5.3 summarizes the breakdown of bioremediation treatment methods using Superfund between 1982 and 1999.

5 Metabolisms and Pathways

5.1 Biotransformation, Biodegradation, Metabolism, and Catabolism

Throughout this text but also in additional publications, the use of degradation and transformation assumes that the author's intent is for interchange. When referring to the two terms, there lies a major difference. The purpose of transformation is to

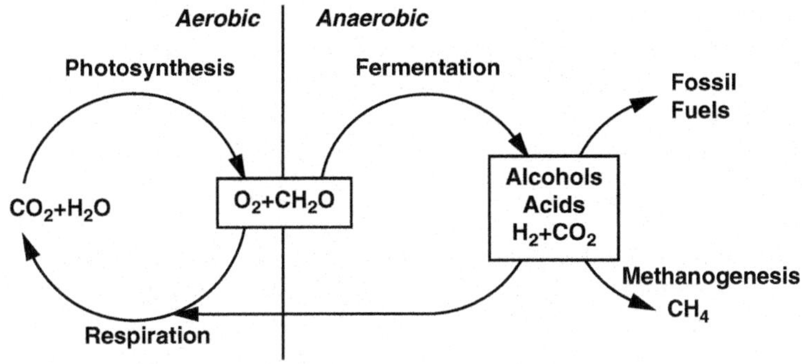

Fig. 5.2 Biochemical processes of C and O [21]

reduce potential human or ecological consequences associated with the presence of contaminants within the system, which typically translates to a change to the contaminant's chemical structure. Aitken and Long [20] recall from previous literature that degradation "is a series of catabolic transformations which lead to intermediary or central metabolic pathways in microorganisms." In addition, it can be described as the "breakdown of organic compounds through biotransformation into less complex metabolites or through mineralization into inorganic minerals, H_2O, CO_2, CH_4 [20]." One can consider the biochemical pathways by reviewing Fig. 5.2. Figure 5.2 demonstrates that degradation reduces over a given period of time. These differences must be dually noted when describing the major differences in the types of processes.

In addition to both transformation and degradation, there are two additional terms involving metabolism. First, catabolism has a relationship with metabolic processes for the purpose of generating new chemical structures or the maintenance of energy by means of using enzyme-catalyzed reactions [7]. Catabolism is different than most metabolic systems because the primary reason for degradation may or may not be the production of energy. Therefore, the microorganisms chosen for bioaugmentation to degrade chemical constituents must be clearly understood.

5.2 Oxidation-Reduction and Respiratory Processes

One of the most important tasks for microorganisms is oxidation-reduction reactions. Oxidation is a chemical process where constituents lose electrons to oxygen atoms, while reduction is gain of electrons and loss of oxygen atoms. This reaction is important for the degradation contaminants within soil, groundwater, or wastewater treatment. Oxidation-reduction reactions are necessary to degrade aliphatic hydrocarbons and chlorinated volatile organic compounds [22].

In addition, microorganisms degrade organics. There are three major ways in which this occurs. First, aerobic respiration is the process where microorganisms use electron acceptors such as molecular oxygen to degrade organic compounds (electron donors) to produce carbon dioxide, water, and other compounds [22]. Aerobic respiration typically uses aerobic microbes. Equation (5.1) summarizes aerobic respiration:

$$\text{Electron donor (organic substrate)} + O_2\,\text{(electron acceptor)} \rightarrow \text{biomass} + CO_2 + H_2O + \text{metabolites} + \text{energy} \quad (5.1)$$

Figure 5.3 provides a diagram of aerobic degradation inside soils.

Second, fermentative respiration is a process similar to aerobic respiration without the presence of molecular oxygen. Unlike aerobic respiration, the electron acceptors in fermentative respiration do not use molecular oxygen, but bound forms such as nitrate (NO_3^-). Equation (5.2) expresses fermentative respiration with nitrate as an electron acceptor. Please note that the products for fermentative respiration using nitrate form nitrogen gas:

$$\text{Electron donor (organic substrate)} + NO_3^-\,\text{(electron acceptor)} \rightarrow \text{biomass} + CO_2 + H_2O + N_2 + \text{metabolites} + \text{energy} \quad (5.2)$$

Finally, anaerobic respiration degrades organics into carbon dioxide, water, other constituents, and various gases. Anaerobic respiration is important to mention because many bioaugmentation processes degradation constituents such as PCE and heavy metal. Because of their chemical nature, many of these processes can occur by means of anaerobic respiration [22]. To illustrate this example, consider the reduction of sulfur by sulfate-reducing bacteria. In this equation,

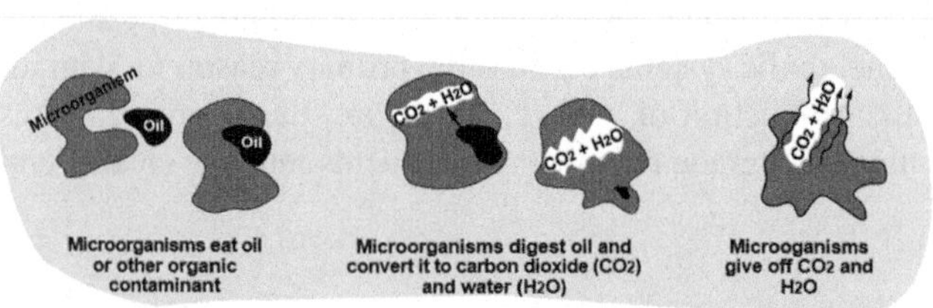

Fig. 5.3 Aerobic degradation within soils [23]

anaerobic respiring bacteria are able to take sulfur dioxide and reduce it to hydrogen sulfide:

$$\text{Electron donor (organic donor)} + SO_2{}^{-}\text{(electron acceptor)} + H^{+} \rightarrow \text{biomass} + CO_2 + H_2O + H_2S \text{ metabolites} + \text{energy} \tag{5.3}$$

In order to decide the best method of treatment, a stark distinction must be made between aerobic, fermentative, and anaerobic respiration through the perspective of microorganisms. For example, facultative microorganisms are the only microorganism able to thrive in both anaerobic and aerobic conditions [22]. Therefore, the microorganism chosen is contingent on the conditions of the environment where bioaugmentation is to take place.

5.3 Kinetics

Otten et al. [24] states that there are three major important kinetic equations that describe the microorganism degradation. These equations are speed dependent. The first equation (Eq. 5.4) holds all variables constant:

$$v = v_{mas}(c_l/K_M + c_l) \tag{5.4}$$

with:

v conversion speed (mg/kg ds/d)
v_{max} maximum degradation speed (mg/kg ds/d)
c_l variable factor concentration (mg/L)
K_M conversion speed $v_{max}/2$ (mg/L)

The remaining two equations are based on order reaction kinetics. The second assumes c_l is much greater than the K_M [24]:

$$v = v_{mas}(c_l/K_M + c_l) = v_{max} \tag{5.5}$$

The third is defines c_l is much less than K_M:

$$v = v_{max}(c_l/K_M + c_l) = K'c_l \tag{5.6}$$

These equations are significant in describing the way contaminants are reduced within the soil. When using these equations, the environmental conditions must be taken into an account when attempting to predict how microorganisms fare with contaminant degradation.

5.4 Growth and Non-growth Metabolism

As previously stated, bioremediation is contingent upon environmental conditions present within the system. While bioaugmentation reconfigures the current status of the environment to degrade xenobiotics present in the system, the major function of contaminants is the use for growth. There are five major factors which influence the biodegradability of substances for the purpose of growth: component structure within the mixture, presence of indigenous microorganisms, environmental conditions meeting the required standards for growth, and the substrate concentration and availability [25].

Growth can either happen during aerobic conditions, or when the proper electron acceptors such as nitrates, sulfates, and iron are present when oxygen is absent. During anaerobic conditions, the use of electron acceptors is more preferable due to the low solubility of oxygen within water. Even though adjustments are made to the environment there are still several mitigating circumstances. Indigenous populations may be unable to directly use constituents present within the system, typically because the concentration of contaminants maybe insufficient for use. However, there are situations when those low concentrations still threaten human or ecological safety. These series of conditions constitute under the category known as non-growth metabolism and require additional treatment [25].

Non-growth metabolism is an important facet in removing organic pollutants within the system. In this instance, the activity potential is not based not on the pollutant. This is a very important distinction from growth metabolism because the pollution presence determines the type of microorganisms required for growth (specific microorganisms can use specific components within contaminants as electron acceptors). Yet there are two benefits of non-growth metabolism—first, microorganisms degrade components without using them for growth. Second, non-growth metabolism can use any group of microorganisms [25].

Along with metabolism, there are three major types of transformations—uncatalyzed, non-enzyme catalyzed, and enzyme catalyzed. First, uncatalyzed transformation is when microorganisms grow microbial cells or supply enzyme activity on non-growing cells. The problem with uncatalyzed transformation is stoichiometric consumption or the need for microorganisms to renew components during reactors before completing any future transformations. Any uncatalyzed transformation can also lead to what is known as suicide enzyme inactivation where activity is loss to reaction-enzyme encounters [25].

Second, non-enzyme catalyzed transformation alter without the use of enzymes. This type of transformation requires the removal of these compounds from organics that can have human and ecological effects. An example of non-enzyme catalyzed transformation is the dehalogenation of halogenated organic compounds. Many organic contaminants consist of halogens, a series of elements located in the seventh column of the periodic table. Halogens have a charge of $^{-}1$. Chlorine (Cl^{-}) is a typical halide attaching to an organic compound. Coenzymes such as F_{430}

found within methanogenic bacteria and vitamin B_{12} are examples of non-enzyme catalyzed transformations [25].

Third, enzymatic catalyzed transformations have two major categories—primary and secondary metabolism. Primary metabolism produces enzymes by growing on a substrate. Examples of primary enzymatic catalyzed transformation are the oxidation of chlorinated aliphatic compounds by methanogenic bacteria on toluene and phenol, and the conversion of methane into methanol by the methane monooxygenases enzyme found in methanogenic bacteria. These water soluble enzymes (sMMO) can also be used fulfill the criteria of catalyzing chlorinated aliphatic compounds. During secondary metabolism, bacteria and fungi release metabolites during the stationary phase of batch culture. *Phanerochaete chrysosporium* and other types of wood-rot fungi are examples of secondary enzymatic catalyzed transformation [25].

Using white rot fungi for bioremediation is advantageous because it can withstand high concentration of toxic pollutants, a very important factor when degrading lignin. Lignin is a polymer having constituents of phenylpropane subunits with β–O ether and carbon-carbon chains [26]. White rot fungi also posses extracellular enzymes that are capable of dividing mirrored and non-mirrored structured carbon-carbon and carbon-oxygen bonds within lignin structures simultaneously producing alternative oxidants for the purpose of oxidizing other constituents. Finally, white rot fungi can use LiPs (lactases, lignin peroxidases), a series of extracellular proteins to oxidize PAH down to simpler quinine-products. Research has concluded that the genus *P. cryptosporidium* exemplifies this activity [26].

5.5 Soil and Groundwater Contamination Equations

There are many equations to describe soil and groundwater contamination. While what is discussed emphasizes soil contamination, these principles can be applied to groundwater. In general, contaminants are either present in pure form or as a solution. They can flow vertically in the soil or remain trapped within smaller particles. The main mechanism for transport is mobile phase convective transport. Mobile phase convective transport is the relationship between movement of gas and the pressure gradient. This mechanism drives materials within the soil. There are three parameters that control mobile phase convective transport—permeability, velocity of the water, and density. These parameters are contingent on the soil characteristics. This relationship is summarized by Darcy's Law:

$$V_w = -k_w{}^* \, d\rho/dx = -k_w \, x \, dH/dx \qquad (5.7)$$

where:

V_w = flow velocity of pore water (m/day)
Kk = permeability coefficient of water (m/day)

$d\rho/dx$ = pressure gradient in the x-direction
dH/dx = hydraulic gradient in the x-direction

In addition, contaminants will float, sink to a saturated zone, or gather under an aquifer. Floating occurs when the density of the contaminant is less than water. Because the density is less than water, the contaminant remains on top of the water surface. When the density of the contaminant is greater than the density of water, the contaminant sinks to the saturated zone. Finally, contaminants have the potential to gather under an aquifer. This depends on how contaminants interact within the soil and the relationship between the solid, water, and gaseous phases of the soil. For the purposes of the processes between water and the solid material, an appropriate relationship can be seen through the development of the adsorption isotherm equation [23]. The adsorption isotherm describes the ability of organic compound contaminants to bind within the soil:

$$q = K_d C_w \tag{5.8}$$

where

K_d = adsorption coefficient (dm^3/kg); K_d is proportionate to the organic carbon present within the soil
C_w = concentration of the contamination in the water phase (mg/dm^3)
q = content level of contaminant in soil

However, there are times when contaminants cannot be removed from the soil. Sometimes they can form residuals. This is important to understand because many compounds may have a higher residual but lower degradable potential. The degradable potential affects the presence and capabilities of microorganisms within the system. Mahro et al. [25] explained the difference between pollutants compound mixtures and their potential for biodegradation. Cassidy and Irvin [27] explained that 2,3,6,7-tetrachlorobenzodium is an example of a non-biodegradable contaminant. In other situations, a shift in environmental conditions can change the potential degradability of pollutants within the system, forming toxin intermediates. The accessibility of microbial communities to pollutants within the system (bioavailability) can also affect the degradation potential of pollutants. One of the effects of bioavailability is contaminant transformation within soils. This reduces the availability of contaminants for microorganisms [27].

Subsequent problems can also be seen during pollutant biodegradation processes, specifically during mass transfer. Mass transfer evaluates how particle size increases diffusibility within a soil medium. When particles are distributed in the soil medium, particle size permits microorganism access. In addition, many soil pores are filled with the associated contaminants [27], making it for them to reach the microorganisms. The mass transfer evaluation is summarized in Eq. (5.9) [28]:

$$N = -D_p * \delta C/\delta x \tag{5.9}$$

where

N = mass flux of pollutant
D_p = diffusion coefficient
$\delta C/\delta x$ = change of concentration to change in time

The significance of this equation is such that typical values for the diffusion coefficient are between 9×10^{-12} and -2×10^{-21} m^2/s.

6 Treated Media

6.1 Groundwater Contamination

Groundwater moves through the pore spaces of the aquifer and contaminants are interchanged and dissolved by diffusion within non-flowing entities. These non-flowing entities become 'sinks' for contaminants such as non-aqueous-phase liquids (NAPL) and organic matter. These contaminants enter into an aquifer and are controlled based on the way materials are released. An example of an NAPL is a leaking fuel. The following is the process of NAPL formation. When water flows through an aquifer, it becomes a source of contaminant transfer, spreading contaminants throughout the media. Aquifers can be a mechanism that transfers contaminants throughout an aquifer Therefore groundwater contamination is not only problematic within the inside of aquifer, but also outside an aquifer [29].

Another type of contaminant, dense non-aqueous phase liquids (DNAPLs), is transmitted through the soils ultimately reaching the water table. Since these constituents are higher in density as compared to water, they remain mobile and eventually settle in low-permeable locations where they form a pool. The remaining DNAPLs sit on the outskirts in soils. Increasing the concentration of DNAPLs within an aquifer ultimately forms plumes [31]. Figure 5.4 provides a diagram of this process happening.

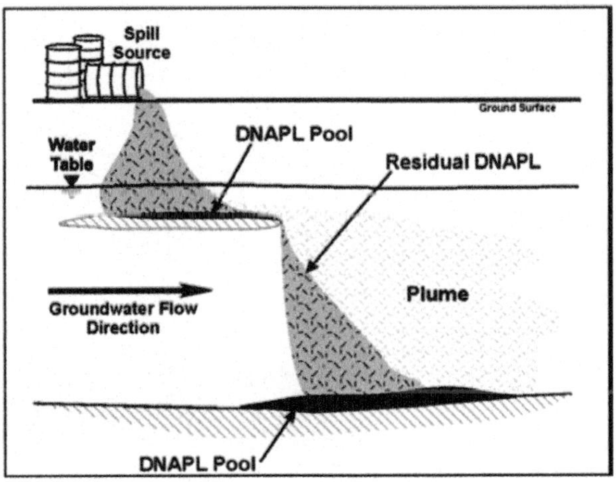

Fig. 5.4 DNAPL plume formation process from a chemical spill [30]

When treating contaminated groundwater, there are two major processes. First, adsorption a thermodynamic property, binds contaminants to a specific binding site, specifically organic matter. Coupled with adsorption is the reverse, desorption, removing contaminants away from the binding site in groundwater. These are very essential for bioaugmentation to retrieve contaminants prior to their entrapment. Traditionally, absorption encompasses the relationship between the aqueous and sorbed phases. As the aqueous phase concentration increases, the sorbed-phase mass increases until reaching the maximum value known as 'equilibrium partitioning' [32]. A discussion of contaminants and their degradability is presented in Sects. 6.2 and 7.

The most common form of groundwater treatment is in situ, or direct treatment at a contaminated site. In situ treatment of groundwater occurs by air sparging or extraction. Air sparging is the addition of air into the contaminated aquifer to remove the contamination, while the use of an extraction [23] integrates electron acceptors such as oxygen and nitrate and nutrients with microorganisms and then re-inserted back into the area of contamination by using either infiltration galleries or injection wells. Constant extraction and reinjection of water back into the system removes constituents within the groundwater [19, 33, 34]. Figures 5.5 and 5.6 provide visual diagrams of in situ treatment of contaminated ground water.

In situ treatment methods of groundwater contamination involving microorganisms take advantage of aerobic and anaerobic conditions in the water. The appropriate addition of microorganisms can degrade methyl *tert*-butyl ether (MTBE) [35, 36] from gasoline. Table 5.4 describes the microorganisms completing in situ groundwater treatment.

One of the more recent in situ methods of treatment is biological enhanced reductive dechlorination (ERD). ERD is the direct application of bacteria that are capable of dechlorinating contaminants such as trichloroethylene and tetrachloroethylene into ethane and ethane. ERD adds a substrate that promotes bacterial growth under anaerobic conditions for treatment near the contaminated medium and production of hydrogen. Hydrogen atoms replace chlorine atoms inside of the compound reducing it to ethene. However, if not monitored ERD has the potential

Fig. 5.5 In situ bioremediation process [19]

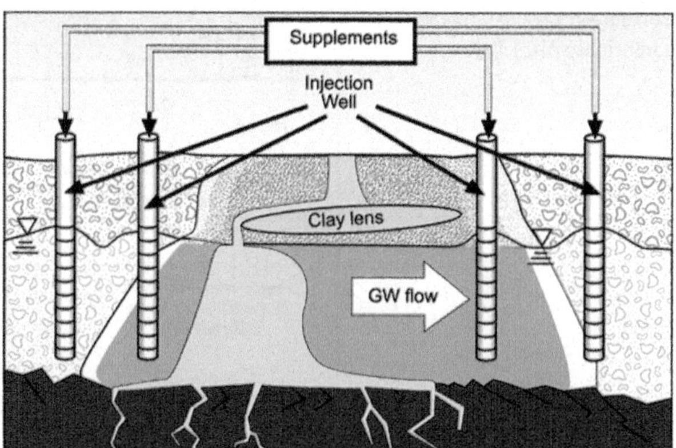

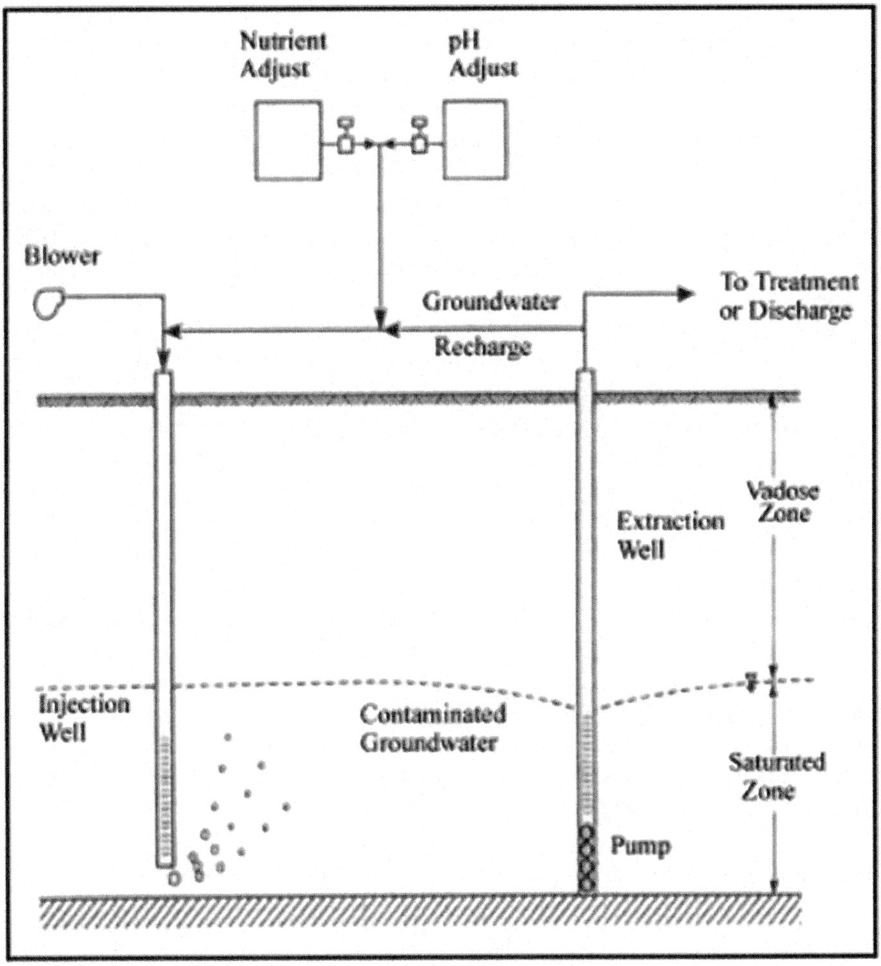

Fig. 5.6 Method for treating contaminated groundwater [34]

Table 5.4 Microorganisms capable of removing petroleum hydrocarbons [37]

Bacteria	Yeast and Fungi
Achromobacter	*Aspergillus*
Acinetobacter	*Candida*
Alcaligenes	*Cladosporium*
Arthrobacter	*Penicillium*
Bacillus	*Rhodotorula*
Brevibacterium	*Sporobolomyces*
Cornybacterium	*Trichoderma*
Flavobacterium	
Nocardia	
Pseudimonas	
Vibrio	

of creating harmful vinyl chlorides (VC) or *cis*-dichloroethene (*cis*-DCE) in the process. Therefore, applying strains such as *Dehalococcoides ethenenogenes* is important to reduce the potential of creating harmful compounds so that the treatment can continue [38]. Literature has shown an increasing interest in using this method. Justicia-Leon et al. [39] achieved complete degradation of chloroform to non-harmful forms within contaminated groundwater samples by using Dehalobacter strains with an abundance of bacteria ($1.6 \times \pm 0.9 \times 10^4$/mL) coupled with 10 mM bicarbonate. Husserl and Hughes [40] applied *Arthrobacter* sp. JBH1 in laboratory studies to completely mineralize nitroglycerin from contaminated groundwater.

An important factor to consider in bioremediation treatment of groundwater is how to appropriately design something that can be applied directly into the groundwater. Additional influences on design include the hydraulic conductivity, capability of waste to be degraded, and the location at which the contamination occur [33]. Overall, there are several advantages and disadvantages of using in situ groundwater remediation. A summary of these advantages and disadvantages are available and summarized in Table 5.5.

Groundwater remediation has been applied in literature. Lin et al. [41] examined the addition of nitrogen as a nutrient to increase removal of benzene, toluene, ethylbenzene and xylene (BTEX) within an oxygen-releasing reactive barrier (ORRB). Marzorati et al. [42] found that lactase as a carbon source increases a consortium that reduces 1,2-dichloroethane within an aquifer. Other consortiums present within groundwater include using *Dehalococcodies* for removing perchloroethylene and trichloroethylene (PCE and TCE) [43], *Acetobacterium* sp. using zero-valent iron within reactive permeable layers to mineralize hexahydro-1,3,5-trinitro-1,3,5-triazine (RDX) [44].

Table 5.5 Advantages and disadvantages of in situ groundwater treatment [33]

Advantages	Disadvantages
Remediates contaminants that are adsorbed onto or trapped within the geologic materials of which the aquifer is composed along with contaminants dissolved in groundwater.	Injection wells and/or infiltration galleries may become plugged by microbial growth or mineral precipitation.
Application involves equipment that is widely available and easy to install.	High concentrations (TPH greater than 50,000 ppm) of low solubility constituents may be toxic and/or not bioavailable.
Creates minimal disruption and/or disturbance to on-going site activities.	Difficult to implement in low-permeability aquifers.
Time required for subsurface remediation may be shorter than other approaches (e.g., pump-and-treat).	Re-injection wells or infiltration galleries may require permits or may be prohibited. Some states require permit for air injection.
Generally recognized as being less costly than other remedial options.	May require continuous monitoring and maintenance.
Can be combined with other technologies (e.g., bioventing, SVE) to enhance site remediation.	Remediation may only occur in more permeable layer or channels within the aquifer.
In many cases this technique does not produce waste products that must be disposed.	

Treatment of groundwater can also occur by ex situ applications. One of the more common ex situ applications of groundwater is thermal treatment or the process of applying heat to aquifer material or microorganisms. Thermal treatment has been used to treat trichloroethylene (TCE) by dechlorinated microorganisms [45] and heated microorganisms [46]. Another method of ex situ applications uses strains to enhance the degradation of contaminants (cometabolism), a method that has been applied in various circumstances [33]. McCarty et al. [31] utilized a modified strain Burkholderia cepacia $PR1_{301}$ with phenol as a carbon source, applying air stripped groundwater into an aquifer to remove TCE.

6.2 Soil Contamination

Organic soils consist of 5–60 % organic material and contain amino acids, proteins, lipids, and carbohydrates. This material remains unchanged (non-humic), or changed due to weather, physical, or chemical processes (humic). The difference between humic and non-humic substances is determined by color—darker colored soils are humic [20].

Organic soils contamination occurs because of material stability, pH, nutrient availability, and mean cell residence time or the time the time contaminants remain in the soil. There are three major factors that inhibit organic soils. First, organic soil properties inhibit contaminant analysis within the soil. Second, the contaminants within soils are diverse in their degradability—in other words, counterpart soils such as mineral soils may experience higher degradability as compare to others. Finally, measurements of organic growth and production of carbon dioxide can be made possible through analytical data [20].

Whenever contaminants are present within organic soils, they will either enter into water particles in the soil or in zones generated by particles within the soil. Chemical characteristics such as molecular weight and vapor pressure determine their ability to dissolve within water depending on the availability and value to the microorganism. Chemical composition also becomes significant for the environment's ability to withstand these contaminants especially with primary degradation by microorganisms in the aqueous phase (see Sect. 6.1) [20].

Additional interactions of organic materials involve a relationship with the soils. Whenever soils are present, biodegradation increases the entrapment of materials within the soil matrix. Alternative organics—soil relations include sorption which not only separates microorganisms from the contaminants but also removes them from the aqueous forms. This creates a change within the degrading capability of microorganisms. When sorption occurs towards humic substances, many contaminants reduce microorganism availability. This is an important facet within the soil since contaminants during sorption withheld in the soil. Subsequent interactions such as cation exchange can also have a profound effect on contamination removal [20].

Similar to the treatment of groundwater, soil treatment is either employed by using in situ and ex situ processes. In situ soil treatment occurs by bioventing or slurry-phase lagoon aeration. Bioventing introduces air by extraction for the purpose of reducing contaminants within soils. Slurry-phase lagoon aeration uses a lagoon as a place to combine air and soil for constituent removal [19].

There are several historical in situ soil bioremediation applications. Microorganisms combining with nutrients and chemicals were applied to treat formaldehyde in Ukiah, California. Other treatments include phenol in Michigan, methylene chloride, *n*-butyl alcohol, dimethylaniline, and acetone in Waldwick, New Jersey, chromium sludge, chemical and electroplating wastes at the Kelly Air Force Base Waste Disposal Site in Texas, and underground leak of jet fuel at Eglin Air Force Base in Florida [47]. Cutright [48] was capable of removing 96 % of polycyclic aromatic hydrocarbons (PAHs) from soils.

As an alternative towards in situ treatment, soils can be removed by ex situ applications. For soil contamination land treatment transforms contaminated constituents to other areas that have a more viable microorganism growth. Land treatment transports engineering designed reactor with clay and liners, and include proper places for drainage and irrigation. To advocate microbial growth within land treatment, fertilizers and animal manures are applied as nutrients and carbon sources using traveling gun sprinkling systems [49]. Fertilizers in the form of sewage sludge are also applied to treat PAHs [50]. Other ex situ methods include composting using wood chips for bulking and hay and manure to increase organic materials and biopiles where treated soils are housed within mounds having received additional constituents and air circulation [19].

Finally, slurry-phased treatment develops a liquid constituent where soils, sediment, and sludge are mixed to maintain contact between the constituents and microbes [19]. Figure 5.7 provides a process-flow diagram of a slurry-phased treatment method. Shailaja et al. [51] use a bioslurry-phased reactor to degrade 90 % di-ethylhexyl phthalate (DEHP), a byproduct of polyvinyl chloride (PVC) plastics. Slurry systems have been used for treating PAH as seen by Valentin et al. [52]. Slurry systems can use white-rot fungus (*Bjerkandera adusta BOS55*), whereas Gamati et al. [53], Cassidy and Hudak [54], and Lewis [55] applied the technique but also use other consortiums. Quintero et al. [56] treated 90 % of the pesticide gamma-hexachlorocyclohexane (gamma-HCH) within 50 days using an anaerobic slurry reactor. Other enhancements of bioreactor slurry were used by Partovinia et al. [57] to degrade *n*-hexadecane, Kao et al. [58] treated total extractable organics using a two-stage bioslurry reactor, pendimethalin [59], 2,4,6-trinitrotoluene [60–62], *n*-dodecane [63], 2-4-dinitrotoluene [64], polychlorinated biphenyls (PCBs) and diesel fuels [65–67], chlorpyrifos [68], and other crude oils [69].

Lebkowska et al. [71] determined that diesel-fuels soils bioaugmented with microorganisms increased treatment by 50 %, while Lee and Kim [72] applied bioleaching using *Acidithiobacillus thiooxidans* along with EDTA-enhanced electrokinetics removing 92.7 % lead (Pb) in the soils. Tang et al. [73] observed a relationship between fertilizer application (20 g Nitrogen/m^2) and treatment

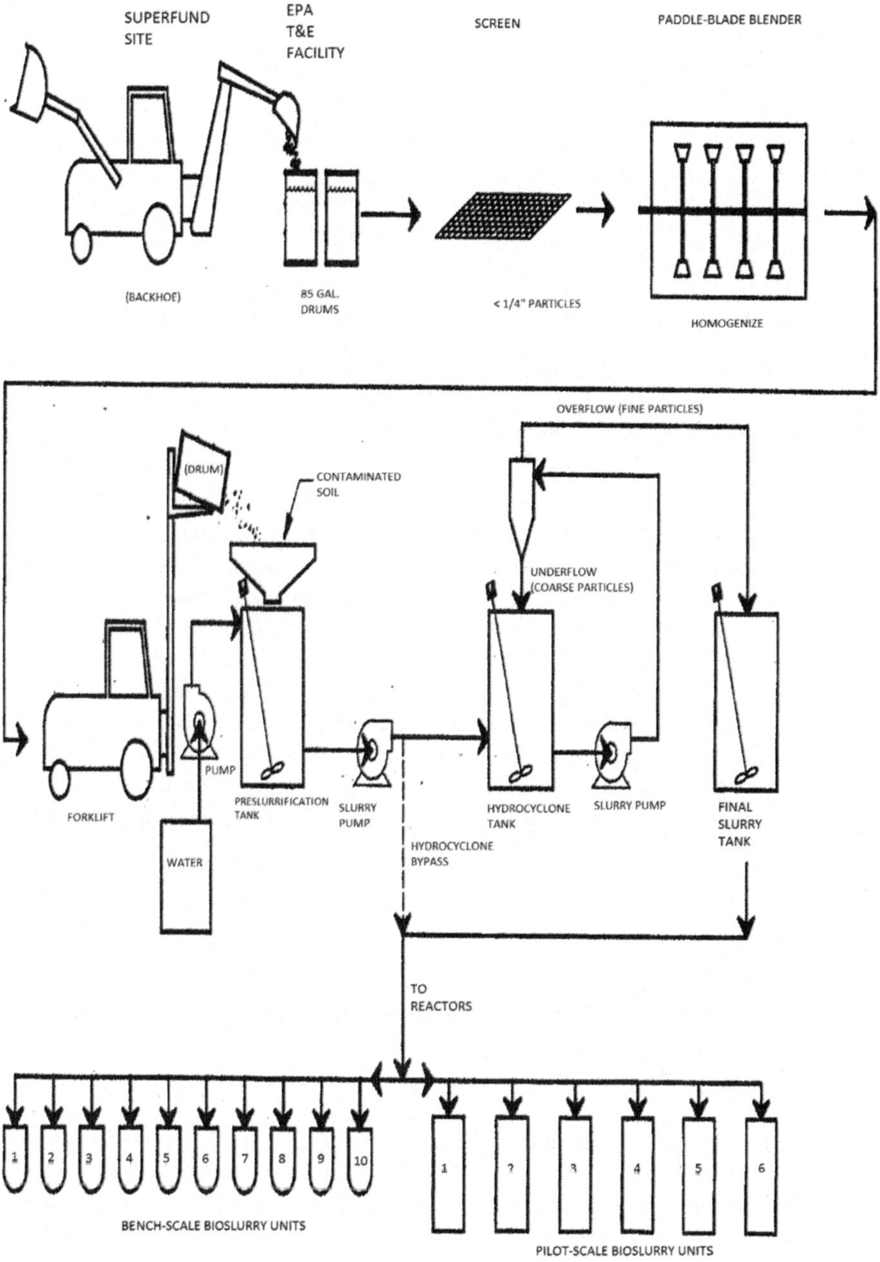

Fig. 5.7 Slurry phased treatment diagram [70]

Table 5.6 Examples of superfund sites using bioremediation technologies [23]

Name of site	Treatment	Contaminants
Applied Environmental Services, NY	Bioventing	Volatile organic compounds (VOCs), semi-volatile organic compounds (SVOCs)
Onalaska Municipal Landfill, WI	Bioventing	VOCs, polyaromatic hydrocarbons (PAHs)
Eielson Air Force Base, AK	Bioventing	VOCs, SVOCs, PAHs
Brown Wood Preserving, FL	Land treatment	PAHs
Vogel Paint & Wax, IA	Land treatment	VOCs
Broderick Wood Products, CO	Land treatment/ bioventing	SVOCs, PAHs, dioxins
Burlington Northern (Somers), MT	Land treatment/in situ bioremediation	SVOCs, PAHs

efficiency of soils with petroleum, where tall fescue showed the highest removal efficiency. Abdulsalam and Omale [74] found that bioaugmentation produced the second highest removal of motor oil (65.4 %) within contaminated soils. Ruberto et al. [75] discovered that bioaugmentation of soils contaminated with diesel oil from Marambio Station in Antarctica had a higher removal efficiency (86 %) as compared with using biostimulation (81 %). Kolwzan [76] observed that while inoculation with *Acinetobacter lwoffi* removed 99.99 % diesel oil, *Pseudomonas putida* has better removal than *Acinetobacter lwoffi* (99.997 %). Also, Perfumo et al. [77] augmented soils with *Geobacillus thermoleovorans T80* and reduce hydrocarbons by 70 %. Andreoni et al. [78] degraded 2,4,6-trichlorophenol (TCP) by using *Alcaligenes eutrophus*. Tables 5.6 and 5.7 indicate the various types of treatment methods used for soil contamination.

6.3 *Hazardous and Industrial Waste Treatment*

6.3.1 Hazardous Waste

Since hazardous wastes contain toxic substances, specifically hydrocarbons, extensive research began in the 1950s and 1960s to consider the relationship of microbial metabolic pathways and the degradation of hydrocarbons within the environment. Extensive research finds that many microbes within the environment of hazardous wastes have a difficulty of completely degrading materials in a timely fashion to prevent adverse effects. Microbes from bioaugmentation become useful in removing harmful hydrocarbons. These microbes use the toxic substances as sources to fulfill metabolic processes or transform various substances unable to be used for the purpose of extracting carbon, energy, or other nutrients [79].

Table 5.7 Several field studies of treatment using bioaugmentation techniques

Treatment process	Description	Location	Compound treated	Percentage removal	Costs	References
Simplot Anaerobic Biological Remediation (SABR)	Removal of contaminated soil to receive microorganisms; add starch as carbon source	Bowers Field, ID	Nitroaromatic compounds, pesticides	99.8 %	$97/yd^3$	[34]
Daramend	Anaerobic, aerobic composting	Cambridge, Ontario	2,4-Dinoseb; atrazine	98.5–100 %	$120/ton	[34]
Xenorem	Composting windrows	Tampa, Florida	Chlordane; DDD; DDT	89–98 %	$192/yd^3$	[34]
Land treatment	Application of lime and cow manure	Burlington Northern Superfund Site, Brainerd/Baxter, MN	Total PAHs	98.2–98.8 %	Not available	[19]
Land treatment	Composting	Dubose Oil Products Co. Superfund Site, Cantonment, FL	Total PAHs	~86.4 %	Not available	[19]
Slurry-phased bioremediation	Slurry-phase bioremediation	Southeastern Wood Preserving Superfund Site, Canton, MS	Total PAHs	67.40 %	Not available	[19]

6.3.2 Industrial Wastewater

Leather tanning, wood preservation, metal finishing, coke, and other industrial processes [80] create wastewater with many harmful products. These products can include surfactants, dyes, oils, phenols, polycylic and heterocyclic aromatic hydrocarbons, heavy metals such chromium (VI) [80–82]. Bioaugmentation is useful in industrial wastewater treatment because it can provide a source of treatment for wastes difficult to be removed. McLaughlin et al. developed parameters which assign specific tasks to microorganisms. These tasks help increase kinetic activity and biological diversity within indigenous species. For example, chlorophenols a subset within a series of organics known as chloroaromatics are most popularly found in water chlorination. Bench test research by McLaughlin et al. [83], provided an initial step towards chlorophenol removal by cultivating *Pseudomonas putida*. Quan et al. [84] found evidence that mixing cultures of bacteria could drive a reduction of 2,4-dichlorophenol within chemical wastewaters.

Other authors find that bioaugmentation useful in industrial wastewater treatment. Bioaugmentation reduce toxicity and lipid content within olive oil wastewater by the use of *Phanerochaete chrysosporidium* [85]. Studies also suggest that strains from bioaugmented bacteria reduce odor during anaerobic transformation, or the biological conversion of methane and carbon dioxide from organic material by methanogenic bacteria. Pandaya [86] concluded bioaugmentation consisting of *Micrococcus*, *Nocardia* and *Pseudomonas* species and photooxidation were suitable in improving treatment of cresol wastewater. Recently, Bhattacharya and Gupta [80] used *Acinetobacter* sp. *B9* to remove Cr(VI), total Cr, and Ni, while Kim et al. [82] developed a consortium consisting of Proteobacteria, Firmicutes, *Bacteroidetes*, *Planctomycetes* and *Deinococcus-Thermus* to significantly improve the treatment of COD, TN, and TP at a tannery wastewater treatment plant.

6.4 Bioaugmentation in Wastewater Treatment Plants

Bioaugmentation provides readily available bacteria to a wastewater treatment plant by culturing. Typically bioaugmentation is used for system start-ups. One of the major advantages of using bioaugmentation in wastewater treatment is that the density of bacteria reduces the production of filamentous bacteria. This is because of the decrease in mixed liquid volatile suspended solid (MLVSS) concentrations and means cell residence times (MCRT). Typical MLVSS concentrations and MCRT are less than 2 mg/L and 2–4 weeks respectively [87].

According to Gerardi [87], there are two major bacteria types available for wastewater treatment bioaugmentation. These bacteria are affected by the nutrients level, dissolved requirements, degradation rates, and substrate available for degradation. First, coli-aerogens are microorganisms that form in fecal waste and the gastrointestinal tract of humans. Second, saprophytic bacteria are a series of microorganisms present in reducing substrate levels. One of the most common

Table 5.8 Treatment methods for various wastewaters

Treatment system/bioaugmented species	Wastewaters	References
Anaerobic sequencing batch reactor (ASBR)	Diary	[88]
Membrane biological reactor (MBR), photocatalysis, ozonation	Bromoamine	[89]
Fluidized bed reactor	1-Napththylamine	[90]
Zeolite-biological aerated	Pyridine Quinolone TOC Ammonium	[91]
Activated sludge	Nitrogen	[92]
Escherichia coli HB101 and *Pseudomonas putida KT2440*	2,4-Dichlorophenoxyactic acid (2,4-D)	[93]
Activated sludge	Nitrification	[94–98]
Mixed bacteria	Municipal wastewater	[99]
Pyridine and quinoline in a sequencing batch reactor (SBR)	Coking wastewater	[100]
Cyanide-degrading microorganisms	Coking wastewater	[101]
3,5-Dinitrobenzoic-degrading strain and bacteria	3,5-Dinitrobenzoic acid (3,5-DNBA)	[102]
Comamonas testosteroni	*p*-Toluenesulfonate (pTS)	[103]
Citrobacter braakii	Anionic surfactants	[104]

degradable constituents using these microorganisms is carbonaceous biochemical oxygen demand (cBOD). This is treated by exoenzyme production. Exoenzyme production of cBOD is treated by first generating lipids, carbohydrates, and proteins to become soluble, and then degraded by both coli-aerogens and the saprophytic bacteria. An example of bacteria within the system is *Cellulomonas*, a series of bacteria that release exoenzymes that break glucose bonds in exchange for cellulose. Cellulose then becomes water soluble to break down water, carbon dioxide, and new bacterial cells [87]. Bacteria can also produce enzymes that can degrade lipids or surfactants that are permissible for the control of foam production [87].

Various authors have used bioaugmentation for various applications of wastewater treatment. Table 5.8 summarizes this research.

6.5 *Phytoremediation*

Phytoremediation uses various techniques to remove constituents from soils, specifically heavy metals. While there are many phytoremediation techniques, there are three major principles applied—the degradation or the uptake of constituents such as trichloroethylene within the root zone of influence, phytoextraction or planting plant species for the purpose of uptake by the plants and removal by harvesting, and containment and immobilization where plants are able to 'hold'

Table 5.9 Description of phytoremediation processes [105]

Process	Definition	Examples of contaminants removed	Plants	Treated medium
Phytoextraction	Growing various plant species for the purpose of uptake and eventual removal by harvesting	Gold (Ag) Cadmium (Cd) Cobalt (Co) Chromium (Cr) Copper (Cu)	*Helianthus annuus* L., *Eichhornia crassipes*, *Myriophyllum spicatum*	Groundwater
Rhizofiltration	Adsorption onto roots or absorption into roots	Lead (Pb) Cadmium (Cd) Nickel (Ni) Zinc (Zn)	*Brassica juncea*, *Thlaspi caerulescens*, *Alyssum wulfenianum*	Groundwater
Phytostabilization	Prevention of constituents by plants from moving further in the media having applied adsorption, absorption, and precipitation	Arsenic (Ar) Cadmium (Cd) Copper (Cu) Mercury (Hg)	*Brassica juncea*, *Agrostis tenuis*, *Andropogon gerardi*	Soil, sediment, sludges
Rhizodegradation	Microorganism growth within soils where the release of various carbon sources from plants or roots	Total petroleum hydrocarbons (TPH) Polycyclic aromatic hydrocarbons (PAH) Pesticides	*Medicago sativa*, *Schizachyrium scoparius*, *Agropyron smithii*	Groundwater
Phytotranspiration	Transformation of contaminants from the soils into roots and transpired into the atmosphere	Chlorinated solvents Arsenic (Ar) Mercury (Hg) Selenium (Se) Insecticides	*Medicago sativa*, *Brassica juncea*, *Arabidopsis thaliana*	Soil, sediment, sludges, groundwater
Phytohydraulics	Uptake of groundwater to prevent further contamination within groundwater	Gasoline Diesel	Cottonwood, poplar trees	Groundwater
Vegetative cover	Cover consisting of soils and plants either with or without engineering techniques	Organics, inorganics	Poplar trees	Groundwater
Riparian covers/ vegetative strips	Plants along streams to reduce contamination	Pesticides	Poplar trees	Groundwater

materials within a given area keeping them from further travel. All forms of phytoremediation are based on these variations [105]. Table 5.9 summarizes the major types of phytoremediation processes, while Fig. 5.8 shows the treatment process.

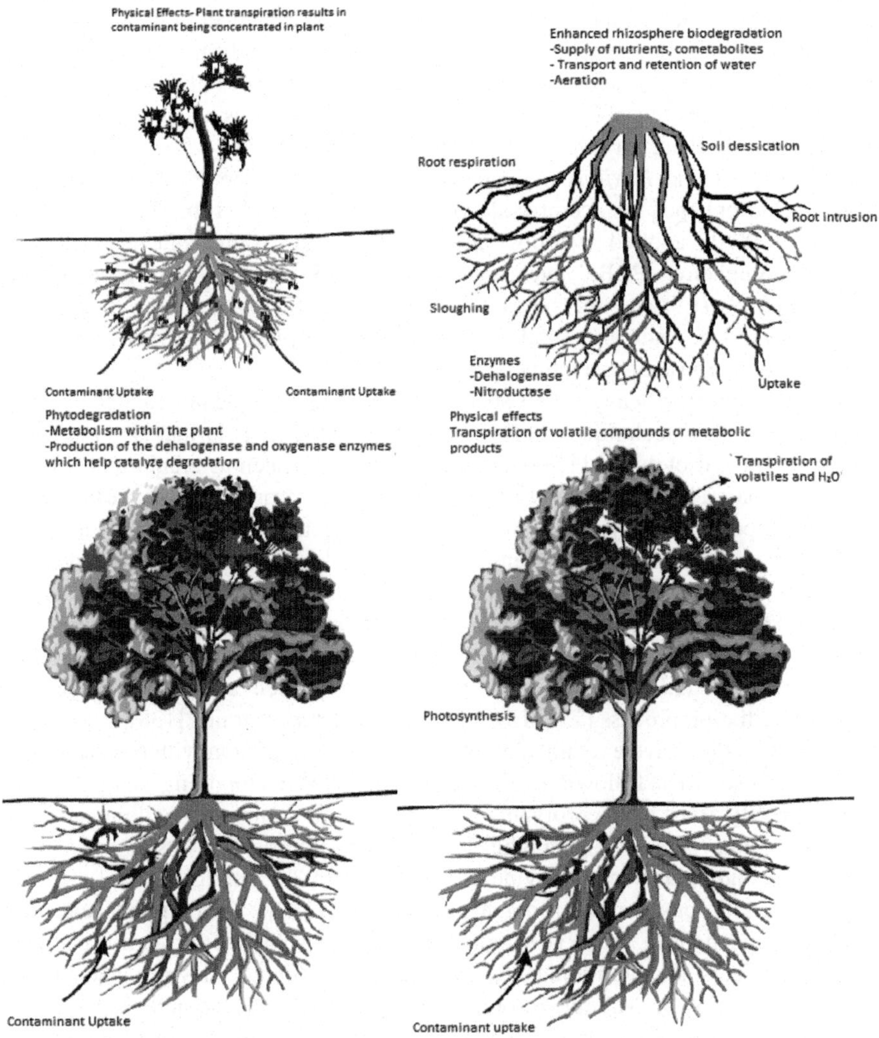

Fig. 5.8 Examples of phytoremediation processes (from top left to bottom right: phytoextraction, rhizodegradation, phytodegradation, phytovolatilization) [105]

There have been various applications of phytoremediation to treating soils. Schwab et al. (1998) use tall fescue and alfalfa where 32–45 % of the mass balance components consisted of naphthalene volatilization [106, 107]. Barber et al. (1995) found that *Lemna gibba* use phenyl-(gr)b-D-glucoside to treating soils containing phenol with a 90 % efficiency [106, 108]. Zablotowicz et al. (1996) reduce the herbicide acifluoren within soils inside the rhizosphere [106, 109]. Siciliano and Germida (1997) found that forage grass species *Bromus biebersteinii*, *Elymus dauricus*, and *Agropyron riparum* [106, 110] and two strains *Pseudomonas aeruginosa R75* and *P. savastoanoi* CB35 effectively treated soils containing

2-chlorobenozic acid. *Pseudomonas SR3* increases the treatment of pentachlorophenol within soils [106, 111]. Boyle and Shann (1998) found that degrading 2,4,5-trichlorophenoxyacetic acid (2,4,5-T) was based on the type of soil [106, 112]. Rivera et al. (1998) removed 100 % 2,4,6-trinitrotoluene (TNT) using phytoremediation [106, 113]. Meharg et al. (1997) studied fungal species—*Suillus variegatus* (*S. variegatus*) and *Paxillus involutus* (*P. involutus*), where *P. involutus* reduce 2,4,6-dichlorophenol as compared with *S. variegatus* [106, 114]. Rhykerd et al. (1998) applied various grasses to treat trinitrotoluene, 2,2′,5,5′-tetrabromobiphenyl (PBB), and chrysene [106, 115]. Conger and Portier (1997) found that *Salix nigra* was capable of removing bentazon from groundwater [106, 116]. Also, Andreazza et al. [117] used oatmeal plant *Avena sativa* L. within soils for the purpose of extracting copper.

For groundwater, heavy metals such as lead or zinc could be remediated by using phytoremediation techniques [118], arsenic [119–122], chlorinated compounds [123–126], trinitrotoluene [127], radioactive materials such as uranium [128], and petroleum based hydrocarbons [129, 130]. In addition, various plants have been used for the purpose of applying phytoremediation techniques—Natarajan et al. [119] apply *Pteris vittata* L. in arsenic removal, Wang et al. [123] utilized dogwood, while Lee and Yang [128] sunflower and bean plants for uranium removal. Yoon et al. [131] applied *Arabidopsis thaliana* to remove 2,6-dinitrotoluene, and El-Gendy [132] poplar trees for removing hydrocarbons. Finally, ethylenediaminetetraacetic acid (EDTA) and other additives have been added to improve the phytoremediation process [133–137] along with bacterial strains [138].

Recent studies have examined combining phytoremediation with bioaugmented microorganisms. The following are three potential reasons for doing so. First, plants can indirectly assist in microorganism degradation of compounds within soils. For example, plant roots act as a conduit for spreading microorganisms across a given space. Second, plants can provide the mechanism for tilling the soil, helping improve conditions for microorganism viability [139]. Third, plants can also provide a mechanism for oxygen transfer, adequate substrate through the release of root exudates, and additional essentials such as nutrients, enzymes, and carbon. Yet a competition of available resources may still arise between native species and bioaugmented species. Therefore, it is important to carefully select appropriate microorganisms that complement with indigenous species. An additional issue is having soil conditions that are not conducive to plant growth. This can be remediated by introducing organics to soils such as biosolids [140].

7 Contaminants

In this section, the series of contaminants discussed are non-aqueous phase density liquids (NAPDLs). NAPDLs are a series of compound contaminants found as mixtures [29]. While there are extensive compounds that can be considered for discussion, three major groups will be discussed—halo-organic compounds, polycyclic aromatic hydrocarbons, and additional compounds such as alkanes, alkenes, and benzene toluene and xylenes (BTEX).

There are two major conditions in which hydrocarbons can be biodegraded—aerobic and anaerobic conditions. As stated in Sect. 5.2, under aerobic conditions, degradation occurs by taking the organic compound and converts into carbon dioxide and water in the presence of oxygen. Equation (5.10) expresses the degradation of toluene (C_7H_8) [141]:

$$C_7H_8 + 9O_2 \rightarrow 7CO_2 + 4H_2O \tag{5.10}$$

This equation states that in order to degrade 1 mg/L of hydrocarbon, 3 mg/L of water are required. In other situations, contaminants cannot be degraded in the presence of oxygen. Therefore, hydrocarbons rely on additional electron acceptors. Within this example, toluene is degraded, by using ferric iron (Fe^{3+}) [30]:

$$C_7H_8 + 36Fe(OH)_3 \rightarrow 7CO_2 + 36Fe^{2+} + 72OH^- + 22H_2O \tag{5.11}$$

In this reaction, ferric iron is reduced from Fe^{3+} to Fe^{+2}, a very essential equation to monitor the overall effects of reduced iron within the system [84].

7.1 Halo-Organic Compounds

Halo-organics are a series of organic pollutants attaching to a series of halogens (typically chlorine). According to author Mohn (2004), halo-organic compounds are the most difficult pollutants handled within the soils. The electronegativity of halo-organic compounds makes the compounds more stable and more difficult to apply bioremediation. In addition, halo-organic compounds are well-known for their ability to be transferred throughout the food chain and can sorb to soil because of hydrophobicity. However, unfamiliarity within natural soils makes them difficult to be removed [142, 143]. There are three halo-organic compounds—trichloroethene (TCE), perchloroethene (PCE), chlorophenols, and polychlorinated biphenyls (PCBs).

TCE compounds are found in textile processing and degreasing machinery. When these materials are improperly removed from the system, they can be easily transferred, affecting ground water and also soils [1]. There are two major methods that remove TCE aerobic co-oxidation by two major enzymes monooxygenases and dioxygenases, and reductive dehalorespiration under anaerobic conditions [20]. TCE are reduced to dichloroethene, vinyl chlorine, and ethane [20]. PCE known as "the fully-chlorinated by unsaturated tetrachloroethene," [1] are removed in a similar fashion, but instead dechlorinate the compounds into 1,2-*cis*-dichloroethene by *Dehalococcoides ethenogens* by using the dechlorination products as electron acceptors [20]. TCE degradation can be summarized in Fig. 5.9.

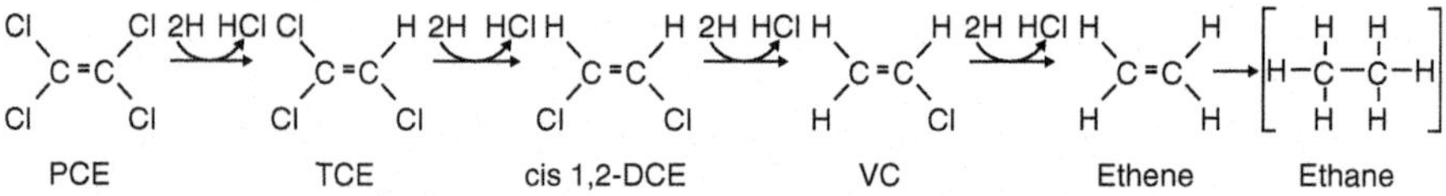

Fig. 5.9 Biological pathway for TCE degradation [30]

Chlorophenols (chlorinated hydrocarbons) are most commonly found in the environment as forms of biocides in agriculture and industry, incineration of organics, and chlorination of water. With the potential problems of these forms released into the environment, there are methods provided to reduce the impact of these pollutants [142, 143].

The final major halo-organic compound, polychlorinated biphenyls (PCBs) are a series of 209 biphenyls. PCBs have a connection of two hydrocarbon structures, renowned for their position and number of chlorine atoms. There are three major properties of PCBs—they are stable, have a high boiling point, and are insoluble in water. PCBs are used in various industrial, thermal, and hydraulic processes. While xenobiotics are beneficial, they continue to be a retardant in degradation. PCBs are infamous for their ability to transform without molecular oxygen in the presence of other conditions such as nitrate, magnesium, iron, and sulfate-reduction. An example of PCBs degradation is the co-oxidation of 2,3-biphenyl dioxygenases. Co-oxidation begins by reducing the regions of the compound that are considered the least chlorinated, followed by the breakdown of the aromatic rings within the PCB, and then hydrolysis [143].

Section 6 of the Toxic Substance Control Act of 1976 develops the proper disposal and handling for soil contamination. In addition, PCBs have been prohibited in the distribution, process, and manufacturing of these contaminants. From previous in situ bioremediation studies, indications towards *Acinetobacter, Athrobacter*, and *Pseudononas putida* have been applied to degrade PCBs in non-laboratory conditions [143].

7.2 *Polycyclic Aromatic Hydrocarbons*

A major application of bioaugmentation is the reduction of polycyclic aromatic hydrocarbons (PAH) found within industrial plant waste. According to Harvey and Thurston, PAH are a series of carcinogenic (cancer causing) chemicals produced by the incineration of organic material to produce petroleum or wood-preservation, from accidental spills or runoff from the landscape [144]. PAH are most commonly formed because pyrolysis or incomplete combustion of organic compounds [145].

The chemical composition of polyaromatic hydrocarbons consists of two or more rings of benzene, becoming insoluble and stable in several forms [146, 147]. PAHs are more stable in form and have three types of angular arrangements (phenanthrene, benzopyrene, pyrene) and two linear arrangements (anthracene, benzanthracene). Angular arrangements become more difficult to biodegrade as compared to linear arrangements [146]. In addition, larger ringed forms of PAHs become more insoluble in water, creating an issue with a bacteria's ability to degrade material [148]. In soil applications, PAHs can be very difficult to treat because of their high hydrophobicity and low bioavailability because they tend to stick to soil colloids [147]. When these pollutants enter into the environment, they remain in their composition until a change is provided by means of processes such as degradation, accumulation, or bioaccumulation [146]. In the environment, PAHs can potentially be mutagenic, carcinogenic, and teratogenic [145].

The typical degradation of polyaromatic hydrocarbons converts pollutants into carbon dioxide or into other products. The most common degradation is modeled through naphthalene since it has the smallest ring structures [146]. Several examples of microorganisms are available to reduce petroleum hydrocarbons and include *Acinetobacter* and *Alcaligens* [147]. The microorganisms metabolize PAHs so they can grow on the PAH substrate. However, if growth is not possible, PAHs are reverted through, or transformed into non-mineral products. Major issues when discussing degradation of PAH is such that these contaminants have a slower rate of contaminant desorption which has an effect on their total degradability within the soil. This is a problem which can provide an issue for availability within soils [20].

According to author Herwijnen et al. [148], bioaugmentation is compiled in nature under various measurements of carbon source limitation due to the constituents of PAHs. Many other authors initially suggest the need of oxygen to degrade unless nitrate is replaced by the terminal electron acceptor. However, many have provided evidence of degradation of PAHs in the presence of sulfur or methane [148]. As shown in the previous section, reductive dechlorination has been recently popularized as a viable method for treating PAHs provided that the potential for developing VC is contained. Nevertheless, native soil microbial species should be considered when attempting to bioaugment. The reduction of contaminants within the soils may require the application of additional substances such as lignocellulosic substances (LS) to foster growth. This might present challenges in fostering healthy microbial groups [147].

7.3 Additional Compounds

There are three additional contaminants that are important to consider concerning biodegradation of chemicals—Figures 5.10a, b and 5.11 provide diagrams of various other major organic compounds present. First, alkanes are a series of hydrocarbons with a formula of C_nH_{2n+2}. Alkane degradation is dependent on their molecular weight—the higher molecular weight contaminants experience a lower degradability because of hydrophobic characteristics that causes them to be less soluble. In addition, many alkanes are branched [149]. The degradation of an alkane can been by considering the degradation pathway of an n-alkane—first, n-alkanes convert into alcohols, then aldehydes and fatty acids, eventually

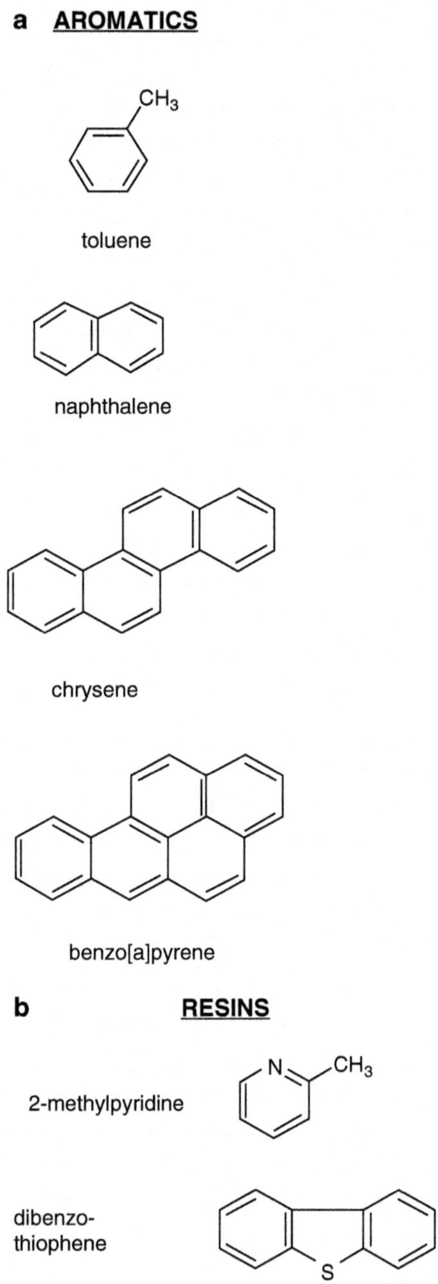

Fig. 5.10 Examples of crude oil organic compounds [37]. (**a**) Aromatics. (**b**) Resins

becoming carbon dioxide. Recently, alkanes have become more of an issue because of recent oil spills. Many petroleum hydrocarbons present during these spills include various alkanes such as normal alkanes (n-alkanes) and cyclic alkanes

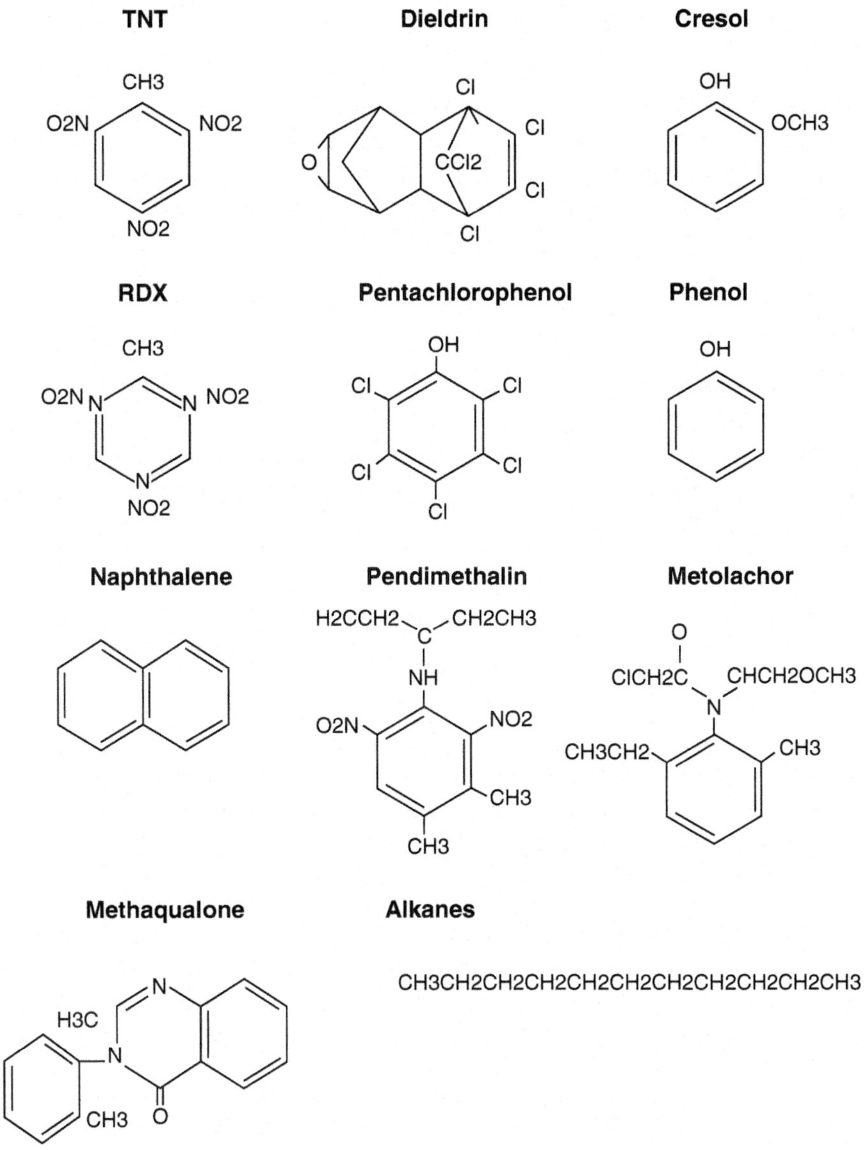

Fig. 5.11 Other organic compounds found treated within bioremediation [152]

(c-alkanes) [150]. Second, alkenes are a series of double bonded hydrocarbons with the formula C_nH_{2n} [149]. Third, aromatics such as benzene, toluene, ethylbenzene, and xylenes (BTEX) are a series of components found within fuels, gasoline production, leaking underground storage tanks, pipelines, and accidental spills [26, 151]. Generally, aromatics are soluble and hazardous components [149].

Degradation by bioaugmentation for each of the aformentioned components begins with the conversion of the compound into fatty acids by the enzymes in the bacterial cell. For alkane hydrocarbons, alkane-oxidizing enzymes are found on *Pseudomonas putida* Gpo1 with ten genes available. Each of the ten genes is able to perform the proper destruction of alkane constituent [29]. On the other hand, unstable BTEX compounds are degraded by opening the aromatic ring [20]. In addition, anaerobic degradation is the best way to mineralize BTEX without the production of nitrates, sulfates, or an additional organics during intermediate steps [153].

8 Case Studies

The following is a series of case studies that applies several case studies involving bioaugmentation. Each of these studies considers the application to wastewater treatment, in situ, groundwater and soil application. These case studies are an addendum to what has been previously discussed earlier in this chapter.

8.1 Full-Scale Coke Wastewater

Coke from steel mills consists of high concentrations of ammonia, phenols, and cyanides. The area of concern for coke wastewater is the pre-denitrification processes that occur under anoxic conditions. This is a process where heterotrophic bacteria convert nitrate and nitrite into nitrogen gas. For this process a high concentration of cyanide is formed. Therefore, treatment is required for effluent to be treated prior to discharge into the environment. Bioaugmentation reduces the potential problems from the effluent [101].

In order to remove cyanide from the wastewater, the authors use *Cryptococcus humicolus* MCN2, a form of cyanide-degrading yeast. Yeast and an unknown species of cyanide-degrading bacteria are applied to the system. According to the research of Park et al. [101], bioaugmentation research is a four step process. First, bacteria and *C. humicolus* are placed in a 5-L fermenter at 25 °C with a maximum containment of 3 L. Second, cultivated material is transferred into 12 tanks with a capacity of 100 L made of polyethylene and cultivated for 7 weeks, where batch cultures were supplied weekly. Third, 1.2 m^3 of the cultures are transferred into four aeration tanks and placed into the system as a batch for 29 days, supplying the batch with nutrients—1 kg/m^3 glucose, 50 g/m^3 KCN, and other nutrients. Finally, the waste treatment process supplies the system with recycled wastewater at a flow rate of 60 m^3/h, a sludge rate of 30 m^3/h, and an inflow 10 m^3/h. Wastewater flow provided into the system consisted of 14 mg/L concentration of cyanide [101].

The authors report that cyanide increase the values of 6 mg/L within the first 6 days of operation and reduce activated sludge to 400 mg/L due to poor settling within the tank. In addition, the study reports that the failure of cyanide reduction within coke wastewater treatment is related to the inability of biodegraded free cyanide within the system. This is because cyanide forms a stable component with iron, a very prominent form of cyanide within the system. A major problem with iron cyanide is the high toxicity of microorganisms. According to research, cyanide is reduced through a carbon source with the system. When the system receives a carbon source, cyanide removal is improved to 78 %, twice the efficiency without carbon presence. This research is separated from the analysis made within the full-scale treatment done in 48 h batch experiments with 20 mg/L of free and nickel cyanides, 100 mg/L cobalt cyanide, and 3 g/L glucose [101].

8.2 Reduction of Polycyclic Aromatic Hydrocarbons

Herwijnen et al. [148] considered this application to a given value of contaminated soil with inoculated bacteria growing in restricted conditions, a kind of culture known as a chemostat culture. The study conducted by the authors considered bacteria cultures from the genus *Sphingomonas* and *Mycobacterium* for their ability to grow and convert limited carbon sources, specifically flourene, phenanthrene, and pyrene. Cultures were supplied with either a combination of easily utilized carbon and PAH. The conclusion provided from the experiment indicates that the removal efficiency of the aforementioned carbon sources recorded 99 % [148].

In addition, Harvey and Thurston [144] consider the application of ligninolytic fungi for practical use in reducing polyaromatic hydrocarbons. The purpose of using ligninolytic fungi as an alternative to bacteria is because the organism has a mechanism which can provide ability to perform the first step of PAH degradation [144].

8.3 Exxon Valdez

In March 1989, 41 million liters of crude oil spilled about 800 km of shoreline in Prince William Sound, Alaska. This infamous oil spill became widely known as *Exxon Valdez*. Following the major incident, many scientists became interested in the degradation of these toxins and how they affected not only the soils, but also their overall effect on the wildlife within the region [154]. Figure 5.12 is a graphic depicting the location of the spill.

According to research, this developed an increase of toxicity within these beach sediments. However, when soil samples were taken in 1992, only 0.2 % of soils across a 4800 km region remained within the soil, which during the accident indicated a measurement of 16 %. In addition, 8 % of quadrants sampled, ¾ of the samples were reduced. According to the authors, this was due to geological conditions with the area—the section of Prince William Sound consisted of boulder

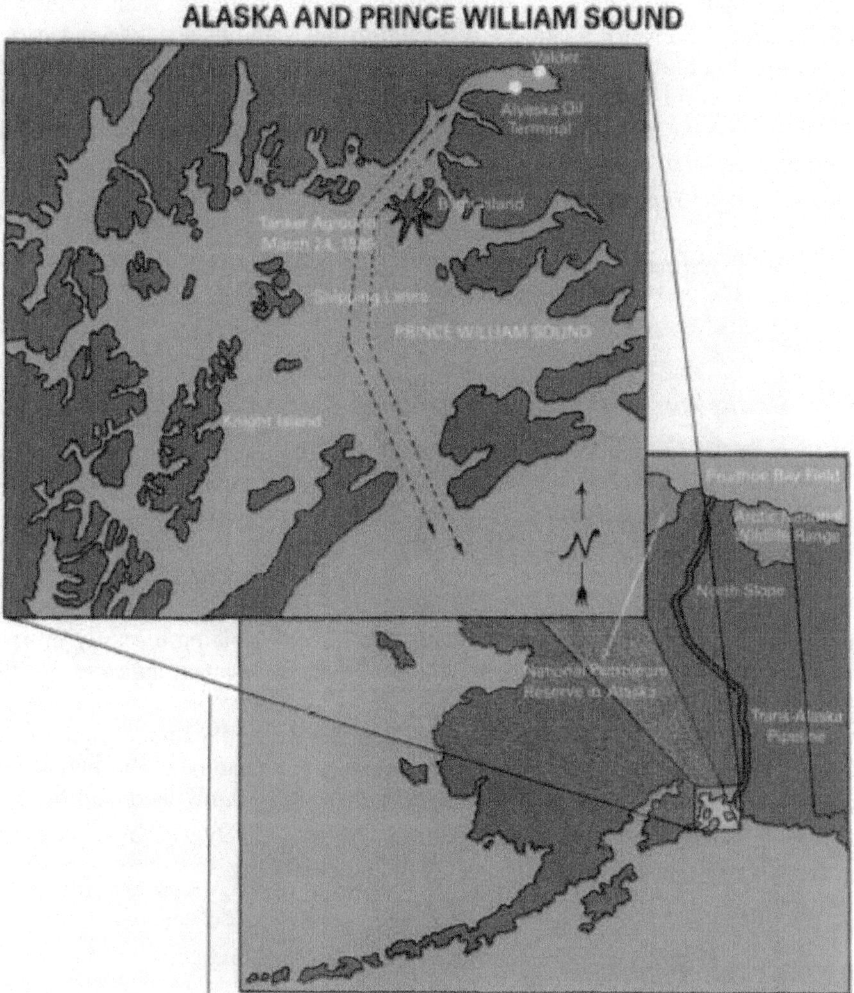

Fig. 5.12 Map of Alaska and Prince William Sound, site of Exxon Valdez oil spill [155]

and cobble shorelines, reducing any erosion and weathering. Finally, additional measurements concluded that in regions with mussel concentrations noticed that "a fraction of 1 % [of hydrocarbons] were readily available. This is due to the fact that mussels or bivalves are able to concentrate contaminants in their tissues which ultimately reduced the concentration of contaminants in the soils [154]."

8.4 Fowler Beach, DE

In response to Exxon Valdez, an experiment was conducted to further understand the combination of inorganics and bioaugmentation in its response to removing contaminants on mixed sand and gravel beaches. The authors experiment on Fowler Beach, Delaware used 15 plots with an area of 36 m^2 on crude oil from Nigeria. The four plots have four treatment conditions—nutrients only, water soluble nutrients, a combination of nutrients and bacteria which was a mix of Delaware Sea water, crude oil, and nutrients in a 55-gal drum, and an untreated plot. Sea water is applied daily, while microbes are applied once a week [156].

The results conclude that nutrients show a more consistent reduction of PAHs and alkanes, specifically linear reduction. But inocculum of bacteria reduce components similar to the non-treated plots. The difference noticed is in the path taken, which means contaminants take longer to degrade within soils. While natural degradation provides a similar reduction at the end of the experiment, the increase can have an impact on reducing hydrocarbons. However, the experiment indicates possible explanations which could affect bacterial maintenance within the system. First, environmental conditions could have a profound effect on cultures which are developed in the laboratory, or natural lag growth at specific times could retard or maximize the possible amount of contaminant reduction to take place by microorganisms [156].

8.5 Benzene, Toluene, Ethylbenzene, Xylene

From the case study, BTEX and additional hydrocarbons were found at a depth of 5 ft. Within the case study, 27 inoculation wells, both shallow and deep, supplied oxygen by a compressor. Inside of the wells, microorganisms from the *Bacillus*, *Pseudomonas*, *Serratia*, *Azobacter* genre are applied to the system. The experiment considered the effectiveness of the innocula in removing BTEX within the soils. The experiment lasted 15 months [157].

Throughout the duration of the experiment, data collection is completed in two phases: a first 6-month observation, and then a post-6 month term which the wells are developed from high yields of water that are produced. Groundwater results did not prove a reduction in their hydrocarbon concentration during the 15 year project period. In fact, research indicates that the population introduced to the well did not survive the experiment, which has been due to the lack of food availability or a potential toxic concentration [157].

8.6 *P. chrysosporium ATCC 24725*

Following successful laboratory experiments which are tests conducted for 3 weeks with soil from timber treatment indicated an 80 % reduction in chlorophenol in 15 weeks of research. Therefore, research is applied to a contaminated site in Finland, in which fungal inoculation is produced and then applied to the contamination. Four treatment beds with a volume of 3000 m^3 are excavated and mixed with the inoculated material and surround within a high density polyethylene (HDPE) line. Additional oxygen is provided by blowers by means of a series of perforated pipes. Ten soils are taken and analyzed and the results conclude that there is a reduction of chlorophenols over the period of 24 months of observation, which is lower than the 0.02 mg/kg reduction limit as required by the maximum contaminant limit for chlorophenol in Finland [158].

8.7 *Bioremediation of Petroleum-Contaminated Soil*

Pearce et al. [159] provided a study considering bioremediation of contaminated soils 50 m from a stream, received the application of a 500-L solution of bacteria every 2 weeks. From the study, Pearce recorded reduction of TPHC (total petroleum hydrocarbon levels) by 90 % during a 169 day time period, a result that showed the effectiveness of bioaugmentation [159].

8.8 *Pseudomonas stutzeri KC*

Pseudomonas stutzeri KC is a denitrifying bacteria which is used for the purpose of reducing carbon tetrachloride into carbon dioxide. Dybas et al. [160, 161] developed an in situ *bioremediation* study considering the feasibility of bioaugmentation based on two factors—to maintain and adjust pH for the formation and transfer of carbon tetrachloride; and to compare the laboratory-cultured species with the indigenous populations already placed within the soils in order to refute arguments suggesting bioaugmented species of bacteria can be successful in contaminant removal in both field and laboratory studies [160, 161].

The site chosen for carbon tetrachloride reduction is found at an aquifer in Schoolcraft, Michigan. At the site, the contaminant is located 12–23 m below the ground. Two studies are compiled at the site. The original study consisted of a laboratory study which consists of a 99 % purity of carbon tetrachloride sample. This sample is analyzed using batch and column studies [160].

Within the groundwater samples, *Pseudomonas stutzeri KC* strains are added at the initial pH of 8.2 with 100 mM sodium carbonate and sterile CO_2 gas. KC strains are analyzed everyday for 1–4 day intervals and then at 14 days, pH adjustment was made to 7.5 by carbon dioxide addition and monitored for 7 more days. In addition, nine columns were generated to analyze the ability of the strain to be able to remove carbon tetrachloride. The columns consisted of the following constituents: for 71 days 20–30 μg/L of groundwater is exchanged in the system. Next, supplements of 0.6 mM NaOH, 0.6 mM acetate, 0.1 mM potassium phosphate exchanged with the water. A second series of columns receive 4.7×10^{10} strain KC cells in 1 mL of 100 mM potassium phosphate buffer. Finally, a set of columns receive 200 mg/L thimersol. The result indicates that those received during the exchange inoculated KC strains remove the most μg of carbon tetrafloride (4.00 μg) as compared to those with thimersol (0.80) [160].

Following this experiment, the authors went to the same site and considered an in situ application within the aquifer site. There are six phases in which the study was conducted. Pre-examination parameters included proper assessment of the site and groundwater core sampling. Phase A was prepared by the adjusting of pH to 7.5–8.0. This was achieved by pumping ground water and providing acetate, phosphorus, and sodium hydroxide. Phase B adds the supply of microorganisms. Phase C applied ground water and the strand of *Pseudomonas stutzeri KC* cells, while Phase D adds additional acetate, phosphorus, and other nutrients into the system. Following experimentation, post-testing was completed to determine the success of contamination reduction [161].

The results from the study indicated a 60 % reduction of carbon tetrachloride in the system. However, the authors explained that there is a decrease in efficiency whenever acetate is lost in the system. Also, it was found that carbon tetrachloride was reduced 50 % for every 70 % of nitration removed [161]. Interesting to note, from this historical location, Tenney et al. [162] applied a three-dimension computer model based on historical data to design methods to completely degrade constituents—these include pumping water and well dimensions [162].

8.9 *Burkholeria cerpacia* PR_{301}

Burkholeria cepacia PR_{301} are microorganisms applied in situations when indigenous populations are unable to use constituents for growth purposes. In addition, the microorganism has the ability to transform oxygen from its molecular form for degradation by means of oxygenase enzyme. The following case study is conducted with the use of these microorganisms for that particular purpose [163].

The city of Wichita, KS found evidence of CAH (chlorinated aliphatic hydrocarbons)—TCE, DCE, and PAC approximately 4.5 m below the surface. The initial concentrations found were 125 μg/L TCE, and 99 μg/L for two forms of DCE. Therefore, research was conducted by constructing one 15-cm injection well and 5-cm PVC piping installed within the system. The purpose of installing the piping was to receive dissolved oxygen of 15 mg/L, nutrients, and *Burkholderia cepacia*. Additional piping provided six monitoring wells for making observations throughout the experiment [163].

Testing was completed in six phases—specific application of biotreatment is done after approximately 30 days from the initial field test construction. The first application continuously applies microorganisms to the system for 4 days, where concentration of cells reaches 10^9 cells/mL. The results indicated that TCE and t-DCE levels were near 0 μg/L—however, the initial conditions were less than 25. Nevertheless, c-DCE levels dropped immediately within the first day of application (150 to <25 μg/L). According to research it was noted that the reduction rate was approximately 94.5 μg/L h [163].

Following a 22 day recovery phase, c-DCE increased significant from concentration <25 μg/L to approximately 240 μg/L. When the application of the second biotreatment occurred, the concentration returned to approximately 150 μg/L. The second biotreatment consisted of the application of 10^8 cells/mL reducing c-DCE contaminants to below detection levels. In addition, TCE and t-DCE contaminants approached zero in concentration within the system [163].

Final analysis were conducted during a 15 day phase known as a post-test. This test reduced the number of microorganisms within the system to determine the amount of contaminants remaining in the system. With the reduction of microorganisms, contaminants increased significantly for c-DCE, reaching a concentration of approximately 170 μg/L [163].

8.10 Chermaya Oil Pollution Restoration

The Chernaya River in the Moscow region known as Lukhovitsy was polluted in 1995. The cause for the pollution was from the proximity of the Lukhovitsy oil bulk plant to the river which received leakage from the plant. During the year, oil thickens to approximately 5 mm in the river tributary, which therefore required a reduction of the amount of oil present within the system [164].

A series of microorganisms known as *Rhodococcus* sp. were applied to the site because of their ability to degrade oil within a system. An application of the microorganisms was prepared in a salt solution consisting of nitrogen, phosphorus, and potassium with concentrations ranging from 0.005 to 0.02 %. The mixture was then sprinkled onto a 1 m^2 surface area twice with a break period of 14 days. The authors discovered that the hydrocarbon concentration significantly decreased from 440 to 0.04 mg/mL, a reduction of 99.001 %. In addition, cations and anions such as Ca^{+2}, Mg^{+2}, Cl^-, and Na^+ also experienced major reduction in their concentration during their application [164].

9 Cost

As engineers solve problems involving soil and groundwater remediation, it becomes imperative to discuss the parameters associated with cost. This is important since the effectiveness of any process treatment would rely on an investment of funds. According to the American Society of Civil Engineers, when it comes to developing cost analysis, there are two major considerations that must be applied to the system—capital investments which are costs associated with the design, equipment, setup, startup, and removal of structures, or all associated costs with the design process of the system. This contrasts to expenses which are a series of costs associated with entities such as operation, maintenance, and additional building expenses [165]. Typical analyses are done in ratio by relating old project costs to new projects. The following equation displays this relationship:

$$\mathrm{Cost_n}/\mathrm{Cost_o} = (\mathrm{Size_n}/\mathrm{Size_o})^{\mathrm{exp}} \tag{5.12}$$

where

$\mathrm{Cost_n}$ and $\mathrm{Cost_o}$ represent the costs for both new and old projects respectively
$\mathrm{Size_n}$ and $\mathrm{Size_o}$ represent the size of both old and new projects
Exp = additional equipment required aboveground (typically 0.7)

In regards to bioaugmentation, associated costs for design may seem impractical because remediation does not operate through many mechanisms. However, as seen in previous case studies, equipment such as blowers are not used for the diffusion of oxygen within the system. Additional costs could occur with other forms of containers.

Additional costs as previously mentioned, include primary equipment. There are two typical methods applied for the analysis of cost. The first method, known as the Lang method, is the sum of the equipment used in the system with an application of a factor between 3.1 and 4.74 following the summation of all the equipment. Another method, known as the Hand method, applies the Hand factor which is used for the overall costs of each piece of equipment used. For cost related to engineering design, typical applications consider a fraction of the overall construction cost present for the given situation, which may be duly noted as reporting higher value than costs for typical design projects [165].

Expenses for entities such as utilities, operation, maintenance, and additional expenses, are calculated using different rubrics. Typical utility expenses for bioaugmentation involve the use of electricity to power blowers; therefore one must calculate the total amount of power expenses for blowers each year. One can suspect that the estimated costs for maintenance can be approximately 4 % of all costs invested within the project. Finally, additional costs calculated for the purpose of operation can be completed based upon information provided or assigned a dollar amount strictly for non-calculated costs [165]. To summarize, bioremediation

Table 5.10 Cost factors affecting bioremediation treatment efficiency [166]

Cost factor	Comments
Concentration of contaminants	Higher concentrations increase the time needed for treatment.
Presence of higher-molecular-weight organics	Higher molecular-weight organics (e.g., PAHs) tend to increase treatment time.
Area or volume requiring treatment	Increased area or contaminated volume requires increased capital expenditures. However, unit costs typically decrease with increased volume to be remediated.
Depth of contamination	Deeper contamination increases the amount of well drilling required for bioventing or excavation requirements for ex situ applications.
Complex geology	Complex interbedding of high and low permeability layers can be difficult to treat, possibly increasing the density of wells needed or the length of treatment time for bioventing or increased mixing requirements and treatment time for ex situ applications.
Low soil permeability	Treating low permeability soils decreases the radius of influence of in situ technologies, requiring an increase in the treatment density for in situ treatment.
Presence of recalcitrant contaminants	Relatively recalcitrant contaminants may require increased treatment time or alternative treatment strategies, or may preclude biological treatment entirely.
Presence of halogenated organics	Halogenated contaminants may require anaerobic pretreatment to reduce the level of chlorination, and may require increased off-gas control measures.

processes such as bioaugmentation are the least expensive of all land remediation techniques. Table 5.10 discusses some of the factors as to why expenses could be increased for treatment for example of contaminated soils.

In situ bioremediation costs have varied across the board. According to the Environmental Protection Agency, 22 bioremediation processes for both in situ groundwater and in situ and ex situ soil are categorized for the treatment of superfund sites. It has been estimated that the costs range from $48,700 to 26,810,000, where the cost differentiation has been contingent upon the amount of soil that is needed to be treated and also the type of treatment process that has been used. For example, the $26 million site in Texas, French Limited Superfund Site treats carcinogenic Volatile Organic Compounds (cVOCs), Semi Volatile Organic Compounds (SVOCs), Volatile Organic Compounds (VOCs), PAHs, and PCBs. This is compared to treating Petroleum Hydrocarbons (PHC) at Glasgow Air Base UST which costs only $60,000 [19]. Table 5.11 provides a summary of various treatment techniques of Superfund sites that used bioremediation.

Table 5.11 Examples of sites treated using bioremediation [19]

Site name	Cleanup program	Treatment application	State	Status	Contaminants	Start year	Technology cost ($) (source)	Volume treated (cy)	Unit cost ($/cy)	Comments
Harve Air Force Station, Remove Abandoned USTs	Superfund	Ex situ Bioremediation (Soil)-Land treatment	MT	Complete	PHC	1992	$48,700	1786	27.3	Soil plowing and tilling
Brown Wood Preserving Superfund Site	Superfund	Ex Situ (Soil)-Land Treatment	FL	Complete	PAHs	1989	$635,000	8100	78.4	Constructed lined treatment; contaminant concentrations
Dubose Oil Products Co. Superfund Site	Superfund	Ex Situ (Soil)-Land Treatment	FL	Complete	BTEX, cVOCs, Other SVOCs, Other VOCs	1993	$4,990,000	13,137	380	Composting treatment system with leachate collection, inoculant generation, vacuum extraction, wastewater treatment
French Limited Superfund Site	Superfund	Ex Situ (Soil)-Slurry Phase	TX	Complete	cVOCs, Other SVOCs, Other VOCs, PAHs, PCBs	1992	$26,810,000	300,000	89.4	Extremely large volume
Bonneville Power Administration Superfund Site	Superfund	Ex Situ (Soil)-Solid Phase	WA	Complete	PAHs, Other SVOCs	1995	$1,280,000	1048	1220	Included extensive technology demonstration activities
Dover AFB, Area 6	Superfund	In Situ Bioremediation (Soil)-Bioventing	DE	Complete	cVOCs, Heavy metals	1996	$551,000	1667	331	Direct injection of air and propane; cometabolic aerobic; pilot test

(continued)

Table 5.11 (continued)

Site name	Cleanup program	Treatment application	State	Status	Contaminants	Start year	Technology cost ($) (source)	Volume treated (cy)	Unit cost ($/cy)	Comments
Edwards AFB	Superfund	In Situ Bioremediation (Groundwater)	CA	Complete	cVOcs	1995	$445,000	1517	293	Recirculation between two aquifer systems; aerobic
Pinellas Northeast Site, Anaerobic Bioremediation	RCRA CA	In Situ Bioremediation (Groundwater)	FL	Complete	cVOCs	1997	$359,000	1238	290	Recirculation with the addition of benzoate, lactate, and methanol; anaerobic; intended to supplement pump-and-treat system

10 Problems with Bioaugmentation

Bioaugmentation has been notorious as exemplary for the purpose of assisting in the removal of contaminants from the soils and groundwater. Nevertheless, there lies a very difficult challenge in regards to the proper use of the treatment process. Skeptics argue that bioaugmentation is not permissible due to the ecology considered. Devinny and Chang quote van Veen et al.—"Introduced bacteria may therefore have great difficulties in finding soil itches in which to survive…and it is extremely hard to predict the final effects." Many of the approving literature supporting the use of bioaugmentation have been merely applied to studies which have been conducted in a laboratory [7].

For example, Taylor and Hanna conducted laboratory research involving the methanotrophic *Methylosinus trichosporium*, a degrader for chlorinated ethenes, for the purpose of preparation for a treatability study in the vicinity of NASA Kennedy Space Center. The study concluded that removal reached 70 % *cis*-DCE and 51 % TCE removal in a 72 h time period [167]. Aamand et al. considered *Arthrobacter* in the application of treating collected soils from a gasworks site in Esbjerg, Denmark. The study determined that in one tar sample presented, 65 % of ^{14}C-phenathrene is mineralized to CO_2 within a month [168]. Imamura et al. concluded that an inducer-free microbe JM1 exponentially reduced TCE over a period of several hours as compared to samples without JM1 at various concentrations [169]. While these experiments indicate great examples of effectiveness of cultures within the laboratory, as seen throughout the text, many of the bioaugmentation studies either do not have numerous examples or the results indicate negligible differences between what is seen by microorganisms present in the system.

Four reasons for failure have been explained by authors Devinny and Chang [9]. The importance of environmental conditions is established throughout the course of the text and also case study research. Factors such as temperature, pH, the amount of water present within the soil, and the concentration of oxygen present must be developed appropriately for each of the various microorganisms. The authors also argued that the effectiveness of laboratory research as previously prevented indicates the success rate of simply reducing the concentration without having designing the appropriate conditions necessary. In addition, many of the cultures which are applied for soil bioremediation are grown in liquid, as compared to those bacteria already grown accustomed to the soils [9].

In relation to the environmental conditions, the economic viability is another consideration. Beyond the environmental conditions that are typical of soil and ground applications, additional parameters within the soil include predators, competitors, and parasites. Reviewing studies compiled by previous authors, inoculated species of microorganism applied in situ are reported dying in situations when the conditions were not sterile. This is exemplified when the microorganism *Pseudomonas* had difficulty in the removal of dichlorophenol. Other examples in literature include microbial species, those unable to adapt to the current situations available [9].

Also, any type of degradation of contaminant within the soil is contingent upon the substrate amount. Whenever levels do not reach acceptable amounts, microorganisms depend upon the use of that specific type of contaminant may be unable to grow. Also, the authors mention that there are situations when microorganisms degrade materials which are not the intended contaminants, or are unable to complete the process without the presence of other contaminants. In addition, many of the prevailing problems include the structural complexity of the contaminant which may require the application of additional species in order to complete degradation. The amount of concentration affect the microorganisms ability to due toxicity which may occur at high levels [9].

Finally, the contaminants may not be unable to reach the microorganisms due to the inability for dispersal within media such as soils. There are cases when soils are thickened and tightly hold onto contaminants which become a challenge for microorganisms to penetrate through. This is be best explained by poor mixture of soils. These reasons create a setback in regards to the current structure in bioaugmentation. However, the acceptance of drawbacks to this treatment should not deter future research towards developing solutions to these common problems [9].

11 Ecological Risk Assessment

Ecological risk assessment is generated whenever there is a presence of a pathway concerning potential contamination concentration. There are three major steps within ecological risk assessment—first, the formation of problem, which is a prepared hypothesis examining cause and potential affects from human activities affecting the ecology of the system in question. Second, an analysis is made considering the problems created and the potential hazards formed. Finally, the situation is evaluated based upon risk, or the probability of the event causing ecological damage to the system [170].

Prior to understand ecological risk assessment, one must consider the term toxicity. Toxicity is the consideration of both exposure and dosage. Exposure is defined as the concentration of material either in the air, water, or soil which makes contact with a given organism, while dosage is the frequency that the organism is exposed to the contaminant including the amount. With toxicity, it becomes imperative that ecological risk assessment is completed in its totality. There are two major types of toxicity that are considered within bioremediation—human and animal toxicity and ecotoxicology [1].

Human and animal toxicity analyses are made on the consideration of mortality. First, acute toxicity is the amount of contaminant exposure over a brief period time which causes death to the organism. Chronic toxicity is the contact between the microorganism and the contaminant for a longer period of time. This exposure is typically a smaller dosage which causes adverse health effects over a long period of time, which is more realistic and more efforts are applied by those in the study of bioaugmentation and additional bioremediation techniques [1].

An exemplary case study for ecological risk assessment involved the Exxon Valdez spill in Alaska 1989. For the purposes of soil contamination, once one considers the concentration of materials presense, it will be compared with standards of what is a non-hazardous level of the contaminants. Known as benchmarks, this system of classification is prevalent to the end purpose of assessed land, considers the levels within the system, and provides an accepted level of soils not tested for contamination to generate the appropriate responses from those soils. For example, in tests involving petroleum hydrocarbons, the American Petroleum Institute stated that concentrations are not severe if they are less than 10,000 mg/kg of grease or other total petroleum hydrocarbons [170].

In addition, Efroymson et al. [171] applied the notions described using ecological risk assessment to make analytical conclusions regarding the ecotoxicology of petroleum products. According to the authors, screening benchmarks are adapted based upon the comparison between concentrations that are deemed hazardous as compared to those which are not hazardous. Also known as ecological soil screening levels, benchmarks are produced on three major factors: to conclude whether the specific amount of material affects the overall production of given ecology—for example the ability to reproduce the given benchmark and proper merging of soils and microorganisms to develop a specific result. Additional parameters necessary for measurement include the type of soil and microorganisms, time in which soil has been in contact with the material, and constituents comprising the petroleum. The purpose of completing an ecological risk assessment based upon screening benchmarks is to provide an overall sense of the petroleum hydrocarbon components present in a given soil and also assess whether concentrations are low enough to present adverse effects on the overall soil. This can ultimately determine whether a site has the ability to fulfill requirements prior to site closure, as outlined by RCRA. For the purpose of bioaugmentation, making benchmarks can determine the toxicity of current organisms present within a given soil and therefore could suggest whether those toxic sites require bioaugmentation treatment [171].

It is very important to keep various experiments and analysis techniques uniform because toxicity levels can vary for many of the different contaminants. Also, a significant difference between measurements can have a profound effect over whether application of bioaugmented microorganisms is going to be successful within a system [171]. Nevertheless, considering proper methodology of the ecological risk assessment can allow for an understanding of the potential problems and develop possible solutions using bioaugmentation.

12 Conclusions

A proper analysis of bioaugmentation techniques as applied to various media allows for the versatility of the application for different purposes for waste treatment. The challenges facing current supporters of this treatment continue to be its inability to be able to withstand harsh conditions outside of sterile laboratories. Nevertheless, as Forsyth et al. explained, bioaugmentation does have a sense of security; after all, it only requires the application of microorganisms which have been completing degradation work for millions of years [5].

12.1 Recommendations for Improving Bioaugmentation

In response to many critics of bioaugmentation, it should be warranted that the development of laboratory cultures for soil, groundwater, and wastewater treatment applications can be effective if certain measures are taken. First, proper analysis of the media, such as an ecological risk assessment study can be completed determining specific benchmarks for contaminants. If benchmarks are applied properly, then one can determine which species are able to withstand concentrations of a given contaminant. Second, proper maintenance of media in which microorganisms will be contained can provide further life expectancy of microorganisms within the system. Third, use of bioaugmentation in conjunction with other practices, such as physical-chemical practices or phytoremediation can allow for contaminant concentration to be suitable for microorganism presence and also relieve stressors placed on the applied species. Finally, continual trial-and-error in situ practices can develop better understanding of microorganisms within the soil. If bioaugmentation successes and failures are consistently based only on laboratory applications, then continual criticism will arise of the technique outside of the laboratory.

Glossary of Terms Related to Bioaugmentation

Bioaugmentation A subset of biotechnology that consists of supplying microorganisms to the soil or water for the sole purpose of removing a given contaminant.

Biodegradation The mineralization of compounds into inorganic minerals, carbon dioxide, water, and methane.

Bioremediation The removal of contaminants from the environment by using biological processes.

Biotransformation A change of a contaminant's chemical structure to reduce potential human or ecological consequences related to its presence.

Catabolism The generation of new chemical structures or the maintenance of energy by means of using enzyme-catalyzed reactions.

Ecological risk assessment Prepared whenever there is a presence of a pathway or a potential contaminant concentration.

Ex situ Any treatment process that requires the extraction of the contaminated medium for further treatment.

Halo-organic compound An organic compound consisting of a halogen. Halogens are a series of elements located in the seventh column of the periodic table.

Hazardous Waste A substance classified as being ignitable, corrosive, reactive, or toxic.

Industrial Waste A substance produced within any given industry. These industries include leather tanning, textile production, wood preservation, metal finishing, coke, and other industrial processes that create wastewater with many harmful products. Usually, industrial waste products are harmful and can include surfactants, dyes, oils, phenols, polycylic and heterocyclic aromatic hydrocarbons, and heavy metals.

In situ Direct application of a treatment process to the contaminated area.

Metabolism The breakdown of compounds by microorganisms into inorganic minerals, water, carbon dioxide, and methane.

Pathway The degradation processes of a compound.

Polycyclic Aromatic Hydrocarbons Carcinogenic (cancer causing) chemicals produced by the incineration of organic material to produce petroleum or wood-preservation, from accidental spills or runoff from the landscape.

Toxicity The concentration of material either in the air, water, or soil which makes contact with a given organism, or the frequency that an organism is exposed to the contaminant including the amount.

References

1. Philip, J. C., Banforth, S. M., Singleton, I., & Atlas, R. M. (2005). Environmental pollution and restoration: A role of bioremediation. In R. M. Atlas & J. Philip (Eds.), *Bioremediation: Applied microbial solutions for real-world environmental cleanup*. Washington, DC: ASM Press.
2. US Environmental Protection Authority. (2005). *Soil bioremediation* (EPA 589/05). http://www.epa.sa.gov.au/xstd_files/Site%20contamination/Guideline/guide_soil.pdf. (6 April 2012).

3. Singh, A., & Ward, O. P. (2004). Biotechnology and bioremediation—An overview. In A. Singh & O. P. Ward (Eds.), *Biodegradation and bioremediation*. Berlin, Germany: Springer.
4. Scragg, A. (2005). *Environmental biotechnology* (2nd ed.). New York: Oxford University Press.
5. Forsynth, J. V., Tsao, M. Y., & Bleam, R. D. (1995). Bioaugmentation: When is augmentation needed? In R. E. Hinchee, J. Fredrickson, & B. C. Alleman (Eds.), *Bioaugmentation for site remediation*. Columbus, OH: Battelle.
6. Ronald, C. S. (1993, April). Site characterization requirements. In *Bioremediation of hazardous waste sites: Practical applications to implementation* (EPA/600/K-93/002). Washington, DC.
7. Wackett, L. P., & Hershberger, C. D. (2001). *Biocatalysis and biodegradation: Microbial transformation of organic compounds*. Washington, DC: ASM Press.
8. Vogel, T. M, & Walter, M. V. (2002). Bioaugmentation. In *Manual of environmental microbiology* (2nd ed.). Washington, DC: ASM Press.
9. Devinny, J. S., & Chang, S. H. (2000). Bioaugmentation for soil remediation. In D. L. Wise, D. J. Trantolo, E. J. Cichon, H. I. Inyang, & U. Stottmeister (Eds.), *Bioremediation of contaminated soils*. New York: Marcel Dekker.
10. Flathman, P. (1993, April). Background on bioremediation. In *Bioremediation of hazardous waste sites: Practical applications to implementation* (EPA/600/K-93/002). Washington, DC.
11. De Rosa, C., & Hans, H. (2003). The impact of 20 years of risk assessment on public health. *Human and Ecological Risk Assessment, 9*(5), 1219–1228.
12. Lupo, T. D., & Harris, L. L. (1998). How the law drives site cleanups. In S. K. Sikdar & R. L. Irvine (Eds.), *Bioremediation: Principles and practice, volume I: Fundamentals and applications*. Lancaster, England: Technomic.
13. US Environmental Protection Agency. (2012). Toxic Substances Act: Summary of TSCA: Law, regulations, guidance, and dockets. http://www.epa.gov/lawsregs/laws/tsca.html. (7 April 2012).
14. US Environmental Protection Agency. (2011). CERCLA overview. http://www.epa.gov/superfund/policy/cercla.htm. (7 April 2012).
15. US Environmental Protection Agency. (2011). SARA overview. http://www.epa.gov/superfund/policy/sara.htm. (2 April 2012).
16. US Environmental Protection Agency. (1990, January). *Handbook of* in situ *treatment of hazardous waste-contaminated sites* (EPA/540/2-90/002). Cincinnati, OH.
17. US Environmental Protection Agency. (2011, March). *Beneficial effects of the superfund program* (EPA EP-W-07-037). Washington, DC.
18. US Environmental Protection Agency. (2012). Introduction to hazard ranking system. http://www.epa.gov/superfund/programs/npl_hrs/hrsint.htm. (2 April 2012).
19. US Environmental Protection Agency. (2001, September). *Use of bioremediation at superfund sites* (EPA 542-R-01-019). Washington, DC.
20. Aitken, M. D., & Long, T. C. (2004). Biotransformation, biodegradation, and bioremediation of polycyclic aromatic hydrocarbons. In A. Singh & O. P. Ward (Eds.), *Biodegradation and bioremediation*. Berlin, Germany: Springer.
21. Flathman, P. (1993, April) Background on bioremediation. In *Bioremediation of hazardous waste sites: Practical applications to implementation* (EPA/600/K-93/002). Washington, DC.
22. Nyer, E. K., & Boettcher, G. (2001). In situ bioremediation (chap. 11). In E. K. Nyer, P. L. Palmer, E. P. Carman, G. Boettcher, J. M. Bedessem, F. Lenzo, et al. (Eds.), *In situ treatment technology* (2nd ed.). Boca Raton, FL: CRC Press.
23. Environmental Protection Agency. (2012). *A citizen's guide to bioremediation* (EPA 542-F-96-007). http://www.epa.gov/tio/download/citizens/bioremediation.pdf. (6 April 2012).

24. Otten, A. M., Alphenaar, A., Pijls, C., Spuij, C., & De Wit, H. (1997). *In situ soil remediation*. Norwell, MA: Kluwer.
25. Mahro, B., Muller, R., & Kasche, V. (2001). Bioavailability—The key factor of soil bioremediation. In R. Stegmann, G. Brunner, W. Clamnzo, & G. Matz (Eds.), *Treatment of contaminated soil: Fundamentals analysis, application*. Berlin, Germany: Springer.
26. Reddy, C. A., & Matthew, Z. (2001). Bioremediation potential of white fungi. In G. M. Gadd (Ed.), *Fungi in bioremediation*. Cambridge, England: Cambridge University Press.
27. Cassidy, D., & Irvine, R. (1998). Interactions between organic contaminants and soil affecting bioavailability. In S. K. Sikdar & R. L. Irvine (Eds.), *Bioremediation: Principles and practice, volume I. Fundamentals and applications*. Lancaster, England: Technomic.
28. Wick, L. Y., & Springael, H. H. (2013). Bacterial strategies to improve the bioavailability of hydrophobic organic pollutants. In R. Stegmann, G. Brunner, W. Clamnzo, & G. Matz (Eds.), *Treatment of contaminate soil: Fundamentals analysis, application*. Berlin, Germany: Spring.
29. Adeel, Z., Mercer, J. W., & Faust, C. R. (2000). Models for describing multiphase flow and transport of contaminants (chap. 1). In J. J. Kaluarachchi (Ed.), *Groundwater contamination by organic pollutants: Analysis and remediation*. Reston, VA: American Society of Civil Engineers.
30. US Environmental Protection Agency. (2004, September). *Demonstration of biodegradation of DNAPL through biostimulation and bioaugmentation at Launch Complex 34 in Cape Canaveral Air Force Station, Florida* (EPA 540/R-007). Cincinnati, OH.
31. McCarty, P. L., Hopkins, G. D., Munakata-Marr, J., Mathwaon, V. C., Dolan, M. E., Dion, L. B., et al. (1998, April). *Bioaugmentation with Burkholderia cepacia PR1301 for* in situ *bioremediation of trichloroethylene contaminated groundwater* (EPA/600/S-98/001). Gulf Breeze, FL.
32. Suthersan, S. S., & Payne, F. C. (2005). *In situ remediation engineering*. Boca Raton, FL: CRC Press.
33. US Environmental Protection Agency. (2004, May). In-situ groundwater bioremediation (chap. 10). In *How to evaluate alternative cleanup technologies for underground storage tank sites: A guide for corrective action plan reviewers* (EPA 510-R-04-002), Washington, DC.
34. Frazar, C. (2000). *The bioremediation and phytoremediation of pesticide-contaminated sites*. Washington, DC: Environmental Protection Agency.
35. Stocking, A., Deeb, R., Flores, A., Stringfellow, W., Talley, J., Brownell, R., et al. (2000). Bioremediation of MTBE: A review from a practical perspective. *Biodegradation, 11*(2–3), 187–201.
36. Prince, R. (2000). Biodegradation of methyl tertiary-butyl ether (MTBE) and other fuel oxygenates. *Critical Reviews in Microbiology, 26*(3), 163–178.
37. Zhu, X., Venosa, A. D., Suidan, M. T., & Lee, K. (2001). *Guidelines for the bioremediation of marine shorelines and freshwater wetlands*. http://www2.epa.gov/sites/production/files/2013-07/documents/guidelines_for_the_bioremediation_of_marine_shorelines_and_freshwater_wetlands.pdf. (6 April 2012).
38. Stroo, H. F., & Ward, C. H. (2010). In situ *remediation of chlorinated solvent plumes*. New York: Springer. http://books.google.com/books?id=7INqhn7WQuwC&pg=PA321&dq=37.%09In+Situ+Remediation+of+Chlorinated+Solvent+Plumes&hl=en&sa=X&ei=3cZU9eMMcHy8AIIf7YGYCA&ved=0CDEQ6AEwAA#v=onepage&q=37.%09In%20Situ%20Remediation%20of%20Chlorinated%20Solvent%20Plumes&f=false. (6 June 2014).
39. Justice-Leon, S. D., Higgins, S., Mack, E. E., Griffiths, D. R., Tang, S., Edwards, E. A., et al. (2014). Bioaugmentation with distinct Dehalobacter strains achieves chloroform detoxification in microorganisms. *Environmental Science & Technology, 48*(3), 1851–1858.
40. Husserl, J., & Hughes, J. B. (2013). Biodegradation of nitroglycerin in porous media and potential for bioaugmentation with Arthrobacter sp. strain JBH1. *Chemosphere, 92*(6), 721–724.

41. Lin, C. W., Chen, L. H., I, Y. P., & Lai, C. (2010). Microbial communities and biodegradation in lab-scale BTEX-contaminated groundwater remediation using an oxygen-releasing reactive barrier. *Bioprocess and Biosystems Engineering, 33*(3), 383–391.
42. Marzorati, M., Balloi, A., de Ferra, F., Corallo, L., Carpani, G., Wittebolle, L., et al. (2010). Bacterial diversity and reductive dehalogenase redundancy in a 1,2-dichloroethane-degrading bacterial consortium enriched from a contaminated aquifer. *Microbial Cell Factories, 9*:12.
43. Vainberg, S., Condee, C. W., & Steffan, R. J. (2009). Large-scale production of bacterial consortia for remediation of chlorinated solvent-contaminated groundwater. *Journal of Industrial Microbiology and Biotechnology, 36*(9), 1189–1197.
44. Oh, B., & Alvarez, P. (2002). Hexahydro-1,3,5-trinitro-1,3,5-triazine (RDX) degradation in biologically-active iron columns. *Water, Air, & Soil Pollution, 141*(1–4), 325–335.
45. Friis, A. K., Albrechtsen, H. J., Cox, E., & Bjerg, P. L. (2006). The need for bioaugmentation after thermal treatment of a TCE-contaminated aquifer: Laboratory experiments. *Journal of Contaminant Hydrology, 88*(3–4), 235–248.
46. Friis, A., Albrechtsen, H., Heron, G., & Bjerg, P. (2005). Redox processes and release of organic matter after thermal treatment of a TCE-contaminated aquifer. *Environmental Science & Technology, 39*(15), 5787–5795.
47. Environmental Protection Agency. (1990, January). *Handbook on* in situ *treatment of hazardous waste-contaminated soils* (EPA/540/2-90/002). Cincinnati, OH.
48. Cutright, T. (1995). A feasible approach to the bioremediation of contaminated soil – From lab-scale to field-test. *Fresenius Environmental Bulletin, 4*(2), 67–73.
49. Pope, D. F., & Matthews, J. E. (1993, April). Soil treatment: Land treatment. In *Bioremediation of hazardous waste sites: Practical applications to implementation* (EPA/600/K-93/002). Washington, DC.
50. Paraiba, L. C., Queiroz, S. C. N., de Souza, D. R. C., & Saito, M. L. (2011). Risk simulation of soil contamination by polycyclic aromatic hydrocarbons from sewage sludge used as fertilizers. *Journal of the Brazilian Chemical Society, 22*(6).
51. Shailaja, S., Mohan, S. V., Krishna, M. R., & Sarma, P. N. (2008). Degradation of di-ethylhexyl phthalate (DEHP) in bioslurry phase reactor and identification of metabolites by HPLC and MS. *International Biodeterioration and Biodegradation, 62*(2), 143–152.
52. Valentin, L., Lu-Chau, T. A., Lopez, C., Feijoo, G., Moreira, M. T., & Lema, J. M. (2007). Biodegradation of dibenzothiophene, fluoranthene, pyrene and chrysene in a soil slurry reactor by the white-rot fungus Bjerkandera sp. BOS55. *Process Biochemistry, 42*(4), 641–648.
53. Gamati, S., Gosselin, C., Bergeron, E., Chenier, M., Truong, T., & Bisaillon, J. (1999). New plug flow slurry bioreactor for polycyclic aromatic hydrocarbon degradation. In *Third Americana Pan-American Environmental Technology Trade Show and Conference*, 24–26 March 1999, Montreal, Canada.
54. Cassidy, D., & Hudak, A. (2002). Microorganism selection and performance in bioslurry reactors treating PAH-contaminated soil. *Environment Technology, 23*(9), 1033–1042.
55. Lewis, R. F. (1993). Site demonstration of slurry-phase biodegradation of PAH contaminated soil. *Journal of the Air and Waste Management Association, 43*(4), 503–508.
56. Quintero, J., Moreira, M., Lema, J., & Feijoo, G. (2006). An anaerobic bioreactor allows the efficient degradation of HCH isomers in soil slurry. *Chemosphere, 63*(6), 1005–1013.
57. Partovinia, A., Naeimpoor, F., & Hejazi, P. (2010). Carbon content reduction in a model reluctant clayey soil: Slurry phase n-hexadecane bioremediation. *Journal of Hazardous Materials, 181*(1–3), 133–139.
58. Kao, C. M., Chen, S. C., Liu, J. K., & Wu, M. J. (2001). Evaluation of TCDD biodegradability under different redox conditions. *Chemosphere, 44*(6), 1447–1454.
59. Ramakrishna, M., Mohan, S., Venkata, S., Shailaja, S., Narashima, R., & Sharma, P. N. (2008). Identification of metabolites during biodegradation of pendimethalin in bioslurry reactor. *Journal of Hazardous Materials, 151*(2–3), 658–661.

60. Newcombe, D. A., & Crawford, R. L. (2007). Transformation and fate of 2,4,6-trinitrotoluene (TNT) in anaerobic bioslurry reactors under various aeration schemes: Implications for the decontamination of soils. *Biodegradation, 18*(6), 741–754.
61. Arienzo, M. (2000). Degradation of 2,4,6-trinitrotoluene in water and soil slurry utilizing a calcium peroxide compound. *Chemosphere, 40*(4), 331–337.
62. Shen, C. F., Guiot, S. R., Thiboutot, S., Ampleman, G., & Hawari, J. (1997). Fate of explosives and their metabolites in bioslurry treatment processes. *Biodegradation, 8*(5), 339–347.
63. Okuda, T., Alcantara-Garduno, M. E., Suzuki, M., Matsui, C., Kose, T., Nishijima, W., et al. (2007). Enhancement of biodegradation of oil adsorbed on fine soils in a bioslurry reactor. *Chemosphere, 68*(2), 281–286.
64. Daprato, R. C., Zhang, C. L., Spain, J. C., & Hughes, J. B. (2005). Modeling aerobic bioremediation of 2,4-dinitrotoluene in a bioslurry reactor. *Environmental Engineering Science, 22*(5), 676–688.
65. Hudak, A. J., & Cassidy, D. P. (2004). Stimulating in-soil rhamnolipid production in a bioslurry reactor by limiting nitrogen. *Biotechnology and Bioengineering, 88*(7), 861–868.
66. Cassidy, D. P., & Hudak, A. J. (2001). Microorganism selection and biosurfactant production in a continuously and periodically operated bioslurry reactor. *Journal of Hazardous Materials, 84*(2–3), 253–264.
67. Cassidy, D. R. (2001). Biological surfactant production in a biological slurry reactor treating diesel fuel contaminated soil. *Water Environment Research, 73*(1), 87–94.
68. Mohan, S. V., Sirisha, K., Rao, N. C., Sarma, P. N., & Reddy, S. J. (2004). Degradation of chlorpyrifos contaminated soil by bioslurry reactor operated in sequencing batch mode: Bioprocess monitoring. *Journal of Hazardous Materials, 116*(1–2), 39–48.
69. Kuyukina, M. S., Ivshina, I. B., Ritchkova, M. I., Philp, J. C., Cunningham, C. J., & Christofi, N. (2003). Bioremediation of crude oil-contaminated soil using slurry-phase biological treatment and land farming techniques. *Soil and Sediment Contamination, 12*(1), 85–99.
70. Nelson, C., & Brenner, R. (1993, April). Reactors for treatment of solid, liquid, and gaseous phases. In *Bioremediation of hazardous waste sites: Practical applications to implementation* (EPA/600/K-93/002). Washington, DC.
71. Lebkowska, M., Zborowska, E., Karwowska, E., Miaskiewicz-Peska, E., Muszynski, A., Tabernacka, A., et al. (2011). Bioremediation of soil polluted with fuels by sequential multiple injection of native microorganisms: Field-scale processes in Poland. *Ecological Engineering, 37*(11), 1895–1900.
72. Lee, K., & Kim, K. (2010). Heavy metal removal from shooting range soil by hybrid electrokinetics with bacteria and enhancing agents. *Environmental Science and Technology, 44*(24), 9482–9487.
73. Tang, J. C., Wang, R. G., Niu, X. W., Wang, M., Chu, H. R., & Zhou, Q. X. (2010). Characterisation of the rhizoremediation of petroleum-contaminated soil: Effect of different influencing factors. *Biogeosciences, 7*(12), 3961–3969.
74. Abdulsalam, S., & Omale, A. B. (2009). Comparison of biostimulation and bioaugmentation techniques for the remediation of used motor oil contaminated soil. *Brazilian Archives of Biology and Technology, 52*(3), 747–754.
75. Ruberto, L., Dias, R., Lo Balbo, A., Vazquez, S. C., Hermandez, E. A., & Mac Cormack, W. P. (2009). Influence of nutrients addition and bioaugmentation on the hydrocarbon biodegradation of a chronically contaminated Antarctic soil. *Journal of Applied Microbiology, 106*(4), 1101–1110.
76. Kolwzan, B. (2008). Assessment and choice of inoculants for the bioremediation of soil contaminated with petroleum products. *Ochrona Środowiska, 30*(4), 3–14.
77. Perfumo, A., Banat, I. M., Marchant, R., & Vezzulli, L. (2007). Thermally enhanced approaches for bioremediation of hydrocarbon-contaminated soils. *Chemosphere, 66*(1), 179–184.

78. Andreoni, V., Baggi, G., Colombo, M., Cavalca, L., Zangrossi, M., & Bernasconi, S. (1998). Degradation of 2,4,6-trichlorophenol by a specialized organism and by indigenous soil microflora: Bioaugmentation and self-remediability for soil restoration. *Letters in Applied Microbiology, 27*(2), 86–92.
79. Atlas, R. M. (1993). Bioaugmentation to enhance bioremediation. In M. A. Levin & M. A. Gealt (Eds.), *Biotreatment of industrial and hazardous waste*. New York: McGraw-Hill.
80. Bhattacharya, A., & Gupta, A. (2013). Evaluation of Acinetobacter sp. B9 for Cr (VI) resistance and detoxification with potential application in bioremediation of heavy-metals-rich industrial wastewater. *Environmental Science and Pollution Research, 20*(9), 6628–6637.
81. Zhang, J., Wen, D., Zhao, C., & Tang, X. (2014). Bioaugmentation accelerates the shift of bacterial community structure against shock load: A case study of coking wastewater treatment by zeolite-sequencing batch reactor. *Applied Microbiology and Biotechnology, 98*(2), 863–873.
82. Kim, I. N., Ekpeghere, K., Ha, S. Y., Kim, S. H., Kim, B. S., Song, B., et al. (2013). An eco-friendly treatment of tannery wastewater using bioaugmentation with a novel microbial consortium. *Journal of Environmental Science and Health. Part A Toxic/Hazardous Substances and Environmental Engineering, 48*(13), 1732–1739.
83. McLaughlin, H., Farrell, A., & Quilty, B. (2006). Bioaugmentation of activated sludge with two *Pseudomonas putida* strains for the degradation of 4-chlorophenol. *Journal of Environmental Science and Health. Part A Toxic/Hazardous Substances and Environmental Engineering, 41*(5), 763–777.
84. Quan, X. C., Hanchang, S. L., Hong, W., & Jianlong, Q. Y. (2004). Removal of 2,4-dichlorophenol in a conventional activated sludge system through bioaugmentation. *Process Biochemistry, 39*(11), 1701–1707.
85. Dhouib, A., Ellouz, M., Aloui, F., & Sayadi, S. (2006). Effect of bioaugmentation of activated sludge with white rot fungi on olive mill wastewater detoxification. *Letters in Applied Microbiology, 42*(4), 405–411.
86. Pandya, M. T. (2007). Treatment of industrial wastewater using photooxidation and bioaugmentation technology. *Water Science and Technology, 56*(7), 117–124.
87. Gerardi, M. H. (2006). *Wastewater bacteria*. Hoboken, NJ: Wiley.
88. Schneider, I. D., & Topalova, Y. I. (2011). Effect of bioaugmentation on anaerobic wastewater treatment in the dairy industry. *Journal of Dairy Science, 94*(9), 4389–4397.
89. Qu, Y., Yang, Q., Zhou, J., Gou, M., Xing, L., & Ma, F. (2009). Combined MBR with photocatalysis/ozonation for bromoamine acid removal. *Applied Biochemistry and Biotechnology, 159*(3), 664–672.
90. Ro, K., Babcock, R., & Stenstrom, M. (1997). Demonstration of bioaugmentation in a fluidized-bed process treating 1-naphthylamine. *Water Research, 31*(7), 1687–1693.
91. Bai, Y., Sun, Q., Sun, R., Wen, D., & Tang, X. (2011). Bioaugmentation and adsorption treatment of coking wastewater containing pyridine and quinoline using zeolite-biological aerated filters. *Environmental Science and Technology, 45*(5), 1940–1948.
92. Jittawattanarat, R., Kostarelos, K., & Khan, E. (2007). Immobilized cell augmented activated sludge process for enhanced nitrogen removal from wastewater. *Water Environment Research, 79*(11), 2325–2335.
93. Tsutsui, H., Anami, Y., Matsuda, M., Inoue, D., Sei, K., Soda, S., et al. (2010). Transfer of plasmid pJP4 from Escherichia coli and Pseudomonas putida to bacteria in activated sludge developed under different sludge retention times. *Journal of Bioscience and Bioengineering, 110*(6), 684–689.
94. Krhutkova, O., Novak, L., Pachmanova, L., Benakova, A., Wanner, J., & Kos, M. (2006). In situ bioaugmentation of nitrification in the regeneration zone: Practical application and experiences at full-scale plants. *Water Science and Technology, 53*(12), 39–46.
95. Leu, S., & Stenstrom, M. K. (2010). Bioaugmentation to improve nitrification in activated sludge treatment. *Water Environment Research, 82*(6), 524–535.

96. Flores, A., Nisola, G. M., Cho, E., Gwon, E., Kim, H., Lee, C., et al. (2007). Bioaugmented sulfur-oxidizing denitrification system with Alcaligenes defragrans B21 for high nitrate containing wastewater treatment. *Bioprocess and Biosystems Engineering, 30*(3), 197–205.
97. Head, M., & Oleszkiewicz, J. (2004). Bioaugmentation for nitrification at cold temperatures. *Water Research, 38*(3), 523–530.
98. Plaza, E., Trela, J., & Hultman, B. (2001). Impact of seeding with nitrifying bacteria on nitrification process efficiency. *Water Science and Technology, 43*(1), 155–163.
99. Guo, J., Wang, J., Cui, D., Wang, L., Ma, F., Chang, C., et al. (2010). Application of bioaugmentation in the rapid start-up and stable operation of biological processes for municipal wastewater treatment at low temperatures. *Bioresource Technology, 101*(17), 6622–6629.
100. Bai, Y., Sun, Q., Zhao, C., Wen, D., & Tang, X. (2010). Bioaugmentation treatment for coking wastewater containing pyridine and quinoline in a sequencing batch reactor. *Applied Microbiology and Biotechnology, 87*(5), 1943–1951.
101. Park, D., Lee, D. S., Kim, Y. M., & Park, J. M. (2008). Bioaugmentation of cyanide-degrading microorganisms in a full-scale cokes wastewater treatment facility. *Bioresource Technology, 99*(6), 2092–2096.
102. Li, M., Peng, L., Ji, Z., Xu, J., & Li, S. (2008). Establishment and characterization of dual-species biofilms formed from a 3,5-dinitrobenzoic-degrading strain and bacteria with high biofilm-forming capabilities. *FEMS Microbiology Letters, 278*(1), 15–21.
103. Bokhamy, M., Deront, M., Adler, N., & Peringer, P. (1997). Survival and activity of Comamonas testosteroni in mixed population. *Water Research, 31*(11), 2802–2810.
104. Dhouib, A., Hamad, N., Hassairi, I., & Sayadi, S. (2003). Degradation of anionic surfactants by Citrobacter braakii. *Process Biochemistry, 38*(8), 1245–1250.
105. US Environmental Protection Agency. (2000, February). *Introduction to phytoremediation* (EPA/600/R-99/107). Washington, DC.
106. US Environmental Protection Agency. (2012). *Phytoremediation resource guide* (EPA 542-B-99-003, June 1999). http://www.epa.gov/tio/download/remed/phytoresgude.pdf. (7 April 2012).
107. Schwab, A., Al-Assi, A., & Banks, M. (1998). Adsorption of naphthalene onto plant roots. *Journal of Environmental Quality, 27*(1), 220–224.
108. Barber, J., Sharma, H., Ensley, H., Polito, M., & Thomas, D. (1995). Detoxification of phenol by the aquatic angiosperm, Lemna-Gibba. *Chemosphere, 31*(6), 3567–3574.
109. Zablotowicz, R., Locke, M., & Hoagland, R. (1997). Aromatic nitroreduction of acifluorfen in soils, rhizospheres, and pure cultures of rhizobacteria. *Phytoremediation of Soil and Water Contaminants, 664*, 38–53.
110. Siciliano, S., & Germida, J. (1997). Bacterial inoculants of forage grasses that enhance degradation of 2-chlorobenzoic acid in soil. *Environmental Toxicology and Chemistry, 16* (6), 1098–1104.
111. Pfender, W. (1996). Bioremediation bacteria to protect plants in pentachlorophenol-contaminated soil. *Journal of Environmental Quality, 25*(6), 1256–1260.
112. Boyle, J., & Shann, J. (1998). The influence of planting and soil characteristics on mineralization of 2,4,5-T in rhizosphere soil. *Journal of Environmental Quality, 27*(3), 704–709.
113. Rivera, R., Medina, V., Larson, S., & McCutcheon, S. (1998). Phytotreatment of TNT-contaminated groundwater. *Journal of Soil Contamination, 7*(4), 511–529.
114. Meharg, A., Cairney, J., & Maguire, N. (1997). Mineralization of 2,4-dichlorophenol by ectomycorrhizal fungi in axenic culture and in symbiosis with pine. *Chemosphere, 34*(12), 2495–2504.
115. Rhykerd, R. L., Hallmark, M. T., & Munster, C. L. (1998). A field facility for phytoremediation research. *The 1998 ASAE Annual International Meeting*, ASAE, Orlando, FL.
116. Conger, R. M., & Portier, R. J. (1997). Phytoremediation experimentation with the herbicide bentazon. *Remediation Journal, 7*(2), 19–37.

117. Andreazza, R., Okeke, B. C., Lambais, M. R., Bortolon, L., Bastos de Melo, G. W., & de Oliveira Camargo, F. A. (2010). Bacterial stimulation of copper phytoaccumulation by bioaugmentation with rhizosphere bacteria. *Chemosphere, 81*(9), 1149–1154.
118. Nouri, J., Lorestani, B., Yousefi, N., Khorasani, N., Hasani, A. H., Seif, F., et al. (2011). Phytoremediation potential of native plants grown in the vicinity of Ahangaran lead-zinc mine (Hamedan, Iran). *Environmental Earth Sciences, 62*(3), 639–644.
119. Natarajan, S., Stamps, R. H., Ma, L. Q., Saha, U. K., Hernandez, D., Cai, Y., et al. (2011). Phytoremediation of arsenic-contaminated groundwater using arsenic hyperaccumulator Pteris vittata L.: Effects of frond harvesting regimes and arsenic levels in refill water. *Journal of Hazardous Materials, 185*(2–3), 983–989.
120. Govindaswamy, S., Schupp, D. A., & Rock, S. A. (2011). Batch and continuous removal of arsenic using hyacinth roots. *International Journal of Phytoremediation, 13*(6), 513–527.
121. Jiang, Y., Purchase, D., Jones, H., & Garelick, H. (2011). Technical note: Effects of arsenate (AS5+) on growth and production of glutathione (GSH) and phytochelatins (PCS) in Chlorella Vulgaris. *International Journal of Phytoremediation, 13*(8), 834–844.
122. Wang, S., & Zhao, X. (2009). On the potential of biological treatment for arsenic contaminated soils and groundwater. *Journal of Environmental Management, 90*(8), 2367–2376.
123. Wang, C., Ma, X., & Walsh, M. P. (2011). Competitive uptake and phytomonitoring of chlorinated contaminant mixtures by Redosier Dogwood (Cornus sericea). *International Journal of Phytoremediation, 13*(4), 333–344.
124. Weyens, N., Taghavi, S., Barac, T., van der Lelie, D., Boulet, J., Artois, T., et al. (2009). Bacteria associated with oak and ash on a TCE-contaminated site: Characterization of isolates with potential to avoid evapotranspiration of TCE. *Environmental Science and Pollution Research, 16*(7), 830–843.
125. Gopalakrishnan, G., Werth, C. J., & Negri, M. C. (2009). Mass recovery methods for trichloroethylene in plant tissue. *Environmental Toxicology and Chemistry, 28*(6), 1185–1190.
126. James, C. A., Xin, G., Doty, S. L., Muiznieks, I., Newman, L., & Strand, S. E. (2009). A mass balance study of the phytoremediation of perchloroethylene-contaminated groundwater. *Environmental Pollution, 157*(8–9), 2564–2569.
127. Rylott, E. L., Budarina, M. V., Barker, A., Lorenz, A., Strand, S. E., & Bruce, N. C. (2011). Engineering plants for the phytoremediation of RDX in the presence of the co-contaminating explosive TNT. *The New Phytologist, 192*(2), 405–413.
128. Lee, M., & Yang, M. (2010). Rhizofiltration using sunflower (Helianthus annuus L.) and bean (Phaseolus vulgaris L. var. vulgaris) to remediate uranium contaminated groundwater. *Journal of Hazardous Materials, 173*(1–3), 589–596.
129. Cook, R. L., Landmeyer, J. E., Atkinson, B., Messier, J., & Nichols, E. G. (2010). Field note: Successful establishment of a phytoremediation system at a petroleum hydrocarbon contaminated shallow aquifer: Trends, trials, and tribulations. *International Journal of Phytoremediation, 12*(7), 716–732.
130. Arnold, C. W., Parfitt, D. G., & Kaltreider, M. (2007). Phytovolatilization of oxygenated gasoline-impacted groundwater at an underground storage tank site via conifers. *International Journal of Phytoremediation, 9*(1–3), 53–69.
131. Yoon, J. M., Oliver, D. J., & Shanks, J. V. (2007). Phytotoxicity and phytoremediation of 2,6-dinitrotoluene using a model plant, Arabidopsis thaliana. *Chemosphere, 68*(6), 1050–1057.
132. El-Gendy, A. S., Svingos, S., Brice, D., Garretson, J. H., & Schnoor, J. (2009). Assessments of the efficacy of a long-term application of a phytoremediation system using hybrid poplar trees at former oil tank farm sites. *Water Environment Research, 81*(5), 486–498.
133. Zhao, S., Lian, F., & Duo, L. (2011). EDTA-assisted phytoextraction of heavy metals by turfgrass from municipal solid waste compost using permeable barriers and associated potential leaching risk. *Bioresource Technology, 102*(2), 621–626.

134. Lai, H. Y., & Chen, Z. S. (2005). The EDTA effect on phytoextraction of single and combined metals-contaminated soils using rainbow pink (Dianthus chinensis). *Chemosphere, 60*(8), 1062–1071.
135. Saifullah, Ghafoor, A., & Qadir, M. (2009). Lead phytoextraction by wheat in response to the EdTA application method. *International Journal of Phytoremediation, 11*(3), 268–282.
136. Lai, H., & Chen, Z. (2007). Multi-dose applying EDTA to decrease the potential groundwater contamination using rainbow pink (Dianthus chinensis) for enhanced phytoextraction. *Desalination, 210*(1–3), 236–247.
137. Neugschwandtner, R. W., Tlustos, P., Komarek, M., & Szakova, J. (2008). Phytoextraction of Pb and Cd from a contaminated agricultural soil using different EDTA application regimes: Laboratory versus field scale measures of efficiency. *Geoderma, 144*(3–4), 446–454.
138. Van Aken, B., Correa, P. A., & Schnoor, J. L. (2010). Phytoremediation of polychlorinated biphenyls: New trends and promises. *Environmental Science and Technology, 44*(8), 2767–2776.
139. Secher, C., Lollier, M., Jézéquel, K., Cornu, J. Y., Amalric, L., & Lebeau, T. (2013). Decontamination of a polychloroinated biphenyls-contaminated soil by photoremediation-assisted bioaugmentation. *Biodegradation, 24*(4), 549–562.
140. Masciandaro, G., Macci, C., Peruzzi, E., Ceccanti, B., & Doni, S. (2013). Organic matter–microorganism–plant in soil bioremediation: A synergic approach. *Reviews in Environmental Science and Bio/Technology, 12*(4), 399–419.
141. Borden, J. (2008). Natural bioattenuation of anaerobic hydrocarbons and chlorinated solvents in groundwater (chap. 4). In J. J. Kaluarachchi (Ed.), *Groundwater contamination by organic pollutants: Analysis and remediation*. Reston, VA: American Society of Civil Engineers.
142. Mohn, W. W. (2004). Biodegradation and bioremediation of halogenated organic compounds. In A. Singh & O. P. Ward (Eds.), *Biodegradation and bioremediation*. Berlin, Germany: Springer.
143. Chawla, R. C., Liou, R., Johnson, J. H., & Tharakan, J. P. (2000). Biodegradation of PCBs in aqueous soil system. In D. L. Wise, D. J. Trantolo, E. J. Cichon, H. I. Inyang, & U. Stottmeister (Eds.), *Bioremediation of contaminated soils*. New York: Marcel Dekker.
144. Harvey, P. J., & Thurston, C. F. (2001). The biochemistry of ligninolytic fungi. In G. M. Gadd (Ed.), *Fungi in bioremediation*. Cambridge, UK: Cambridge University Press.
145. Baneshi, M. M., Kalantary, R. R., Jafari, A. J., Nasseri, S., Jaafarzadeh, N., & Esrafili, A. (2014). Effect of bioaugmentation to enhance phytoremediation for removal of phenanthrene and pyrene from soil with Sorghum and Onobrychis sativa. *Journal of Environmental Health Science and Engineering, 12*(1), 24.
146. Coates, J. D. (2004). Anaerobic biodegradation of hydrocarbons. In A. Singh & O. P. Ward (Eds.), *Biodegradation and bioremediation*. Berlin, Germany: Springer.
147. Lladó, S., Covino, S., Solanas, A. M., Viñas, M., Petruccioli, M., & D'annibale, A. (2013). Comparative assessment of bioremediation approaches to highly recalcitrant PAH degradation in a real industrial polluted soil. *Journal of Hazardous Materials, 248*, 407–414.
148. Herwijnen, R., van Joffe, B., Ryngaert, A., Hausner, M., Springael, D., Govers, H. A. K., et al. (2006). Effect of bioaugmentation and supplementary carbon sources on degradation of polycyclic aromatic hydrocarbons by a soil-derived culture. *FEMS Microbiology Ecology, 55*(1), 122–135.
149. Mulligan, C. N. (2002). *Environmental biotreatment: Technologies: Technologies for air, water, and wastes*. Rockville, MD: ABS Consulting.
150. Fukuhara, Y., Horii, S., Matsuno, T., Matsumiya, Y., Mukai, M., & Kubo, M. (2013). Distribution of hydrocarbon-degrading bacteria in the soil environment and their contribution to bioremediation. *Applied Biochemistry and Biotechnology, 170*(2), 329–339.
151. Tsai, S. L., Lin, C. W., Wu, C. H., & Shen, C. M. (2013). Kinetics of xenobiotic biodegradation by the Pseudomonas sp. YATO 411 strain in suspension and cell-immobilized beads. *Journal of the Taiwan Institute of Chemical Engineers, 44*(2), 303–309.

152. US Environmental Protection Agency. (1998, April). *An analysis of composting as an environmental remediation technology* (EPA530-R-98-008). Washington, DC.
153. Xin, B. P., Wu, C. H., Wu, C. H., & Lin, C. W. (2013). Bioaugmented remediation of high concentration BTEX-contaminated groundwater by permeable reactive barrier with immobilized bead. *Journal of Hazardous Materials, 244*, 765–772.
154. Weins, J. (2007). Applying ecological risk assessment to environmental accidents: Harlequin ducks and the *Exxon Valdez* spill. *Bioscience, 57*(9), 769–777.
155. US Environmental Protection Agency. (1989, August). *Alaska oil spill bioremediation project* (EPA/600/8-89/073). Washington, DC.
156. Venosa, A. D., Suidan, M. T., Haines, J. R., Wrenn, B. A., Strogmeier, K. L., Eberhart, B. L., et al. (1995). Field bioremediation study: Spilled crude oil on Fowler Beach, Delaware. In R. E. Hinchee, J. Fredrickson, & B. C. Alleman (Eds.), *Bioaugmentation for site remediation*. Columbus, OH: Battelle.
157. Maxwell, C. R., & Baqai, H. A. (1995). Remediation of petroleum hydrocarbons by inoculation with laboratory-cultured microorganisms. In R. E. Hinchee, J. Fredrickson, & B. C. Alleman (Eds.), *Bioaugmentation for site remediation*. Columbus, OH: Battelle.
158. Holroyd, M. L., & Caunt, P. (1995). Large-scale soil bioremediation using white-rot fungi. In R. E. Hinchee, J. Fredrickson, & B. C. Alleman (Eds.), *Bioaugmentation for site remediation*. Columbus, OH: Battelle.
159. Pearce, K., Snyman, H. G., Oellermann, R. A., & Gerber, A. (1995). Bioremediation of petroleum-contamination soil. In R. E. Hinchee, J. Fredrickson, & B. C. Alleman (Eds.), *Bioaugmentation for site remediation*. Columbus, OH: Battelle.
160. Dybas, M. J., Tatara, G. M., Knoll, W. H., Mayotte, T. J., & Criddle, C. S. (1995). Niche adjustment for bioaugmentation with *Pseudomonas* sp. strain KC. In R. E. Hinchee, J. Fredrickson, & B. C. Alleman (Eds.), *Bioaugmentation for site remediation*. Columbus, OH: Battelle.
161. Dybas, M. J., Bezborodinikov, S., Voice, T., Wiggert, D. C., Davies, S., Tiedje, J., et al. (1997). Evaluation of bioaugmentation to remediate an aquifer contaminated with carbon tetrachloride. In situ *and on-site bioremediation Volume 4: Papers from the Fourth International* In Situ *and On-Site Bioremediation Symposium New Orleans, April 28–May 1, 1997*. Columbus, OH: Battelle.
162. Tenney, C., Lastoskie, C., & Dybas, M. (2004). A reactor model for pulsed pumping groundwater remediation. *Water Research, 38*(18), 3869–3880.
163. Bourquin, A. W., Mosteller, D. C., Olsen, R. L., Smith, M. J., & Reardon, K. F. (1997). Aerobic remediation of TCE-contaminated groundwater: bioaugmentation with *Burkholderia Cepacia* PR1$_{301}$. *In Situ and On-Site Bioremediation Symposium New Orleans, April 28–May 1, 1997*. Columbus, OH: Battelle.
164. Murygaia, V., Arinbasarov, M. U., Korotayeva, E. V., Stolyarova, A. V., & Peterson, L. R. (1997). Pilot clean-up of Chernaya River (Moscow Region) from oil pollution. *In Situ and On-Site Bioremediation Symposium New Orleans, April 28–May 1, 1997*. Columbus, OH: Battelle.
165. Hyman, M., & Ryan Dupont, R. (2000). *Groundwater and soil remediation: Process design and cost estimation of proven technologies*. Reston, VA: American Society of Civil Engineers.
166. Goetz, J., & Brenner, R. (2002, September). *Application, performance, and costs of biotreatment technologies for contaminated soils* (EPA/600/R-03/037). Cincinnati, OH.
167. Taylor, R. T., & Hanna, M. L. (1995). Laboratory treatability studies for resting-cell in situ microbial filter bioremediation. In R. E. Hinchee, J. Fredrickson, & B. C. Alleman (Eds.), *Bioaugmentation for site remediation*. Columbus, OH: Battelle.
168. Aamand, J., Bruntse, G., Jepsen, M., Jorgensen, C., & Jensen, B. K. (1995). Degradation of PAHs in soil by indigenous and inoculated bacteria. In R. E. Hinchee, J. Fredrickson, & B. C. Alleman (Eds.), *Bioaugmentation for site remediation*. Columbus, OH: Battelle.

169. Imamura, T., Kozabi, S., Kuriyama, A., Kawaguchi, M., Touge, Y., Yano, T., et al. (1997). Inducer-free microbe for TCE degradation and feasibility study in bioaugmentation. In Situ *and On-Site Bioremediation Symposium New Orleans, April 28–May 1, 1997*. Columbus, OH: Battelle.
170. US Environmental Protection Agency. (1998, September). *Guidelines for ecological risk assessment* (EPA/630/R-95/002F). Washington, DC.
171. Efroymson, R., Sample, B. E., & Peterson, M. J. (2004). Ecotoxicity test data for total petroleum hydrocarbons in soil: Plants and soil-dwelling invertebrates. *Human and Ecological Risk Assessment, 10*(2), 207–231.

Chapter 6
Wastewater Renovation by Flotation for Water Pollution Control

Nazih K. Shammas

Contents

Abstract Planning and management of water resources is more and more being based on the renovation and utilization of wastewater for use in agricultural and landscape irrigation as well as in industrial production. Conventional wastewater treatment, even when it is economically feasible, is costly because of biosolids handling and tertiary sedimentation tanks. The fact that (DAF) dissolved air flotation (with 3–5 min of detention time) can replace sedimentation (with 2–3 h of detention time) for clarification has been overlooked by environmental engineers for many decades. Modern high rate DAF clarifiers have advanced to such an extent that they could overshadow the conventional settling clarifiers in wastewater treatment. DAF hydraulic loadings increased from 1 to 2.5 L/m^2/s and for a triple

N.K. Shammas, Ph.D. (✉)
Lenox Institute of Water Technology, Lenox, MA, USA

Krofta Engineering Corporation, Lenox, MA, USA

277 Pleasant Street, Unit 311, Pasadena, CA 91101, USA
e-mail: N.K.Shammas@Gmail.com; NazihShammas@AOL.com

L.K. Wang, C.T. Yang, and M.-H.S. Wang (eds.), *Advances in Water Resources Management*, Handbook of Environmental Engineering, Volume 16,
DOI 10.1007/978-3-319-22924-9_6

stacked unit to 7.5 L/m^2/s; the detention time decreased from 30 to 3 min; air dissolving is improved and now requires only 10 s retention time in the air dissolving tube instead of the previous 60 s. This chapter discusses the field application of the DAF process for primary wastewater clarification, secondary separation of mixed liquor from activated sludge and the design and operation parameters for a two-stage DAF operation. The performance of the compact units in treatment of various wastewater effluents illustrates the system's usefulness for wastewater reclamation, recycling and reuse.

Keywords Wastewater reclamation and reuse • Dissolved air flotation • DAF • Primary DAF • Secondary DAF • Two-stage DAF

1 Introduction

Planning and management of water resources is more and more being based on the renovation and utilization of wastewater for use in agricultural and landscape irrigation as well as in industrial production. Conventional wastewater treatment, even when it is economically feasible, is costly because of biosolids handling and tertiary sedimentation tanks. The fact that (DAF) dissolved air flotation (with 3–5 min of detention time) can replace sedimentation (with 2–3 h of detention time) for clarification has been overlooked by environmental engineers for many decades. Modern high rate DAF clarifiers have advanced to such an extent that they could overshadow the conventional settling clarifiers in wastewater treatment. DAF hydraulic loadings increased from 1 to 2.5 L/m^2/s and for a triple stacked unit to 7.5 L/m^2/s; the detention time decreased from 30 to 3 min; air dissolving is improved and now requires only 10 s retention time in the air dissolving tube instead of the previous 60 s [1, 2]; the modern DAF unit smaller size and weight allows for inexpensive construction and flexibility of erection using prefabricated units; and the availability of excellent flocculating chemicals gives a high stability of operation and high clarification degree [3–6].

Flotation has been widely used for industrial waste treatment. Typical industrial wastes treated by flotation are oily wastewaters [7], pulp and paper mill wastes [8, 9], slaughterhouse wastes [10], poultry processing wastewater [11], hazardous wastes [12], food and dairy processed wastewater [13] and seafood processing wastewater [14]. However, although flotation has also been used extensively for waste activated sludge (WAS) thickening [15–17] yet only limited applications are available in wastewater clarification [18–20] or Wastewater reuse [21, 22].

Conventional biological treatment [23, 24] is the predominant selection for municipal wastewater treatment plants. Whenever upgrading is deemed to be necessary, the remedial action in most cases is to merely enlarge the size of existing facilities. In both instances the above choices may not be practical or economical. Examples of such cases are:

1. Seasonal operations where huge population swelling occurs during religious festivities, in sea and mountain resorts, and in ski areas [25].
2. Recovery and treatment of wastes generated from agricultural crops, especially those that are seasonal in nature [26, 27].
3. Chemical and toxic wastes that are not amenable to biological treatment [7, 8, 28].
4. Emergency situations which demand immediate action. Such events cannot be remediated by biological means.
5. Control and reduction of combined sewer overflows (CSOs), especially where space is at a premium [28, 29].
6. Wastewater renovation and tertiary treatment for water cycling and reuse [25, 30].
7. Retrofitting of treatment plants for upgrading or plant capacity expansions [31–33].

Harleman et al. [29] and Harleman and Murcott [32] consider chemically-enhanced primary treatment (CEPT) and chemical secondary treatment (CST) to be significant innovations in wastewater treatment technology. They conclude that CEPT can remove over 80 % TSS and 60 % BOD compared to 60 % TSS and 35 % BOD for conventional primary treatment. They also indicate that CST can achieve higher removals of up to 90 % TSS and 80 % BOD compared to biological secondary treatment (BST) of 85 % TSS and 85 % BOD. On the other hand, Firmin [34] who reviewed the performance of CEPT plants in North America and Scandinavia has found that their actual removal efficiencies and effluent quality are highly variable and specific to given wastewaters. He concluded that CEPT will remove 65–80 % TSS and only 40–50 % BOD.

The goal of this chapter is to illustrate and explain how the flotation technology can be employed to further enhance the CEPT concept as a feasible alternative or an upgrade to conventional wastewater treatment processes. An innovative physicochemical treatment system, based on flotation technology developed by the Lenox Institute of Water Technology [35–37], is being used for this purpose. Furthermore, in order to illustrate the universal applicability of this system to various wastewaters, the system was pilot tested at different geographical locations in seven states and on different wastewaters. This multitasked testing program allowed the evaluation of flotation cells design criteria and the effect of wastewater variability on removal efficiencies and final effluent quality.

The technical feasibility of the DAF process for primary and secondary flotation in wastewater treatment is illustrated by the following applications [31, 38–40]:

(a) Applications of one-stage flotation (F-Cell)

1. Primary clarification of raw wastewater in Indiana
2. Clarification of primary settled effluent in Maine
3. Secondary flotation of mixed liquor from an activated sludge aeration tank in Maryland
4. Full-scale operation of an upgraded conventional activated sludge plant in Texas

(b) Applications of two-stage flotation (F-Cell) and flotation/filtration (FF-Cell)

1. Wastewater I in Massachusetts
2. Wastewater II in Arkansas
3. Wastewater III in Alabama

2 DAF Pilot Plant for Single-Stage Operations

The DAF pilot plant used in single-stage applications is shown in Fig. 6.1. It is a self-contained, skid mounted flotation clarifier capable of treating 163 m^3/d (30 gpm) wastewater flow. The flotation tank is 1.2 m (4 ft) in diameter with

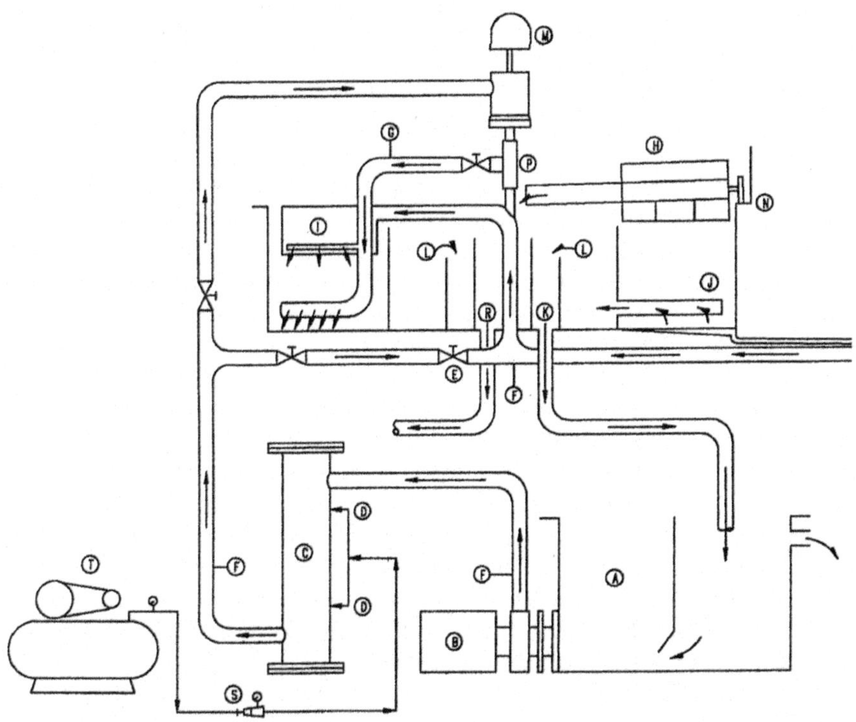

(A) Holding Tank (100 Gal.)
(B) Pressure Pump (30 GPM @ 60 PSI)
(C) Air Dissolving Tube
(D) Air Meters (+/- 1 SCFM)
(E) Pressure Release Valve (Globe Valve)
(F) Chemical Feed Fittings
(G) Sampling Valve
(H) Spiral Scoop (Float Removal)
(I) Distribution Duct (Raw Water Inlet)
(J) Clarified Water Extraction Pipes
(K) Clarified Water Outlet
(L) Level Control Weir
(M) Rotary Electrical Contact
(N) Gear Motor
(P) Rotary Pipe
(R) Floated Material Outlet
(S) Air Regulator
(T) Compressor

Fig. 6.1 The flotation cell (F-Cell) flow diagram [31]

041–0.46 m (16–18 in.) effective water depth. The system is complete with feed pump, pressure pump, compressor, air dissolving tube, collection tank and chemical feed equipment. The flotation tank is equipped with a spiral scoop for collection and removal of floated sludge.

The inlet, outlet and sludge removal mechanisms are contained in the central rotating section. This section and the spiral scoop rotate around the tank at a speed synchronized with the flow. The system is operated in the recycle flow pressurization mode, whereby a portion of clarified effluent (30–40 %) is continuously recycled from the collection tank to the air dissolving tube under 400 kPa (60 psi) pressure. After pressure release, the aerated water is mixed with the influent flow just before the inlet to the distribution duct that moves with the same velocity, but in opposite direction to the incoming flow thus creating a quiescent state in the flotation chamber. Fine bubbles generated in this manner attach to the suspended particles and float them to the surface. The spiral scoop takes up the floated biosolids, pouring them into the stationary center section where they are discharged by gravity. Clarified water is removed by extraction pipes which are attached to the moving center section and discharged into the collection tank.

Wiper blades attached to the moving distribution duct scrape the bottom and the sides of the tank and discharge settled sludge into the built-in sump, for periodic purging. The variable speed gear motor drives the rotating elements and scoop. Electrical current for the gear motor feeds from a rotary contact mounted on the central shaft.

3 Clarification of Raw Wastewater

The results of the pilot applications indicated that the DAF clarifier (F-CELL) achieved good performance in both primary and secondary flotation. Table 6.1 shows that an average removal of 67 % in suspended solids from raw wastewater is feasible without any chemical addition. Raw wastewater was pumped to the F-CELL from the Kirklin wastewater plant immediately after the grinder. When the flotation cell was run with polymer addition to the incoming flow, no significant

Table 6.1 Clarification of raw wastewater by flotation at Kirklin WWTP, Kirklin, IN [31]

	Total suspended solids		
Composite sample	Influent (mg/L)	Effluent (mg/L)	Removal (%)
1	97	27	72
2	226	129	43
3	298	68	77
4	415	197	53
5	238	54	77
6	300	50	83
Average	262	87	67

improvement in removal was obtained. One of the collected composite samples was tested for both BOD and ammonia-N. Although the BOD removal was not very high (28 % removal from 188 mg/L down to 136 mg/L), the Ammonia was almost completely removed. Ammonia-N concentration decreased from 31.4 mg/L in the influent to 0.41 mg/L in the effluent i.e. 99 % removal.

4 Clarification of Primary Effluent

The F-CELL was installed at the Bangor, Maine Pollution Abatement Facility to evaluate the DAF performance in the clarification of the plant's primary effluent. Trials were initially run on the plant effluent without chemical treatment and then with alum treatment only, however, these tests produced poor results. Further tests were made with alum and sodium aluminate and no polymer, but with similar results. Optimum chemical addition was found to consist of 15 mg/L alum, 10 mg/L sodium aluminate and 1 mg/L anionic polymer (Nalco 2 P0 462). Table 6.2 indicates that suspended solids concentration in the effluent from primary settling can be further reduced to an average value below 20 mg/L by flotation. The average suspended solids concentration was reduced from 73 to 18 mg/L, a reduction of 72 %. The averaged biosolids consistency was 2.5 % with a range of 1.7–3.8 % solids. Approximate biosolids flow was 0.5 % of wastewater feed. Six composite samples were analyzed for BOD. The feed BOD ranged between 114 and 164 mg/L with an average value of 128 mg/L, while the effluent BOD ranged between 34 and 67 mg/L with an average value of 45 mg/L. The average BOD removal was 65 %.

5 Secondary Separation of Mixed Liquor from Activated Sludge

The application for secondary flotation of biological mixed liquor from an activated sludge aeration tank was carried out at Patapsco wastewater treatment plant, Baltimore, MD. Data from the application (see Table 6.3) indicated that polymer (Secodyne 1533) addition was not necessary. Biosolids consistency and solids capture did not improve with polymer addition. The biological activated sludge flocs floated easily and quickly, achieving an average biosolids consistency of 4 % solids and an average solids capture of 96 %. The high solids content of the DAF float compared to 0.4–0.7 % concentration from a sedimentation tank, reduces the volume of the recycle flow to the aeration basin and consequently increases the hydraulic capacity of the basin and its detention time. This potential of secondary flotation can be utilized for improving the treatment efficiency of an existing overloaded activated sludge plant or to handle additional wastewater flow.

Table 6.2 Clarification of primary settling tank effluent by flotation at Bangor WWTP, Bangor, ME [31]

Sample	Total suspended solids Influent (mg/L)	Effluent (mg/L)	Removal (%)	Sludge concentration (%)
1	70	26	63	1.9
2	67	36	46	2.0
3	172	10	94	2.0
4	140	16	89	2.6
5	118	14	88	2.5
6	100	16	84	2.6
7	96	20	79	2.5
8	88	13	85	2.6
9	75	19	75	2.4
10	81	18	78	2.6
11	52	8	85	–
12	55	11	80	–
13	64	12	81	2.9
14	52	16	69	3.2
15	60	29	52	3.8
16	59	18	70	2.3
17	59	25	58	2.4
18	29	14	52	1.7
19	44	19	57	1.9
20	47	16	66	2.3
21	55	16	71	1.9
22	56	18	68	2.8
23	48	23	52	3.1
Average	73	18	72	2.5

Table 6.3 Secondary clarification by flotation at Patapsco Activated Sludge WWTP, Baltimore, MD [31]

Composite sample	Total suspended solids Influent (mg/L)	Effluent (mg/L)	Removal (%)	Sludge concentration (%)	Polymer dosage (mg/L)
1	4240	220	95	4.0	0
2	3660	200	95	3.5	3
3	4550	180	96	4.7	3
4	4260	140	97	3.6	2
5	5525	270	95	4.3	2
6	5610	260	95	4.2	0
7	6040	160	97	4.1	1
8	6040	210	97	4.1	0
9	4050	120	97	3.3	1
Average	4886	196	96	4.0	1.3

6 Full-Scale Operation of an Upgraded Activated Sludge Plant

The activated sludge treatment plant at a paper mill in Lufkin, TX treats 68,200 m^3/d (18 MGD) of wastewater. The plant was designed to produce a final effluent with BOD and TSS that would not exceed 20 mg/L. However, several expansions resulted in poor effluent quality and borderline permit compliance, particularly during the periods of peak BOD loading. The first alternative to solve the plant's problems namely increasing the aeration time by adding another aeration basin of the same size was not a viable option. The company did not have enough land space and the capital expenditure for this conventional option is high. The alternative decision was the use of a 16.8 m (55 ft) DAF F-CELL (see Fig. 6.2) as a secondary clarifier that would be installed in front of the final sedimentation tanks and has a capacity to handle 30,000 $m^{3/}$d (8 MGD) of flow. This was accomplished at only 12 % of the cost of the conventional expansion project estimate. A top view of the F-CELL is shown in Fig. 6.3 and the cell details are shown in Fig. 6.4.

The sludge return to the aeration basin from the flotation cell at 2 % consistency is five times thicker than the 0.4 % sludge return from the final settling tanks (see Fig. 6.2). The resulting reduction in the volume of recycle to the aeration basin by 9500 m^3/d (2.5 MGD) provides an extra 10 % hydraulic capacity for aeration. The solids removed from the 30,000 m^3/d flow processed by the flotation cell reduced the solids flowing to the final clarifiers by at least 30 % that no violations of the discharge limits have occurred since installation. The net results were reduced solids loading to the final clarifiers, increased hydraulic capacity and retention time of aeration basin, threefold increase in overall concentration of biosolids, more active recycled sludge, better effluent quality and no biosolids bulking problems.

7 DAF Pilot Plant for Two-Stage Operations

The DAF pilot plant (flotation/filtration FF-CELL) used in the following applications is shown in Fig. 6.5 and its details in Fig. 6.6. The system, mounted on a mobile trailer, consists of a combination of a primary dissolved oxygen flotation (DAF) process [31] and a secondary duel media filtration unit [25]. The inlet, outlet, and sludge removal mechanisms of both clarifiers are contained in the central rotating carriage. Raw wastewater mixed with flocculating agents enters the flotation unit through a manifold on the side of the carriage. Because the speed of the carriage is set to match the velocity of the incoming flow, the velocity of the incoming wastewater is effectively zero, enhancing the flotation process.

Flotation is achieved with the introduction of pressurized aerated recycled effluent through air dissolving tube (ADT). When this water is released, its pressure drops, and the dissolved air comes out as microscopic bubbles which attach

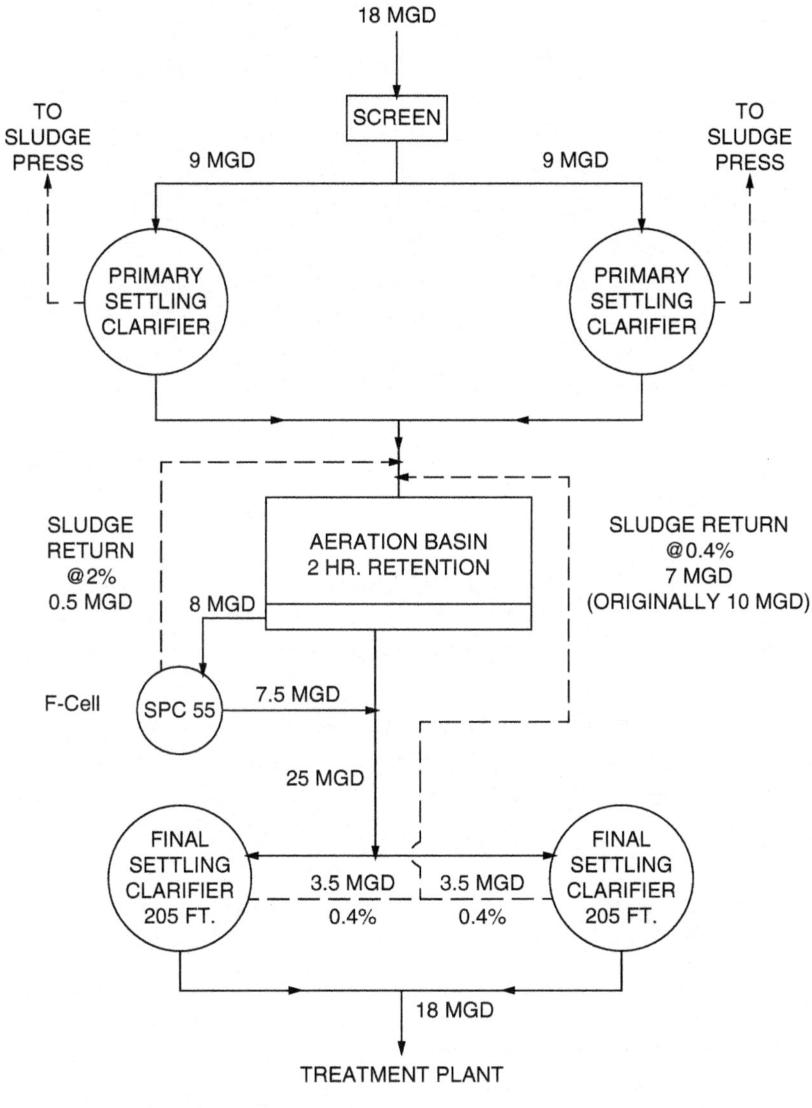

Fig. 6.2 Process flow diagram of upgrade activated sludge plant by using DAF F-Cell [31]

themselves to floc particles, causing them to float to the surface where they are removed by the rotating spiral scoop and discharged into the central sludge or biosolids collector. The clarified water from the flotation process then passes down through the anthracite—sand filtration media located at the bottom of the unit. The filtration compartment is composed of multiple separate sections which are individually backwashed while the rest of the filter is on line. The backwash water recycled from the filter is mixed with the influent water at the inlet. Thus backwash

Fig. 6.3 Top view of flotation cell (F-Cell) [35]

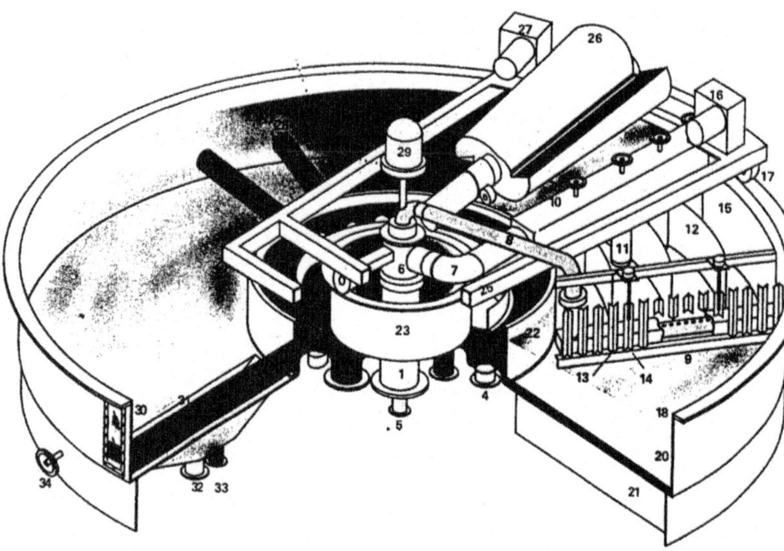

1) RAW WATER INLET
2) CLARIFIED WATER OUTLET
3) FLOATED SLUDGE OUTLET
4) CLARIFIED WATER RECYCLE OUTLET
5) PRESSURIZED WATER INLET
6) ROTARY JOINT
7) RUBBER PIPE CONNECTION
8) PRESSURIZED WATER PIPING
9) PRESSURIZED WATER DISTRIBUTION HEADER
10) RAW WATER DISTRIBUTION HEADER
11) DISTRIBUTION HEADER OUTLET PIPES
12) FLOW CONTROL CHANNELS
13) TURBULENCE REDUCTION BAFFLES
14) ADJUSTABLE HEIGHT BAFFLE ATTACHMENT
15) FLOW CONTROL CHANNEL OUTER WALL
16) ROTATING CARRIAGE GEARMOTOR DRIVE
17) CARRIAGE DRIVE WHEEL
18) WHEEL SUPPORT RIM
20) TANK WALL
21) TANK FLOOR SUPPORT STRUCTURE
22) ROTATING CLARIFIED WATER CONTAINMENT WALL
23) SLUDGE WELL
24) LEVEL CONTROL OVERFLOW WEIR
25) ROTATING CARRIAGE STRUCTURE
26) REVOLVING SPIRAL SCOOP
27) SPIRAL SCOOP GEARMOTOR DRIVE
28) CLARIFIED WATER EXTRACTION PIPES
29) ELECTRICAL SLIP RING
30) TANK WINDOW
31) SEDIMENT REMOVAL SUMP
32) FINAL DRAIN
33) SEDIMENT PURGE OUTLET
34) LEVEL CONTROL ADJUSTMENT HANDWHEEL

Fig. 6.4 Details of the flotation cell (F-Cell) [35]

Fig. 6.5 The flotation/filtration cell (FF-Cell) [35]

discharge is eliminated, and the only discharge is the float, thickened biosolids (2–3 % solids concentration) suitable for direct handling, thus also eliminating the need for biosolids thickening.

The pilot plant had a nominal flow rate of 150 L/min (40 gpm). The main characteristics of the units are listed in Table 6.4. Wastewater and air flow rates, ADT pressure, pH, and temperature were monitored continuously. Relevant parameters including TSS, COD, BOD, total phosphorous, and total Kjeldahl nitrogen (TKN) were analyzed according to Standard Methods [41].

8 Design and Operational Parameters for the Two-Stage Operation

The pilot plant was operated over a broad range of operational parameters, including various flow rates (75–170 L/min) and recycle ratios (5–60 %) as well as different coagulant and flocculant combinations and doses (up to 200 and 10 mg/L respectively). In addition to providing clear-cut evidence about the effectiveness of the multistage flotation system for treating various municipal wastewaters, the pilot study also afforded the opportunity to optimize both coagulant and flocculant dosages as well as relevant operational parameters. However since the wastewater quality at each location is different, it was expected and actually found that chemical additions did vary. Results of required chemical doses as well as optimized operational parameters, including hydraulic loading and recycle ratios, are summarized in Table 6.5.

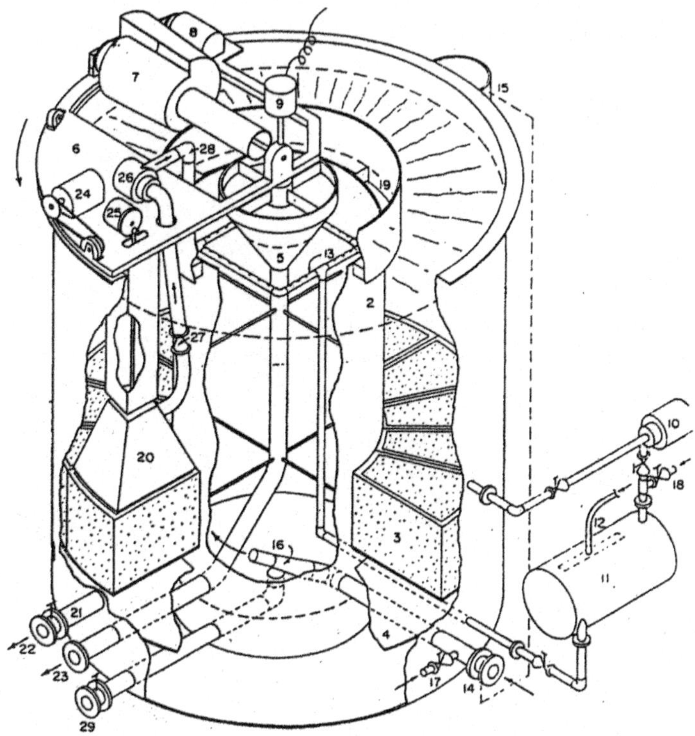

1. Outside tank
2. Inside flocculation tank
3. Sandbed assembly with screen
4. Tank bottom
5. Sludge collection funnel
6. Moveable carriage assembly
7. Spiral scoop
8. Scoop variable speed drive
9. Electrical rotary contact
10. Pressure pump
11. Air dissolving tube
12. Compressed air addition point
13. Aerated water distribution pipes
14. Raw water inlet regulating valve
15. Tank level control sensor
16. Raw water inlet jet nozzle
17. Coagulant addition point
18. Polyelectrolyte addition point
19. Deflector ring into flotation tank
20. Backwash hood assembly
21. Clarified water pipeline
22. Clarified water flow regulating valve
23. Floated sludge discharge pipe
24. Main carriage drive
25. Motor to lift backwash hood assembly
26. Backwash suction pump
27. Check valve (backflow preventor)
28. Dirty backwash water recycle pipe
29. Drain line

Fig. 6.6 Details of flotation/filtration cell (FF-Cell) [35]

8.1 Hydraulic Loading

The flotation-enhanced treatment system, in addition to improving treatment efficiency, is shown (Table 6.5) to be able to allow hydraulic loading rates (130–200 L/min/m^2) 7–10 times higher than the conventional primary treatment rate of 20 L/min/m^2. The consequence is a corresponding reduction in space requirement from one seventh to one tenth of the space needed by sedimentation clarifiers. This saving in area requirement has a dominant impact on the final cost of the system.

Table 6.4 Characteristics of pilot plant units [38]

Parameter	First stage F-Cell	Second stage FF-Cell
Nominal capacity, L/min	150	150
Cell diameter, m	1.2	1.5
Cell depth, m	0.6	0.9
Air feed, L/min	2.4	2.4
Air pressure, bar	6	6.5
Sludge scoop speed, Rev./min	2	2
Backwash rate, L/min	Not applicable	38
Backwash time, s	Not applicable	45

Table 6.5 Operational parameters of pilot plant [38]

Parameter	Wastewater I WWTP Massachusetts	Wastewater II WWTP Arkansas	Wastewater III WWTP Alabama
Hydraulic loading, L/min/m^2			
First stage	202	202	202
Second stage	135	130	130
Recycle ratio, %			
First stage	30	24	24
Second stage	30	20	20
Chemical dosage, mg/L			
First stage			
Magnifloc 496 C	4.6	3	4
Second stage			
Magnifloc 496 C	2.4	–	–
Magnifloc 1849 A	–	1	1
Ferric sulfate	83	–	–
Alum	–	8.3	4.1

This treatment system based on flotation technology exhibited a similar decided advantage over chemically-enhanced primary treatment. According to Harleman and Murcott [32], CEPT allows overflow rates 2–3 times higher than conventional clarifiers. Consequently, hydraulic loadings for flotation clarifiers using chemical addition can be tripled or quadrupled over those used for chemically-enhanced sedimentation clarifiers.

8.2 *Chemical Requirements*

Cationic polymers used alone were found to be very effective in the primary stage. The optimum chemical dose was found to be 3–5 mg/L (Table 6.5). In the second stage, a combination of 2.4 mg/L cationic polymer and 83 mg/L ferric sulfate was required for Wastewater I. However, for the other two wastewaters the best combination was using 1 mg/L anionic polymer with 4–8 mg/L alum (Table 6.5). These chemical requirements compare favorably with chemical additions practiced at CEPT plants. Firmin [34] assessed chemically-enhanced primary treatment in the United States and reported chemical doses varying from 22 mg/L ferric chloride and 0.25 mg/L polymer at Hyperion wastewater treatment plant in Los Angeles to 250 mg/L alum and 6.5 mg/L polymer at Tacoma treatment plant in Washington.

The estimated cost of chemicals used in the pilot plant applications in terms of 2014 U.S. Dollars [42] were \$0.075, \$0.053, and \$0.043 per m^3 of wastewater treated at locations I, II, and III respectively. The overall average cost was 5.70 cents/m^3.

8.3 Removal Rates

The pilot plant testing results are shown in Tables 6.6, 6.7, and 6.8 for Wastewaters I, II, and III respectively. Average concentrations of TSS, COD, BOD, and total phosphorous and Kjeldahl nitrogen (TKN) are cited for plant influent, first stage effluent after primary flotation in the F-Cell, and final effluent after the secondary flotation–filtration unit FF-Cell. Also shown in same tables are

Table 6.7 Testing results—Wastewater II (WWTP, Arkansas) [38]

Parameter	TSS	COD	BOD	Total P	TKN
Concentration, mg/L					
Influent	253	197	123	1.8	78
First stage effluent	34	96	68	0.9	30
Final effluent	1	37	20	0.1	18
Removal, %					
First stage	87	51	45	50	62
Second stage	97	62	71	89	40
Total	99	81	84	94	77

Table 6.6 Testing results—Wastewater I (WWTP, Massachusetts) [38]

Parameter	TSS	COD	BOD	Total P	TKN
Concentration, mg/L					
Influent	250	435	252	4.5	32
First stage effluent	43	168	89	3.6	26
Final effluent	1	94	40	1	20
Removal, %					
First stage	83	61	65	20	19
Second stage	95	44	55	72	23
Total	99	78	84	78	38

Table 6.8 Testing results—Wastewater III (WWTP, Alabama) [38]

Parameter	TSS	COD	BOD	Total P	TKN
Concentration, mg/L					
Influent	183	133	92	1.6	44
First stage effluent	14	53	10	0.7	22
Final effluent	1	24	8	0.1	19
Removal, %					
First stage	92	60	89	56	50
Second stage	93	55	20	86	14
Total	99	82	91	94	57

Table 6.9 Summary of removal rates [38]

Parameter	Removal, % Primary first stage	Removal, % Total
TSS	87	99
COD	57	80
BOD	66	87
Total P	42	89
TKN	44	57

the computed average values of % removals in both primary and secondary cells as well as in the system as a whole.

Significant reductions in all quality parameters were achieved for all wastewaters and at all different locations. TSS concentration in every case was reduced to non detectable levels (less than 1 mg/L) resulting in practically complete suspended solids removal (greater than 99 %). COD values were reduced by 51–61 % in the primary stage and by 44–62 % in the secondary unit, with overall reduction of 78–82 % (Tables 6.6, 6.7, and 6.8). BOD concentrations were also reduced by 45–89 % in the F-Cell and by 20–71 % in the FF-Cell, with overall reduction of 84–91 %. Good nutrients removal was also accomplished. Total phosphorous reduction reached 94 % (Tables 6.7 and 6.8) and total Kjeldahl nitrogen was reduced by up to 77 % (Table 6.7).

Table 6.9 summarizes the results obtained from the pilot plant study at all various locations. The primary single-stage DAF unit was able to achieve overall average removal efficiencies of 87 % total suspended solids, 66 % BOD, 57 % COD, 42 % total phosphorous, and 44 % total Kjeldahl nitrogen. The two-stage DAF was able to achieve excellent removal efficiencies of over 99 % TSS, 87 % BOD, 80 % COD, 89 % total phosphorous, and 57 % TKN.

Table 6.10 presents a summary of the performance of different wastewater treatment processes [1, 32, 34, 43–46] together in comparison with the results of the application of the flotation-enhanced pilot plant system. The data demonstrate the superior effectiveness of flotation based treatment systems which are capable of removing 87–99 % TSS and 66–87 % BOD as opposed to 60–90 % and 35–80 % respectively, for other wastewater treatment processes. When this high performance

Table 6.10 Comparison of various treatment systems [38]

Parameter	TSS removal (%)	BOD removal (%)	Hydraulic loading (L/min/m^2)	Metal salt dosage (mg/L)	Polymer dosage (mg/L)
Conventional primary	60	35	20	–	–
CEPT	65–80	40–60	40–50	20–250	0.14–6.5
CST	90	80	50	150–160	0.2–2.8
Primary flotation	87	66	200	–	3–4.6
Secondary flotation	99	87	130	4–83	1–2.4

is coupled with the units high hydraulic loading rates that are seven to tenfold greater in comparison to conventional primary systems and three to fourfold greater as opposed to recent experience with chemically-enhanced primary treatment, it can be concluded that the innovative flotation-enhanced wastewater treatment system can be used to replace or upgrade conventional physicochemical treatment facilities with a greater flexibility, less land use, and at a saving of 30–50 % in cost.

The ramification of this is that combining low installation and operational costs with affordable water user rates, would allow municipalities to finance plant construction, through local or international financial institutions, without having to resort to investing its own or asking for grant money. Opting for the extremely compact, modular, and low cost DAF treatment system allows the building of a plant that is profitable. Financial groups consider municipalities the most stable lenders, which make them attractive to long-term investors. If a municipality proposes to upgrade an existing plant or construct and operate a new wastewater treatment plant with a profit, long-term investors will provide the means. CEPT utilizing DAF in lieu of sedimentation is not only a proven technology; it is also representative of a new way of thinking for wastewater management and a new paradigm for treatment plant design [47–52] to serve for the renovation of wastewater for reuse.

Appendix: United States Yearly Average Cost Index for Utilities U.S. Army Corps of Engineers [42]

Year	Index	Year	Index
1967	100	1989	383.14
1968	104.83	1990	386.75
1969	112.17	1991	392.35
1970	119.75	1992	399.07
1971	131.73	1993	410.63
1972	141.94	1994	424.91

(continued)

Year	Index	Year	Index
1973	149.36	1995	439.72
1974	170.45	1996	445.58
1975	190.49	1997	454.99
1976	202.61	1998	459.40
1977	215.84	1999	460.16
1978	235.78	2000	468.05
1979	257.20	2001	472.18
1980	277.60	2002	486.16
1981	302.25	2003	497.40
1982	320.13	2004	563.38
1983	330.82	2005	605.47
1984	341.06	2006	645.52
1985	346.12	2007	681.88
1986	347.33	2008	741.36
1987	353.35	2009	699.70
1988	369.45	2010	720.80
		2011	758.79
		2012	769.30
		2013	776.44
		2014	790.52
		2015	803.83

References

1. Wang, L. K., Hung, Y. T., & Shammas, N. K. (Eds.). (2005). *Physicochemical treatment processes*. Totowa, NJ: The Humana Press, Inc.
2. Wang, L. K., Hung, Y. T., & Shammas, N. K. (Eds.). (2007). *Biosolids treatment processes*. Totowa, NJ: The Humana Press, Inc.
3. Wang, L. K., Shammas, N. K., Selke, W. A., & Aulenbach, D. B. (2010). *Flotation technology* (p. 680). Totowa, NJ: The Humana Press, Inc.
4. Zabel, T. (1985). The advantages of dissolved air flotation for water treatment. *Journal of American Water Works Association, 77*, 42.
5. Krofta, M. (1990). *Flotation replaces sedimentation in water and effluent clarification* (Report). Lenox, MA: Krofta Engineering Corp.
6. Janssens, J. G. (1992). Developments in coagulation, flocculation and dissolved air flotation. *Water/Engineering & Management, 139*, 26.
7. Eckenfielder, W. W. (1989). *Industrial water pollution control*. New York: McGraw-Hill.
8. Krofta, M., & Wang, L. K. (1989). Total closing of paper mills with reclamation and deinking installations. In *Proc. 43rd Ind. Waste Conf.* (p. 535). Chelsea, MI: Lewis.
9. Barton, C. A., Byrd, J. F., Peterson, R. C., Walter, J. H., & Woodruff, P. R. (1968). A total systems approach to pollution control at a pulp and paper mill. *Journal of the Water Pollution Control Federation, 40*, 1471.
10. Cooper, R. N., & Denmead, C. F. (1979). Chemical treatment of slaughterhouse wastes with protein recovery. *Journal of the Water Pollution Control Federation, 51*, 1017.
11. Woodard, F. E., Sproul, O. J., Hall, M. W., & Ghosh, M. M. (1972). Abatement of pollution from a poultry processing plant. *Journal of the Water Pollution Control Federation, 44*, 1909.

12. Wang, L. K., & Wang, M. H. S. (1980, October 11–13). Control of hazardous wastes in petroleum refining industry. Technical Paper (No. LIR/10-80/2) presented at the *6th Annual Convention of the Chinese-American Academic and Professional Association*, New York.
13. Ross, C. C., & Valentine, G. E., Jr. (2014). *Coupled biological/dissolved air flotation processes for treatment of food and dairy processed wastewater*. Environmental Treatment Systems, Inc., http://www.etsenvironmental.com/white-papers/coupled-biological
14. Krofta, M., Wang, L. K., & Pollman, C. D. (1989). Treatment of seafood processing wastewater by dissolved air flotation carbon adsorption and free chlorination. In *Proc. 43rd Ind. Waste Conf.* (p. 535). Chelsea, MI: Lewis.
15. Wang, L. K., Shammas, N. K., Selke, W. A., & Aulenbach, D. B. (2007). Flotation thickening. In L. K. Wang, N. K. Shammas, & Y. T. Hung (Eds.), *Biosolids treatment processes*. Totowa, NJ: The Humana Press, Inc.
16. Metcalf & Eddy, Inc. (2003). *Wastewater engineering treatment and reuse* (4th ed.). New York: McGraw Hill.
17. Cizinska, S., Matejo, V., Wase, C., Kiasson, Y., Krejci, J., & Daihainmar, G. (1992). Thickening of waste activated sludge by biological flotation. *Water Research, 26*, 139.
18. Krofta, M., Guss, D., & Wang, L. K. (1988). Development of low-cost flotation technology and systems for wastewater treatment. In *Proc. 42nd Ind. Waste Conf.* (p. 185). Chelsea, MI: Lewis.
19. Bratby, J. R. (1982). Treatment of raw wastewater overflows by dissolved air flotation. *Journal of the Water Pollution Control Federation, 54*, 1558.
20. Mennell, M., Merrill, D., & Jorden, R. M. (1974). Treatment of primary effluent by lime precipitation and dissolved air flotation. *Journal of the Water Pollution Control Federation, 46*, 2471.
21. Ksenofontov, B. S., & Ivanov, M. V. (2013). A novel multistage kinetic modeling of flotation for wastewater treatment. *Water Science & Technology, 68*, 807–812, IWA Publishing.
22. Metcalf & Eddy. (2007). In T. Asano, S. L. Burton, H. L. Leverenz, P. Tsuchihashi, & G. Tchobanoglous (Eds.), *Water reuse*. New York: McGraw Hill.
23. Wang, L. K., Pereira, N. C., Hung, Y. T., & Shammas, N. K. (Eds.). (2009). *Biological treatment processes*. Totowa, NJ: The Humana Press, Inc.
24. Wang, L. K., Shammas, N. K., & Hung, Y. T. (Eds.). (2009). *Advanced biological treatment processes*. Totowa, NJ: The Humana Press, Inc.
25. Shammas, N. K., & Krofta, M. (1994). A compact flotation–filtration treatment unit for wastewater reuse. In *Proc. Water Reuse Symposium, AWWA* (pp. 97–109). Dallas, TX, February 27–March 2.
26. Schneider, I. A. H., Neto, V. M., Soares, A., Rech, R. L., & Rubio, J. (1995). Primary treatment of soybean protein bearing effluent by dissolved air flotation and by sedimentation. *Water Research, IAWQ, 29*(1), 69–75.
27. Cooper, R. N., & Denmead, C. F. (1979). Chemical treatment of slaughterhouse wastes with protein recovery. *Journal of the Water Pollution Control Federation, 51*, 1017.
28. Wang, L. K. (1991). Water and wastewater treatment using advanced dissolved air flotation. In *The 1991 Annual Conference of the Korea Society of Water Pollution Research and Control*, Seoul, Korea, February 22–23.
29. Harleman, D. R. F., Wolman, L. M. G., & Curll, D. B., III. (1992). Boston harbor cleanup plan can be improved. In *Paper winner of the 1992 Better Government Competition*, Boston, MA, December.
30. U.S. EPA. (2014). *Water recycling and reuse: The environmental benefits*, http://www.epa.gov/region9/water/recycling
31. Shammas, N. K., & DeWitt, N. (1992). Flotation: A viable alternative to sedimentation in wastewater treatment plants. In *Water Environment Federation 65th Annual Conf., Proc. Liquid Treatment Process Symposium* (pp. 223–232). New Orleans, LA, September 20–24.
32. Harleman, D. R. F., & Murcott, S. (1991). Recent experience with chemically-enhanced primary treatment in North America and Scandinavia. *Journal of New England Water Pollution Control Association, 25*(2), 111–125.

33. Rad, H., & Cross, J. T. (1990). Chemically-assisted primary treatment: A viable alternative to upgrading overloaded treatment plants. In *63rd Annual WPCF Conf.*, Washington, DC.
34. Firmin, A. C. (1993). South Essex Sewerage District assessment of chemically-enhanced primary treatment. *Journal of New England Water Environment Association, 27*(1), 61–82.
35. Krofta, M., & Wang, L. K. (2000). *Flotation engineering* (Technical Manual No. Lenox/1-06-2000/368). Lenox, MA: Lenox Institute of Water Technology.
36. Krofta, M., & Wang, L. K. (1999). *Flotation and related adsorptive bubble separation processes* (Technical Manual No. Lenox 7-25-1999/348, 4th ed.). Lenox, MA: Lenox Institute of Water Technology.
37. Wang, L. K., Fahey, E. M., & Wu, Z. (2005). Dissolved air flotation. In L. K. Wang, Y. T. Hung, & N. K. Shammas (Eds.), *Physicochemical treatment processes*. Totowa, NJ: The Humana Press, Inc.
38. Shammas, N. K. (1997). Physicochemically-enhanced pollutants separation in wastewater treatment. In *Proc. International Conference: Rehabilitation and Development of Civil Engineering Infrastructure Systems – Upgrading of Water and Wastewater Treatment Facilities*, organized by The American University of Beirut and University of Michigan, Beirut, Lebanon, June 9–11.
39. Krofta, M., Miskovic, D., Shammas, N. K., & Burgess, D. (1994). Pilot-scale applications of a primary–secondary flotation system on three municipal wastewaters. In *Specialist Conference on Flotation Processes in Water and Sludge Treatment*, Orlando, FL, April 26–28.
40. Krofta, M., Miskovic, D., Shammas, N. K., Burgess, D., & Lampman, L. (1994). An innovative multiple stage flotation–filtration low cost municipal wastewater treatment system. In *IAWQ 17th Biennial International Conference*, Budapest, Hungary, July 24–30.
41. APHA. (2005). *Standard methods for the examination of water and wastewater* (latest edition). Washington, DC: APHA, AWWA, and WEF.
42. U.S. ACE. (2014). *Civil works construction cost index system manual*, 110-2-1304. U.S. Army Corps of Engineers, Tables, Washington, DC, p. 44, PDF file is available on the Internet at http://www.nww.usace.army.mil/cost
43. WPCF. (1985). *Clarifier design*. Manual of Practice FD-8, Water Pollut. Control Fed.
44. Ekama, G. A., Barnard, G. L., Gunthert, F. W., Krebs, P., McCorquodale, J. A., Parker, D. S., & Wahlberg, E. J. (Eds.). (1997). *Secondary settling tanks*. London: IWA Publishing.
45. Wang, L. K., Hung, Y. T., & Shammas, N. K. (Eds.). (2006). *Advanced physicochemical treatment*. Totowa, NJ: The Humana Press, Inc.
46. Asano, T., & Bahri, A. (2011). *Global challenges to wastewater reclamation and reuse*, www.worldwaterweek.org/documents
47. Lincoln Wastewater System. (2012). *Wastewater reclamation and reuse*. Blue Ribbon Presentation, Department of Public Works and Utilities, http://lincoln.ne.gov/city/pworks/waste/index.htm, April 24.
48. Aerofloat 100 DAF. (2013). *Aerofloat DAF solution for treatment of production wastewater*, www.aerofloat.com.au, March 28.
49. Water Action Hub. (2014). *Effluent management/wastewater reclamation/reuse*, Website: http://wateractionhub.org/action_areas/view/18
50. Environmental Treatment Systems, Inc. (2014). *Dissolved air flotation*, http://www.etsenvironmental.com/products-and-services/dissolved-air-flotation
51. Dissolved Air Flotation Corporation. (2014). *What is a dissolved air flotation clarifier?* http://www.dafcorp.com/aboutdaf, Kaukauna, WI.
52. Hydro Flow Technologies. (2014). *Dissolved air flotation theory and operation*, http://www.dissolved-air-flotation.hydroflotech.com/Engineering%20Data/Dissolved%20Air%20Flotation%20Theory%20of%20Operation.htm

Chapter 7
Determination of Reaeration Coefficient of Saline Receiving Water for Water Quality Management

Ching-Gung Wen, Jao-Fuan Kao, Chii Cherng Liaw, Mu-Hao S. Wang, and Lawrence K. Wang

Contents

Abstract A practical method of determining reaeration coefficients would greatly aid design engineers in determining the degree of wastewater treatment required for a proposed effluent discharge. Many previous tidal river and estuary studies emphasized mainly the effects of flow conditions (such as velocity, water depth, turbulent intensity, hydraulic gradient, etc.) and temperature on stream aeration, but

C.-G. Wen, Ph.D., P.E. (✉) • J.-F. Kao, Ph.D., P.E. • C.C. Liaw, M.E., P.E.
Department of Environmental Engineering, College of Engineering, National Cheng Kung University, Tainan, Taiwan
e-mail: t15250@mail.ncku.edu.tw; ckwen33@yahoo.com.tw

M.-H.S. Wang, Ph.D., P.E. • L.K. Wang, Ph.D., P.E.
Department of Environmental Engineering, College of Engineering, National Cheng Kung University, Tainan, Taiwan

Lenox Institute of Water Technology Newtonville, NY, USA

L.K. Wang, C.T. Yang, and M.-H.S. Wang (eds.), *Advances in Water Resources Management*, Handbook of Environmental Engineering, Volume 16,
DOI 10.1007/978-3-319-22924-9_7

the effect of salts was not seriously considered. In this research new mathematical models of reaeration coefficient considering the effect of salts have been developed for water quality analysis of tidal rivers and estuaries. The reaeration coefficient in saline water, K_{2s} (day^{-1}, base e) at any chloride concentration, chlorinity, or salinity, and at 20 °C, can be expressed by the authors' NCKU (National Cheng Kung University) equations:

$$K_{2s} = K_{2f} \exp(0.0127 \text{ Chlorinity})$$

$$K_{2s} = K_{2f} \exp(0.0000127 \text{ Chloride})$$

$$K_{2s} = K_{2f} \exp(0.007 \text{ Salinity})$$

where

K_{2s} = reaeration coefficient of saline water, day^{-1}
K_{2f} = reaeration coefficient of fresh water, day^{-1}
Chlorinity = chlorinity of receiving water, g/L
Chloride = chloride concentration of receiving water, mg/L
Salinity = salinity of receiving water, 0/00, or ppt, or parts per thousand

In this research, the step-by-step experimental procedures, experimental apparatus, and statistical analysis of the mathematical models are fully presented.

Keywords Water resources • Stream pollution • Saline water • Reaeration coefficient • Reaeration • Dissolved oxygen • Tidal river • Estuary • Chlorinity • Chloride • Salinity • Receiving water

Nomenclature

a_o	A regression coefficient (dimensionless)
a_1	A regression coefficient (dimensionless)
a_2	A regression coefficient (dimensionless)
a_n	A regression coefficient (dimensionless)
A	Area of gas–liquid interface (L^2)
b_o	A regression coefficient (dimensionless)
b_1	A regression coefficient (dimensionless)
b_2	A regression coefficient (dimensionless)
b_n	A regression coefficient (dimensionless)
B	A coefficient (dimensionless)
C	Concentration of chloride (M/L^3)
C_A	A coefficient (dimensionless)
C_B	A coefficient (dimensionless)
C_O	Dissolved oxygen concentration at the beginning of aeration test, mg/L (M/L^3)

C_S	Concentration of saturated gas in liquid or concentration of saturated dissolved oxygen in water (M/L^3)
C'	Concentration of gas in liquid, or concentration of dissolved oxygen in water at any time t(M/L^3)
d	A constant (dimensionless)
D	Oxygen deficit in water (M/L^3)
D_L	Molecular diffusivity of oxygen in water, ft^2/day (L^2/t)
D_{LO}	Molecular diffusivity of oxygen in fresh water, ft^2/day (L^2/t)
E	Dissipation energy per unit mass of flow, ft^2/s^3 (L^2/t^3)
E_e	A coefficient (dimensionless)
f	A symbol of mathematical function (dimensionless)
f_1	A symbol of mathematical function (dimensionless)
f_2	A symbol of mathematical function (dimensionless)
f_3	A symbol of mathematical function (dimensionless)
f_4	A symbol of mathematical function (dimensionless)
f_5	A symbol of mathematical function (dimensionless)
f_6	A symbol of mathematical function (dimensionless)
F	A coefficient (dimensionless)
F_r	Froude number (dimensionless)
H	Stream water depth, ft (L)
K_2	Reaeration coefficient, day^{-1} (t^{-1})
K_{2f}	Reaeration coefficient of fresh water at 20 °C(t^{-1})
K_{2s}	Reaeration coefficient of saline water at 20 °C (t^{-1})
K_L	Physical mass-transfer coefficient (L/t)
P	Mass density of fluid, lb/ft^3 (M/L^3)
p	Atmospheric pressure, mm Hg
P_s	Saturated steam pressure, mm Hg
R^2	Coefficient of determination for regression equation (dimensionless)
s	Hydraulic gradient (dimensionless)
S	Surface tension, dyne/cm (F/L)
S_a	Salinity, percent (dimensionless)
S_{eq}	Equivalent surface tension, dyne/cm (F/L)
t	Reaeration time (t)
t_o	Time zero, or beginning of aeration test (t)
T	Temperature, °C (T)
U	Velocity of stream flow, ft/day (L/t)
U_f	Friction velocity, ft/s (L/t)
U_i	First order tensor of velocity
U_j	Second order tensor of velocity
u	Absolute viscosity
v	Kinetic viscosity
V	Liquid volume (L^3)

X_i Distance in i direction (L)
X_j Distance in j direction (L)
Y Ratio of K_{2S} to K_{2f} (dimensionless)

1 Introduction

The ability of a body of water to assimilate organic waste is dependent on the oxygen absorption characteristics of the water. A practical method of determining reaeration coefficients would greatly aid the design engineer in determining the degree of wastewater treatment required for a proposed effluent discharge. There are many intrinsic stream factors which affect the reaeration coefficient of a natural receiving water. The simultaneous action of stream velocity, temperature, water depth, hydraulic gradient, turbulent intensity, and salts in water further complicates the ecological system. Many previous estuary studies emphasized on the effects of flow conditions (such as velocity, hydraulic gradient, etc.) and temperature on stream aeration, but the effect of salts was not seriously considered. Concentrations of salts in estuaries are usually high; therefore, its effect on reaeration coefficient must not be overlooked.

In this study, historical investigations of reaeration, gas adsorption, salinity, and salt pollution by various researchers have been reviewed and discussed [1–43], and extensive experiments have been conducted to determine the effect of salinity on the reaeration coefficient in receiving waters. Since the reviewed mathematical models involve the use of the US customary units, or the international SI units, the readers are referred to the literatures [44, 45] for detailed conversation factors. Technical terminologies related to this chapter [45–48] have been summarized in Glossary section, for the convenience of the readers. Based on the step-by-step experimental procedures, and statistical analysis, new reaeration coefficient models are developed to predict how the salinity will affect the reaeration coefficient in receiving waters.

2 Reaeration and Gas Adsorption

Streeter and Phelps [22] suggested that atmospheric reaeration is a first-order reaction, and is described by the equation:

$$dD/dt = -K_2D \tag{7.1}$$

where

D = oxygen deficit in water, mg/L
K_2 = reaeration coefficient, day^{-1}
t = reaeration time, day

Based on Streeter and Phelps' reaeration formula [22], Velz [28] developed a standard curve to estimate reaeration. Nemerow [17] used dissolved oxygen (DO) data and Streeter–Phelps formula to calculate reaeration coefficient. Tsivoglou et al. [24–26] directly measured the reaeration coefficient of natural streams by releasing a gaseous tracer [12]. The reaeration coefficient is usually estimated by the prediction models which had been the subject of many studies and investigations in the 1960s. The earliest and most famous model is O'Connor–Dobbins' Model [18], in which, Danckwerts Surface Renewal Model and Prandtl maxing Length Theory were introduced. The initial O'Connor–Dobbins' Model for isotropic and non-isotropic turbulent flow are presented as Eqs. 7.2 and 7.3 respectively.

$$K_2 = (D_L U)^{0.5}/H^{1.5} \tag{7.2}$$

$$K_2 = 11.04 {D_L}^{0.5} s^{0.25}/H^{1.25} \tag{7.3}$$

where

D_L = the molecular diffusivity of oxygen in water, ft^2/day
H = the stream water depth, ft
U = the velocity of stream flow, ft/day
s = hydraulic gradient, dimensionless

O'Connor and Dobbins [18] explained that the differences between the values predicted by Eqs. 7.2 and 7.3 were, in most cases, insignificant, and proposed that the former be used universally. O'Connor and Dobbins model is applicable to (a) moderately deep to deep channels; (b) less than 30 ft, but greater than or equal to 0.5 ft. of stream depth [18].

Churchill et al. [5] and Wang et al. [21–31] used Multiple Regression Techniques to relate the observed reaeration coefficients with various hydraulic parameters, and proposed an empirical prediction model which is a function of mean velocity of stream flow and mean depth. Isaacs and Gaudy [11] proposed an equation which is the same as Churchill's equation except the regression coefficient and exponential of velocity and depth. The equations proposed by Wang et al. [29–31] are slightly different. Dobbins [9] experimentally derived a dimensional equation which was:

$$K_2 = 2.76 C_A F {E_e}^{0.375} \mathrm{Coth}(B' {E_e}^{0.125}/{C_B}^{1.5} H) {C_B}^{-1.5} H^{-1} \tag{7.4}$$

where

$C_A = 1.0 + F_r$
F_r = Froude number

$C_B = 0.9 + (F_r)^{0.5}$
$F = 9.68 + 0.054(T - 20)$
$T =$ temperature (°C)
$B' = 0.976 + 0.0137(30 - T)^{1.5}$
$E_e = 30(s)U$
K_2 has units of 1/day when U and H are in FPS units.

Krenkel and Orlob [12] proposed an empirical equation which is

$$K_2 = 2.624\left(10^{-4}\right)E^{0.408}H^{-0.66} \tag{7.5}$$

where

E = dissipation energy per unit mass of flow, ft^2/s^3
H = depth, ft
K_2 = reaeration coefficient at 20 °C

Thackston and Krenkel [23] postulated that the rate of surface renewal could be related to the vertical eddy diffusion coefficient at the surface, and proposed a model:

$$K_2 = 0.000288\left(1 + {F_r}^{0.25}\right)U_f H^{-1} \tag{7.6}$$

where

U_f = friction velocity, ft/s,
H = depth, ft.

Parkhurst and Pomeroy [19] performed an experimental investigation of the reaeration coefficients in sewers, and proposed the following model for water temperature at 20 °C:

$$K_2 = 0.1917(1 + 0.17F_r)(SU)^{0.375}H^{-1} \tag{7.7}$$

K_2 has the unit of 1/day with the other terms in FPS system. Parkhurst and Pomeroy's model was developed from data collected in 12 sewers and in natural streams [19].

After using a tracer to measure actual reaeration coefficient, Tsivoglou et al. [24–27] proposed that the reaeration coefficient be proportional to the water surface elevation change in the stream reach. The other predicting models of reaeration coefficient include Negulescu's model [16], Bennett's model [37], Cadwallader's model [3], and Lau's model [14], etc. Wilson and Macleod [37] statistically tested 16 models with a set of 400 experimental measurements, and fond that Dobbins model [9, 10] and Parkhurst's model [19], gave the test general fit to all the data investigated. Brown [2] tested the lack of fit for O'Connor–Dobbins' model, Churchill's model, Isaac's models and other models which are only the functions of stream velocity and depth, and indicated that all models are lack of fit at 5 % significant level.

The rate of gas (say oxygen) adsorbed into agitated liquids without chemical reaction is expressed by [1]:

$$dC'/dt = K_L A\, V^{-1}(C_S - C') \tag{7.8}$$

where

K_L = the physical mass-transfer coefficient, L/t
C_S = the concentration of saturated gas in liquid, M/L^3
C' = the concentration of gas in liquid, M/L^3
A = the area of gas–liquid interface, L^2
V = liquid volume, L^3

In a natural stream, V/A is approximately equal to average depth (H) of stream; therefore, Eq. 7.8 becomes.

$$dC'/dt = K_L H^{-1}(C_S - C') = K_L H^{-1} D \tag{7.9}$$

Comparing Eqs. 7.1 and 7.9, one concludes that reaeration coefficient equals the mass transfer coefficient divided by stream depth. In chemical engineering, there are many models describing the mass transfer coefficient, such as film model, still surface model, surface renewal model [7], Lamont's (or eddy cell) model [13], Levich's model [8, 20], and Kramei-Oishi's model [4]. Lamont's model and Kramei-Oishi's model are presented as Eqs. 7.10 and 7.11 respectively:

$$K_L = 0.41\left(vD_L{}^{-1}\right)^{-0.5}(vE)^{0.25} \tag{7.10}$$

$$K_L = 0.32 D_L{}^{0.5} U_f{}^{1.5} p^{0.5} S_{eq}{}^{-0.5} \tag{7.11}$$

where

v = kinetic viscosity
p = mass density of fluid
S_{eq} = equivalent surface tension

All parameters in Eq. 7.10 are in CGS units, while in Eq. 7.11 are in MKS units. Later on, Prasher [20], proved that Lamont's model is more suitable for stream and stirred tank.

3 Effect of Salinity on Reaeration Coefficient

Reaeration coefficient is affected by temperature, molecular diffusivity of oxygen in water, mass density of water, viscosity, surface tension, dissipation energy per unit mass of fluid, water depth, velocity, hydraulic gradients, etc. If the water temperature is kept constant, reaeration coefficient can be expressed as follows:

$$K_2 = f(D_L, p, u, S, E, H, U_f, s) \tag{7.12}$$

The velocity, water depth, and. hydraulic gradient are independent of salinity. Maa and Tsay [15] concluded that the relationships between salinity and viscosity, density, or surface tension at 20 °C are

$$\text{Log } u = 0.40698 + 0.00235\left({S_a}^{1.45}\right) \tag{7.13a}$$

$$p = 62.286 + 0.489(S_a) \tag{7.13b}$$

$$S = 72.626 + 0.216\left({S_a}^{1.1516}\right) \tag{7.13c}$$

where

u = viscosity, lb/ft/h
S_a = percent salinity
p = mass density of saline water, lb/ft^3
S = surface tension of saline water, dyne/cm

In Eq. 7.13, u, p, and S are all functions of S_a, and can be expressed by Eq. 7.14:

$$u = f_1(S_a) \tag{7.14a}$$

$$p = f_2(S_a) \tag{7.14b}$$

$$s = f_3(S_a) \tag{7.14c}$$

where

f_1, f_2, f_3 = symbols that represent mathematical functions.

Ratclife [21] experimentally verified that the molecular diffusivity of oxygen in saline water (D_L) is

$$D_L = D_{LO}(1 - dS_a) \tag{7.15}$$

where d is a constant; D_{LO} and D_L are the molecular diffusivity of oxygen in fresh water and saline water respectively. D_{LO} and d are independent of S_a; therefore, Eq. 7.15 can be expressed by

$$D_L = f_4(S_a) \tag{7.16}$$

The dissipation energy per unit mass of fluid (E) can be expressed by Eq. 7.17.

$$E = up^{-1}\left[\left(\partial U_i/\partial x_j + \partial U_j/\partial x_i\right)\left(\partial U_j/\partial x_i\right)\right]^{0.5} \tag{7.17}$$

where U_i and U_j are first order tensor and second order tensor, respectively, of velocities; X_i and X_j are distances in i direction and j direction, respectively.

Since there is a little effect of salinity on velocity gradients, (i.e. $\partial U_i/\partial x_j$ and $\partial U_j/\partial x_i$), the dissipation energy can be rewritten by

$$E = f_5(u, p) = f_6(S_a) \tag{7.18}$$

Combined Eqs. 7.14, 7.16, and 7.18, reaeration coefficient is a function of S_a, if temperature is constant, and can be expressed by Eq. 7.19.

$$K_2 = f(S_a) \tag{7.19}$$

The function $f(S_a)$ is very difficult to be derived theoretically, but can be developed by a polynomial regression analysis. Since the salinity of estuary water is proportional to the concentration of chloride (C), K_2 is also a function of C. Let Y be the estimated ratio of the reaeration coefficient of saline water (K_{2s}) at 20 °C and a chloride concentration C to the reaeration coefficient of fresh water (K_{2f}) at 20 °C and zero chloride concentration. The polynomial regression equation of Y and C is

$$Y = K_{2s}/K_{2f} = a_0 + a_1C + a_2C^2 + \ldots\ldots. + a_nC^n \tag{7.20}$$

or

$$Y = K_{2s}/K_{2f} = \exp(b_0 + b_1C + b_2C^2 + \ldots.. + b_nC^n) \tag{7.21}$$

where $a_0, a_1, a_2, \ldots\ldots a_n$, and $b_0, b_1, b_2, \ldots\ldots b_n$, are regression coefficients. Since K_{2s} is equal to K_{2f} at $C = 0$, from Eqs. 7.20 and 7.21, we obtain $a_0 = 1$ and $b_0 = 0$. Therefore, Eqs. 7.20 and 7.21 become

$$Y = 1 + a_1C + a_2C^2 + \ldots\ldots\ldots\ldots + a_nC^n \tag{7.22}$$

$$Y = \exp\left(b_1C + b_2C^2 + \ldots\ldots\ldots\ldots + b_nC^n\right) \tag{7.23}$$

A practical method of determining reaeration coefficients would greatly aid design engineers in determining the degree of wastewater treatment required for a proposed effluent discharge. The tidal river and estuary studies shall emphasize not only the effects of flow conditions (such as velocity, water depth, turbulent intensity, hydraulic gradient, etc.) and temperature on stream aeration, but also the effect of salts. In a research conducted by National Cheng Kung University, Taiwan, and the Lenox Institute of Water Technology (formerly Lenox Institute for Research), USA, a new mathematical model of reaeration coefficient considering the effect of salts has been developed for water quality analysis in tidal rivers and estuaries. The reaeration coefficient in (day^{-1}, base e) at any chloride concentration C (g/L) and 20 °C can be expressed by

$$K_{20} = K_{2f}\exp(0.0127C) \tag{7.24}$$

in which K_{2f} is the reaeration coefficient in fresh water at 20 °C. The correlation coefficient of K_{2s}/K_{2f} and C is 0.88. The experimental evaluation of this important mathematical model presented in Sects. 4–8. National Cheng Kung University, Stevens Institute of Technology and Lenox Institute of Water Technology have jointly conducted many research projects on river and estuary studies [29–36, 38].

In a recent study, a mathematical model describing the effect of salinity on reaeration coefficients of receiving waters, such as tidal rivers and estuaries, has been introduced [36]. This technical paper presents the step-by-step experimental procedures, experimental apparatus, and statistical analysis of the mathematical model. It is concluded that the model is valid and acceptable for tidal river and estuary studies.

4 Experimental Apparatus

The experimental apparatus shown in Fig. 7.1 was designed by the authors for use in this investigation. All tests were conducted in a temperature controlled room at 20 ± 1 °C. The water temperature in plastic container was controlled at 20 ± 0.2 °C

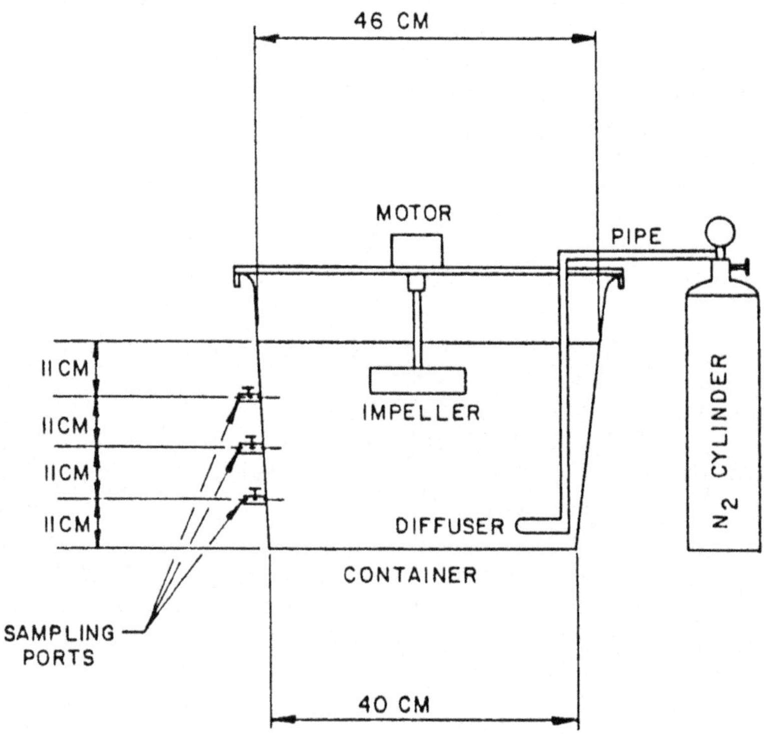

Fig. 7.1 Experimental apparatus

during the experimental period. The components of experimental apparatus used in this study are described below:

a. Plastic Container—The container for saline water testing was a 60-L plastic tank having 44 cm height, 46 cm inside diameter at top and 40 cm inside diameter at bottom. There were three 0.8 cm diameter sampling port on the container wall at different heights.
b. Surface Aeration Devices—Aeration devices consisted of a motor and an impeller. The angular velocities of motors controlled in this experiment were 3.3, 10, 19, 28, 56, 86, and 106 rpm (revolutions per minute). The impeller was 18 cm in width and 2.3 cm in depth, and was under the same depth from water surface for each run of the experiment in order to keep the same rate of aeration.
c. Nitrogen Diffuser—A rubber tubing was used to disperse nitrogen gas throughout the water in the container for deaeration. The nitrogen gas was supplied from a cylinder through a dual gauge and needle valve arrangement for regulation of flow.
d. Reagents—Salt for preparation of saline water was supplied by Taiwan Salt Works, and contained 99.9 % NaCl. Other reagents for titration of dissolved oxygen by Winkler Method and determination of sodium chloride Mercuric Nitrate Method were all supplied by MERCK chemical Inc., Germany, and were all Reagent Grade. The Winkler Method and Mercuric Nitrate Method are documented elsewhere [49].

5 Experimental Procedure

The following are general procedures used for all reaeration experiments:

a. Fill the plastic container with 54 L stilled water and adjust water temperature to 20 °C with a heat exchanger.
b. Deaerate the water with nitrogen gas until dissolved oxygen (DO) in water is decreased to 1.5 mg/L.
c. Remove nitrogen gas bubbling device, then start the motor to agitate the water for 10–30 min at one of the desired rotating speeds of the motor.
d. Record the aeration time and take samples to analyze dissolved oxygen concentration according to Standard Methods [49]. The interval time of sampling ranges from 3 to 12 h
e. Calculate aeration coefficient by the following formula

$$K_2 = [\ell n(C_s - C_o) - \ell n(C_s - C')]/(t - t_o)$$

where

C_S = the concentration of saturated dissolved oxygen, mg/L
= $0.0678\ (P - P_s)(1 - 0.01C)/(35 + T)$, for water temperature ranging from 0 to 30 °C

P = Atmospheric pressure, mm Hg
P_s = Saturated steam pressure, mm Hg, at temperature T (°C) and chloride concentration C (g/L)
C_o = Dissolved oxygen concentration at time zero (t_o), mg/L
C′ = Dissolved oxygen concentration at any time (t), mg/L
t_o = Time zero f or beginning of aeration
t = Aeration time

f. Repeat 4 times from procedure a to procedure e.
g. Change the concentration of chloride by an increment of about 5000 mg/L until total concentration of chloride being about 25,000 mg/L, and
h. Repeat the procedure a to procedure f using a new motor rotating speed in Procedure c.

6 Experimental Results

One hundred and sixty-eight experiments were completed. Average reaeration coefficient (K_2) determined under various rotating speeds and chloride concentrations are listed in Table 7.1. Standard deviations and coefficients of variation are also listed in the table. The ratio of reaeration coefficient at any chloride concentration (C) and 20 °C to that at zero chloride concentration and 20 °C is defined as

$$Y = K_{2S}/\,K_{sf}$$

Table 7.2 documents the values of this ratio determined under various experimental conditions. It should be noted that the saturated dissolved oxygen concentration (C_s) in water at various chloride concentrations (C), temperatures (T) and pressures (P and Ps) can also be easily calculated using a mathematical model recently developed by Wang and Elmore [30].

7 Analysis of Results

7.1 Model Selection

It is difficult to derive the relationship of reaeration coefficient and chloride concentration physically and mathematically. The only feasible method is statistical regression analysis. The general regression equations are shown in Eqs. 7.22 and 7.23. From these equations, one can choose the best fit of regression equation. Initially the first order equations are tried:

$$\text{Linear Model}: \quad Y = 1 + a_1C \tag{7.25}$$

Table 7.1 Experimental data of reaeration experiments

Chloride concentration, g/L	Rotating speed, rpm	Reaeration coefficient study at 20 °C		
		Mean K_2, day^{-1}	Standard deviation	Coefficient of variation, %
0	3.3	0.3871	0.0645	16.7
0	9.6	0.6527	0.0701	10.7
0	19	0.9416	0.0797	8.5
0	28	1.0664	0.0473	4.4
0	56	1.6371	0.0586	3.6
0	86	2.2364	0.1352	6.0
0	106	2.6380	0.0892	3.4
5.602	3.3	0.4327	0.0193	4.5
5.602	9.6	0.7071	0.0339	4.8
5.602	19	0.9622	0.0474	4.9
5.602	28	1.1459	0.0718	6.3
5.602	56	1.7627	0.1184	6.7
5.602	86	2.5194	0.0586	2.3
5.602	106	3.0229	0.0976	3.2
9.291	3.3	0.4022	0.0637	15.8
9.291	9.6	0.6956	0.0466	6.7
9.291	19	1.0206	0.0263	2.6
9.291	28	1.2327	0.0104	0.8
9.291	56	1.6943	0.0964	5.7
9.291	86	2.6400	0.1388	5.2
9.291	106	3.1813	0.0596	1.8
14.250	3.3	0.4727	0.0448	9.5
14.250	9.6	0.7620	0.0317	4.2
14.250	19	1.0051	0.1052	10.5
14.250	28	1.2844	0.0379	2.9
14.250	56	1.8101	0.0603	3.3
14.250	86	2.8021	0.0732	2.6
14.250	106	3.3566	0.1383	4.1
20.745	3.3	0.5152	0.0870	1.7
20.745	9.6	0.7999	0.0259	3.2
20.745	19	1.1222	0.0547	4.9
20.745	28	1.3603	0.0583	4.3
20.745	56	2.1255	0.0566	2.7
20.745	86	3.2567	0.0731	2.3
20.745	106	3.6051	0.1300	3.6
24.471	3.3	0.5345	0.0266	5.0
24.471	9.6	0.7777	0.0249	3.2
24.471	19	1.1346	0.0512	4.5
24.471	28	1.4565	0.0378	2.5
24.471	56	2.0039	0.0193	1.0
24.471	86	3.6275	0.1814	5.0
24.471	106	3.8435	0.0939	2.4

Table 7.2 Ratio (Y) of reaeration coefficient at any chloride concentration to that at zero chloride concentration, both at 20 °C

	Values of Y ratio at various chloride concentrations, g/L					
Rotation speeds, rpm	0	5.602	9.291	14.250	20.745	24.471
106	1	1.1132	1.2059	1.2724	1.3667	1.4570
86	1	1.0939	1.1805	1.2529	1.4562	1.6220
56	1	1.0767	1.0349	1.1057	1.2983	1.2241
28	1	1.0439	1.5181	1.2067	1.2780	1.3684
19	1	1.0219	1.0839	1.0673	1.1918	1.2050
10	1	1.0833	1.0656	1.1675	1.2255	1.1914
3.3	1	1.1178	1.0390	1.2211	1.3309	1.3807

$$\text{Exponential Model}: \mathrm{Y} = \exp(\mathrm{b_1C}) \tag{7.26}$$

With the experimental data in Table 7.2, regression equations of Eqs. 7.25 and 7.26 were determined by the least squares method, and expressed as follows:

$$\text{Linear Model}: \quad \mathrm{Y} = 1 + 0.0146\mathrm{C} \tag{7.27}$$

$$\text{Exponential Model}: \mathrm{Y} = \exp(0.0126\mathrm{C}) \tag{7.28}$$

The observed data and regression equations are both shown in Figs. 7.2 and 7.3. The coefficient of determinations, R^2, of Eqs. 7.27 and 7.28 are 0.738 and 0.771 respectively. The sums square of error between observed data and estimated values of Eqs. 7.27 and 7.28 are 0.2498 and 0.1476 respectively.

The second order of Eqs. 7.22 and 7.23 was also determined.

$$\text{Linear Model}: \quad \mathrm{Y} = 1 + \mathrm{a_1C} + \mathrm{a_2C^2} \tag{7.29}$$

$$\text{Exponential Model}: \mathrm{Y} = \exp(\mathrm{b_1C} + \mathrm{b_2C^2}) \tag{7.30}$$

It was found that the multiple determination coefficients of Eqs. 7.29 and 7.30 were not better than that of Eqs. 7.27 and 7.28, respectively. In other words, there are no improvements in predicting Y values even the second order C^2 values are taken into consideration in the regression equations. Accordingly the best equation chosen as the correction model of reaeration coefficient for chloride modification is the first order Exponential Model, or Eq. 7.28.

7.2 *Analysis of Variance*

In order to test whether the regression coefficient of Exponential Model, Eq. 7.28, is equal to zero or not, analysis of variance has been conducted. From the

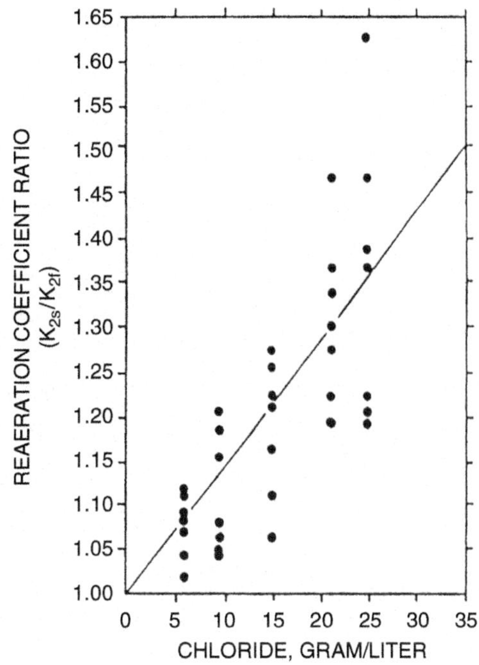

Fig. 7.2 Linear model

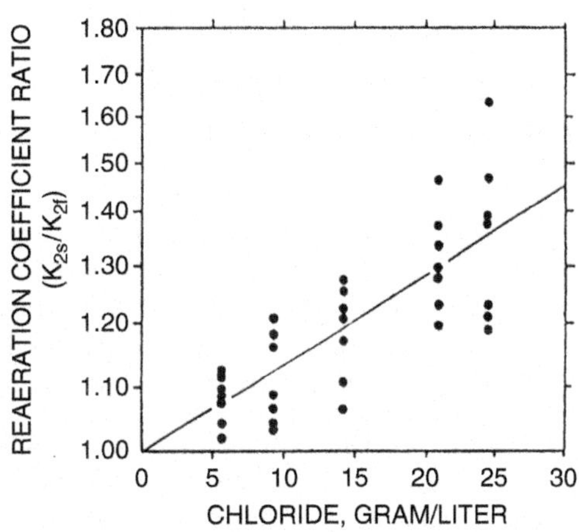

Fig. 7.3 Exponential model

observed data, the results of variance analysis are calculated. Since the calculated F value (415.8) exceeds F (1, 41, 0.05) = 4.08, the regression coefficient of data population is not equal to zero, the regression equation can be accepted under 5 % significance level.

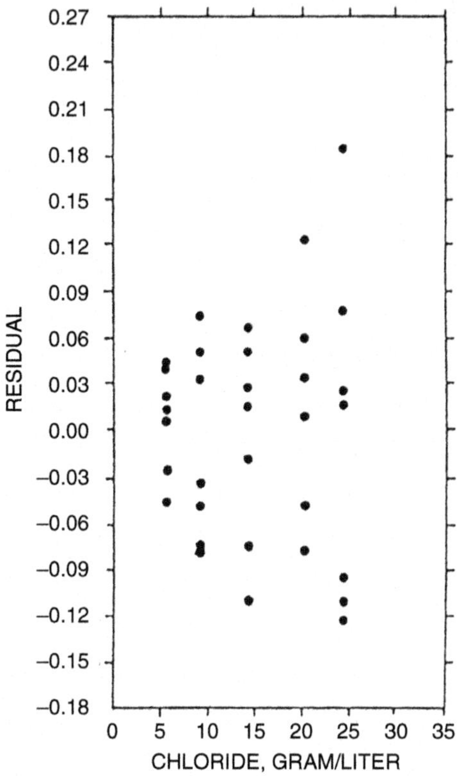

Fig. 7.4 Residual distribution of exponential model

7.3 Examination of Residuals

In performing the regression analysis certain assumptions are made about the errors. The usual assumptions are that the errors are independent, have zero mean and a constant variance, and follow a normal distribution. Thus, if the fitted model is correct, the residuals should exhibit tendencies that tend to confirm the aforementioned assumptions or at least f should not exhibit a denial of the assumptions.

a. Test a Constant Variance: Plot residuals versus chloride in Fig. 7.4. From Fig. 7.4, it is seen that the variance is not a constant, and the regression coefficients must be solved by weighted least squares. Using 1/x as the weighting factor and using the least squares method the following Exponential Model is derived

$$Y = \exp(0.0127C) \tag{7.31}$$

b. Test the Randomness and Independence of Residues: The run test indicates no evidence of time dependent non-randomness of residuals.

c. The Test of Normal Distribution: The distribution of residuals was tested by Kolmogorov–Smirnov's test. Since the maximum absolute value between empirical cumulative probability and the probability of normal distribution is 0.173 which is less than 0.23 for 35 samples under 5 % significance level, the hypothesis of normal distribution of residual can be accepted.

8 Conclusions

The reaeration coefficient of an estuary increases with increasing chloride concentration. The reaeration coefficient, K_{2S} (day^{-1}, base e) at any chloride concentration C (g/L) and at 20 °C can be expressed by

$$K_{2S} = K_{2f}\exp(0.0127C) \tag{7.24}$$

in which K_{2f} is reaeration coefficient in fresh water at 20 °C, The relative coefficient of K_{2s}/K_{2f} and C is 0.88.

9 Related Scientific Developments and Examples

Reaeration process is the process of oxygen exchange between the atmosphere and a water body in contact with the atmosphere, and is modeled as the product of reaeration coefficient multiplied by the difference between dissolved oxygen saturation and the actual dissolved oxygen concentration [48], that is:

$$F_C = K_2(C_s - C) = (K_L/H)(C_s - C) \tag{7.32}$$

where

F_c = rate or flux of dissolved oxygen across the water body, $M/L^3/T$
C = dissolved oxygen concentration, M/L^3
C_s = saturation dissolved oxygen concentration, M/L^3
K_2 = reaeration coefficient, 1/T
H = water depth, L
K_L = surface transfer coefficient, L/T.

Reaeration coefficient is a mass transfer coefficient (K_2) in the reaeration process. In this study, historical investigations of reaeration, gas adsorption, salinity, and salt pollution by various researchers have been reviewed and discussed [1–42], and extensive experiments have been conducted to determine the effect of salinity on the reaeration coefficient in receiving waters. Salinity is the total concentration of all ionic constituents present in water. This is comprised mostly by chloride ions and sodium ions. Seawater may have other ions, such as potassium ions, magnesium ions, sulfate ions, etc. Salinity is traditionally defined as the total solids in water after all carbonates have been converted to oxides, all bromide and iodide have been replaced by chloride, and all organic matter has been oxidized. The new scale used to define salinity is based on the electrical conductivity of seawater relative to a specified solution of potassium chloride ($KC\ell$) and water (H_2O). The scale is dimensionless and the traditional dimensions of parts per thousand (i.e., mg/g of solution) no longer applies. The unit of salinity is ppt, or parts per thousand, or g/L [48].

Chlorinity is defined in relation to salinity as follows:

$$\text{Salinity} = 1.80655(\text{Chlorinity}) \tag{7.33}$$

Although chlorinity is not equivalent to chloride concentration, the factor for translating a chloride determination in seawater to include bromide, for example, is only 1.0045 based on the molecular weights and the relative amounts of the two ions. Therefore, for practical purposes, chloride (in mg/g of solution, or g/L of solution) is nearly equal to chlorinity in seawater. Or 1 g/L of chloride concentration = 1 g/L of chlorinity in seawater [48].

Combination of Eqs. 7.24 and 7.33 gives Eq. 7.34:

$$K_{2S} = K_{2f}\exp(0.007\ \text{Salinity}) \tag{7.34}$$

In which Salinity is in the unit of ppt (parts per thousand, or g/L).

Equation 7.24 can also be rewritten as Eq. 7.35 using chlorinity instead of chloride concentration:

$$K_{2S} = K_{2f}\exp(0.0127\ \text{Chlorinity}) \tag{7.35}$$

where Chlorinity is in the unit of g/L.

9.1 Determination of Chlorinity and Salinity

A saline water has a chloride concentration of 20,000 mg/L. What is the saline water's salinity?

Solution

Chloride concentration = 20,000 mg/L = 20 g/L

Chlorinity = 20 g/L

Salinity = 1.8066 × Chlorinity = 1.8066 × 20 = 36.1 ppt

9.2 Determination of Chloride Concentration and Reaeration Coefficient in Saline Water

Assume the reaeration coefficient (base e) in fresh water at 20 °C is 0.23 day^{-1}. A saline water has a salinity of 36.1 ppt. Determine the saline water's chloride concentration in mg/L and its reaeration coefficient (base e) at 20 °C.

Solution
Salinity = 36.1 ppt = 1.8066 Chlorinity
Chlorinity = 20 g/L = C
Chloride concentration = 20 g/L = 20,000 mg/L

$$
\begin{aligned}
K_{2S} &= K_{2f} \exp(0.007\ \text{Salinity}) \\
&= 0.23 \exp(0.007 \times 36.1) \\
&= 0.23 \exp(0.253) \\
&= 0.23 \times 2.71828^{0.253} \\
&= 0.297\ \ \text{day}^{-1}(\text{base e})\ \text{at}\ 20^{\circ}\text{C}
\end{aligned}
$$

Or

$$
\begin{aligned}
K_{2S} &= K_{2f} \exp(0.0127C) \\
&= 0.23 \exp(0.0127 \times 20) \\
&= 0.23 \exp(0.254) \\
&= 0.23 \times 2.71828^{0.254} \\
&= 0.297\,\text{day}^{-1}(\text{base e})\ \text{at}\ 20^{\circ}\text{C}.
\end{aligned}
$$

Acknowledgement This investigation was supported by a research grant from the National Science Council of the Republic of China, Taiwan, to which the authors wish to express their sincere thanks.

Glossary

Adsorption (a) A physicochemical, passive and reversible process that attracts and adsorbs molecules of a gas, liquid, or dissolved substance to a surface of adsorbent, (b) The interaction of a analyte with the surface of a matrix.

Aeration A gas transfer process that allows for the absorption of gas (such as oxygen) by liquid (such as water or wastewater).

Chloride It is one of the major inorganic anions, Or negative ions, in saltwater and freshwater. It originates from the dissociation of salts, such as sodium chloride or calcium chloride, in water. Chlorides are binary compounds of chlorine. 1 g/L of chloride concentration = 1 g/L of chlorinity = 1.80655 ppt (or g/L) of salinity in seawater.

Chlorinity Chlorinity is defined in relation to salinity as follows: Salinity = 1.80655 (Chlorinity). Although chlorinity is not equivalent to chloride concentration, the factor for translating a chloride determination in seawater to include bromide, for example, is only 1.0045 based on the molecular weights and the relative amounts of the two ions. Therefore, for practical purposes, chloride (in mg/g of solution, or g/L of solution) is nearly equal to chlorinity in seawater. Or 1 g/L of chloride concentration = 1 g/L of chlorinity in seawater. For wastewater, a knowledge of the ions responsible for the solution's electrical conductivity is necessary to correct for the ions impact on oxygen solubility and use of the tabular value or the equation is inappropriate unless the relative composition of the wastewater is similar o seawater.

Dissolved oxygen (DO) The amount of oxygen dissolved in water. The amount is usually expressed in parts per million (ppm) or milligrams per liter (mg/L).

Dynamic viscosity It is the property of a fluid whereby it tends to resist relative motion within itself. It is the shear stress, i.e., the tangential force on unit area, between two infinite horizontal planes at unit distance apart, one of which is fixed while the other moves with unit velocity. In other words, it is the shear stress divided by the velocity gradient, i.e., $(N/m^2)/(m/s/m) = N\ s/m^2$.

Fresh water Water that contains less than 1000 mg/L of dissolved solids; generally, more than 500 mg/L of dissolved solids is undesirable for drinking and many industrial uses.

Froude number (F_r) It is a numerical quantity used in open-channel flow studies or in cases in which the free surface plays an essential role in influencing motion. It is given by the equation of $F_r = V^2/(gL)$ where V = the characteristic velocity; g = the gravitational constant; and L = the characteristic linear dimension.

Gas adsorption An adsorption process that attracts and adsorbs a gas to a liquid (such as water), or a solid (such as granular activated carbon).

Kinematic viscosity It is the dynamic viscosity of a fluid divided by its density, i.e., W/m^2.

Mass transfer coefficient (a) A constant of proportionality that is specific to an individual compound and is used in a mass transfer expression to determine equilibrium conditions between two phases, (b) Mass transfer coefficients are determined experimentally; the units will depend on the nature of the mathematical expression and the phase transfer.

Molecular diffusion A physical–chemical process whereby mobile compounds (dissolved or suspended in another compound) move from areas of high concentration to areas of low concentration.

Molecular diffusion coefficient of oxygen (a) A coefficient (D_L) for diffusion of oxygen into water in a natural aquatic system; (b) The molecular diffusion coefficient (D_L) of oxygen at 20 °C = 1.76×10^{-4} m^2/day. (c) The molecular diffusion coefficient (D_L) of oxygen at any temperature = $(1.76 \times 10^{-4}\ m^2/day) \times (1.037^{T-20})$.

NCKU (National Cheng Kung University) equations They are reaeration coefficient equations developed by National Cheng Kung University, Taiwan, showing the effect of salinity on receiving water's reaeration coefficient. The NCKU equations are modeled by $K_{2s} = K_{2f}$ exp (0.0127 Chlorinity); $K_{2s} = K_{2f}$ exp (0.0000127 Chloride); and $K_{2s} = K_{2f}$ exp (0.007 Salinity); in which K_{2s}: reaeration coefficient of saline water, day^{-1}; K_{2f} = reaeration coefficient of fresh water, day^{-1}; Chlorinity = chlorinity of receiving water, g/L; Chloride = chloride concentration of receiving water, mg/L; and Salinity = salinity of receiving water, o/oo, or ppt, or parts per thousand.

Nitrate nitrogen A common way to report nitrate concentration (expressed as nitrate-nitrogen or NO_3^--N).

Nitrification A process of formation of nitrate nitrogen from reduced inorganic nitrogen compounds, such as ammonia nitrogen. Nitrification in the natural environment is carried out primarily by autotrophic bacteria.

Nitrite nitrogen A common way to report nitrite concentration (expressed as nitrate-nitrogen or NO_2^--N).

Non-tidal stream/river A stream/river which water level and flow direction will not fluctuate and will not be affected by the action of lunar and solar forces upon the rotating Earth.

Oxygen-sag curve See dissolved oxygen sag curve.

Photosynthesis The conversion of light energy to chemical energy. At night, this process reverses: plants and algae suck oxygen out of the water.

Reaeration (a) The physical chemical reaction by which oxygen is absorbed back into water, (b) An aeration process by which oxygen in air is absorbed back into natural water, such as stream water and lake water, (c) A natural process of oxygen exchange between the atmosphere and a natural water body in contact with the atmosphere. Typically, the net transfer of oxygen is from the atmosphere and into the water, since dissolved oxygen levels in most natural waters are below saturation. When photosynthesis produces supersaturated dissolved oxygen levels, however, the net transfer is back into the atmosphere, (d) Reaeration process is modeled as the product of reaeration coefficient multiplied by the difference between dissolved oxygen saturation and the actual dissolved oxygen concentration, that is: $F_c = K_2(C_s - C) = (K_L/K)(C_s - C)$. Here F_c = rate or flux of dissolved oxygen across the water body, $M/L^3/T$; C = dissolved oxygen concentration, M/L^3; C_s = saturation dissolved oxygen concentration, M/L^3; K_2 = reaeration coefficient, 1/T; H = water depth, L; and K_L = surface transfer coefficient, L/T.

Reaeration coefficient A mass transfer coefficient (K_2) in reaeration process. See reaeration and mass transfer coefficient.

Reaeration rate (a) The rate at which oxygen is absorbed back into water. This is dependent, among other things, upon turbulence intensity, temperature, and the water depth, (b) The reaeration rate is defined as the rate of dissolved oxygen across the water body $F_c = K_2(C_s - C)$. Here F_c = rate or flux of dissolved oxygen across the water body, $M/L^3/T$; C = dissolved oxygen concentration, M/L^3; C_s = saturation dissolved oxygen concentration, M/L^3; K_2 = reaeration coefficient, 1/T.

Reaeration rate coefficient See reaeration coefficient.

Receiving waters A river, lake, ocean, stream, or other bodies of water into which wastewater or treated effluent is discharged.

Regression analysis (a) A statistical method in which an empirical relationship between and independent variable ad a dependent variable can be determined such that the average tendency of the observed values and the average tendency of the predictive values given by the empirical equation may be identical, (b) A statistical method attempts to determine the best mathematical model, given the available data, to describe a dependent variable as a function of an independent variable or, in the case of multiple regression analysis, more than one independent variable.

Regression coefficient (a) A derived coefficient in regression analysis that expresses the change in the dependent variable associated with a change in one or more independent variables, (b) It is referred to as the slope of the relationship between the variables, as the derivative in bivariate analysis, or as the partial derivative in multiple regression.

Saline water Water that contains significant amounts of dissolved solids. Here are our parameters for saline water: freshwater, less than 1000 mg/L; slightly saline water, from 1000 to 3000 mg/L; moderately saline water, from 3000 to 10,000 mg/L; and highly saline water, from 10,000 to 35,000 mg/L.

Salinity (a) Salinity is the total concentration of all ionic constituents present in a water. This is comprised mostly by chloride ions and sodium ions. Seawater may have other ions, such as potassium ions, magnesium ions, sulfate ions, etc.; (b) Salinity is defined in relation to chlorinity as follows: Salinity = 1.80655 (Chlorinity); (c) Salinity is traditionally defined as the total solids in water after all carbonates have been converted to oxides, all bromide and iodide have been replaced by chloride, and all organic matter has been oxidized. The new scale used to define salinity is based on the electrical conductivity of seawater relative to a specified solution of potassium chloride (KCl) and water (H_2O). The scale is dimensionless and the traditional dimensions of parts per thousand (i.e., mg/g of solution) no longer applies. The unit of salinity is ppt, or parts per thousand, or g/L.

Surface transfer coefficient (a) A mass transfer coefficient (K_L) which governs the rate for transferring dissolved oxygen across the water surface, L/T (b) A mass transfer coefficient which is defined as $K_L = (K_2)H$. Here K_2 = reaeration coefficient, 1/T; and H = water depth, L. See mass transfer coefficient, reaeration coefficient, and reaeration.

References

1. Bird, R. B., & Stewart, W. E. (I960). *Lightfoot EN transport phenomena*. New York: Wiley.
2. Brown, L. C. (1974). Statistical evaluation of reaeration prediction equations. *Journal of the Environmental Engineering Division, ASCE, 100*(EE5), 1051–1068.
3. Cadwallader, T. E., & McDonnell, A. J. (1969). A multivariate analysis of reaeration data. *Water Research, 3*, 731–742.
4. Chung, D. K., & Mills, A. F. (1976). Experimental study of gas adsorption into turbulent falling films of water and ethylene glycol water mixtures. *International Journal of Heat and Mass Transfer, 19*, 51–59. Pergamon Press.
5. Churchill, M. A., Elnore, H. L., & Buckingham, R. A. (1962). The prediction of stream reaeration rates. *International Journal of Air and Water Pollution, 6*, 467–504.
6. Danckwerts, P. V. (1951). Significance of liquid-film coefficients in gas adsorption. *Industrial and Engineering Chemistry, 43*(6), 1460–1467.
7. Danckwerts, P. V. (1970). *Gas–liquid reactions* (pp. 98–103). Copyright in Taiwan.
8. Davies, J. T. (1972). *Turbulence phenomena* (p. 187). New York: Academic Press.
9. Dobbins, W. E. (1964). BOD and oxygen relationships in streams. *Journal of the Sanitary Engineering Division, ASCE, 90*(SA3), 53–78.
10. Dobbins, W. E. (1965). Discussion of BOD and oxygen relationships in streams. *Journal of the Sanitary Engineering Division, ASCE, 91*(SA5), 49–55.
11. Isaacs, W. P., & Gaudy, A. F., Jr. (1968). Atmospheric oxygenation in a simulated stream. *Journal of the Sanitary Engineering Division, ASCE, 94*(SA2), 319–344.
12. Krenkel, P. A., & Orlob, G. T. (1962). Turbulent diffusion and the reaeration coefficient. *Journal of the Sanitary Engineering Division, ASCE, 88*(SA2), 53–83.
13. Lamont, J. C., & Scott, D. S. (1970). An eddy cell model of mass transfer into the surface of a turbulent liquid. *AIChE Journal, 16*(4), 513–519.
14. Lau, L. Y. (1972). Prediction equation for reaeration in open channel flow. *Journal of the Sanitary Engineering Division, ASCE, 98*(SA6), 1063–1068.
15. Maa, J. R., & Tsay, S. Y. (1972). *Studies of a modified multistage flash distillation process for producing fresh water and concentrated brine* (p. 45). Taiwan: National Science Council Engineering Science Research Center.
16. Negulescu, M., & Rojanski, V. (1969). Recent research to determine reaeration coefficient. *Water Research, 3*, 189–202.
17. Nemerow, N. L. (1971). *Theories and practices of industrial waste treatments*. New York: Addison-Wesley.
18. O'Connor, D. J., & Dobbins, W. E. (1958). The mechanism of reaeration in natural streams. *Transactions of the American Society of Civil Engineers, 123*, 641–666.
19. Parkhurst, J. D., & Pomeroy, R. D. (1972). Oxygen adsorption in streams. *Journal of the Sanitary Engineering Division, ASCE, 98*(SA1), 101–124.
20. Prasher, B. D., & Wills, G. B. (1973). Mass transfer in an agitated vessel. *Industrial and Engineering Chemistry Process Design and Development, 12*(3), 351–353.
21. Ratclife, G. A., & Holdcroft, J. G. (1963). Diffusivities of gases in aqueous electrolyte solution. *Transactions of the Institution of Chemical Engineers, 41*, 315.
22. Streeter, H. W., & Phelps, E. B. (1926). The rate of atmospheric reaeration of sewage polluted streams. *Transactions of the American Society of Civil Engineers, 89*, 1351.
23. Thackston, E. L., & Krenkel, P. A. (1969). Reaeration prediction in natural streams. *Journal of the Sanitary Engineering Division, ASCE, 95*(SA1), 65–94.
24. Tsivoglou, E. C., O'Connell, R. L., Walter, C. M., Godsil, P. J., & Logsdon, G. S. (1972). Tracer measurement of stream reaeration, II. Field studies. *Journal of the Water Pollution Control Federation, 40*, 2.
25. Tsivoglou, E. C., Cohen, J. B., Shearer, S. D., & Godsil, P. J. (1968). Tracer measurement of stream reaeration, II. Field studies. *Journal of the Water Pollution Control Federation, 40*(2), 285–305.

26. Tsivoglou, E. C., & Wallace, J. R. (1972). *Characterization of stream reaeration capacity.* EPA-R-3-72-012, EP, USA.
27. Tsivoglou, E. C., & Neal, L. A. (1976). Tracer measurement of reaeration, III. Predicting the reaeration capacity of inland streams. *Journal of the Water Pollution Control Federation, 48* (12), 2669–2689.
28. Velz, C. J. (1970). *Applied stream sanitation*. New York: Wiley.
29. Wang, L. K., Wang, M. H. S., Kao, J. F., & Wen, C. G. (1979). *Journal of Environmental Management, 9*, 185–204 (England).
30. Wang, L. K., & Elmore, D. C. (1981). *Computer-aided modeling of water vapor pressure, gas absorption coefficient, and oxygen solubility* (Technical Report No. PB82-118787, p. 137). Springfield, VA: US Dept of Commerce, National Technical Information Service.
31. Wang, M. H. S., Wang, L. K., Kao, J. F., Wen, C. G., & Vielkind, D. (1979). *Journal of Environmental Management, 9*, 165–183 (England).
32. Wen, C. G., Kao, J. F., & Wang, L. K. (1985). *Effect of salinity on reaeration coefficient of receiving waters, Part II: Experimental study*. Lenox Institute of Water Technology (LIWT) Technical Report LIR/11-85/48B.
33. Wen, C. G., Kao, J. F., Wang, L. K., & Wang, M. H. S. (1982). *Journal of Environmental Management, 14*, 17–34 (England).
34. Wen, C. G., Kao, J. F., Wang, L. K., & Wang, M. H. S. (1983). *Civil Engineering for Practicing and Design Engineers, 2*, 425–445.
35. Wen, C. G., Kao, J. F., Wang, L. K., & Wang, M. H. S. (1983). *Civil Engineering for Practicing and Design Engineers, 2*, 537–550.
36. Wen, C. G., Kao, J. F., & Wang, L. K. (1985). *Effect of salinity on reaeration coefficient of receiving waters, Part, I: Mathematical models*. Lenox Institute of Water Technology (LIWT) Technical Report LIR/11-85/48A.
37. Wilson, G. T., & Macleod, N. (1974). A critical appraisal of empirical equations and models for the prediction of the coefficient of reaeration of deoxygenated water. *Water Research, 8*, 341–366.
38. Wang, L. K., Kao, J. F., Wen, C. G., & Liaw, C. C. (1984). Effect of salinity on reaeration coefficient of receiving waters. *Water Science & Technology, 16*(5–7), 139.
39. US EPA. (1985, June). *Rates, constants, and kinetics formulations in surface water quality modeling* (EPA/600/3-85/040). Washington, DC: US Environmental Protection Agency.
40. EPA Victoria. (2008). *An ecological risk assessment of the Lower Wimmera River*. EPA, Victoria, Australia, Publication 1257, October.
41. FAO. (2014). *Site selection for aquaculture: Chemical features of water*. Washington, DC: Fisheries and Aquaculture Department, www.fao.org.
42. Lehigh University. (2000). *Chloride and salinity*. Bethlehem, PA: Lehigh University
43. Gunderson, J. (2014). Is salt the sleeping dragon of wastewater pollutants? *Water Environment and Technology, 26*(4), 17–18.
44. Wang, M. H. S., & Wang, L. K. (1978). Conversion factors for environmental engineers and scientists. *Water & Sewage Works*, R203–R214.
45. Wang, M. H. S., & Wang, L. K. (2014). Glossary and conversion factors for water resources engineers. In: L. K. Wang & C. T. Yang (Eds.), *Modern water resources engineering* (pp. 759–851). New York: Humana Press.
46. Symons, J. M., Bradley, L. C., Jr., & Cleveland, T. C. (Eds.). (2000). *The drinking water dictionary* (p. 506). Denver, CO: American Water Works Association.
47. Wang, L. K. (1974). *Environmental engineering glossary* (p. 439). Buffalo, NY: Calspan Corporation.
48. Wang, M. H. S., & Wang, L. K. (2015). Environmental water engineering glossary. In: C. T. Yang & L. K. Wang (Eds.), *Advances in water resources engineering* (pp. 471–556). New York: Springer.
49. AWWA, APHA, & WEF. (2012). *Standard methods for the examination of water and wastewater* (22nd ed.). Washington, DC: American Publish Health Association, ISBN-10: 0875530133.

Chapter 8
Sensitivity Analysis for Stream Water Quality Management

Ching-Gung Wen, Jao-Fuan Kao, Mu-Hao S. Wang, and Lawrence K. Wang

Contents

Abstract The sensitivity equations of stream water quality parameters are presented, and their practical applications to stream pollution control scientifically illustrated. Non-tidal streams are classified into: (a) clean or slightly polluted swift non-tidal streams, (b) moderately polluted swift non-tidal streams, (c) heavily polluted swift non-tidal streams, (d) clean or slightly polluted, intermediate non-tidal streams, (e) moderately polluted intermediate non-tidal streams, (f) heavily polluted intermediate non-tidal streams, (g) clean or slightly polluted

C.-G. Wen, Ph.D., P.E. (✉) • J.-F. Kao, Ph.D.
Department of Environmental Engineering, College of Engineering, National Cheng Kung University, Tainan, Taiwan
e-mail: t15250@mail.ncku.edu.tw; ckwen33@yahoo.com.tw

M.-H.S. Wang, Ph.D., P.E. • L.K. Wang, Ph.D., P.E.
Department of Environmental Engineering, College of Engineering, National Cheng Kung University, Tainan, Taiwan

Lenox Institute of Water Technology Newtonville, NY, USA

L.K. Wang, C.T. Yang and M.-H.S. Wang (eds.), *Advances in Water Resources Management*, Handbook of Environmental Engineering, Vol. 16,
DOI 10.1007/978-3-319-22924-9_8

slow non-tidal streams, (h) moderately polluted slow non-tidal streams, and (i) heavily polluted slow non-tidal streams.

Tidal streams are classified into: (a) clean or slightly polluted tidal streams, (b) moderately polluted tidal streams, and (c) heavily polluted tidal streams. The characteristics and water quality parameter ranges of different types of receiving streams are presented. The significance of water quality sensitivities and dissolved oxygen deficits for water quality management are systematically identified by the author's mathematical models.

Keywords Water resources • Environmental management • Sensitivity • Dissolved oxygen deficit • Systems analysis • Water quality • Stream pollution control • Non-tidal streams • Tidal streams • River management

Nomenclature

B	Bottom deposit uptake rate, mg/L-day
C	Concentration of dissolved oxygen, mg/L
D	Total dissolved oxygen deficit, mg/L
Do	Initial concentration of dissolved oxygen deficit, mg/L
D_B	Dissolved oxygen deficit caused by bottom deposit uptake, mg/L
D_D	Dissolved oxygen deficit caused by initial DO deficit, mg/L
D_L	Dissolved oxygen deficit caused by BOD, mg/L
D_N	Dissolved oxygen deficit caused by nitrification, mg/L
D_α	Dissolved oxygen deficit reduced by photosynthesis reaction, mg/L
E	Longitudinal dispersion coefficient, km^2/day
J_1	$(U/2E) + (U^2/4E^2 + K_1/E)^{0.5}$
J_2	$(U/2E) - (U^2/4E^2 + K_2/E)^{0.5}$
J_n	$(U/2E) - (U^2/4E^2 + K_n/E)^{0.5}$
K	one of water quality parameters, such as K_1, K_2, K_n, α, E or B
K_1	Deoxygenation coefficient (base e), day^{-1}
K_2	Reaeration coefficient (base e), day^{-1}
K_n	Nitrification rate coefficient (base e), day^{-1}
K_r	BOD removal rate constant (base e), day^{-1}
L	Concentration of remaining carbonaceous biochemical oxygen demand (CBOD), mg/L
L_o	Initial concentration of remaining CBOD, mg/L
m_1	$(U^2 + 4K_1E)^{-0.5}$
m_2	$(U^2 + 4K_2E)^{-0.5}$
m_3	$(U^2 + 4K_nE)^{-0.5}$
N	Concentration of ammonia nitrogen, mg/L
N_O	Initial concentration of ammonia nitrogen, mg/L
n_1	$\partial J_1/\partial E = -(J_1/E + 2\,K_1 m_1)$

n_2	$\partial J_2/\partial E = -(J_2/E + 2\ K_2 m_2)$
n_n	$\partial J_n/\partial E = -(J_n/E + 2\ K_n m_n)$
$S_{C,K}$	Sensitivity of C to K for non-tidal streams
$S_{C,K,\ t}$	Sensitivity of C to K for tidal streams
$S_{D,\ B}$	Sensitivity of D to B for non-tidal streams, day
$S_{D,\ K1}$	Sensitivity of D to K_1 for non-tidal streams, mg-day/L
$S_{D,\ K2}$	Sensitivity of D to K_2 for non-tidal streams, mg-day/L
$S_{D,\ Kn}$	Sensitivity of D to K_2 for non-tidal streams, mg-day/L
$S_{D,\ \alpha}$	Sensitivity of D to α for non-tidal streams, day
$S_{D,\ B,\ t}$	Sensitivity of D to B for tidal streams, day
$S_{D,\ E,\ t}$	Sensitivity of D to E for tidal streams, mg-day/L-km^2
$S_{D,\ K1,\ t}$	Sensitivity of D to K_1 for tidal streams, mg-day/L
$S_{D,\ K2,\ t}$	Sensitivity of D to K_2 for tidal streams, mg-day/L
$S_{D,\ Kn,\ t}$	Sensitivity of D to K_n for tidal streams, mg-day/L
$S_{D,\ \alpha,\ t}$	Sensitivity of D to α for tidal streams, day
$S_{D,\ K}$	Sensitivity of D to K for non-tidal streams, day
$S_{D,\ K,\ t}$	Sensitivity of D to K for tidal streams
$S_{L,\ K1}$	Sensitivity of L to K1 for non-tidal streams, mg-day/L
$S_{L,\ K1,\ t}$	Sensitivity of L to K1 for tidal streams, mg-day/L
$S_{L,\ E,\ t}$	Sensitivity of L to E for tidal streams
t	Flow time of pollutant, day
U	Mean velocity of streams, Km/day
X	The downstream distance from the point of effluent discharge, km
α	Photosynthesis rate, mg/L-day

1 Introduction

The authors have developed practical formulas for calculating the following sensitivities of critical water quality parameters: (a) the sensitivities of biochemical oxygen demand (L) to deoxygenation coefficient (K_1) and to longitudinal dispersion coefficient (E), and (b) the sensitivities of dissolved oxygen deficit (D) to deoxygenation coefficient, longitudinal dispersion coefficient, reaeration coefficient (K_2), photosynthesis rate (α), and bottom deposit uptake rate (B). It was also concluded that the sensitivity of dissolved oxygen (C) to any water quality parameter ($S_{C,K}$ or $S_{C,K,t}$) is equal to the negative sensitivity of dissolved oxygen deficit ($S_{D,k}$ or $S_{D,K,t}$) specific parameter (K).

$$S_{C,K} = -S_{D,K} \tag{8.1}$$

$$S_{C,K,t} = -S_{D,K,t} \tag{8.2}$$

where K in the subscripts represents any one of the water quality parameters, K_1, K_2, K_n, E, α or B; t in the subscripts stands for tidal streams. $S_{D,k}$, $S_{D,K,t}$, $S_{C,K}$ and $S_{C,K,t}$ are all defined clearly in the NOMENCLATURE section.

In this chapter, the sensitivity equations of stream water quality parameters are summarized and presented. Practical applications of the sensitivity equations are proposed and illustrated in detail. Receiving waters are classified according to the hydraulic characteristics and the degree of pollution as follows:

Non-tidal Receiving Streams.;

a. clean or slightly polluted swift non-tidal streams
b. moderately polluted swift non-tidal streams
c. heavily polluted swift non-tidal streams
d. clean or slightly polluted intermediate non-tidal streams
e. moderately polluted intermediate non-tidal streams
f. heavily polluted intermediate non-tidal streams
g. clean or slightly polluted slow non-tidal streams
h. moderately polluted slow non-tidal streams
i. heavily polluted slow non-tidal streams

Tidal Receiving Streams;

a. clean or slightly polluted tidal streams
b. moderately polluted tidal streams
c. heavily polluted tidal streams

The hydraulic characteristics and the water quality parameters of the aforementioned [1] types of receiving streams are scientifically assigned. The sensitivities of water quality (L, D, and C) to various water quality parameters (K_1, K_2, K_n, E, α, and B) are systematically analyzed and compared with one another for determination of relative environmental significance. Recent development in stream pollution control and the sensitivities of water quality parameters is introduced [1–17]. Finally important conclusions are drawn for water quality control in receiving streams.

2 Water Quality Models and Sensitivity Equations

The formulas of sensitivities have been theoretically derived from four steady state water quality models. The derivation of the sensitivity formulas can be found elsewhere [2], The steady state water quality models, two for non-tidal streams and two for tidal streams, are summarized in Appendix 1. The derived sensitivity equations for non-tidal streams and tidal streams are summarized in Appendices 2 and 3, respectively [15]. All terms are defined clearly in the NOMENCLATURE section.

3 Significance of Sensitivities for Non-tidal Streams

When a realistic set of consistent stream water quality parameter values need to be developed and/or identified, the sensitivities of these water quality parameters must be evaluated and discussed. Though the sensitivities are different from stream to stream, and from distance to distance within a stream, they can be generalized and compared with one another under specified stream conditions. The non-tidal streams are classified into swift, intermediate, and slow streams, according to their hydraulic characteristics. Each type of stream is further classified into clean or slightly polluted, moderately polluted, and heavily polluted streams according to their degrees of pollution.

The water quality of common receiving streams can be generalized as follows: (a) the initial concentration of remaining ultimate carbonaceous biochemical oxygen demand (CBOD) ranges from 3 to 30 mg/L; (b) the initial concentration of dissolved oxygen deficit (D) ranges from 1 to 8 mg/L; (c) the deoxygenation coefficient (K_1, base e) ranges from 0.1 to 6.0 day^{-1}; (d) the reaeration coefficient (K_2, base e) ranges from 0.1 to 20 day^{-1}; (e) the bottom deposit uptake rate (B) ranges from 0 to 2 mg/L-day; and (f) the photosynthesis rate (α) ranges from 0 to 2 mg/L-day. The sensitivities of various stream water quality parameters are discussed according to the classification and ranges of parameters stated previously.

Of the five most important water quality parameters for non-tidal streams (K_1, K_2, K_n, α, and B), only K_1 affects the Biochemical Oxygen Demand Model vindicated in Eq. 8.14, Appendix 1, and the sensitivity of L to K_1 ($S_{L,K1}$, indicated in Eq. 8.18, Appendix 2). Accordingly $S_{L,K1}$ is the only and the most important sensitivity term for biochemical oxygen demand. No evaluation and further discussion on other sensitivity terms are attempted in this research.

The total dissolved oxygen deficit for non-tidal streams (D, indicated in Eq. 8.15, Appendix 1) can be divided into five terms as follows:

$$D = D_L + D_N + D_B + D_\alpha + D_D \tag{8.3}$$

where

D_L = dissolved oxygen deficit caused by biochemical oxygen demand, mg/L
$$= K_1 L_O (K_2 - K_1)^{-1}[\exp(-K_1 t) - \exp(-K_2 t)] \tag{8.4}$$

D_N = dissolved oxygen deficit caused by nitrification, mg/L
$$= K_n N_O (K_2 - K_n)^{-1}[\exp(-K_n t) - \exp(-K_2 t)] \tag{8.5}$$

D_B = dissolved oxygen deficit caused by bottom deposit uptake, mg/L
$$= B(K_2)^{-1}[1 - \exp(-K_2 t)] \tag{8.6}$$

D_α = dissolved oxygen deficit reduced by photosynthesis reaction, mg/L
$$= -\alpha[1 - \exp(-K_2 t)] \tag{8.7}$$

D_D = dissolved oxygen deficit caused by initial DO deficit, mg/L
$$= D_O[\exp(-K_2 t)] \tag{8.8}$$

Accordingly D, D_L, D_N, D_B, D_α, D_D and all sensitivity terms $S_{D,K}$ will be analyzed and discussed in detail in Sect. 3.

The sensitivity of dissolved oxygen to a water quality parameter $S_{C,K}$ is simply the negative sensitivity of dissolved oxygen deficit to the same water quality parameter $S_{D,K}$ (Eq. 8.1); therefore $S_{C,K}$ is not included in the systems analysis.

3.1 *Swift Non-tidal Streams*

Non-tidal swift streams are generally shallow, rocky and/or sandy with mean stream velocity ranging from 0.3 to 1 m/s. Since the stream velocity is high, plankton blooms are difficult to grow, and the photosynthesis reaction almost can be neglected. The sensitivities of stream water quality parameters are discussed in the subsequent sections.

3.1.1 Clean or Slightly Polluted Swift Non-tidal Streams

Since stream water is clean or slightly polluted, its CBOD concentration is usually under 5 mg/L; the deoxygenation coefficient (K_1, base e) is under 0.23 day^{-1}; the reaeration coefficient (K_2, base e), which is proportional to velocity, is in the range from 5 to 20 day^{-1} or even higher. The stream bed is sandy or rocky; therefore, the bottom deposit uptake rate (B) is very small and can be neglected. Ammonia nitrogen concentration should be very low because of good stream water quality; therefore, the nitrification can be neglected. Let the initial concentration of remaining CBOD (L_O) be equal to 3 mg/L, the initial concentration of dissolved oxygen deficit (D_O) be equal to 1 mg/L, K_1 (base e) be equal to 0.15 day^{-1}, and bottom deposit uptake rate (B), photosynthesis rate (α), initial concentration of ammonia nitrogen (N_O), and nitrification coefficient (K_n) all be equal to zero. K_2 values (base e), however, are assigned to be equal to 5, 10 and 20 day^{-1}. Substituting those values into Eqs. 8.15, and 8.19–8.22 (See Appendices 1 and 2) for calculating the sensitivities and dissolved oxygen deficits under various flow times (t), one can obtain the results shown in Fig. 8.1. The sensitivity of dissolved oxygen deficit to reaeration coefficient ($S_{D,K2}$) is negative for this kind of river. It is important to know that sensitivity is a slope of water quality to a stream parameter, thus, may be positive or negative. Each sensitivity should be evaluated in accordance with its absolute value. Therefore the absolute value of the sensitivity $|S_{D,K2}|$ is plotted in Fig. 8.1. From the figure it can be seen that the sensitivity of D to K_1 ($S_{D,K1}$) is greater than the absolute value of $S_{D,K2}$. However, the value of dissolved oxygen deficit due to the term (D_D) is larger than that of due to deoxygenation term (D_L). The sensitivities of $S_{D,K1}$ and $|S_{D,K2}|$ decrease when K_2 increases. $S_{D,K1}$ increases with increasing flow time (t). When t increases, $|S_{D,K2}|$ increases to a peak point and then decreases. From above discussion, one can see that K_1 and K_2 are the two most important stream parameters for clean or slightly polluted swift non-tidal streams.

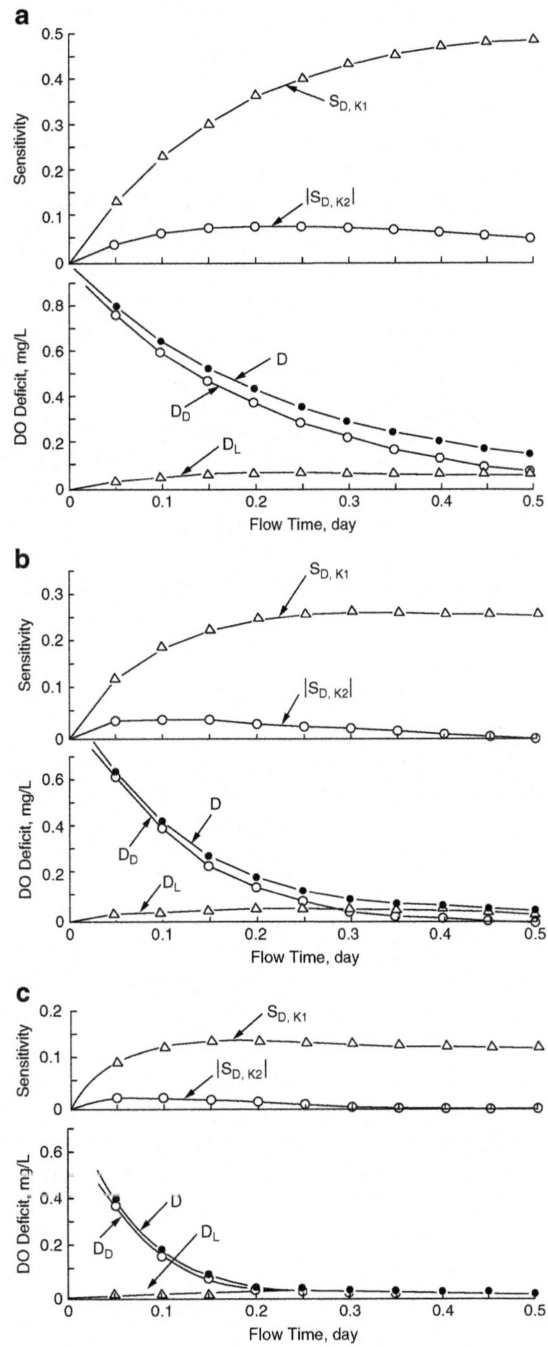

Fig. 8.1 (**a**) The sensitivities and dissolved oxygen deficits for clean or slightly polluted swift non-tidal stream.($K_2 = 5$ 1/day). (**b**) The sensitivities and dissolved oxygen deficits for clean or slightly polluted swift non-tidal stream ($K_2 = 10$ 1/day). (**c**) The sensitivities and dissolved oxygen deficits for clean or slightly polluted swift non-tidal stream ($K_2 = 20$ 1/day)

3.1.2 Moderately Polluted Swift Non-tidal Streams

For this kind of stream water, the initial concentration of remaining carbonaceous biochemical oxygen demand (L_O) is about 10 mg/L; initial dissolved oxygen deficit (D_O) ranges from 1 to 3 mg/L; deoxygenation coefficient (K_1) ranges from 0.2 to 0.5 day^{-1} (base e); the range of reaeration coefficient (K_2) is same as that for clean or slightly polluted swift streams; the bottom deposit uptake rate (B) is under 0.1 mg/L-day. Let $L_O = 10$ mg/L, $D_O = 3$ mg/L, $K_1 = 0.345$ day^{-1}, and neglect the photosynthesis and nitrification effects (i.e., B = 0.1 mg/L-day, and $\alpha = K_n = N_O = 0$), one can plot Fig. 8.2 with Eqs. 8.15, and 8.19–8.22. Figure 8.2 indicates that the $S_{D,K1}$ is the most sensitive. The sensitivities of dissolved oxygen deficit to reaeration coefficient $|S_{D,K2}|$, to nitrification coefficient ($S_{D,Kn}$), and to bottom deposit uptake rate ($S_{D,B}$) are relatively insensitive. All sensitivities rapidly reduce with increasing K_2 values from 5 to 20 day^{-1}. Figure 8.2 also indicates that the most part of dissolved oxygen deficit is contributed by D_D term when the flow time (t) is short. The percentages of dissolved oxygen deficit due to nitrification (D_N) and bottom deposit uptake (D_B), are very small. In conclusion, the nitrification and bottom deposit uptake can be neglected, and K_1 and K_2 must be carefully and accurately measured for the moderately polluted swift non-tidal streams.

3.1.3 Heavily Polluted Swift, Non-tidal Streams

Generally for this type of streams the remaining CBOD of the stream is about 20 mg/L, or more. Wastewaters which discharge into the receiving stream of this type can even be untreated wastes. The deoxygenation coefficient (K_1, base e) is greater than 0.5 day^{-1}. The dissolved oxygen concentration of the stream water is very low. Initial dissolved oxygen deficit (D_O) ranges from 3 to 8 mg/L. There are sludge banks in the slow water segments of the stream. The range of bottom deposit uptake rate (B) is about from 0.1 to 1.0 mg/L-day. When the stream is heavily polluted, the nitrification bacteria cannot compete with the saprophyta; therefore, the nitrification effect can be neglected. Let $L_O = 20$ mg/L, $D_O = 5$ mg/L, $K_1 = 1.0$ day^{-1} (base e), $K_n = N_O = \alpha = 0$, and B = 0.5 mg/L-day. Substituting those values into Eqs. 8.15, and 8.19–8.22 for calculating sensitivities and dissolved oxygen deficits under various flow times (t), one can then plot Fig. 8.3. The figure indicates that the $S_{D,K1}$ is the most sensitive among the sensitivities evaluated. The $|S_{D,K2}|$ is less sensitive than $S_{D,K1}$. The $S_{D,B}$, b is very inert especially under the condition of high K_2 value. All sensitivities rapidly decline when K_2 increases from 5 to 20 day^{-1}. Figure 8.3 also indicates that the most part of dissolved oxygen deficit is contributed by D_D term when t value is small. The percentage of dissolved oxygen deficit due to bottom deposit uptake (D_B) is so small that it can be neglected.

In summation, the order of sensitivity (from very sensitive to insensitive) for swift streams is in the order of $S_{D,K1}$, $|S_{D,K2}|$, $S_{D,B}$ and $S_{D,Kn}$. The major sources or

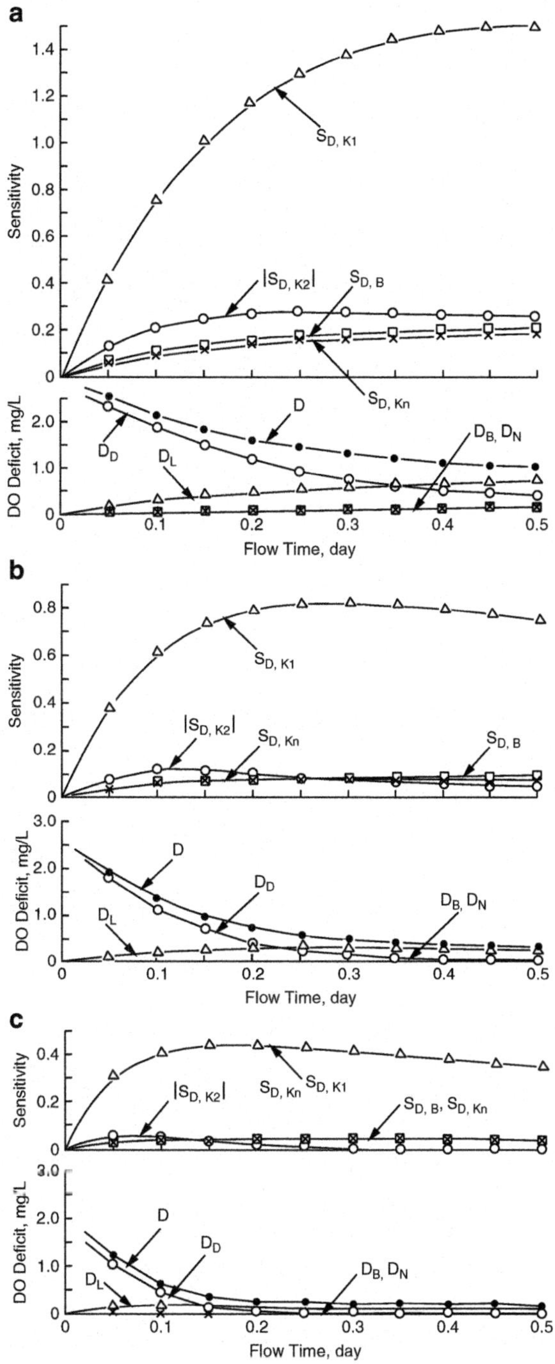

Fig. 8.2 (**a**) The sensitivities and dissolved oxygen deficits for moderately polluted swift non-tidal stream ($K_2 = 5$ 1/day). (**b**) The sensitivities and dissolved oxygen deficits for moderately polluted swift non-tidal stream ($K_2 = 10$ 1/day). (**c**) The sensitivities and dissolved oxygen deficits for moderately polluted swift non-tidal stream ($K_2 = 20$ 1/day)

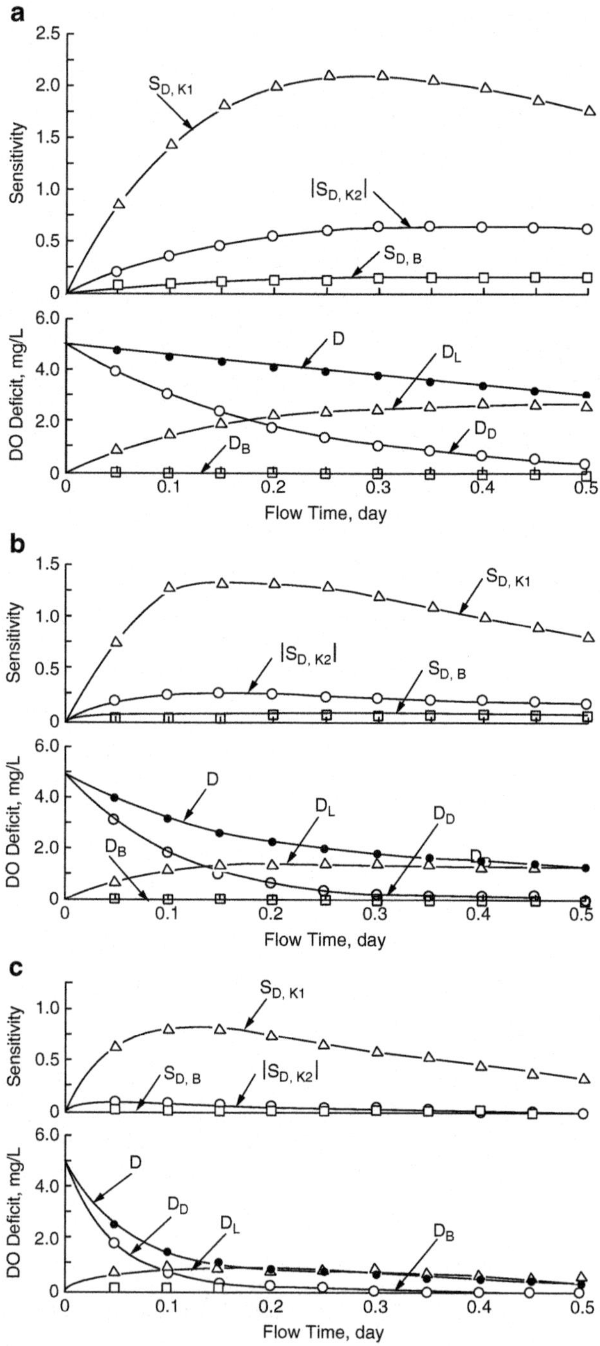

Fig. 8.3 (**a**) The sensitivities and dissolved oxygen deficits for heavily polluted swift non-tidal stream ($K_2 = 5$ 1/day). (**b**) The sensitivities and dissolved oxygen deficits for heavily polluted swift non-tidal stream ($K_2 = 10$ 1/day). (**c**) The sensitivities and dissolved oxygen deficits for heavily polluted swift non-tidal stream ($K_2 = 20$ 1/day)

sinks of dissolved oxygen deficit (D) are deoxygenation and initial DO deficit, the percentage of dissolved oxygen deficit due to bottom deposit uptake (D_B) and nitrification (D_N) is not more than 2 %. Therefore, the parameters of K_1 and K_2 are the two important stream parameters which should be carefully determined. The parameters of B and K_n may be roughly measured or omitted for the swift streams.

3.2 Intermediate Non-tidal Streams

The mean velocity of intermediate non-tidal streams is generally in the range of 0.1 to 0.3 m/s. This kind of stream has a sandy or silty bed. Usually the bed would be sandy if mean velocity is higher than 0.2 m/s. If the mean velocity is lower than 0.2 m/s and the stream water is also polluted, the sludge will be settled in the bed, and the bottom deposit uptake will be significant. In this section, the sensitivities of intermediate non-tidal streams are analyzed and are discussed in following three conditions: clean or slightly polluted, moderately polluted and heavily polluted.

3.2.1 Clean or Slightly Polluted Intermediate Non-tidal Streams

When a stream is clean or slightly polluted, nitrification coefficient (K_n) and ammonia nitrogen concentration (N_O) of the stream water are almost equal to zero, thus, the nitrification can be neglected. The bottom deposit uptake and photosynthesis are also insignificant. The range of reaeration coefficient (K_1) which is a function of velocity and depth, is 0.25 to 15 day^{-1}. The flow through time (t) is longer than that of swift streams. Based on the above discussion, it is reasonable to assume that $L_O = 3$ mg/L, $D_O = 1$ mg/L, $K_1 = 0.15$ day^{-1}, $B = 0.1$ mg/L-day, $\alpha = 0.2$ mg/L-day, and $K_n = N_O = 0$. K_2 is assigned to be 0.5, 2.5 and 10 day^{-1}.

The sensitivities and the dissolved oxygen deficit at various flow times can be determined by substituting the L_O, D_O, K_1, B, α, K_n, N_O, and K_2 values into Eqs. 8.15, and 8.19–8.22. The results are plotted as Fig. 8.4 from which we can see that:

$$S_{D,K1} > S_{D,B} > |S_{D,K2}|$$

in the order of decreasing sensitivity.

All sensitivities decline when K_2 value increases. The dissolved oxygen deficit due to bottom deposit uptake is relatively small.

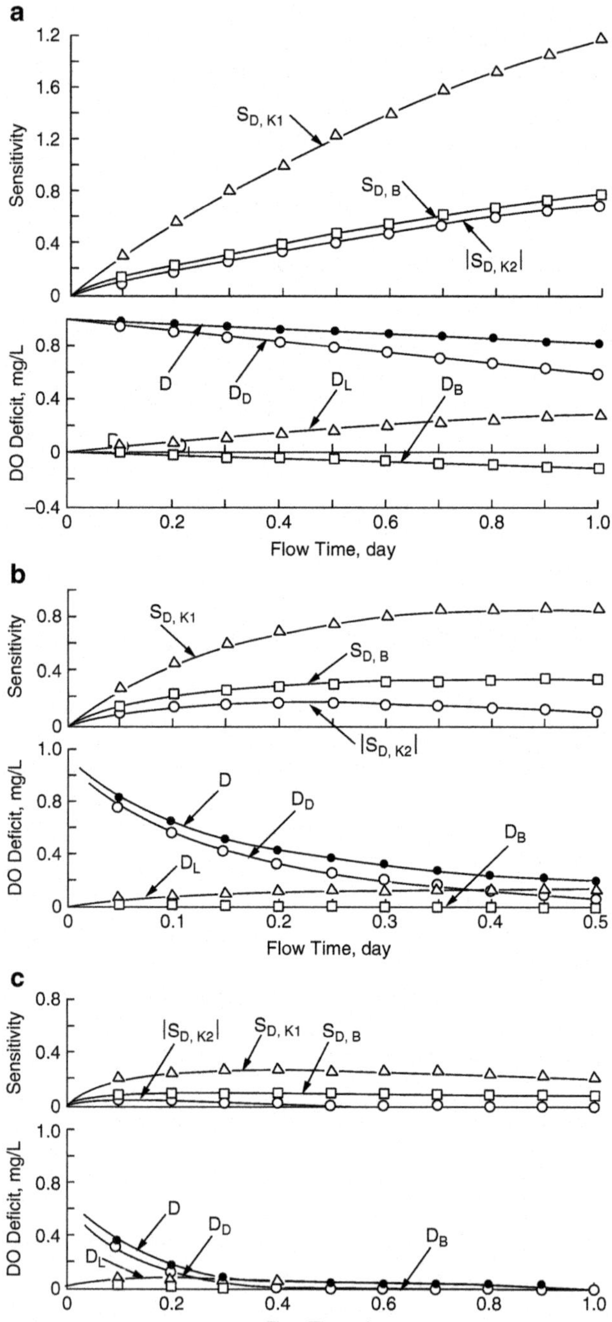

Fig. 8.4 (**a**) The sensitivities and dissolved oxygen deficits for clean or slightly polluted intermediate non-tidal stream ($K_2 = 0.5$ 1/day). (**b**) The sensitivities and dissolved oxygen deficits for clean or slightly polluted intermediate non-tidal stream ($K_2 = 2.5$ 1/day). (**c**) The sensitivities and dissolved oxygen deficits for clean or slightly polluted intermediate non-tidal stream ($K_2 = 10$ 1/day)

3.2.2 Moderately Polluted Intermediate Non-tidal Streams

Nitrification may or may not exist under these stream conditions. If nitrification occurs in the stream water, the reactions follow:

$$2NH_3 + 3O_2 \rightarrow 2HNO_2 + 2H_2O \tag{8.9a}$$

$$2HNO_2 + O_2 \rightarrow 2HNO_3 \tag{8.9b}$$

The overall reaction is:

$$2NH_3 + 4O_2 \rightarrow 2HNO_3 + 2H_2O \tag{8.10}$$

The reactions of Eqs. 8.8 and 8.9 obey the first order reaction. The ranges of these two reaction coefficients are 0.01 to 0.50 day^{-1} (base e) and 0.50 to 2.00 day^{-1} (base e), respectively. [1] It is clear that the nitrification is controlled mainly by oxidation of NH_3 to HNO_2. The rate of oxygen uptake per unit of ammonia in oxidation ranges from 3 to 4 mg of O_2/mg of NH_3. Therefore, the range of nitrification coefficient in Appendices 1–3 is from 0.03 (i.e. 0.01 × 3) to 2.0 (i.e., 0.5 × 4) day^{-1} (base e).

The bottom deposit uptake of the stream under the stated environmental conditions is significant, and ranges from 0.1 to 1.0 mg/L-day [14]. Let $B = 0.5$ mg/L-day, $\alpha = 0$, $L_O = 10$ mg/L, $D_O = 3$ mg/L, $K_1 = 0.345$ day^{-1} (base e), $N_O = 1$ mg/L, $K_n = 0.345$ day^{-1} (base e), $K_2 = 0.5$, 1.5, and 10 day^{-1} (base e), and substitute these values into Eqs. 8.15, and 8.19–8.22. The sensitivities and the dissolved oxygen deficits at various flow times (t) are determined and illustrated in Fig. 8.5.

From Fig. 8.5, one can understand that the sensitivities of $S_{D,Kn}$ and $S_{D,B}$ are very small in comparison with the sensitivities of $S_{D,K1}$ and $S_{D,K2}$ when K_2 is 5 day^{-1}. The sensitivities of $S_{D,K2}$, $S_{D,Kn}$ and $S_{D,B}$ a1l approach to zero when K_2 is as high as 10 day^{-1}. All sensitivities rapidly decline when K_2 increases. The dissolved oxygen deficits due to nitrification (D_N) and bottom deposit uptake (D_B) are relatively small, under the conditions of high K_2 value, but the latter (D_B) is significant when K_2 value is as low as 0.5 day^{-1}.

3.2.3 Heavily Polluted Intermediate Non-tidal Streams

The point sources of pollution which discharge into this type of stream are mostly untreated wastes. The deoxygenation coefficient (K_1) is higher than 0.5 day^{-1} (base e). The concentration of dissolved oxygen of the stream water is usually very low, the initial concentration of dissolved oxygen deficit (D_O) ranges from 3 to 8 mg/L. There are sludge blankets in the slow water segment of the stream. The range of bottom deposit uptake rate (B) is about 0.1 to 1.5 mg/L-day. Since the stream is heavily polluted, the bacteria cannot compete with the saprophyta, therefore, the nitrification can be neglected. Let $L_O = 20$ mg/L, $D_O = 5$ mg/L,

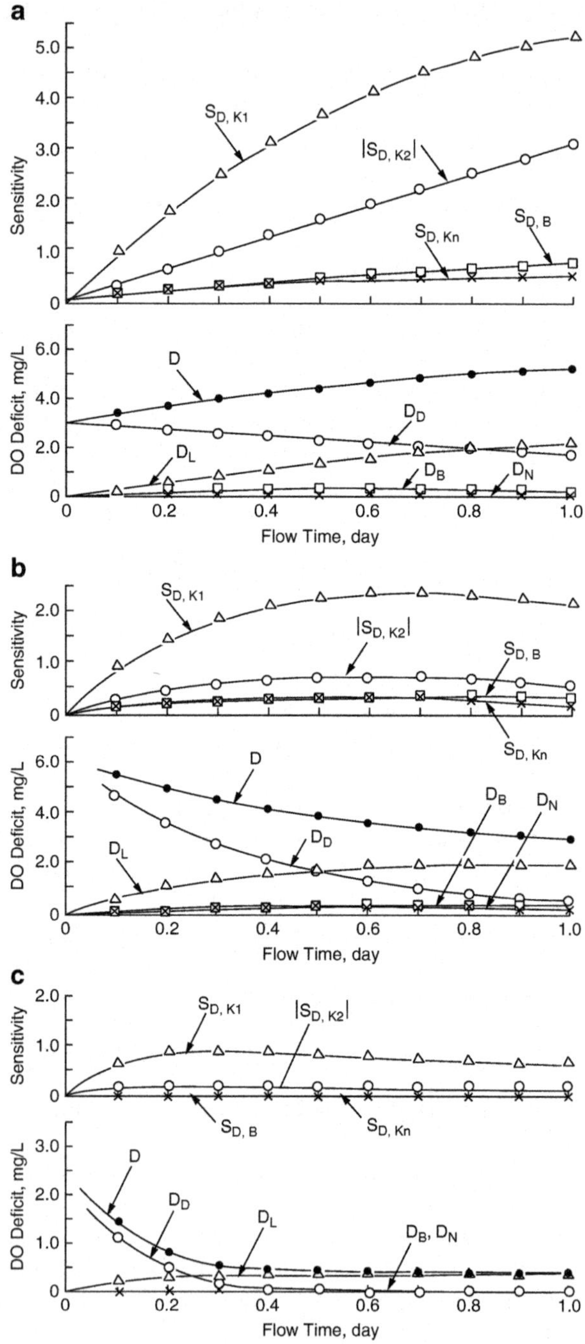

Fig. 8.5 (**a**) The sensitivities and dissolved oxygen deficits for moderately polluted intermediate non-tidal stream ($K_2 = 0.5$ 1/day). (**b**) The sensitivities and dissolved oxygen deficits for moderately polluted intermediate non-tidal stream ($K_2 = 2.5$ 1/day). (**c**) The sensitivities and dissolved oxygen deficits for moderately polluted intermediate non-tidal stream ($K_2 = 10$ 1/day)

$B = 1$ mg/L-day, $\alpha = 0$, $K_1 = 1.0$ day^{-1} (base e), $K_n = N_O = 0$, and $K_2 = 0.5$, 2.5 and 10 day^{-1} (base e). The calculated sensitivities and the dissolved oxygen deficits at various times (t) are illustrated in Fig. 8.6. The sensitivities at $K_2 = 0.5$ day^{-1} are not shown in Fig. 8.6 because of the low values. The stream water is in an anaerobic condition, and the existing models cannot be properly applied.

From Fig. 8.6, it can be understood that the $S_{D,K1}$ larger than $|S_{D,K2}|$, and $|S_{D,K2}|$ is larger than $S_{D,B}$. All sensitivities' decline when K_2 increases. The dissolved oxygen deficit due to bottom deposit uptake (D_B) is very low.

In summation, for the intermediate non-tidal streams, the sensitivity term of $S_{D,K1}$ is the most sensitive, $|S_{D,K2}|$ is next, $S_{D,B}$ and $S_{D,N}$ are relatively insensitive. The major sources or sinks of dissolved oxygen deficit are deoxygenation and initial dissolved oxygen deficit. The percentage of dissolved oxygen deficit due to bottom deposit uptake (D_B) and nitrification (D_N) is usually small. Therefore, the stream parameters of K_1 and K_2 are important for stream pollution control. B and K_n, however, are not significant.

3.3 Slow Non-tidal Streams

The mean velocities of slow non-tidal streams are generally under 0.1 m/s. For this kind of stream, the depth is deep, the reaeration coefficient (K_2) is very low ranging from 0.05 to 0.6 day^{-1} (base e). The sensitivities of this kind of stream are discussed in three pollutional conditions.

3.3.1 Clean or Slightly Polluted Slow Non-tidal Streams

When stream water is clean or slightly polluted, its nitrification can be neglected. The bottom deposit uptake and photosynthesis reactions do exist, but are very insignificant. Let $L_O = 3$ mg/L, $D_O = 1$ mg/L, $K_1 = 0.15$ day^{-1} (base e), $K_n = N_O = 0$, $B = 0.2$ mg/L-day, $\alpha = 0.4$ mg/L-day, and $K_2 = 0.2$ and 0.6 day^{-1} (base e). By substituting these values into Eqs. 8.15, and 8.19–8.22, one can determine the sensitivities and dissolved oxygen deficits at various flow times (t) and graphically illustrate the results in Fig. 8.7. From the figure, one can conclude that the $S_{D,K1}$ is the most sensitive, and $|S_{D,K2}|$ and $S_{D,B}$ (or $|S_{D,\alpha}|$) very sensitive too. The dissolved oxygen deficits due to combined action of bottom deposit uptake and photosynthesis ($D_B + D_\alpha$) are very significant, especially when K_2 value is low. Therefore, the photosynthesis and the bottom deposit uptake play an important role for clean or slightly polluted slow non-tidal streams.

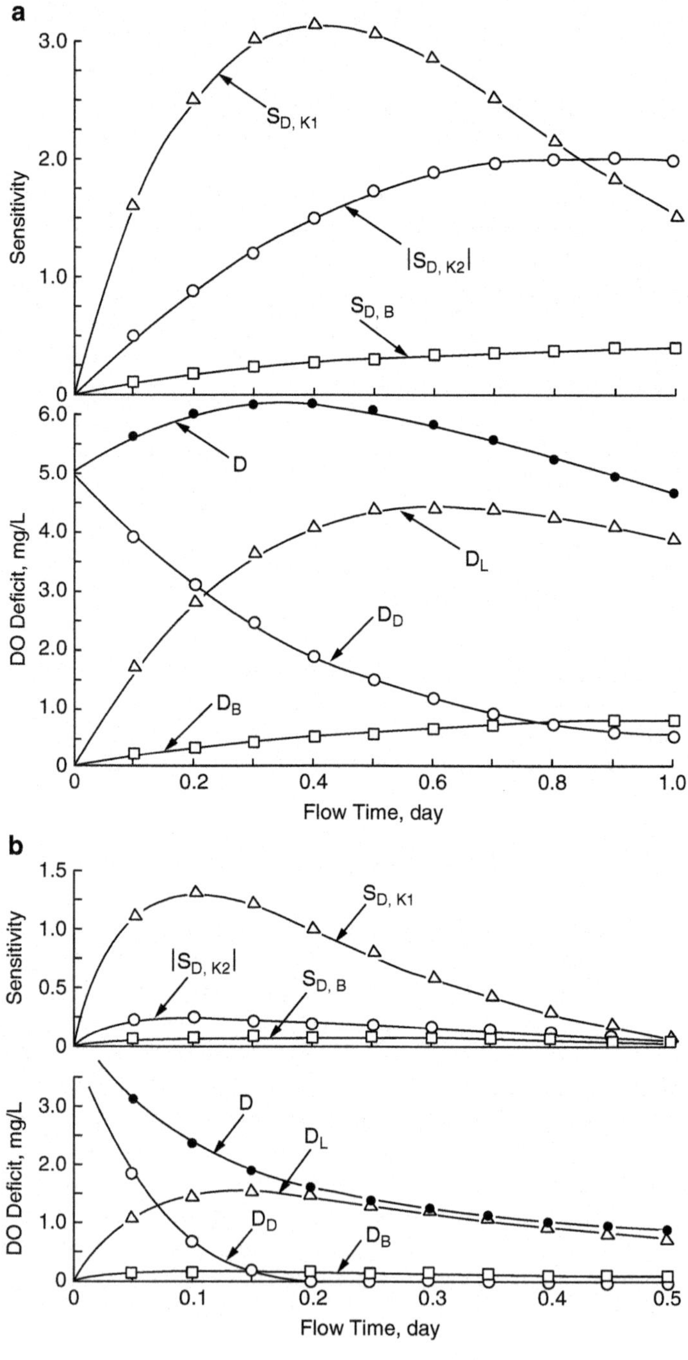

Fig. 8.6 (**a**) The sensitivities and dissolved oxygen deficits for heavily polluted intermediate non-tidal stream ($K_2 = 0.5$ 1/day). (**b**) The sensitivities and dissolved oxygen deficits for heavily polluted intermediate non-tidal stream ($K_2 = 2.5$ 1/day)

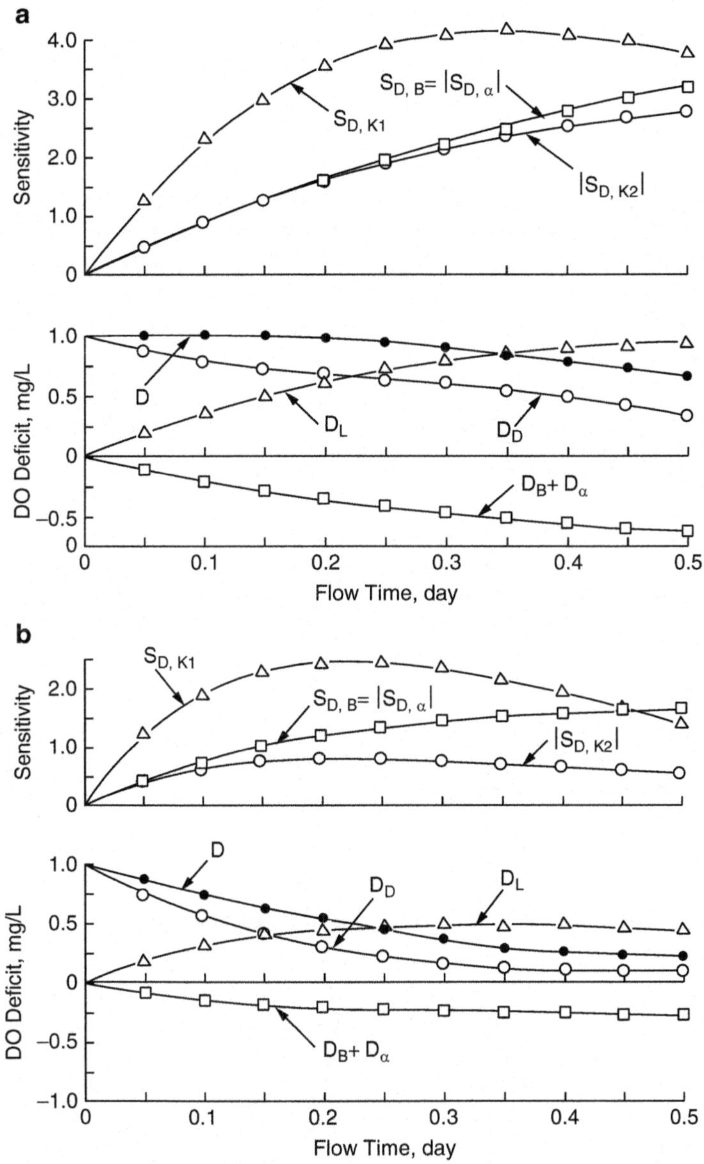

Fig. 8.7 (**a**) The sensitivities and dissolved oxygen deficits for clean or slightly polluted slow non-tidal stream ($K_2 = 0.2$ 1/day). (**b**) The sensitivities and dissolved oxygen deficits for clean or slightly polluted slow non-tidal stream ($K_2 = 0.6$ 1/day)

3.3.2 Moderately Polluted Slow Non-tidal Streams

Nitrification may or may not exist in a moderately polluted non-tidal stream. If nitrification occurs, the range of nitrification coefficient K_n is from 0.03 to

2.0 day^{-1} (base e). The bottom deposit uptake and photosynthesis are significant. Usually the bottom deposit uptake rate (B) ranges from 0.2 to 2 mg/L-day, and the photosynthesis rate ranges from 0.1 to 2.0 mg/L-day. Assuming $L_O = 10$ mg/L, $D_O = 3$ mg/L, $K_1 = 0.345$ day^{-1}, $K_n = 0.345$ day^{-1}, $K_2 = 0.2$ and 0.6 day^{-1} (base e), $\alpha = 2$ mg/L-day, $N_O = 1$ mg/L, and $B = 1$ mg/L-day, one can determine all sensitivities and dissolved oxygen deficit concentrations with Eqs. 8.15, and 8.19–8.22. Figure 8.8 illustrates the calculated results. From Fig. 8.8, one can see that the sensitivities of $S_{D,K1}$ and $|S_{D,K2}|$ are very high, whereas the $S_{D,Kn}$ and $S_{D,B}$ (or $|S_{D,\alpha}|$) are comparatively low. The dissolved oxygen deficit due to the combined effects of bottom deposit uptake and photosynthesis ($D_B + D_\alpha$) is significant, but that due to nitrification (D_N) is relative insignificant. The bottom deposit uptake and photosynthesis cannot be omitted in this type of streams.

3.3.3 Heavily Polluted Slow Non-tidal Streams

When a slow non-tidal stream is heavily polluted, K_1 is very high, which ranges from 0.5 to 5 day^{-1}, but the reaeration coefficient (K_2) is very low due to slow stream velocity. The dissolved oxygen is close to zero for the most part of the stream reach; therefore, the water quality models are difficult to be applied. The authors make no attempt to analyze the sensitivity at present. More research in this area is needed.

Section 4 starting below, presents only the significance of sensitivity analysis for three types of tidal streams. Important water quality models, nomenclature and references are also included in this Part.

4 Significance of Sensitivities for Tidal Streams

For tidal streams, the velocity is very slow and the water depth is relatively deep; therefore, the reaeration coefficient is smaller than that of non-tidal streams, thus K_2 ranges from 0.1 to 1.0 day^{-1} generally. Since the stream velocity is very low, suspended sludges are settled in the stream bed, and the algae blooms are dominant in the water. Therefore, bottom deposit uptake, and photosynthesis always occur in tidal streams. In addition, nitrification always occurs in tidal streams. The dispersion coefficient (E) ranges from 1 to 60 km^2/day. The sensitivities of water quality model for tidal streams should be controlled mainly according to the pollution loading, but not according to the stream velocity, because the range of velocity of tidal streams is narrow.

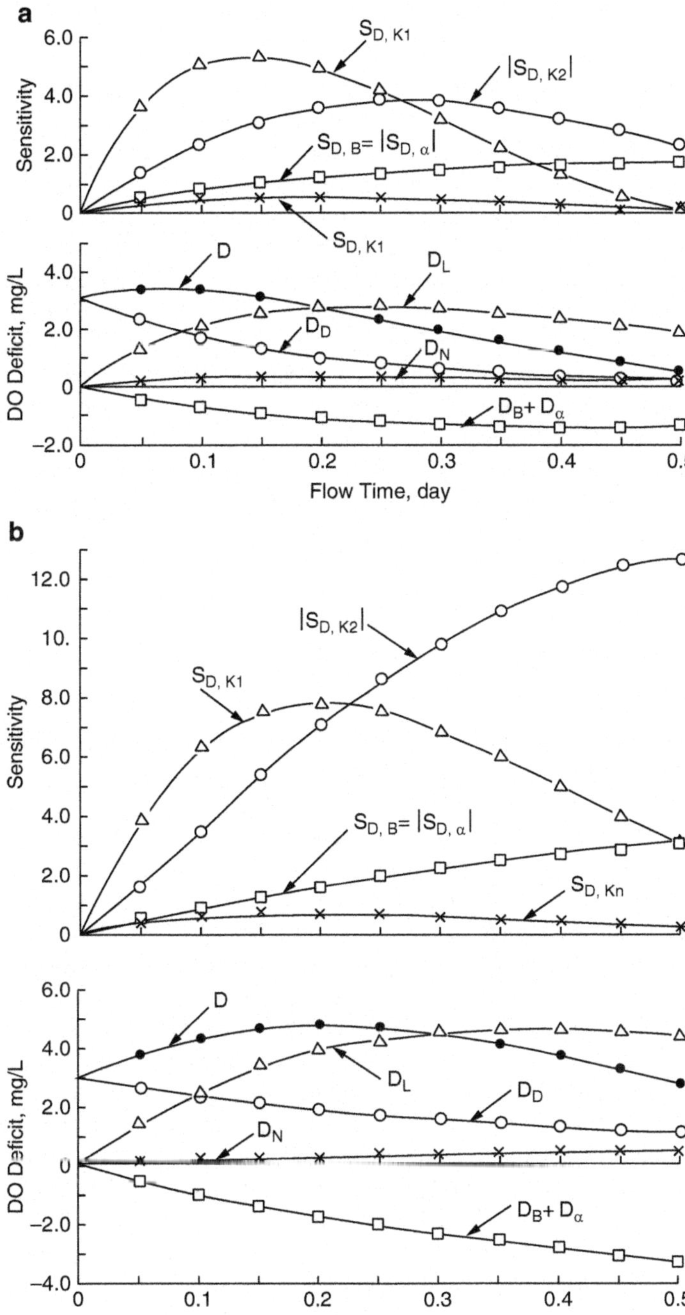

Fig. 8.8 (**a**) The sensitivities and dissolved oxygen deficits for moderately polluted slow non-tidal stream ($K_2 = 0.2$ 1/day). (**b**) The sensitivities and dissolved oxygen deficits for moderately polluted slow non-tidal stream ($K_2 = 0.6$ 1/day)

4.1 Clean or Slightly Polluted Tidal Streams

It is reasonable to assume the following figures for the major water quality parameters of clean or slightly polluted tidal streams: $L_O = 3$ mg/L, $D_O = 1$ mg/L, $K_1 = 0.15$ day^{-1} (base e), $K_n = N_O = B = 0$, $\alpha = 0.1$ mg/L-day, $K_2 = 0.4$ to 0.8 day^{-1} (base e), and U (Mean velocity of fresh water) = 0.5 km/day. Two values each are assigned for K_2 and U for the purpose of illustration. By substituting these values into Eqs. 8.17 and 8.25–8.29, one can determine the sensitivities and dissolved oxygen deficit and illustrate the results in Fig. 8.9. From the figure, we can see that $S_{D,K1,t}$ and $S_{D,K2,t}$ are extremely sensitive, $S_{D,\alpha,t}$ is moderately sensitive, and $S_{D,E,t}$ is relatively insensitive. All sensitivities increase in absolute values when freshwater velocity (U) increases, and rapidly decrease when K_2 increases. $S_{D,K1,t}$ is more sensitive than $S_{D,K2,t}$ when K_2 value is low, but is less sensitive than $S_{D,K2,t}$ when the K_2 value is high. The dissolved oxygen deficit is mainly contributed by initial DO deficit (D_D), deoxygenation (D_L), and combined action of photosynthesis and bottom deposit uptake (D_B and D_α). Therefore, the photosynthesis and deposit uptake cannot be neglected.

4.2 Moderately Polluted Tidal Streams

When sea water intrudes into a tidal stream, the concentration of remaining CBOD and initial dissolved oxygen deficit (D_O) are lower than that of a non-tidal stream for the same pollution loading. Let $L_O = 8$ mg/L, $D_O = 3$ mg/L, $K_1 = K_n = 0.345$ day^{-1} (base e), B = 1 mg/L-day, $\alpha = 2$ mg/L-day, E = 10 km^2/day, $N_O = 1$ mg/L, $K_2 = 0.4$ and 0.8 day^{-1} (base e), and U = 0.5 and 5 km/day. Substituting those values into Eqs. 8.17, and 8.25–8.29, one can obtain the sensitivities and dissolved oxygen deficits at various distances. All calculated results are presented in Fig. 8.10. Dissolved oxygen deficit declines to zero after 8 km from the river mouth at U = 0.5 km/day, thus the water quality models cannot be applied, and the sensitivities need not to be analyzed under this condition. $S_{D,K2,t}$ is extremely sensitive, while $S_{D,Kn,t}$ is very insensitive. The dissolved oxygen deficit associated with the nitrification (D_N) is also insignificant, thus the nitrification can be neglected. The sensitivity terms $S_{D,B,t}$ and ($|S_{D,\alpha,t}|$), are low when compared with $S_{D,K1,t}$. The dissolved oxygen deficit associated with the net action of photosynthesis and bottom deposit uptake ($D_\alpha + D_B$) is significant, thus the photosynthesis and bottom deposit uptake must be analyzed carefully for stream water quality control. Another sensitivity term, $S_{D,E,t}$ is in the same magnitude of $S_{D,K1,t}$; thus the longitudinal dispersion coefficient should not Be overlooked.

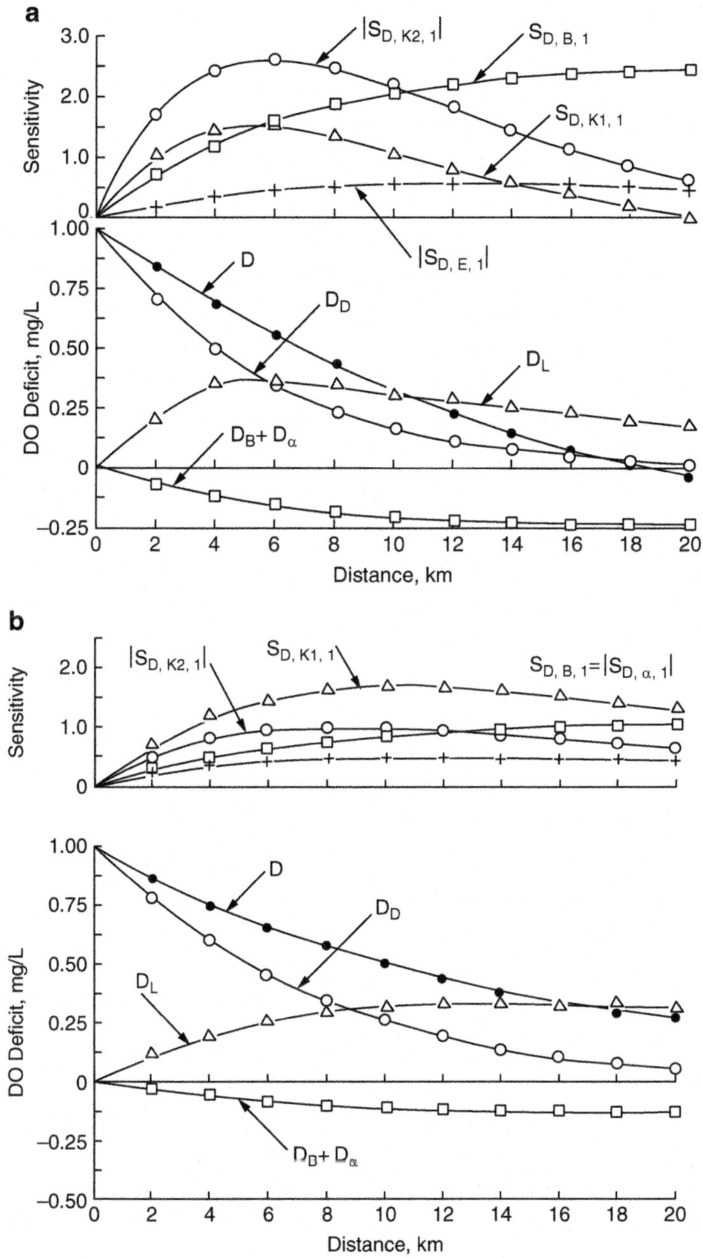

Fig. 8.9 (**a**) The sensitivities and dissolved oxygen deficits for clean or slightly polluted tidal stream ($K_2 = 0.4$ 1/day and $U = 0$ km/day). (**b**) The sensitivities and dissolved oxygen deficits for clean or slightly polluted tidal stream ($K_2 = 0.8$ 1/day and $U = 5$ km/day). (**c**) The sensitivities and dissolved oxygen deficits for clean or slightly polluted tidal stream ($K_2 = 0.8$ 1/day and $U = 0.5$ km/day). (**d**) The sensitivities and dissolved oxygen deficits for clean or slightly polluted tidal stream ($K_2 = 0.4$ 1/day and $U = 5$ km/day)

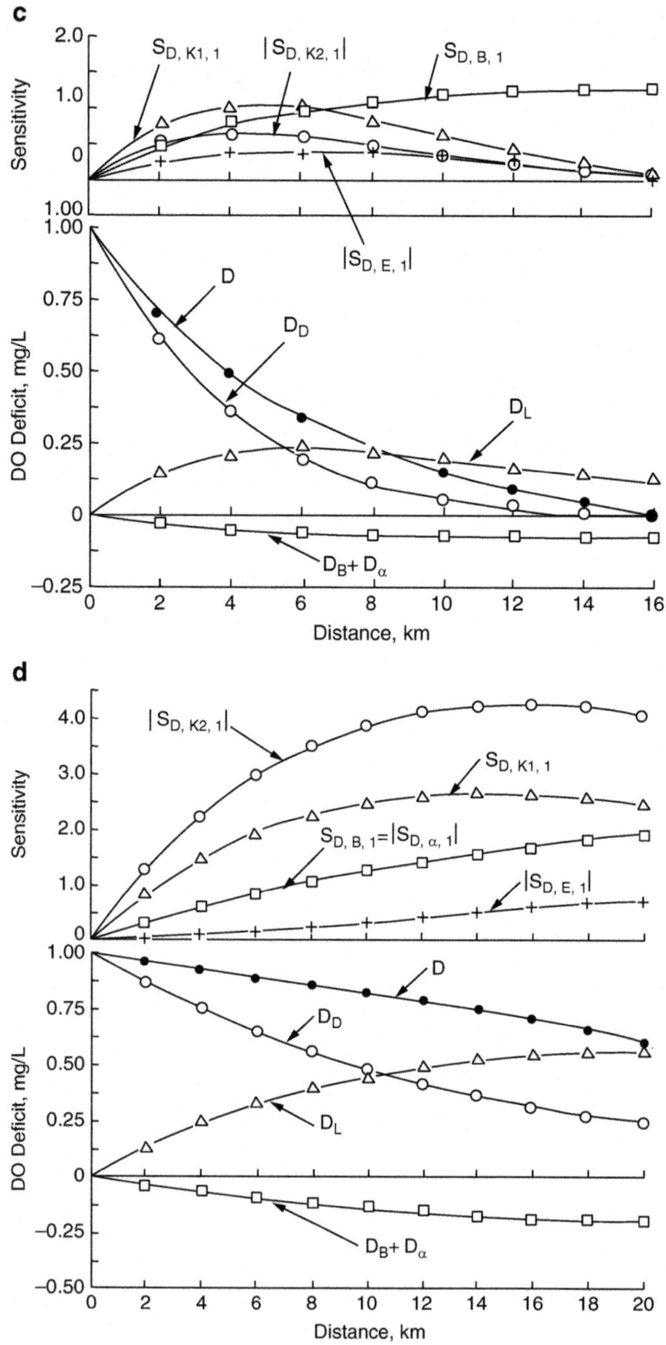

Fig. 8.9 (continued)

4.3 Heavily-Polluted Tidal Streams

The following water quality may be considered for heavily polluted tidal streams: $L_O = 15$ mg/L, $D_O = 5$ mg/L, $B = 3$ mg/L-day, $\alpha = 1$ mg/L-day, $K_1 = 1.0$ day^{-1} (base e), $K_2 = 0.4$ to 0.8 day^{-1} (base e), $K_n = 0$ day^{-1}, $N_O = 0$ mg/L, and $E = 10$ km^2/day. Choosing two K_2 values, 0.4 and 0.8 day^{-1}, and two U values, 0.5 and 5 km/day, and then substituting these values and L_O, D_O, B, α, K_1, K_n, N_O, and E values into Eqs. 8.17 and 8.25–8.29. The calculated sensitivities and

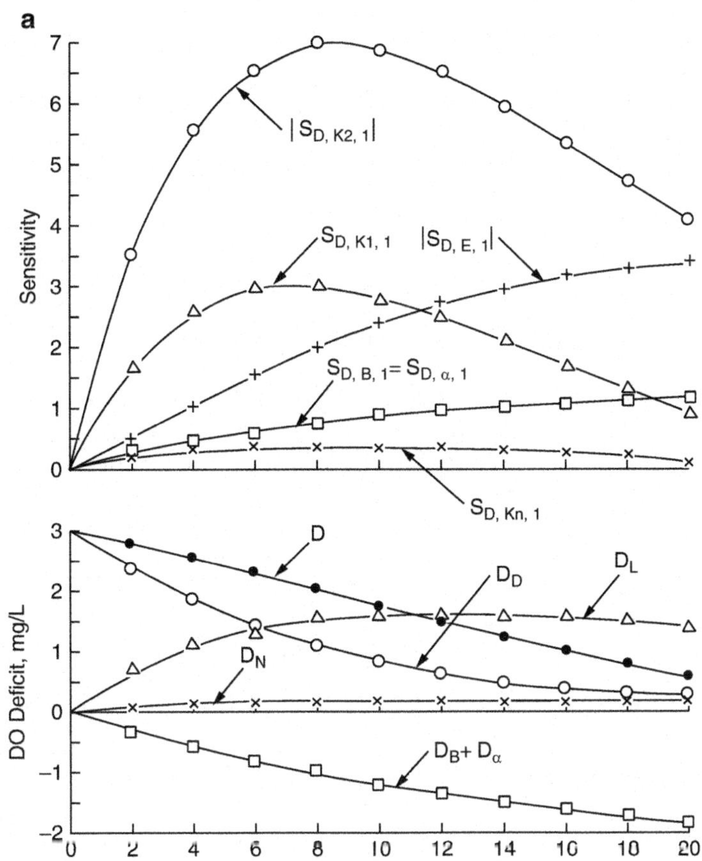

Fig. 8.10 (**a**) The sensitivities and dissolved oxygen deficits for moderately polluted tidal stream ($K_2 = 0.8$ 1/day and $U = 5$ km/day). (**b**) The sensitivities and dissolved oxygen deficits for moderately polluted tidal stream ($K_2 = 0.4$ 1/day and $U = 5$ km/day). (**c**) The sensitivities and dissolved oxygen deficits for moderately polluted tidal stream ($K_2 = 0.8$ 1/day and $U = 0.5$ km/day). (**d**) The sensitivities and dissolved oxygen deficits for moderately polluted tidal stream ($K_2 = 0.4$ 1/day and $U = 0.5$ km/day)

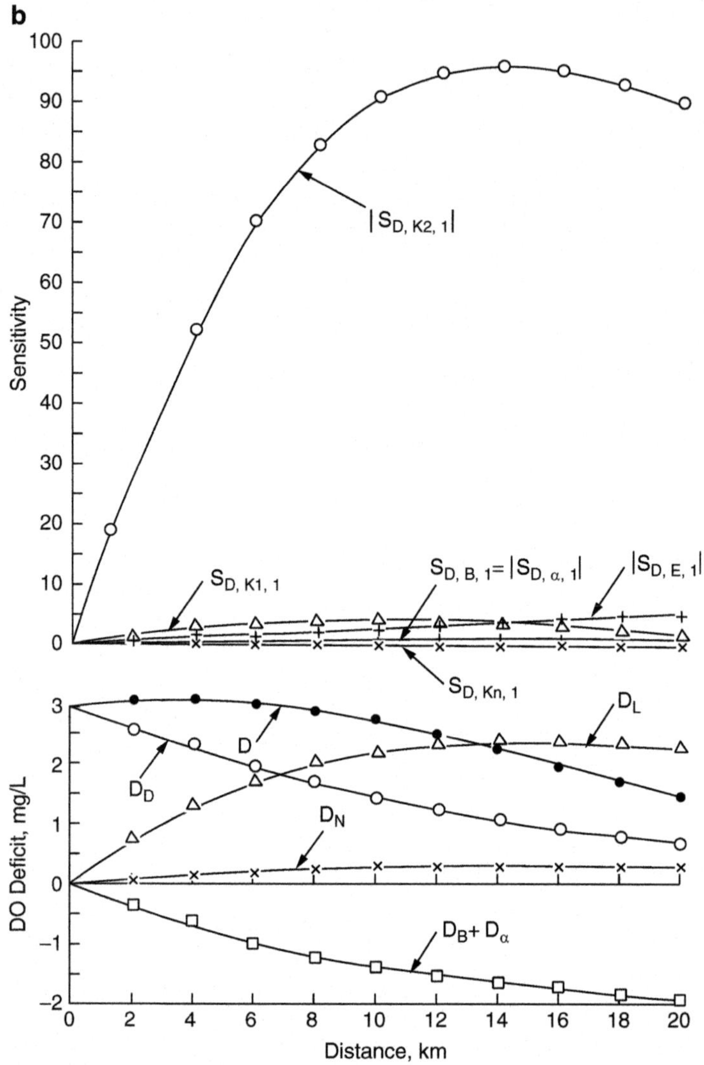

Fig. 8.10 (continued)

dissolved oxygen deficits are partially illustrated in Fig. 8.11. It should be noted that the sensitivities at $K_2 = 0.4\ \text{day}^{-1}$ are not shown in the figure. This is due to the fact that the total dissolved oxygen deficit at various X values is so high that the dissolved oxygen reduces to almost zero; in turn, the water quality models cannot be employed. It can also be seen that the sensitivity term $S_{D,K2,t}$ is extremely high,

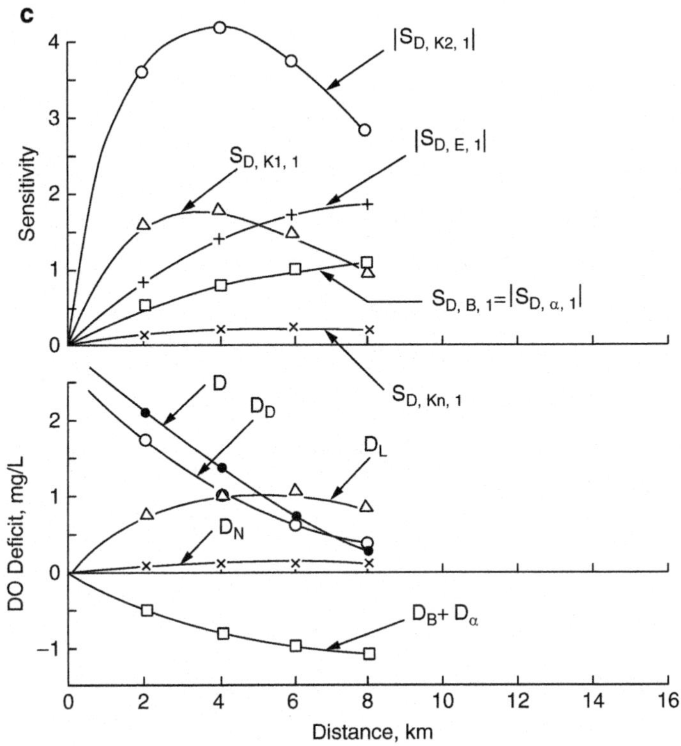

Fig. 8.10 (continued)

and $S_{D,E,t}$ is comparatively sensitive than $S_{D,K1,t}$ and $S_{D,B,t}$ (or $|S_{D,\alpha,t}|$). The dissolved oxygen deficit associated with photosynthesis and bottom deposit deficit ($D_\alpha + D_B$) is significant, thus cannot be overlooked.

5 Discussions and Recommendation

This research divides the non-tidal streams into nine categories, and divides the tidal streams into three categories as follows, and presents the average hydraulic and water quality characteristics of each stream category:

Non-tidal Receiving Streams:

a. clean or slightly polluted swift non-tidal streams
b. moderately polluted swift non-tidal streams
c. heavily polluted swift non-tidal streams
d. clean or slightly polluted intermediate non-tidal streams
e. moderately polluted intermediate non-tidal streams

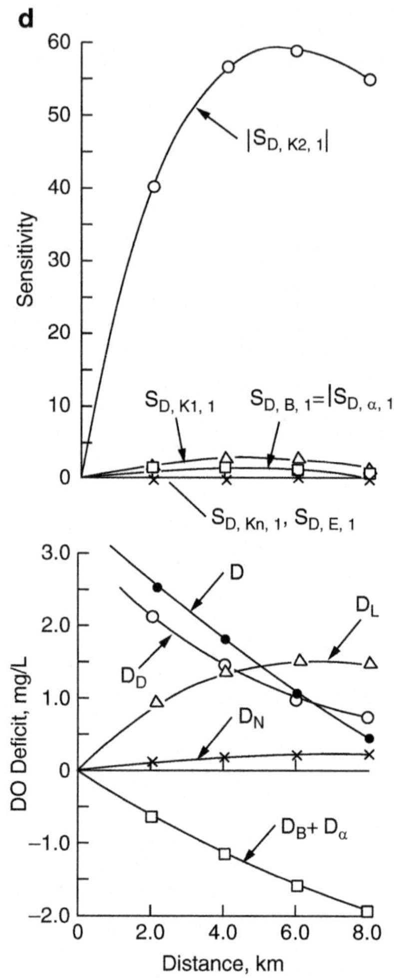

Fig. 8.10 (continued)

f. heavily polluted intermediate non-tidal streams
g. clean or slightly polluted slow non-tidal streams
h. moderately polluted slow non-tidal streams
i. heavily polluted slow non-tidal streams

Tidal Receiving Streams:

a. clean or slightly polluted tidal streams
b. moderately polluted tidal streams
c. heavily polluted tidal streams

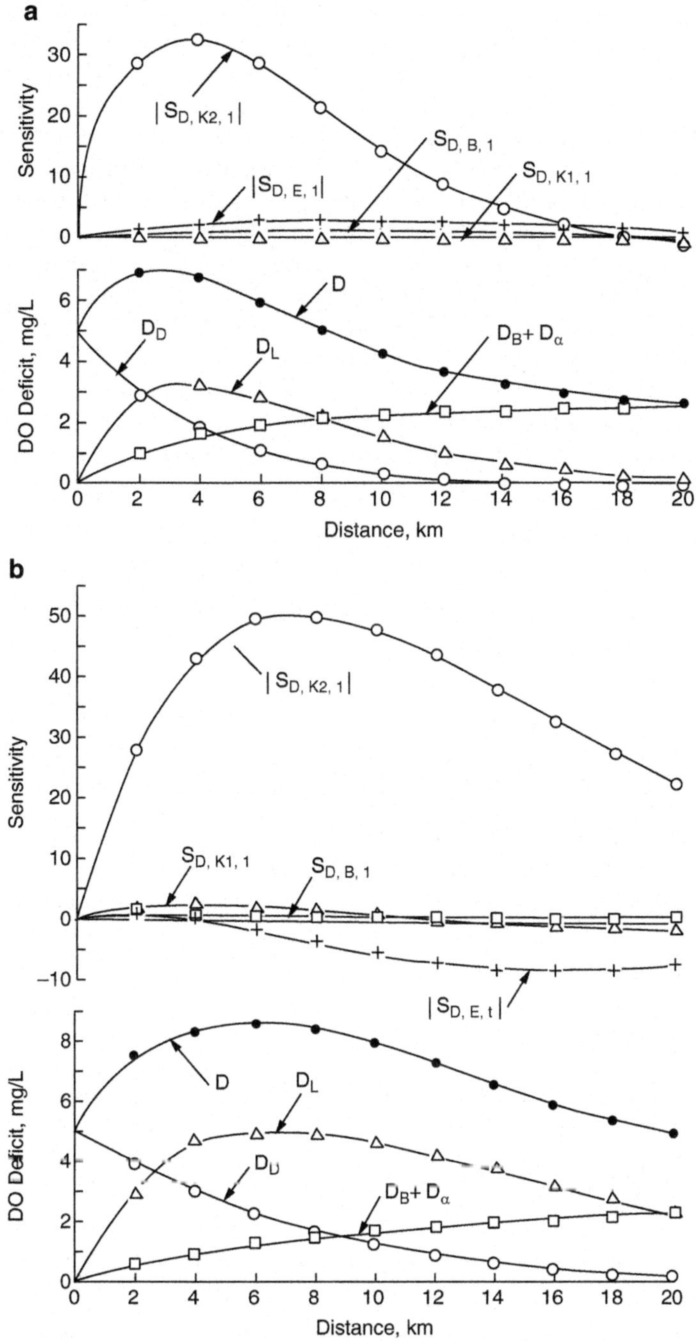

Fig. 8.11 (**a**) The sensitivities and dissolved oxygen deficits for heavily polluted tidal stream ($K_2 = 0.8$ 1/day and $U = 5$ km/day). (**b**) The sensitivities and dissolved oxygen deficits for heavily polluted tidal stream ($K_2 = 0.8$ 1/day and $U = 5$ km/day)

The sensitivity analysis can be a very useful scientific tool for stream water quality management. The results of this research identify the significance of water quality sensitivities and dissolved oxygen deficits systematically and graphically for each stream category. Knowing the category or type of a target stream (tidal or non-tidal; swift, intermediate or slow stream velocity; clean, moderately polluted, or heavily polluted), an environmental water resources engineer can determine/understand the sensitivity of each water quality/quantity parameter, and in turn, can better manage his/her stream pollution control projects.

Sensitivity analyses have been used by environmental water resources engineers extensively in recent years for environmental risk and decision analysis [18–21], ecological investigations [22–25], aquaculture site selection [26] and waterway management [27]. Its other applications may be further explored.

Further research in the area of sensitivity analysis of tidal streams may consider the effect of salinity because the salinity affects the reaeration coefficient [15–17, 28–30], in accordance with the following NCKU (National Cheng Kung University) equations.

$$K_{2s} = K_{2f} \exp\left(0.0127 \text{ Chlorinity}\right) \quad (8.11)$$

$$K_{2s} = K_{2f} \exp\left(0.0000127 \text{ Chloride}\right) \quad (8.12)$$

$$K_{2s} = K_{2f} \exp\left(0.007 \text{ Salinity}\right) \quad (8.13)$$

where

K_{2s} = reaeration coefficient of saline water, day^{-1}
K_{2f} = reaeration coefficient of fresh water, day^{-1}
Chlorinity = chlorinity of receiving water, g/L
Chloride = chloride concentration of receiving water, mg/L
Salinity = salinity of receiving water, ‰, or ppt, or parts per thousand

Glossary [31–35]

Ammonia nitrogen A common way to report ammonia concentration (expressed as ammonia-nitrogen).

Ammonification A process of formation of ammonia nitrogen from reduced organic nitrogen compounds.

Biological oxidation A process by which living organisms in the presence of oxygen convert organic matter into a more stable or a mineral form.

Carbonaceous Containing carbon and derived from organic substances such as coal, coconut shells, and organic waste.

Denitrification A biochemical process of conversion of nitrite nitrogen and nitrate nitrogen to molecular nitrogen, nitrogen dioxide, or a mixture of these two gases, under reducing conditions in the absence of free dissolved oxygen.

Deoxygenation It is a process for depletion of the dissolved oxygen in a liquid either under natural conditions associated with the biochemical oxidation of organic matter present or by addition of chemical reducing agents.

Deposit Material left in a new position by a transporting agent such as earth quake, gravity, human activity, ice, water current, or wind.

Dissolved gases The sum of gaseous components, such as oxygen, nitrogen, carbon dioxide, methane, hydrogen sulfide, etc. that are dissolved in water.

Dissolved oxygen (DO) The concentration of oxygen dissolved in water, which is often expressed in units of mg/L.

Dissolved oxygen deficit (D) The difference between the dissolved oxygen saturation concentration (C_s) and actual dissolved oxygen concentration at time t (Q) in a receiving water (such as river) at some downstream distance away from the point of waste discharge ($D = C_s - C$). See dissolved oxygen deficit and dissolved sag curve.

Dissolved oxygen sag curve (DO sage curve) A stream water quality curve that represents the profile of dissolved oxygen concentration along the course of a stream resulting from deoxygenation associated with biochemical oxidation of organic matter and reoxygenation through the absorption of atmospheric oxygen and biological photosynthesis. Also called oxygen sag curve.

Dissolved oxygen saturation concentration (C_s) The maximum concentration (mg/L) of dissolved oxygen in water under specific water temperature, pressure and salinity.

Dissolved solids The constituents in water that can pass through a 0.45-μm pore-diameter filter.

Initial dissolved oxygen deficit (D_0) The difference between the dissolved oxygen saturation concentration (C_s) and actual dissolved oxygen concentration (C) in a receiving water (river or lake) at the point of waste discharge ($D_0 = C_s - C$). See dissolved oxygen deficit.

NCKU (National Cheng Kung University) equations They are reaeration coefficient equations developed by National Cheng Kung University, Taiwan, showing the effect of salinity on receiving water's reaeration coefficient. The NCKU equations are modeled by $K_{2s} = K_{2f}$ exp (0.0127 Chlorinity); $K_{2s} = K_{2f}$ exp (0.0000127 Chloride); and $K_{2s} = K_{2f}$ exp (0.007 Salinity); in which K_{2s} = reaeration coefficient of saline water, day^{-1}; K_{2f} = reaeration coefficient of fresh water, day^{-1}; Chlorinity = chlorinity of receiving water, g/L; Chloride = chloride concentration of receiving water, mg/L; and Salinity = salinity of receiving water, ‰, or ppt, or parts per thousand.

Nitrate nitrogen A common way to report nitrate concentration (expressed as nitrogen).

Nitrification A process of formation of nitrate nitrogen from reduced inorganic nitrogen compounds, such as ammonia nitrogen. Nitrification in the natural environment is carried out primarily by autotrophic bacteria.

Nitrite nitrogen A common way to report nitrite concentration (expressed as nitrogen).

Non-tidal stream/river A stream/river which water level and flow direction will not fluctuate and will not be affected by the action of lunar and solar forces upon the rotating earth.

Oxygen-sag curve See dissolved oxygen sag curve.

Photosynthesis The conversion of light energy to chemical energy. At night, this process reverses: plants and algae suck oxygen out of the water.

Reaeration (a) The physical chemical reaction by which oxygen is absorbed back into water, (b) An aeration process by which oxygen in air is absorbed back into natural water, such as stream water and lake water, (c) A natural process of oxygen exchange between the atmosphere and a natural water body in contact with the atmosphere. Typically, the net transfer of oxygen is from the atmosphere and into the water, since dissolved oxygen levels in most natural waters are below saturation. When photosynthesis produces supersaturated dissolved oxygen levels, however, the net transfer is back into the atmosphere, (d) Reaeration process is modeled as the product of reaeration coefficient multiplied by the difference between dissolved oxygen saturation and the actual dissolved oxygen concentration, that is: $F_c = K_2\ (C_s - C) = (K_L/H)\ (C_s - C)$. Here F_c = rate or flux of dissolved oxygen across the water body. $M/L^3/T$; C = dissolved oxygen concentration, M/L^3, C_s = saturation dissolved oxygen concentration, M/L^3, K_2 = reaeration coefficient, $1/T$, H = water depth, L, K_L = surface transfer coefficient, L/T.

Reaeration coefficient A mass transfer coefficient (K_2) in reaeration process. See reaeration and mass transfer coefficient.

Reaeration rate (a) The rate at which oxygen is absorbed back into water. This is dependent, among other things, upon turbulence intensity, temperature, and the water depth, (b) The reaeration rate is defined as the rate of dissolved oxygen across the water body $F_c = K_2\ (C_s - C)$. Here F_c = rate or flux of dissolved oxygen across the water body, $M/L^3/T$; C = dissolved oxygen concentration, M/L^3; C_s = saturation dissolved oxygen concentration, M/L^3; K_2 = reaeration coefficient, $1/T$.

Reaeration rate coefficient See reaeration coefficient.

Receiving waters (a) A river, lake, ocean, stream, or other bodies of water into which wastewater or treated effluent is discharged; (b) A distinct water body that receives run off, or wastewater discharges, such as streams, rivers, lakes, estuaries and oceans.

Saline water intrusion The movement of saline groundwater into a formerly freshwater aquifer as a result of pumping in that aquifer usually near coastal areas where the source of saline water is the nearby ocean.

Sensitivity (a) In analytical testing, the lowest practical detection level; (b) In microbiological testing, the likelihood that the test result will be positive when the target organism is present, (c) In water resources engineering, the smallest changes of certain physical parameters that will affect hydraulic or hydrological model's solutions.

Sensitivity analysis (a) A mathematical analysis of the sensitivity of the dependent variable in a mathematical expression as a function of variations in the value of any independent variables or coefficients associated with the independent variables, (b) A mathematical analysis which determines how much the value of Y is affected by changes in the values of a and b.

Tidal Pertaining to periodic water level fluctuations due to the action of lunar (moon) and solar (sun) forces upon the rotating Earth.

Tidal current A water current brought about or caused by tidal forces.

Tidal stream/river A stream/river which is affected by tidal current and its water level and flow direction fluctuate due to the action of lunar and solar forces upon the rotating Earth.

Appendix 1: Water Quality Models

For Non-tidal Streams

1. Biochemical Oxygen Demand (L) Model

$$L = L_O[\exp(-K_1t)] \tag{8.14}$$

2. Dissolved Oxygen Deficit (D) Model

$$\begin{aligned} D = {} & K_1L_O(K_2 - K_1)^{-1}[\exp(-K_1t) - \exp(-K_2t)] \\ & + K_nN_O(K_2 - K_n)^{-1}[\exp(-K_nt) - \exp(-K_2t)] \\ & + (B - \alpha)(K_2)^{-1}[1 - \exp(-K_2t)] + D_O\exp(-K_2t) \end{aligned} \tag{8.15}$$

For Tidal Streams

1. Biochemical Oxygen Demand (L) Model

$$L = L_O[\exp(J_1X)] \tag{8.16}$$

2. Dissolved Oxygen Deficit (D) Model

$$\begin{aligned} D = {} & K_1L_O(K_2 - K_1)^{-1}[\exp\,(J_1X) - \exp\,(J_2X)] \\ & + K_nN_O(K_2 - K_n)^{-1}[\exp\,(J_nX) - \exp\,(J_2X)] \\ & + (B - \alpha)\ K_2^{-1}[1 - \exp\,(J_2X)\ ] + D_o\exp\,(J_2X) \end{aligned} \tag{8.17}$$

Appendix 2: Sensitivity Formulas for Non-tidal Streams

1. The sensitivity of L to K_1:

$$S_{L,\,K1} = -L_Ot[\exp(-K_1t)] \tag{8.18}$$

2. The sensitivity of D to K_1:

$$\begin{aligned} S_{D,\,K1} = {} & K_2L_o(K_2 - K_1)^{-2}[\exp(-K_1t) - \exp(-K_2t)\ ] \\ & - K_1L_Ot(K_2 - K_1)^{-1}\exp(-K_1t) \end{aligned} \tag{8.19}$$

3. The sensitivity of D to K_n:

$$\begin{aligned} S_{D,\,Kn} = {} & K_2N_O(K_2 - K_n)^{-2}[\exp(-K_nt) - \exp(-K_2t)] \\ & - K_nN_Ot(K_2 - K_n)^{-1}\exp(-K_nt) \end{aligned} \tag{8.20}$$

4. The sensitivities of D to α and B:

$$S_{D,\,B} = -S_{D,\alpha} = (K_2)^{-1}[1 - \exp(-K_2t)] \tag{8.21}$$

5. The sensitivity of D to K_2:

$$\begin{aligned} S_{D,\,K2} = {} & -K_1L_O(K_2 - K_1)^{-2}[\exp(-K_1t) - \exp(-K_2t)] \\ & + K_1L_Ot(K_2 - K_1)^{-1}\exp(-K_2t) \\ & - K_nN_O(K_2 - K_n)^{-2}[\exp(-K_nt) - \exp(-K_2t)] \\ & + K_nN_Ot(K_2 - K_n)^{-1}\exp(-K_2t) \\ & + (\alpha - B)(K_2)^{-2}[1 - \exp(-K_2t)] \\ & - (\alpha/K_2 - B/K_2 + D_O)\ t\ [\exp(-K_2t)] \end{aligned} \tag{8.22}$$

6. The sensitivity of DO to K:

$$S_{C,\,K} = -S_{D,\,K} \tag{8.23}$$

Appendix 3: Sensitivity Formulas for Tidal Streams

1. The sensitivity of L to K_1:

$$S_{L,\ Kl,\ t} = -L_O m_1 X[\exp(J_1X)] \tag{8.24}$$

2. The sensitivity of D to K_1:

$$\begin{aligned} S_{D,\ Kl,\ t} = {} & K_2Lo(K_2 - K_1)^{-2}[\exp(J_1X) - \exp(J_2X)] \\ & - K_1m_1XL_O(K_2 - K_1)^{-1}\exp(J_1X) \end{aligned} \tag{8.25}$$

3. The sensitivity of D to K_n:

$$\begin{aligned} S_{D,\ Kn,\ t} = {} & K_2N_O(K_2 - K_1)^{-2}[\exp(J_nX) - \exp(J_2X)] \\ & - K_nN_Ot(K_2 - K_n)^{-1}\exp(-K_nt) \end{aligned} \tag{8.26}$$

4. The sensitivities of D to α and B:

$$S_{D,\ B,\ t} = -S_{D,\alpha,\ t} = (K_2)^{-1}[1 - \exp(J_2X)] \tag{8.27}$$

5. The sensitivity of D to E:

$$\begin{aligned} S_{D,\ E,\ t} = {} & K_1L_OX(K_2 - K_1)^{-1}[n_1\exp(J_1X) - n_2\exp(J_2X)] \\ & + K_nN_OX(K_2 - K_n)^{-1}[n_n\exp(J_nX) - n_2\exp(J_2X)] \\ & + (\alpha/K_2 - B/K_2 + D_O)n_2X[\exp(J_2X)] \end{aligned} \tag{8.28}$$

6. The sensitivity of D to K_2:

$$\begin{aligned} S_{D,\ K2,\ t} = {} & -K_1L_O(K_2 - K_1)^{-2}[\exp(J_1X) - \exp(J_2X)] \\ & - K_1L_OXm_2(K_2 - K_1)^{-1}\exp(J_2X) \\ & - K_nN_O(K_2 - K_1)^{-2}[\exp(J_nX) - \exp(J_2X)] \\ & - K_nN_OXm_2(K_2 - K_n)^{-1}\exp(J_2X) \\ & + (\alpha - B)(K_2)^{-2}[1 - \exp(J_2X)] \\ & - \left[(\alpha - B)(K_2)^{-1} + D_Om_2X\exp(J_2X)\right] \end{aligned} \tag{8.29}$$

7. The sensitivity of DO to K:

$$S_{C,\ K,\ t} = S_{D,\ K,\ t} \tag{8.30}$$

References

1. Norton, W. R., & Roesner, L. A. (1974). *Computer program: Documentation for the stream quality model QUAL-II*. Washington, DC: US Environmental Protection Agency.
2. Wen, C. G., Kao, J. F., Wang, L. K., & Wang, M. H. S. (1982). Determination of sensitivity of water quality parameters for stream pollution control. *Journal of Environmental Management, 14*, 17–34.
3. Wen, C. G., Kao, J. F., & Wang, L. K. (1980). Mathematical modeling of stream water quality by a new moment method: Theoretical development. *Journal of Environmental Management, 10*, 1–11.
4. Wen, C. G., Kao, J. F., & Wang, L. K. (1981). Mathematical modeling of stream water quality by a new moment method: Field investigation of a tidal river. *Journal of Environmental Management, 12*, 127–140.
5. Wang, M. H. S., Wang, L. K., Kao, J. F., Wen, C. G., & Vielkind, D. (1979). Computer-aided stream pollution control and management. Part I. *Journal of Environmental Management, 9*, 165–183.
6. Wang, L. K., & Elmore, D. C. (1981). *Computer-aided modeling of water vapor pressure, gas absorption coefficient and oxygen solubility* (Report no. PB82-118787, p. 137). Springfield, VA: US, Department of Commerce, National Technical Information Service.
7. Wang, L. K., & Elmore, D. C. (1981). *Development of a computer plotting program* (Report no. PB81-202558, p. 89). Springfield, VA: US, Department of Commerce, National Technical Information Service.
8. Wang, L. K., Wang, M. H. S., Kao, J. F., & Wen, C. G. (1979). Computer-aided stream pollution control and management. Part II. *Journal of Environmental Management, 9*, 185–204.
9. Wen, C. G., & Kao, J. F. (1978). Mathematical models of dissolved oxygen level of Po-Tzu creek. In Proceedings of Aquatic Environment in Pacific Region, Taipei, Taiwan, Republic of China.
10. Kao, J. F., Wen, C.G., & Cheng, S. S. (1978). Water pollution of Chi-Shui Creek in Taiwan. In International Conference on Water Pollution Control in Developing Countries, Bangkok, Thailand.
11. National Cheng Kung University. (1979). The assimilative: capacity of Kao-Ping River, Environmental Engineering Department. National Cheng Kung University, Tainan, Taiwan, Republic of China, No. 10.
12. National Taiwan University. (1978). Water pollution control: planning of Chang-Kang River. National Taiwan University, Taipei, Taiwan, Republic of China.
13. Taipei Area Sewerage Engineering Department. (1971). Sewerage planning in the greater Taipei area-water pollution survey. Taipei, Taiwan, Republic of China.
14. Thomann, R. V. (1972). *Systems analysis and water quality management*. New York: McGraw-Hill.
15. Wang, L. K., Kao, J. F., Wen, C. G., & Liaw, C. C. (1984). Effect of salinity on reaeration coefficient of receiving waters. *Water Science & Technology, 16*(5–7), 139.
16. US EPA. (1985). Rates, Constants, and Kinetics Formulations in Surface Water Quality Modeling. EPA 600-3-85-040 US Environmental Protection Agency, Washington, DC, June 1985.

17. James, K., Cant, B., & Ryan, T. (2003). Responses of freshwater biota to rising salinity levels and implications for saline water management: A review. *Australian Journal of Botany, 51*, 703–713.
18. US EPA. (1998). Guidelines for ecological risk assessment. EPA-630-R-95-002F, US Environmental Protection Agency, Washington, DC.
19. Ferenc, S. A., & Foran, J. A. (2000). *Multiple stressors ecological risk and impact assessment: Approaches to risk estimation*. Pensacola, FL: SETAC Press.
20. Burgman, M. A. (2005). *Environmental risk and recision analysis: For conservation and natural resource management*. London: Cambridge University Press.
21. EPA-Victoria. (2008). An ecological risk assessment of the lower Wimmera River. Environment Protection Authority, Victoria, Australia. www.vict.gov.au Publication 1257, October.
22. Taylor, L. R. (2013). Ecological sensitivity analysis and aquatic assessment for the proposed poultry abattoir at Funda Mlimi Farms, Mpumalanga. Taylor Environmental CC, White River, South Africa, May 1, 2013.
23. Kassenaar, D. (2012). *Barrie, lovers, and Hewitt Creeks—Ecologically significant ground water recharge area assessment and sensitivity analysis*. Toronto, Ontario, Canada: Earth Fx Inc.
24. Simpson, T. G., Druppel, J. L., Watson, M. C., Benson, B. I., & Mullen, J. D. (2011). *Comparing ecological sensitivity with stream flow rates in the Apalachicola—Chattahoochee-flint River basin*. Athens, GA: The University of Georgia.
25. Harper, E. B., Stella, J. C., & Fremier, A. (2011). Global sensitivity analysis for complex ecological models: A case study of riparian cottonwood population dynamics. *Ecological Applications, 21*(4), 1225–1240.
26. FAO. (2014). Site selection for aquaculture: Chemical features of water. Fisheries and Aquaculture Department, Washington. www.fao.org.
27. Ladson, A. R., & White, L. J. (1999). Development and testing of an index of stream condition for waterway management in Australia. *Freshwater Biology, 41*(2), 453–468.
28. Wen, C. G., Kao, J. F., Wang, L. K., & Wang, M. H. S. (1983). Sensitivity analysis of ecological design parameters in streams. *Civil Engineering for Practicing and Design Engineers, 2*(4), 425–446. 2(6), 537–550.
29. Wen, C. G., Kao, J. F., Wang, L. K., & Liaw, C. C. (1986). Effect of salinity on reaeration coefficient of receiving waters. *Civil Engineering for Practicing and Design Engineers, 5*(1), 1–18. 5(2), 57–66.
30. Wen, C. G., Kao, J. F., Liaw, C. C., Wang, M. H. S., & Wang, L. K. (2015). Determination of reaeration coefficient of saline receiving water for water quality management. In L. K. Wang, C. T. Yang, & M. H. S. Wang (Eds.), *Advances in water resources management*. New York: Springer Science + Business Media.
31. Wang, M. H. S., & Wang, L. K. (2014). Glossary and conversion factors for water resources engineers. In: L. K. Wang & C. T. Yang (Eds.), *Modem water resources engineering*. New York: Humana Press.
32. Symons, J. M., Bradley, L. C., Jr., & Cleveland, T. C. (Eds.). (2000). *The drinking water dictionary* (p. 506). Denver, CO: American Water Works Association.
33. Wang, L. K. (1974). *Environmental engineering glossary* (p. 439). Buffalo, NY: Calspan Corporation.
34. Wang, M. H. S., & Wang, L. K. (2015). Environmental water engineering glossary. In: C. T. Yang & L. K. Wang (Eds.), *Advances in water resources engineering*. New York: Springer.
35. Wang, M. H. S., & Wang, L. K. (1986). Water quality control of tidal rivers and estuaries. In: L. K. Wang & N. C. Pereira (Eds.), *Water resources and natural control processes* (pp. 61–106). Totowa, NJ: Humana Press.

Chapter 9
River Ice Processes

Hung Tao Shen

Contents

H.T. Shen, Ph.D. (✉)
Department of Civil and Environmental Engineering, Wallace H. Coulter School of Engineering, 232 Rowley Laboratories, Clarkson University, 8 Clarkson Avenue, Potsdam, NY 13699-5710, USA
e-mail: HTShen@Clarkson.edu

L.K. Wang, C.T. Yang, and M.-H.S. Wang (eds.), *Advances in Water Resources Management*, Handbook of Environmental Engineering, Volume 16,
DOI 10.1007/978-3-319-22924-9_9

Abstract The presence of ice in rivers is an important aspect to be considered in the development of water resources in cold regions. River ice research has largely been driven by engineering and environmental problems that concern society. Ice formation can affect the design, operation, and maintenance of hydraulic engineering facilities, in addition to issues related to ecological, environmental, and morphological aspects of the river. River ice processes involve complex interactions between hydrodynamic, mechanical, and thermal processes. They are also influenced by weather and hydrologic conditions. This chapter gives a brief overview of river ice processes, followed by discussions on the state of knowledge of these processes from freeze-up to breakup, and sediment transport in rivers under ice conditions.

Keywords Anchor ice • Border ice • Breakup • Frazil ice • Freeze up • Ice dynamics • Ice jams • Ice runs • River ice • Sediment transport

Nomenclature

A	Flow cross sectional area under the cover, m^2
A_s	Mass rate of snowfall over a unit area of water surface, kg m^{-2} s^{-1}
a_0	Mean area of frazil crystal discs, mm
B	Top width of channel, m
B_a	River surface width in the wind direction, m
B_o	Open water width, m
C	Cloud cover in tenths
C_a	Area concentration of surface ice
C_b	Fraction of bed width covered by anchor ice
C_i	Specific heat of ice, 4.1855×10^3 J kg^{-1} °C
C_p	Specific heat of water, 4.215×10^3 J kg^{-1} °C check p. 25?
C_v	Suspended ice concentration
C_v^g	Suspended ice concentration due to thermal growth
C_o	A reference sediment concentration at a near bed reference level z_o, cm
C_w	Wind drag coefficient on water surface
C_z	Chezy's coefficient, m$^{0.5}$ s^{-1}
$C_{a,max}$	Maximum allowable surface ice concentration
C_s	Suspended sediment concentration, mg/L
$c_f = \dfrac{n_b^2 g}{(\alpha_b H')^{1/3}}$	Bed friction coefficient
$c_{iw} = \dfrac{n_i^2 g}{(\alpha_i H')^{1/3}}$	Coefficient of water drag on ice

D	Sediment particle diameter, cm
D_H	Hydraulic diameter, m
$D_* = D_{50}[(s-1)g/\nu^2]^{1/3}$	A dimensionless sediment particle parameter
D_{50}	Median diameter of sediment, cm
d_e	Frazil crystal thickness, mm
d_n	Nominal diameter of under-cover granular ice, cm
d_a	Day number of the year counting from January 1
E	Net volumetric rate of loss of frazil due to exchange with the surface layer and the anchor ice, m^3/s
E_a	Rate of change of surface ice concentration due to sources and sinks, s^{-1}
E_m	Rate of change of surface ice mass per unit area of water surface due to sources and sinks, kg/m
E_o	Eccentricity correction factor of the earth's orbit
e_a	Vapor pressure, mb
e_{an}	Porosity of anchor ice
e_c	Porosity in the accumulation between ice floes
e_f	Porosity of the frazil ice portion of surface ice floes
$e_j = e_c + (1-e_c)e_p$	Overall porosity of the surface ice accumulation
e_p	Porosity of ice floes
e_s	Saturated vapor pressure above the river surface, mb
e_{so}	Saturated vapor pressure corresponding to the river surface temperature, mb
$e_T = C_p\rho(1-C_v)T_w - \rho_i C_v L_i$	Thermal energy of the ice–water mixture per unit volume, J/m^3
e_u	Porosity of the frazil jam
F	Shape factor of under-cover granular ice
$\vec{F}_a$	Wind drag on ice, N/m^2
F_b	Inter-bed particle resistance per unit area, N/m^2
F_{rc}	A critical Froude number beyond which juxtaposition of ice floes cannot occur
F_{rp}	A limiting Froude number for cover progression
$\vec{F}_w$	Water drag on ice, N/m^2
f	Fraction of the width covered by border ice cover
$\vec{G}$	Gravitational force along the water surface slope, N/m^2
g	Gravity, m/s^2
H	Water level, m

$H_T = H_R + \eta$	Total water depth from channel bottom to the water surface, m
$H' = H_R + \eta'$	Water depth beneath the ice layer, m
H_R	Water depth below the reference level, m
H_t	Water depth underneath an equivalent ice–water interface, m
$h_e = h_i + h_f(1 - e_f)$	Equivalent thickness of surface ice floes, m
h_f	Thickness of the frazil accumulation on the underside of surface ice floes, m
h_i	Thickness of the solid portion of surface ice floes, m
$h_{i,p}$	Thickness of the porous portion of the ice cover, m
h_{iws}	Heat exchange coefficient between anchor ice and the substrate flow, W/m^2 °C
h_s	Snow cover thickness on top of the ice cover, m
h_{sa}	Heat transfer coefficient at the snow–air interface, W/m^2 °C
h_{sb}	Substrate flow depth at the bottom of the anchor ice, m
h_{wi}	Heat transfer coefficient between water and ice cover, W/m^2 °C
I_{so}	Solar constant, 1367 W/m^2
K_w	Thermal conductivity of water, W m^{-1} °C^{-1}
k_i	Thermal conductivity of ice cover, W m^{-1} °C^{-1}
k_{ic}	Ice cover roughness height, m
k_n	Decay constant of the ice cover roughness coefficient, day^{-1}
k_s	Thermal conductivity of snow cover, W m^{-1} °C^{-1}
L_i	Latent heat of fusion of ice, 3.3484×10^5 J kg^{-1}
$M_i = \rho_i C_a t_i$	Surface ice mass per unit area, kg/m^2
m	Optical air mass
m_o	Optical air mass at sea level
N_f	Number of frazil particles per unit volume, 1/m^3
$N_u = h_{wi} D_H / K_w$	Nusselt number
n_b, n_i, and n_c	Bed, ice cover, and composite Manning's coefficients, respectively, s ft$^{-1/3}$
n_{io}	Single layer surface ice roughness, s ft$^{-1/3}$
n_o, n_e	Initial and end values of the ice cover roughness coefficient, s ft$^{-1/3}$
P	Internal pressure of pack ice, Pa
$P_r = C_p \mu K_w$	Prandtl number
p_a	Atmospheric pressure, mb

p_b and p_i	Wetted perimeters formed by the channel bed and the ice cover, respectively, m
p_o	Atmospheric pressure at sea level, mb
Q	Water discharge, m^3/s
Q_d^i, Q_s^i	Volumetric rates of suspended and surface ice discharge, respectively, m^3/s
Q_u	Volumetric rate of ice entertainment under the cover at the leading edge, m^3/s
q_f	Heat transfer between water and a frazil particle per unit frazil surface area, J/cm^2 s
q_b	Volumetric rate of bed sediment transport per unit width, m^3/s
q_c	Volumetric rate of undercover ice transport per unit width, m^3/s
$\vec{q}_l$, and $\vec{q}_u$	Unit-width water discharge beneath the ice layer, and in the ice layer, respectively, m^2/s
q_s	Seepage flow in the surface ice layer, m^3/s
$\vec{q}_{ice}$	Unit-width surface ice discharge, m^3/s
R	Hydraulic radius, m
R_a	Rate of change of surface ice concentration due to mechanical redistribution of the ice mass, 1/s
R_{an}	Contribution to the surface ice run from anchor ice release, m^3/s
R_b	Rate of lateral growth of border ice, m/h
R_c	Radius of curvature of the centreline of the ice sheet, m
R_i	Hydraulic radius associated with the ice cover, m
$R_{eH} = \rho U D_H / \mu$	Reynolds number
R_t	Albedo of river surface
$\vec{R}$	Internal ice resistance due to floe-to-floe interactions, N/m^2
RH	Relative humidity in percentage, %
r_s	Reflectivity of long wave radiation from the river surface
S_f	Friction slope of the channel
S_w	Water surface slope
S_5	Cumulative degree-days above −5 °C, °C day
$s = \rho_s/\rho$	The ratio of sediment density and water density
T	Temperature in the cover, °C
T_a	Air temperature, °C
T_{cr}	Supercooled water surface temperature below which static skim ice will form, °C
T_{ak}	Air temperature at 2 m above the surface, in the absolute scale, K

T_m	Freezing point of water, 0 °C
T_s	River surface temperature, °C
T_t	A sediment transport stage parameter
T_{sk}	River surface temperature in the absolute scale, K
T_{sn}	Snow surface temperature, °C
T_w	Cross section-averaged water temperature, °C
$T_{w,s}$	Water surface temperature, °C
t_{eq}	Equilibrium ice jam thickness, defined as the jam thickness in the uniform reach, m
t_f	Thickness of frazil jam, m
t_i	Surface ice layer or ice cover thickness, m
t'_i	Submerged surface ice layer thickness, m
u_{*i}	Shear velocity on the undersurface of the frazil jam, m/s
v_{a2}	Wind velocity at 2 m above the surface, m/s
v_b	Buoyancy velocity of frazil ice, m/s
v_0	Mean volume of frazil crystals, cm^3
v'_z	Vertical component of turbulent fluctuation velocity, m/s
U	Cross-section-averaged flow velocity, m/s
U_s	Width-averaged surface ice velocity, m/s
V_a	Wind velocity, m/s
V_{cb}	Maximum velocity at which a surface ice floe can adhere to the border ice edge, m/s
V_{cp}	Rate of progression of leading edge, m/s
$\hat{V}$	Visibility, km
$\forall$	Volume of anchor ice per unit bed area, m
$\vec{V}_i$	Surface ice velocity, m/s
$\vec{V}_w$	Water current velocity underneath the ice cover, m/s
W_b	Submerged weight of bed materials per unit area, N/m^2
W_i	Width of ice sheet detached from banks, i.e. the distance between hinge cracks, m
x, y, and t	Space and time variables, m and s
α	Probability of deposition of frazil particles reaching the surface layer
α_b	Fraction of the water depth affected by the bed friction
α_i	Fraction of the water depth affected by the ice friction
α_z	Solar latitude, angular elevation of the sun above the true horizon, degrees

β	Coefficient quantifying the rate of re-entrainment of surface ice per unit area, 1/s
γ	Coefficient quantifying the rate of accretion to the bed per unit area, m/s
$\gamma_e = 0.5\rho_i g\left(1 - \frac{\rho_i}{\rho}\right)\left(1 - e_j\right)$	N/m^3
ΔB	Growth of border ice width for a given time step, m
Δ_d	Dune height, m
Δ_i	The ratio $(\rho - \rho_i)/\rho$
$\Delta\phi*$	Surface heat exchange during a time step, J m^{-2}
δ	Solar declination, in radians
ε_a	Emissivity of atmosphere
ε_s	Emissivity of river surface
ξ_i	Bulk extinction coefficient, 1/cm
η	Water surface elevation, m
η'	Elevation of the bottom of surface ice, m
$\widetilde{\eta} = \frac{\Theta}{\Theta_c}$	Ratio of θ to the critical value of θ for incipient motion
$\Theta = \frac{\tau_b}{\Delta\rho g D}$	Dimensionless flow strength for sediment transport
Θ_c	Critical value of Θ for incipient motion
Θ_i	Dimensionless flow strength for cover load transport
Θ_{ci}	Critical flow strength for undercover ice transport
θ_a	Angle between the wind direction and the x-axis, degree
λ_s	Seepage coefficient in the pack ice, m/s
μ	Dynamic viscosity of water, N s/m^2
μ_i	Ice-over-ice friction coefficient
$\hat{\nu}$	Nonlinear shear viscosity of the surface ice run
Π	A stress parameter for ice cover during breakup period, kPa
ρ, ρ_a, ρ_i, and ρ_s	Density of water, air, ice, and sediment, respectively, kg/m^3
σ	Stefan–Boltzmann constant, 5.67×10^{-8} W m^{-2} K^{-4}
σ_f	Flexural strength of the ice cover, kPa
σ_{ij}	Internal ice stress in the surface ice layer, N/m^2
τ_a	Wind drag on the water or ice surface along the channel, N/m^2
τ_{ai}	Wind drag on ice, N/m^2
τ_{aw}	Wind drag on water, N/m^2
τ_b, τ_i	Shear stresses at the channel bottom and the ice–water interface, respectively, N/m^2

τ_s	Shear stresses at the ice–water interface, N/m^2
$\tau^{(a-w)}$	Wind shear stress on water surface, N/m^2
$\tau^{(i-w)}$	Shear stress at the ice–water interface, N/m^2
τ_c	Cohesion in the shear stress of floating surface ice accumulation, N/m^2
τ_g	Weight component of the ice cover along the surface slope, N/m
τ_w	Water drag, N/m^2
$\widetilde{\tau}_i$	Driving stress on the ice cover during breakup period, N/m^2
Φ_i	Dimensionless under-cover ice transport capacity
$\Phi_b = \frac{q_b}{\sqrt{\Delta g D^3}}$	Dimensionless bed load intensity
φ	Internal friction angle of surface ice, degree
ϕ_b	Bed heat flux per unit area, W/m^2
ϕ_{ba}	Atmospheric long-wave radiation reaching the river surface, kJ/m^2 h
ϕ_{br}	Reflected long wave radiation from the river surface, kJ/m^2 h
ϕ_{bs}	Long wave radiation emitted by the water surface, kJ/m^2 h
ϕ_B	Effective back radiation or terrestrial radiation, kJ/m^2 h
ϕ_{cl}	Incoming short wave radiation under clear skies, kJ/m^2 h
ϕ_E	Evaporation heat transfer, kJ/m^2 h
ϕ_H	Sensible heat transfer, kJ/m^2 h
ϕ_{ia}	Heat loss at the air-ice interface, kJ/m^2 h
ϕ_P	Heat transfer due to precipitation on the water surface, kJ/m^2 h
ϕ_{ps}, ϕ_{pz}	Solar radiation on the ice cover surface, and at a depth below the ice surface, kJ/m^2 h
ϕ_{ri}	Incoming solar radiation, kJ/m^2 h
ϕ_{rr}	Solar radiation reflected back to the atmosphere, kJ/m^2 h
ϕ_R	Net solar radiation, kJ/m^2 h
ϕ_{si}	Net rate of heat loss from top and bottom of surface ice floes, kJ/m^2 h
ϕ_{sk}	Rate of heat loss through top and bottom of the suspended ice layer, kJ/m^2 h
ϕ_{so}	Total extraterrestrial solar radiation incident on a horizontal surface, kJ/m^2 h
ϕ_{ss}	Rate of heat gain through top and bottom of the suspended layer, kJ/m^2 h

ϕ_v	Rate of internal heating of the cover due to absorption of penetrated short wave radiation, kJ/m^2 h
ϕ_{wi}	Heat flux between river water and ice cover, kJ/m^2 h
ϕ^*	Total surface heat flux between air and water, ϕ_{wa}, or between ice and air, ϕ_{ia}, kJ/m^2 h
ϕ_s^*	Rate of heat loss at the snow surface, kJ/m^2 h
ψ	Latitude in degrees, north positive, south negative
ω	Hour angle, degrees

1 Introduction

Ice formation can affect the design, operation, and maintenance of hydraulic engineering facilities in cold regions. Major engineering concerns related to river ice are ice jam flooding, hydropower operation, inland navigation, water transfer, and ice effects on environmental, ecological, and morphological conditions. In regions with severe winter conditions, the river regime experiences dramatic changes compared to open water conditions. This section gives an overview of river processes.

River ice phenomena include formation, evolution, transport, accumulation, dissipation, and deterioration of various forms of ice. These phenomena involve complex interactions between hydrodynamic, mechanical, and thermal processes, as well as the ambient hydro-meteorological conditions and channel morphology. Figure 9.1 illustrates various river ice processes that can occur during a winter.

In the beginning of winter, when the water temperature drops to the freezing point, further heat loss will lead to supercooling and frazil ice formation [2, 3]. Subsequently, various types of ice formation can develop in the river. In slow flowing areas, where the turbulence intensity is not strong enough to mix the water or frazil ice crystals formed on the water surface over the depth, skim ice can form in the river even before the cross-section-averaged water temperature drops to the freezing point. Stationary skim ice attached to the river bank is the first appearance of surface ice in the form of static border ice formed on the supercooled water surface. In areas with higher turbulence intensity, the supercooling condition and frazil ice extend over the depth of the flow. Frazil particles grow in size and quantity and develop into flocs. The increase in buoyancy of the frazil flocs may overcome the vertical mixing to form the surface ice run. Matousek [4] provided a useful empirical method for determining the occurrence of various types of ice formation.

Frazil ice is the origin of almost all other forms of river ice during the freeze up period. In supercooled water, frazil is active and can attach to any object. The initiation of frazil in a river is mainly due to the mass exchange of cold seed crystals across the free surface [5]. When mixed in the supercooled turbulent water, frazil

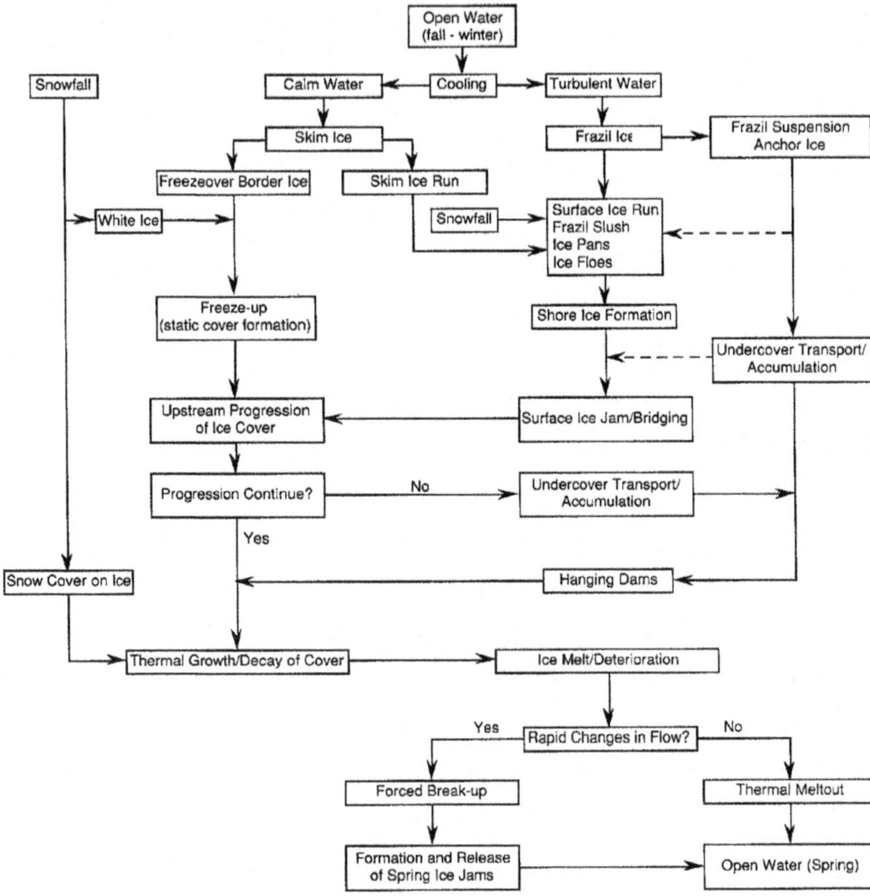

Fig. 9.1 River ice processes [1]

crystals can grow in size, multiply in number, and agglomerate into flocs [3, 6, 7]. Frazil crystals entrained to the channel bottom under supercooled conditions can attach to the bed and underwater objects to form anchor ice. The presence of anchor ice can cause changes in flow resistance, water level, and discharge [8]. The formation of anchor ice also has important biological implications for biota sensitive to freezing or the oxygen supply in the inter-gravel flow [9]. The knowledge on the development of anchor ice in river channels is limited. A few analytical and experimental investigations on this subject have been made [8, 10–12].

Ice flocs that rise to the water surface can form ice pans, which will grow in size and strength due to freezing of interstitial water, thermal thickening, and further accumulation of frazil ice from the suspension. Ice pans may sinter into large floes while travelling along the river. Partial coverage of the water surface by ice pans and floes will result in a reduction of the ice production rate due to the insulating effect of the surface ice [13, 14].

In addition to the border ice formed from skim ice, border ice can form due to the accumulation of surface ice along river banks or the edges of existing border ice. This type of lateral accumulation is limited by the stability of surface ice floes in contact with the existing edge of the border ice. An empirical relationship developed by Michel et al. [15] describes this border ice growth phenomenon. Limited field observations indicated this mode of border ice growth might cease when the local depth-averaged velocity exceeds a critical value of about 0.4 m/s [16, 17].

Jamming or bridging of the surface ice run by congestion initiates the accumulation of incoming surface ice into an ice cover or ice jam. The ice cover or ice jam thickness profile can be approximated by the static balance between the external and internal forces of the floating ice pack [18–24]. Since this type of static ice jam theory ignores the dynamics of ice motion, they cannot determine when and where an ice jam initiates. Numerical models for the dynamic transport of surface ice and ice jam evolution have been developed [25, 26]. Ice jams initiate at locations where ice concentration is high and convergence of ice mass occurs [27, 28].

The rate of progression of the leading edge of the cover depends on the rate of surface ice supply and the thickness of the cover. The flow condition at the leading edge of the ice cover governs the stability of the ice floes arriving from upstream and the limiting condition for cover progression [18]. When the flow velocity is relatively low, incoming ice floes will form a relatively smooth cover of one floe thickness through juxtaposition. When the surface ice consists mainly of frazil slush and loose frazil pans, the juxtaposition mode of cover formation may not be identifiable due to the compression of surface ice elements. At higher velocities, surface ice elements can underturn or submerge at the leading edge to form a thicker ice cover [18, 29]. This mode of ice cover progression is often termed hydraulic thickening or narrow river jam. The progression of the cover will cease when the velocity at the leading edge exceeds a critical condition. At this velocity the incoming surface ice that is swept under at the leading edge, and will be carried downstream contribute to undercover accumulation. This critical condition for the entrainment of surface ice can be represented by a critical Froude number for cover progression, F_{rp}, at the leading edge. In natural rivers the maximum value of the Froude number at the cross section controls the progression limit rather than the cross-section-averaged value. Field observations indicated that this value is about 0.09 [30, 31].

The cover formed by surface accumulation under any hydraulic condition has to be thick enough so that it is capable of withstanding the forces acting on it. These forces include current drag, wind drag, and the weight component of the cover along the water surface. The cover will collapse when the applied forces exceed the resistance provided by the cover strength and the bank shear. Mechanical thickening will occur until the cover reaches a thickness that is capable of withstanding the external forces. This process is often called secondary consolidation or shoving. Thick accumulations or jams formed after shoving are often called wide river jams. The formation of a thin crust near the top of the newly formed cover due to freezing can significantly increase the strength of the cover [32].

When the upstream progression of an ice cover stops at a cross section with high flow velocity, the rapid reach upstream of the cover will remain open during the winter. This open water reach will continue to produce frazil ice to be transported into the ice-covered reach. Since the open water reach has relatively high velocity, the ice entering the covered reach will be in the form of frazil granules. These frazil granules can transport and deposit on the underside of the cover. A localized undercover accumulation is often called a hanging dam or frazil jam. Frazil jams will lead to a rise in both the water level and the ice cover. It can also increase the potential of a breakup jam due to the excessive volume of ice in the channel. The location and thickness of a frazil jam were commonly determined by a critical velocity criterion or its equivalent in the past [33–35]. Shen and Wang [36] developed the theory of ice transport capacity. According to this theory, the ice discharge in the river adjusts itself towards the ice transport capacity through erosion and deposition, similar to the process of bed sediment transport. White and Acone [37] and Mao et al. [38] further confirmed this concept with field studies.

As heat exchange continues over a consolidated ice cover, water-filled voids in the cover will freeze from the free surface downward. This thermal growth will continue beyond the initial cover into the underlying river water. At the beginning of spring, or during a warm spell in the winter, melting of the ice cover will occur with the rise in air temperature. The slight increase in water temperature due to the warm spring runoff can significantly accelerate the thermal erosion of the underside of the cover through the turbulent heat transfer between the river water and the ice cover [39]. The thermal growth and decay of ice covers are relatively well understood and can be analyzed by heat conduction and heat transfer formulations [40, 41]. The existence of snow cover and frazil accumulation on the underside of the cover can affect the growth and decay of the cover [42]. Ice covers with layers of snow ice, snow slush, and black ice may form when repeated heavy snowfalls occur between cold weather spells [43]. During the melting stage the turbulent heat transfer between the river water and the cover can increase due to the formation of ripples on the underside of the cover [44]. Incoming solar radiation starts the decay and melting of the ice cover even before the air temperature rises above freezing. The solar radiation can cause internal melting and the loss of structural integrity of the cover [45, 46].

If the river discharge remains relatively steady, the ice cover will remain stable until the eventual melt out. This type of ice clearance in channels is often called thermal breakup or matured breakup. A more common type of breakup is the pre-matured breakup or mechanical breakup [47, 48]. This type of breakup is the breakup of the cover by hydraulic and mechanical forces associated with rapid changes in river discharge and water level. A mechanical breakup often leads to severe ice runs and ice jams. The pertinent parameters governing mechanical breakup are the increase in stage above the stable ice cover formed during the winter, ice thickness, and degree-days of thaw which provides a measure of ice cover strength [40]. Rainfall is an important factor that causes discharge to increase [27, 49].

2 River Hydraulics

River ice processes are closely related to flow conditions. Formulations for the hydraulics of flow in rivers with ice have been developed.

2.1 Flow with a Floating Ice Cover

Except for very small streams, river ice covers are floating on the water surface. For a river with floating solid ice covers, the Saint-Venant equations can be modified for the ice cover effects [50]:

$$\frac{\partial Q}{\partial x} + \frac{\partial A}{\partial t} = 0 \tag{9.1}$$

and,

$$\rho \frac{\partial Q}{\partial t} + \rho \left(\frac{2Q}{A} \frac{\partial Q}{\partial x} - \frac{Q^2}{A^2} \frac{\partial A}{\partial x} \right) + \rho g A \frac{\partial H}{\partial x} + p_b \tau_b + p_i \tau_i + B_o \tau_a = 0 \tag{9.2}$$

in which, Q = water discharge; A = flow cross sectional area under the cover; x and t are distance and time, respectively; ρ = density of water; H = water surface elevation; B_o = top width; g = gravity; p_b and p_i = wetted perimeters formed by the channel bed and the ice cover, respectively; τ_b and τ_i = shear stresses at the channel bottom and the ice–water interface, respectively; and τ_a = wind drag on the water surface along the channel.

The wind drag can be calculated by

$$\tau_a = \tau_{ai} C_a + \tau_{aw} (1 - C_a) \tag{9.3}$$

in which, C_a = surface ice concentration; τ_{ai} = wind drag on ice; τ_{aw} = wind drag on water. Equation (9.3) assumes the shear stress between the moving ice and water is the same as the wind drag on the top surface of the ice floe. The wind drag on water can be calculated by

$$\vec{\tau}_{aw} = C_w \rho_a \left| \vec{U}_s - \vec{V}_a \right| \left(\vec{U}_s - \vec{V}_a \right) \tag{9.4}$$

in which, C_w = wind drag coefficient; U_s = river surface velocity; m/s, approximated by the mean velocity considering the effect of inter-particle resistance of surface ice; V_a = wind velocity at 10 m above the water surface; m/s; ρ_a = density of air. Hicks [51] found that $C_w = 1.0 \times 10^{-3}$ for wind speed up to 5.0 m/s and increases linearly to a value of 1.5×10^{-3} for wind speed of 15 m/s. The wind drag on the ice surface can be calculated by the same formula except that the value of the drag coefficient is approximately 1.55×10^{-3} [52].

The resistance terms in the momentum equation are generally expressed in terms of the friction slope. The friction slope due to the bed and ice cover can be obtained as

$$S_f = \frac{p_b \tau_b + p_i \tau_i}{\rho g A} \tag{9.5}$$

The friction slope can be expressed in terms of Manning's coefficients of bed, n_b, and cover, n_i [53]. For fully covered reaches, the composite Manning's coefficient, n_c, can be calculated with the Belokon–Sabaneev formula:

$$n_c = \left(\frac{n_b^{3/2} + n_i^{3/2}}{2} \right)^{2/3} \tag{9.6}$$

It is important to note that the cover roughness varies with time over the winter. The time-dependent variations of the cover roughness can be expressed as [54]:

$$n_i = n_e + (n_o - n_e) \cdot e^{-k_n t} \tag{9.7}$$

in which, n_o, n_e, k_n = initial and end values of the cover roughness, and a decay constant, respectively. Either the roughness height of the ice cover, or Chezy's C, may be used instead of the Manning's coefficient to represent the flow resistance.

2.2 *Flow with Surface Ice Run and Ice Jam*

The modified Saint-Venant equations have been further improved for surface ice runs or ice jams when the ice velocity is different from the current velocity [25, 26]. By considering the flow in a river with an upper surface ice layer and a lower water layer, the two-dimensional depth-averaged total water and ice mass conservation equation for the case with a moving surface ice layer can be written as:

$$\frac{\partial \left[\rho H' + \rho t_i'(1 - C_a) + \rho_i t_i C_a\right]}{\partial t} + \nabla \cdot \left(\rho \vec{q}_l + \rho \vec{q}_u + \rho_i \vec{q}_{ice}\right) = 0 \tag{9.8}$$

in which, $H' = H_R + \eta'$ = water depth beneath the ice layer; H_R = water depth below the reference level; η' = elevation of the bottom of ice; t_i = ice layer thickness; t'_i = submerged ice layer thickness; ρ, ρ_i = water and ice density; C_a = surface ice concentration; $\vec{q}_l$ = unit-width water discharge beneath the ice layer; $\vec{q}_u$ = unit-width water discharge in the ice layer; and $\vec{q}_{ice} = \vec{V}_i C_a t_i$ = unit-width ice discharge. Since the ice mass conservation gives $\frac{\partial [\rho_i t_i C_a]}{\partial t} \approx -\nabla \cdot \left(\rho_i \vec{q}_{ice}\right)$, the above equation reduces to:

$$\frac{\partial H_T}{\partial t} + \nabla \cdot \left(\vec{q}_l + \vec{q}_u\right) = \frac{\partial}{\partial t}\left(C_a t_i^{'}\right) \tag{9.9}$$

Therefore, the continuity equation for the total water discharge can be written as:

$$\frac{\partial \eta}{\partial t} + \frac{\partial (q_{tx})}{\partial x} + \frac{\partial (q_{ty})}{\partial y} = \frac{\partial}{\partial t}\left(C_a t_i^{'}\right) \tag{9.10}$$

in which, $H_T = H_R + \eta$ = total water depth from channel bottom to the water surface; η = water surface elevation; $q_{tx} = q_{lx} + q_{ux}$ and $q_{ty} = q_{ly} + q_{uy}$ are components of total unit width water discharge; q_{lx}, q_{ly} = components of the unit width water discharge beneath the ice layer; $q_{ux} = q_{ix} + q_{sx}$, $q_{uy} = q_{iy} + q_{sy}$ are the water discharge in the upper layer; $q_{ix} = u_i(\eta - \eta^{'})(1 - C_a)$ and $q_{iy} = v_i(\eta - \eta^{'})(1 - C_a)$ are components of water discharge carried by ice; u_i, v_i = ice velocity components; and q_{sx}, q_{sy} = components of unit width water discharge in the ice layer relative to the moving ice, or the seepage discharge in stationary ice accumulations.

The momentum equation for hydrodynamics is

$$\frac{\partial q_{tx}}{\partial t} + \frac{\partial}{\partial x}\left(\frac{q_{tx}^2}{H_t}\right) + \frac{\partial}{\partial y}\left(\frac{q_{tx} q_{ty}}{H_t}\right) = \frac{1}{\rho}(\tau_{sx} - \tau_{bx}) + \frac{1}{\rho}\left(\frac{\partial T_{xx}}{\partial x} + \frac{\partial T_{yx}}{\partial y}\right) - gH_t \frac{\partial \eta}{\partial x} \tag{9.11}$$

$$\frac{\partial q_{ty}}{\partial t} + \frac{\partial}{\partial x}\left(\frac{q_{tx} q_{ty}}{H_t}\right) + \frac{\partial}{\partial y}\left(\frac{q_{ty}^2}{H_t}\right) = \frac{1}{\rho}(\tau_{sy} - \tau_{by}) + \frac{1}{\rho}\left(\frac{\partial T_{xy}}{\partial x} + \frac{\partial T_{yy}}{\partial y}\right) - gH_t \frac{\partial \eta}{\partial y} \tag{9.12}$$

in which, $T_{xy} = \varepsilon_{xy}\left(\frac{\partial q_{tx}}{\partial y} + \frac{\partial q_{ty}}{\partial x}\right)$; ε_{xy} are generalized eddy viscosity coefficients; τ_s and τ_b are shear stresses at the ice–water interface and the river bed; and H_t is the water depth underneath an equivalent ice–water interface calculated from:

$$\frac{H_t}{H^{'}} = \left(\frac{q_t}{q_l}\right)^{\frac{3}{5}} \tag{9.13}$$

The seepage flow is estimated by $q_s = \lambda_s \frac{H_s}{1-C_a} S^{\frac{1}{2}}; H_s = (\eta - \eta')(1 - C_a)$ = net water depth in the ice layer; λ_s is a seepage coefficient. Based on field calibrations, Beltaos [55, 56] suggested that λ_s ranges between 1.0 and 2.5 m/s, with an average value of 1.6 m/s. The seepage coefficient increases with the ice particle and void size in the jam. Hence, breakup jams will have a higher seepage coefficient than freeze up jams.

The bed shear stresses can be expressed as:

$$\tau_{bx} = c_f \rho \frac{q_x \left(q_x^2 + q_y^2\right)^{1/2}}{H'^2} \tag{9.14}$$

$$\tau_{by} = c_f \rho \frac{q_y \left(q_x^2 + q_y^2\right)^{1/2}}{H'^2} \tag{9.15}$$

in which, the friction coefficient, c_f, can be expressed in terms of the Manning's coefficients of the bed, n_b, as:

$$c_f = \frac{n_b^2}{(\alpha_b H')^{1/3}} g \tag{9.16}$$

in which, α_b is the fraction of the water depth affected by the bed friction. The shear stress components at the ice–water interface due to the ice and water velocity difference can be written as:

$$\tau_x^{(i-w)} = \rho c_{iw} \left| \vec{V}_i - \vec{V}_w \right| (u_i - V_{wx}) \tag{9.17}$$

$$\tau_y^{(i-w)} = \rho c_{iw} \left| \vec{V}_i - \vec{V}_w \right| (v_i - V_{wy}) \tag{9.18}$$

in which, $\vec{V}_w$ = water current velocity underneath the ice cover; $\vec{V}_i = u_i \vec{i} + v_i \vec{j}$ = ice velocity; and c_{iw} = coefficient of water drag on ice, which varies with ice concentration and ice floe geometry. It can be related to Manning's coefficient of ice layer, n_i, as:

$$c_{iw} = \frac{n_i^2 g}{(\alpha_i H')^{1/3}} \tag{9.19}$$

where, α_i is the fraction of the water depth affected by the ice friction. For a surface ice layer or ice jam of multiple floe thicknesses, the Manning's coefficient n_i appears to vary linearly with the thickness, i.e. $n_i = n_{io}\left(\frac{t_i}{h_e}\right)$, where n_{io} is the single layer ice roughness [26, 57, 58]. The coefficient α_i can be obtained by dividing the flow depth H' into a portion affected by the ice cover resistance, Y_i, and a portion affected by the bed shear stress, Y_b, separated by the zero shear stress plane. For a steady uniform flow

$$\tau_b = \rho g Y_b S \tag{9.20}$$

and

$$C_a \tau^{(i-w)} = \rho g Y_i S \tag{9.21}$$

in which, S is the water surface slope. For a uni-directional flow, the ratio between Y_i and Y_b can be expressed as:

$$\frac{Y_i}{Y_b} = \left[\frac{n_i^2}{n_b^2} \frac{C_a |V_w - V_i|(V_w - V_i)}{|V_w| V_w}\right]^{3/4} \tag{9.22}$$

Hence, the fraction of water depth governed by the bed resistance α_b is

$$\alpha_b = \frac{Y_b}{Y_i + Y_b} = \frac{1}{1 + \left[\frac{n_i^2}{n_b^2} \frac{|V_w - V_i|(V_w - V_i)}{|V_w| V_w}\right]^{3/4}} \tag{9.23}$$

and $\alpha_i = 1 - \alpha_b$. For two-dimensional flows the ice velocity and water velocity may not be in the same direction. The above equations can be extended to:

$$\frac{Y_i}{Y_b} = \left[\frac{n_i^2}{n_b^2} \frac{C_a \left|\vec{V}_w - \vec{V}_i\right| (V_{wx} - u_i)}{\left|\vec{V}_w\right| V_{wx}}\right]^{3/4} \tag{9.24}$$

in the x-direction, and

$$\frac{Y_i}{Y_b} = \left[\frac{n_i^2}{n_b^2} \frac{C_a \left|\vec{V}_w - \vec{V}_i\right| (V_{wy} - v_i)}{\left|\vec{V}_w\right| V_{wy}}\right]^{3/4} \tag{9.25}$$

in the y-direction.

On the open water surface, the surface shear stress due to the wind effect can be expressed as:

$$\tau_x^{(a-w)} = \rho_a C_w V_a^2 \cos \theta_a \tag{9.26}$$

$$\tau_y^{(a-w)} = \rho_a C_w V_a^2 \sin \theta_a \tag{9.27}$$

in which, C_w = wind drag coefficient [59]; V_a = wind velocity at 10 m above the water surface; ρ_a = density of air; θ_a = angle between the wind direction and the x-axis.

For a partially ice-covered water surface, the surface shear stress is assumed to be a linear combination of $\tau^{(a-w)}$ and $\tau^{(i-w)}$:

$$\vec{\tau}_s = (1 - C_a)\vec{\tau}^{(a-w)} + C_a\vec{\tau}^{(i-w)} \tag{9.28}$$

When the surface ice is grounded, whether it is moving or stationary, the condition $\eta - \eta' = \frac{\rho_i}{\rho} t_i$ is no longer valid, and the water discharge in the lower layer is zero. The water mass conservation equation reduces to:

$$\frac{\partial}{\partial t}[(1 - C_a)\Delta\eta] = -\nabla \cdot \vec{q}_u \tag{9.29}$$

in which, $\Delta\eta = \eta + H_R = H$. Hence

$$\frac{\partial H(1 - C_a)}{\partial t} + \frac{\partial q_{tx}}{\partial x} + \frac{\partial q_{ty}}{\partial y} = 0 \tag{9.30}$$

and $q_{lx}, q_{ly} = 0$.

Similarly, the momentum equations become:

$$\frac{\partial q_{tx}}{\partial t} + \frac{\partial}{\partial x}\left(\frac{q_{tx}^2}{H_s}\right) + \frac{\partial}{\partial y}\left(\frac{q_{tx}q_{ty}}{H_s}\right) = \frac{1}{\rho}(\tau_{ix}) + \frac{1}{\rho}\left(\frac{\partial T_{xx}}{\partial x} + \frac{\partial T_{yx}}{\partial y}\right) - gH_s\frac{\partial \eta}{\partial x} \tag{9.31}$$

$$\frac{\partial q_{ty}}{\partial t} + \frac{\partial}{\partial x}\left(\frac{q_{tx}q_{ty}}{H_s}\right) + \frac{\partial}{\partial y}\left(\frac{q_{ty}^2}{H_s}\right) = \frac{1}{\rho}(\tau_{iy}) + \frac{1}{\rho}\left(\frac{\partial T_{xy}}{\partial x} + \frac{\partial T_{yy}}{\partial y}\right) - gH_s\frac{\partial \eta}{\partial y} \tag{9.32}$$

The drag of the seepage flow on ice can be expressed as:

$$\tau_{ix} = -\rho g H_s \frac{q_s q_{sx}}{K_s^2} \tag{9.33}$$

$$\tau_{iy} = -\rho g H_s \frac{q_s q_{sy}}{K_s^2} \tag{9.34}$$

in which, $K_s = \lambda_s H_s/(1 - N)$.

3 Energy Budget

The thermal regime of a river affects the river ice evolution from freeze-up to breakup. It is necessary to understand the energy fluxes across interfaces between the atmosphere, ice cover, and water.

3.1 *Surface Heat Exchange*

A dominating part of the heat exchange occurs at the top surface of the river. The surface heat exchange at the air–water or air–ice interface includes short and long wave radiation, evapo-condensation; sensible heat exchange, and precipitation [40, 60]. The total surface heat loss rate, ϕ*, may be written as

$$\phi^{*} = -\phi_R + \phi_B + \phi_E + \phi_H + \phi_P \tag{9.35}$$

in which, ϕ* = total surface heat flux between air and water, ϕ_{wa}, or between ice and air, ϕ_{ia}; ϕ_R = net short wave radiation, which is the difference between the incoming solar radiation, ϕ_{ri}, and the solar radiation reflected back to the atmosphere, ϕ_{rr}; ϕ_B = effective back radiation or terrestrial radiation, which is the net balance of the atmospheric long-wave radiation reaching the river surface, ϕ_{ba}, the fraction of the atmospheric radiation reflected back by the river surface, ϕ_{br}, and the long wave radiation emitted by the water surface, ϕ_{bs}; ϕ_E = evaporation heat transfer; ϕ_H = conductive or sensible, heat transfer; and ϕ_P = heat transfer due to precipitation. Determination of the surface heat exchange components requires meteorological data that may not be readily available. Some of these components, especially the long wave radiation, ϕ_B, are nonlinear functions of air and water temperatures. The absence of complete meteorological data would necessitate the use of simplified formulation. Linearized surface heat exchange formulas have been commonly used in river ice modeling, especially when applied to long-term simulations.

3.1.1 Solar Radiation

The incoming short wave radiation under clear skies, ϕ_{cl}, can be calculated as the following [61]:

$$\phi_{cl} = (0.99 - 0.17m)\phi_{so} \tag{9.36}$$

$$\phi_{so} = \frac{12}{\pi} I_{so} E_0 [(\omega_1 - \omega_2)\sin\delta\cos\psi + (\sin\omega_1 - \sin\omega_2)\cos\delta\cos\psi] \tag{9.37}$$

$$m = m_0 \frac{p_a}{p_0}, \; m_0 = \left[\sin \alpha_z + 0.15 (\alpha_z + 3.885)^{-1.253} \right]^{-1} \tag{9.38}$$

$$\frac{p_a}{p_o} = \exp(-0.0001184z) \tag{9.39}$$

$$\delta = \frac{23.45\pi}{180} \sin \left[\frac{360}{365} (d_a + 284) \right] \tag{9.40}$$

$$E_o = 1 + 0.033 \cos \left(\frac{2\pi d_a}{365} \right) \tag{9.41}$$

where, ϕ_{so} = total extraterrestrial solar radiation per unit area incident on a horizontal surface, J/m^2; I_{so} = solar constant, 1380 W/m^2 for the winter season; ω = hour angle, which is zero at noon and changes 15° per hour, morning positive and afternoon negative; δ = solar declination, in radians; ψ = latitude in degrees, north positive, south negative; d_a = day number of the year counting from 1 on January 1, February is always assumed to have 28 days; m = optical air mass; p_o = pressure at sea level; m_o = optical air mass at sea level; solar latitude $\alpha_z = 90 - \theta_z$; $\cos \theta_z = \sin \delta \sin \psi + \cos \delta \cos \psi \cos \omega$; and E_o = eccentricity correction factor of the earth's orbit.

Under cloudy skies, the solar radiation reaching the earth surface will be reduced. This effect can be estimated by

$$\phi_{ri} = \phi_{cl} (1 - 0.0065C^2) \tag{9.42}$$

in which, C = cloud cover in tenths, with $C = 0$ for clear skies and $C = 10$ for overcast skies. If the cloud cover data is not available, an estimated value might be used. Letting $\phi_{rr} = R_t \phi_{ri}$, the net solar radiation ϕ_R is

$$\phi_R = (1 - R_t) \phi_{ri} \tag{9.43}$$

in which, R_t = the albedo or reflectivity. For ice surfaces, the albedo depends on the ice type. Bolsenga [62] reported values for various ice conditions on the Great Lakes.

Besides the surface reflection, the solar radiation is absorbed along the depth of ice cover. For ice and snow, the Lambert–Bouguer law gives

$$\phi_{pz} = \phi_{ps} e^{-\xi_i z} \tag{9.44}$$

in which, ϕ_{pz}, ϕ_{ps} = solar radiation at depth z and solar radiation at the surface; and ξ_i = bulk extinction coefficient. The value of ξ_i varies considerably depending on the snow or ice property [63].

3.1.2 Effective Back Radiation

The effective back radiation can be expressed as

$$\phi_B = \phi_{bs} - \phi_{ba} + \phi_{br} \tag{9.45}$$

Among all the heat exchange components, the long wave radiation from the river surface has the largest magnitude. The Stefan–Boltzmann law gives

$$\phi_{bs} = \varepsilon_s \sigma T_{sk}^4 \tag{9.46}$$

in which, ε_s = emissivity of river surface, usually taken as 0.97; σ = Stefan–Boltzmann constant, 5.67×10^{-8} W m^{-2} K^{-4}; and T_{sk} = surface temperature in the absolute scale, K.

The magnitude of the atmospheric radiation is generally larger than the net solar radiation reaching the ground, and is usually the second largest component among the various heat exchange processes. Under cloudy skies, the atmospheric radiation can be calculated by

$$\phi_{ba} = \varepsilon_a \sigma \left(1 + kC^2\right) T_{ak}^4 \tag{9.47}$$

and the reflected long wave radiation is

$$\phi_{br} = r_s \phi_{ba} \tag{9.48}$$

in which, ε_a = emissivity of atmosphere; k = an empirical constant which depends on the cloud condition, typically about 0.0017; T_{ak} = air temperature at 2 m above the surface, in the absolute scale, K; and $r_s = (1 - \varepsilon_s)$, reflectivity of the river surface. The emissivity of atmosphere can be estimated by [64]:

$$\varepsilon_a = 1.08\left[1 - \exp\left(-e_a^{T_{ak}/106}\right)\right] \tag{9.49}$$

where the vapor pressure, in mb, $e_a = \frac{RH}{100} e_s$, RH = relative humidity in percentage, and e_s = saturated vapor pressure.

In summary, the effective back radiation is

$$\phi_B = 0.97\sigma\left[T_{sk}^4 - \varepsilon_a\left(1 + kC^2\right)T_{ak}^4\right] \tag{9.50}$$

3.1.3 Turbulent Heat and Mass Transfer

Both the latent heat transfer and the sensible heat transfer are driven by the turbulent mixing over the river surface. The heat transfer associated with evaporation or condensation when the vapor pressure in the air is greater than that at the surface is given by the Rimsha–Donchenko [65] formula as

$$\phi_E = \frac{4.1855}{8.64}(1.56K_n + 6.08v_{a2})(e_{so} - e_a) \tag{9.51}$$

in which, $K_n = 8.0 + 0.35(T_s - T_a)$; T_a, T_s = air and river (water or ice) surface temperature in °C; e_{so} = saturated vapor pressure corresponding to the surface temperature, mb; e_a = vapor pressure corresponding to air temperature at z = 2 m above the surface, mb; and v_{a2} = wind velocity at 2 m above the surface.

The sensible heat transfer ϕ_H is proportional to the difference between the surface temperature and the air temperature, and is related to ϕ_E by the Bowen ratio. Based on the Rimsha–Donchenko formula above, ϕ_H is

$$\phi_H = \frac{4.1855}{8.64}(k_n + 3.9v_{a2})(T_{sk} - T_{ak}) \tag{9.52}$$

3.1.4 Heat Exchanges Due to Precipitation

The heat loss due to snow falling on the water surface can be estimated by

$$\phi_P = A_s[L_i + C_i(T_w - T_a)] \tag{9.53}$$

The rate of snowfall may be estimated by visibility as

$$A_s = \frac{78.5}{86400}\hat{V}^{-2.375} \tag{9.54}$$

in which, A_s = mass rate of snowfall over a unit area of water surface, kg m^{-2} s^{-1}; L_i = latent heat of fusion of ice, 3.3484×10^5 J kg^{-1}; $C_i = 4.1855 \times 10^3$ Jkg^{-1}°C, specific heat of ice; $\hat{V}$ = visibility in km. When the precipitation is in the form of rainfall, the rate of heat loss, i.e. negative heat gain, can be calculated as $\phi_P = A_pC_p(T_w - T_a)$, with A_p = rainfall rate, and $C_p = 4.215 \times 10^3$ Jkg^{-1}°C, specific heat of water. The rainfall effect is generally small because the only contribution is the due to the specific heat effect associated with the difference in temperature between the water and the rain.

3.1.5 Linear Approximations

The required meteorological data for calculating surface heat exchange components may not be readily available at a project site. Therefore the use of simplified linearized formulas has been popular in river ice engineering. However, the heat exchange coefficients in linear models need to be calibrated. Since the diurnal variation of solar radiation and air temperature could be important to some of the ice processes such as frazil and anchor ice evolution, the linearized expression may include a component proportional to the difference between the river surface temperature T_s and the air temperature T_a, and a temperature independent short wave radiation component [66]:

$$\phi^* = \phi_R + \alpha' + \beta'(T_s - T_a) \tag{9.55}$$

in which, α', β' are empirical constants; ϕ_R = net short wave radiation. If the diurnal variations are not of significant interest, the linear equation can be further simplified to $\phi^* = \alpha^*(T_s - T_a)$.

3.2 Heat Transfer at Ice–Water Interface

The heat transfer from river water to ice at the ice–water interface depends on the water temperature and flow condition, which may be expressed as

$$\phi_{wi} = h_{wi}(T_w - T_m) \tag{9.56}$$

where, ϕ_{wi} = heat flux between river water and ice; h_{wi} = a heat transfer coefficient; T_m = freezing point of water, 0 °C. The heat transfer coefficient can be determined by formulas for turbulent heat transfer in channels. For fully developed turbulent flow, the Dittus–Boelter formula [67] gives

$$N_u = 0.023 R_{eH}{}^{0.8} \mathrm{Pr}^{n_h} \tag{9.57}$$

where, $N_u = h_{wi}D_H/K_w$ = Nusselt number; K_w = thermal conductivity of water, 0.566 W m^{-1} °C^{-1}; D_H = hydraulic diameter; $R_{eH} = \rho U D_H/\mu$ = Reynolds number; $P_r = C_p \mu K_w$ = Prandtl number, 13.25; μ = dynamic viscosity of water; C_p = specific heat of water, 4215 J°C^{-1} kg^{-1}; and n_h = 0.3 or 0.4 depending on whether water temperature is below or above 0 °C.

4 Water Temperature and Ice Concentration

Variations of water temperature, frazil and surface ice discharges, and anchor ice evolution are closely related phenomena. They play important roles in the evolution of river ice. Frazil ice production over the depth of the flow occurs when the water temperature is supercooled. In open water reaches, frazil ice particles in suspension will grow both in size and number due to continuous surface heat loss. Under the influence of the buoyant velocity, some of the frazil particles may rise against turbulent mixing to the water surface to form the surface ice run. Turbulent mixing can also carry frazil particles to the channel bottom contributing to anchor ice growth. In the meantime, the latent heat released due to frazil production tends to raise the water temperature to 0 °C. This recovery of water temperature is enhanced by the insulation effect of the surface ice pieces and the latent heat released due to the thermal growth of anchor ice. After entering an ice covered reach, the ice discharge will travel along the underside of the cover as a cover load [36]. The amount of ice in the cover load also changes with the deposition or erosion of the frazil accumulation, i.e. frazil jams or 'hanging dam', on the underside of the cover.

4.1 Ice Transport

The above discussion showed that river ice process is essentially a thermal-ice transport process. It is important to have a unified analytical framework that includes all components of the thermal and ice processes. Lal and Shen [13] introduced a two-layer formulation for modeling river ice processes. This one-dimensional formulation, which includes all major thermal and ice transport processes and can easily be extended into two-dimensional forms, is used here for the convenience of presentation.

In the two-layer formulation, the ice discharge in the river is considered to consist of surface ice and suspended ice discharges:

$$Q_d^i = AUC_v \tag{9.58}$$

$$Q_s^i = B_o U_s C_a h_e \tag{9.59}$$

in which, Q_d^i, Q_s^i = volumetric rates of suspended and surface ice discharge, respectively; A = flow cross sectional area; $B_o = (1 - f)B$ = open water width; f = fraction of the width covered by border ice cover; C_v = ice concentration in the suspended layer; C_a = area concentration of the surface ice; $h_e = h_i + h_f(1 - e_f)$ = equivalent thickness of surface ice floes; h_i = thickness of the solid portion of surface ice floes; h_f = thickness of the frazil accumulation on the underside of surface ice floes; e_f = porosity of the frazil ice portion of surface ice floes; U = cross-section-averaged flow velocity; and U_s = surface ice velocity. The surface ice transport is affected by ice dynamics, which will affect the ice velocity and concentration. The concentration C_a is limited by a maximum allowable surface ice concentration, $C_{a,max}$, beyond which the surface ice layer will change from a single layer of ice floes to multilayer ice floes. This condition develops when the surface ice run is approaching the jamming condition. The formulation of the surface ice dynamics will be discussed in Sect. 4.4.

4.2 Water Temperature and Suspended Ice Concentration

Besides the production, melting, and mass exchanges between the two layers, the rate of change of the suspended ice concentration, C_v, is also affected by the rate of frazil accretion on anchor ice. The conservation of thermal energy of the ice–water mixture in the suspended layer can be expressed in a Lagrangian form as [68]

$$\frac{De_T}{Dt} = \frac{1}{A}(\phi_{ss} - \phi_{sk}) + \rho_i L_i E \tag{9.60}$$

in which, the material derivative $\frac{D}{Dt} \equiv \frac{\partial}{\partial t} + \vec{U} \cdot \nabla$; $e_T = C_p \rho (1 - C_v) T_w - \rho_i C_v L_i$, thermal energy of the ice–water mixture per unit volume; T_w = cross section-averaged water temperature; ρ and ρ_i = mass density of water and ice, respectively; C_p and L_i = specific heat of water and latent heat of fusion, respectively; ϕ_{ss} = rate of heat gain through top and bottom boundaries; ϕ_{sk} = rate of heat loss through top and bottom boundaries; and E = net volumetric rate of loss of frazil due to exchange with the surface layer and the anchor ice layer.

The change in suspended ice concentration can be caused by both thermal growth and mass exchanges with surface ice and anchor ice, i.e.

$$\frac{DC_v}{Dt} = \frac{DC_v^g}{Dt} - E \tag{9.61}$$

in which, $\frac{DC_v^g}{Dt}$ = rate of increase of suspended ice concentration due to thermal growth. The formation and evolution of frazil have been studied extensively [6]. Hammar and Shen [7] formulated a numerical model for the formation and evolution of frazil ice and the corresponding water temperature change in turbulent channel flows, and validated with observed water temperature during frazil formation in a laboratory flume [2]. The frazil formation is initiated by the mass exchange of seeding crystals at the water surface. Assuming a well-mixed frazil suspension of uniform disc-shaped particles, the rate of change of frazil concentration in a one-dimensional formulation reduces to:

$$\frac{DC_v^g}{Dt} = \frac{1}{\rho_i L_i} q_f a_0 N_f = -\frac{1}{\rho_i L_i} \frac{N_u^f K_w}{d_e} a_0 T_w N_f \tag{9.62}$$

in which, q_f = heat transfer between water and a frazil particle per unit frazil surface area; and N_f = number of frazil particles per unit volume, C_v/v_0; v_0 = mean volume of a frazil crystal; N_u^f = Nusselt number; a_0 = mean area of frazil crystal discs; K_w = thermal conductivity of water, W m^{-1} °C^{-1}; and d_e = frazil crystal thickness.

The net volumetric rate of loss of frazil ice to the surface layer and anchor ice can be formulated as

$$E = \frac{1}{d_w}(\alpha V_b C_v - \beta h_e C_a + \gamma C_v) \tag{9.63}$$

in which, d_w = flow depth; V_b = average rising velocity of suspended ice; and α, β, and γ = mass exchange coefficients at interfaces between the suspended layer, the surface ice layer, and anchor ice. Physically, α = probability of deposition of frazil particles reaching the surface layer; β = coefficient quantifying the rate of re-entrainment of surface ice per unit area, 1/s; and γ = coefficient quantifying the rate of accretion to the bed per unit area, m/s. The exchange coefficients α, β, and γ, are governed by the turbulent intensity and the characteristics of ice particles. Since complete theories on the exchanges at interfaces are not yet available, these coefficients need to be determined empirically.

Frazil and anchor ice growths occur under the supercooling condition. The supercooling condition can be simulated by including the heat exchange between water and ice in the conservation of thermal energy in water. Using Eqs. (9.60) and (9.61) the following water temperature equation can be obtained:

$$\frac{D\left(C_p\rho(1-C_v)T_w\right)}{Dt} = \frac{1}{A}(\phi_{ss} - \phi_{sk}) + \rho_i L_i \frac{DC_g^v}{Dt} \tag{9.64}$$

in which,

$$\begin{aligned}\phi_{ss} - \phi_{sk} = P_b\phi_b - B(1-f)(1-C_a)\phi_{wa} \\ -[(1-f)BC_a + C_bP_b + fB]\cdot h_{wi}(T_w - T_m)\end{aligned} \tag{9.65}$$

where, P_b = wetted perimeter of the river bed ≈B; ϕ_b = bed heat flux per unit area; f = fraction of the width covered by border ice cover; C_b = fraction of bed width covered by anchor ice; and $T_m = 0°C$.

4.3 Surface Ice Concentration

The one-dimensional formulation for surface ice transport in this section assumes the surface ice layer has a single floe thickness, neglecting ice dynamics. Such an assumption may be acceptable when the ice concentration is low with negligible ice-to-ice interaction stress, before approaching the jamming condition. The dynamics of surface ice layer thickening and ice jam development will be discussed in Sect. 4.4. If the dynamics is neglected, the thickness of stationary surface ice accumulations, i.e. ice cover or ice jam, can be approximated by the static ice jam theory presented in Sect. 5.4.

The mass conservation equation for surface ice is:

$$\frac{\partial}{\partial t}[h_eC_aB_0] + \frac{\partial}{\partial x}[h_eC_aB_0U] = S_a - M_a \tag{9.66}$$

in which,

$$S_a - M_a = \frac{B_0 C_a \phi_{si}}{\rho_i L_i} + \alpha V_b B_0 C_v - \beta h_e B_0 C_a + R_{an} \tag{9.67}$$

where, $B_0 = (1 - f)B$; $\phi_{si} = \phi_{ia} - h_{wi}(T_w - T_m)$, net rate of heat loss from top and bottom of surface ice elements; ϕ_{ia} = heat loss at the air–ice interface, and R_{an} = contribution from anchor ice release. When anchor ice detaches from the bed, it will rise to the surface unless in rapids, where the buoyancy cannot overcome the turbulent mixing.

Combining Eqs. (9.64) and (9.66):

$$\frac{DC_a}{Dt} + \frac{C_a}{h}\frac{Dh_e}{Dt} + \frac{C_a}{B_0}\frac{DB_0}{Dt} + C_a\frac{\partial U}{\partial x} = \frac{1}{h}\left[\frac{C_a \phi_{si}}{\rho_i L_i} + \alpha V_b C_v - \beta h_e C_a + \frac{R_{an}}{B_0}\right] \tag{9.68}$$

The heat transfer between surface ice floes and the river water can have an important effect on the thickness of the surface ice, h_e, when the water temperature is significantly different from 0 °C. The rates of change in the solid and frazil ice thickness portions of surface ice floes are:

$$\frac{Dh_i}{Dt} = \frac{\phi_{ia}}{e_f \rho_i L_i} \tag{9.69}$$

and

$$\frac{Dh_f}{Dt} = \frac{(\alpha V_b C_v - \beta h_e C_a)}{(1 - e_f)} - \frac{Dh_i}{Dt} - \frac{h_{wi}(T_w - T_m)}{(1 - e_f)\rho_i L_i} \tag{9.70}$$

Equations (9.69) and (9.70) are valid only for the case when the ice floes have a frazil layer on its underside, a condition that usually exists during freeze up ice runs. Other conditions can be formulated in a similar manner. Using Eqs. (9.69) and (9.70), or their equivalent, Eq. (9.68) can be written as:

$$\frac{DC_a}{Dt} = \frac{(1 - C_a)}{h_e}\left[\alpha V_b C_v - \beta h_e C_a + \frac{R_{an}}{B_o}\right] - \frac{C_a}{B_o}\frac{DB_o}{Dt} - C_a\frac{\partial U_s}{\partial x} \tag{9.71}$$

4.4 Surface Ice Dynamics

The transport of surface ice is a dynamic process and involves external forces such as gravity, wind and water drags, as well as internal ice resistance due to interactions between ice floes. If ice dynamics is neglected, ice jamming conditions cannot be determined analytically from the surface ice transport formulation discussed in Sect. 4.3. Since bank resistance leads to a velocity gradient for ice movement across the channel, the surface ice transport is essentially a two-dimensional phenomenon even in a straight uniform channel. One-dimensional ice dynamic models have limited applicability [25]. Shen et al. [26] developed a two-dimensional model DynaRICE for river ice dynamics and applied it to field conditions for ice jam development and other engineering applications [58]. The basic formulation of river ice dynamics is similar to that of sea ice [69] and lake ice [70]. Considering the surface ice as a continuum, the governing equations for surface ice transport in a Lagrangian form are:

$$M_i \frac{D\vec{V}_i}{Dt} = \vec{R} + \vec{F}_a + \vec{F}_w + \vec{G} + \vec{V}_i E_m \tag{9.72}$$

$$\frac{DM_i}{Dt} + M_i \nabla \cdot \vec{V}_i = E_m \tag{9.73}$$

$$\frac{DC_a}{Dt} + C_a \nabla \cdot \vec{V}_i = E_a - R_a \tag{9.74}$$

in which, x, y and t = space and time variables; $M_i = \rho_i C_a t_i$ = ice mass per unit area; ρ_i, C_a, t_i = density, area concentration, and thickness of the surface ice layer, respectively; $\vec{V}_i = u_i \vec{i} + v_i \vec{j}$, surface ice velocity; D/Dt = material derivative; $\vec{F}_a$ = wind drag on ice; $\vec{F}_w$ = water drag on ice; $\vec{G}$ = gravitational force due to the water surface slope; $\vec{R} = \vec{i}\left[\frac{\partial}{\partial x}(\sigma_{xx} C_a t_i) + \frac{\partial}{\partial y}(\sigma_{xy} C_a t_i)\right] + \vec{j}\left[\frac{\partial}{\partial x}(\sigma_{yx} C_a t_i) + \frac{\partial}{\partial y}(\sigma_{yy} C_a t_i)\right]$ = internal ice resistance due to floe-to-floe interactions; E_m = rate of change of surface ice mass per unit area of water surface due to sources and sinks, including changes in thermodynamic processes and the exchange with frazil suspended over the depth of the flow; E_a = time rate of change of ice concentration due to thermodynamic processes and the exchange with frazil suspended in the flow; and R_a = time rate of change of ice concentration due to mechanical redistribution of the ice mass. A constitutive law is required to describe the ice internal stress. Following Hibler [69] the internal stress was described by a viscous-plastic constitutive law, $\sigma_{ij} = 2\hat{\nu}\dot{\varepsilon}_{ij} + (\hat{\varsigma} - \hat{\nu})\dot{\varepsilon}_k \delta_{ij} - P\delta_{ij}/2$, in which, $\hat{\varsigma}$ and $\hat{\nu}$ = nonlinear bulk and shear viscosity defined as $\hat{\varsigma} = P/2\Delta$ and $\hat{\nu} = \varsigma/e^2$, with $\Delta^2 = D_I^2 + (D_{II}/e)^2$; D_I, D_{II} are first and second invariant strain rates, respectively; $D_I = \dot{\varepsilon}_{xx} + \dot{\varepsilon}_{yy}$, $D_{II} = \left[(\dot{\varepsilon}_{xx} - \dot{\varepsilon}_{yy})^2 + 4\dot{\varepsilon}_{xy}^2\right]^{1/2}$; $\dot{\varepsilon}_{xx} = \partial u_i/\partial x, \dot{\varepsilon}_{yy} = \partial v_i/\partial y$; and $\dot{\varepsilon}_{xy} = (\partial v_i/\partial x + \partial u_i/\partial y)/2$. The principal axes ratio of the elliptical yield curve, $e = 2$. Hibler [69] formulated the pressure term as $P = p^*\left(\frac{t_i}{t_o}\right)\exp[-C(1 - C_a)]$, where t_o is a reference ice thickness. Wake and Rumer [70] replaced the exponential term by a power law as $P = p^*\left(\frac{t_i}{t_o}\right)C_a{}^k$. In both cases, the parameters p^*, C, and k are empirical constants to be calibrated. Based on the formulation of static ice jam theories, Shen et al. [25] formulated the pressure term as:

$$P = \tan^2\left(\frac{\pi}{4} \pm \frac{\varphi}{2}\right)\left(1 - \frac{\rho_i}{\rho}\right)\frac{\rho_i g t_i}{2}\left(\frac{C_a}{C_{a,\max}}\right)^j \tag{9.75}$$

in which, φ = internal friction angle of ice, ≈46°; ρ = water density; $C_{a,max}$ = maximum allowable surface ice concentration, 0.6 is used based on the consideration that the typical porosity of an ice jam is 0.4; and j = an empirical constant, a value of 15 has been used in field applications [26]. The + and − signs are for convergent and divergent states of ice flow, respectively. Since the viscous-plastic law allows the jammed ice accumulation to creep downstream, the constitutive law has been modified to a viscoelastic-plastic model [71]. Bank resistance is a key factor in the jamming of ice runs. Similarly, bed-to-ice friction affects the ice motion when ice is in contact with the bed. Shen et al. [26] used a dynamic Coulomb yield criterion to model the bank and bed frictions for moving ice. Equations (9.72), (9.73), and (9.74) can be used to solve for ice velocity, thickness, and concentration, as well as ice cover and ice jam dynamics.

5 Freeze Up

When the water surface cools to 0 °C, and the heat loss from the river continues, various types of ice appear during the freeze up period. Matousek [4] developed an empirical formulation for determining the conditions for the occurrence of different types of freeze up ice runs.

5.1 *Skim Ice*

According to Matousek [4], the following criteria for skim ice formation may be used:

When $T_w > 0$ °C, skim ice run will occur if $T_{cr} < T_{w,s} < 0$ °C and $v_b > v_z'$;

When $T_w > 0$ °C, a static skim ice cover will form if $T_{w,s} \le T_{cr}$.

in which, T_w = depth-averaged water temperature; $T_{w,s}$ = water surface temperature; v_b = buoyancy velocity of frazil ice, m/s; v_z' = vertical component of turbulent fluctuation velocity; T_{cr} = a supercooled surface temperature below which static skim ice will form. The value of T_{cr} used by Matousek for River Ohre is −1.1 °C.

$$T_{w,s} = T_w - \frac{\phi_{wa}}{1130U + bV_a} \tag{9.76}$$

$$v_b = -0.025T_{w,s} + 0.005 \tag{9.77}$$

$$v_z' = \frac{\sqrt{g}}{5\sqrt{(0.7C_z + 6)C_z}} U \quad (9.78)$$

with,

$b = 15.0$, when $B_a \leq 15.0\,\text{m}$

$b = -0.9 + 5.87\ln B_a$ when $B_a > 15.0$ m

in which, C_z = Chezy's coefficient, $\text{m}^{0.5}\,\text{s}^{-1}$; g = gravity, m/s^2; U = flow velocity, m/s; V_a = wind velocity at an elevation of 2.0 m above water surface; B_a = river surface width in the wind direction, *m*.

When the wind effect is strong, the following equation by Hicks [51] may be used instead of Eq. (9.78):

$$v_z' = \left[\left(C_*^{\frac{1}{2}} g^{\frac{1}{2}} \frac{U n_b}{R^{1/6}} \right)^3 + C_{w*} \left(C_N \frac{\rho_a}{\rho} \right)^{\frac{3}{2}} V_a^3 \right]^{1/3} \quad (9.79)$$

in which, C_* = coefficient which relates bed shear stress to turbulent fluctuation velocity, between 0.2 and 0.3; n_b = river bed Manning's coefficient; R = hydraulic radius; C_{w*} = constant explaining the efficiency of wind energy utilization [51]; C_N = wind drag coefficient, 1.3×10^{-3}; ρ_a = density of air, and ρ = density of water.

5.2 Dynamic Border Ice

Border ice is an important ice process due to its effect on the flow condition, frazil production, ice transport, and ice cover progression. Michel et al. [15] developed the following method based on field observations. The dimensionless relationship for the rate of lateral growth of border ice can be expressed as

$$R_b = 14.1 V_*^{-0.93} C_a^{1.08} \quad (9.80)$$

in which, $R_b = \frac{\rho L_i \Delta B h_e}{\Delta\phi^*}$; $V_* = \frac{u}{V_{cb}}$; u = depth-averaged flow velocity in the open water adjacent to the edge of the existing border ice, m/s; V_{cb} = maximum velocity at which a surface ice floe can adhere to the border ice edge; ΔB = growth of border ice width for a given time step, *m*; h_e = floe thickness; $\Delta\phi^*$ = surface heat exchange during a time step, $\text{J}\,\text{m}^{-2}$; and C_a = area concentration of the surface ice.

Michel et al. [15] gave the physical limits of this equation as:

$V_* < 0.167$, static ice growth occurs,
$V_* > 1.0$, border ice growth is negligible,
$C_a < 0.1$, border ice grows only in thermal mode, set $C_a = 0.1$.

The last condition can be replaced by Matousek's formulation for static thermal border ice growth [4].

The limiting velocity for border ice growth, V_{cb}, is governed by the gravity, and drag and friction forces acting on the surface ice floes, and will vary accordingly. The V_{cb} value is about 0.4 m/s [4, 16, 30].

5.3 Anchor Ice

Frazil crystals entrained to the bottom of the channel in supercooled water can attach to the bed to form anchor ice. Anchor ice may cause significant changes in flow resistance, water level, and discharge [8]. In the upper Niagara River, it was reported that anchor ice could grow from 0.3 to 0.5 m thick in a flow depth of 6 m [72]. Sudden flow reductions of 20–30 % have been observed. In the St. Lawrence River near Montreal, the existence of anchor ice was considered to be the major cause of power production losses at a nearby power station [73]. Anchor ice formation in the tailrace causes significant losses of power production at Manitoba Hydro's Limestone Generating Station on the Nelson River. Anchor ice formation on a granite rock shelf four kilometers downstream from the generation station initiates an ice dam that covers up to 90 % of the channel width, which could cause up to 1.5 m of staging in the tailrace [74].

A flume study on anchor ice evolution and its hydraulic effects in channels with a gravel bed showed that anchor ice formed from the bed attachment of frazil ice [8]. The initial accumulation on a gravel bed roughly starts from half a gravel diameter below the crown level of the gravel. The growth of anchor ice consists of an initial growth stage, followed by a transition stage, then the final stage of continued growth. During the initial stage anchor ice grew in the forms of tails, scales, or balls, depending on the flow velocity and Froude number. During the transition stage, the anchor ice forms were either flattened or released. The release of anchor ice appeared to be dependent on the Reynolds number of the flow. During the final stage of continued growth, the anchor ice grew at a steady rate with respect to the surface heat loss. When the air temperature was constant, this growth rate was nearly constant for a given flow condition. It increased with Froude number and is inversely proportional to the flow depth.

Due to the complexity of anchor ice evolution, more study is needed in order to develop an analytical formulation for anchor ice evolution. A semi-empirical formulation was proposed by Shen et al. [68]. In this formulation, the anchor formation starts from the theoretical bed surface below the crown of bed particles when the river water is supercooled. The rate of growth and decay of anchor ice thickness at the top surface of the anchor ice is:

$$\frac{dh_{an}}{dt} = \frac{1}{1 - e_{an}}\left[\gamma_b C_v + \frac{\phi_{wi}}{\rho_i L_i}\right] \tag{9.81}$$

in which, dh_{an}/dt = growth rate of anchor ice thickness; e_{an} = porosity of anchor ice; and ϕ_{wi} = heat flux from ice to channel flow. The frazil ice accretion rate is set to zero, i.e. $\gamma_b = 0$, when (a) water is no longer supercooled, i.e. $T_w \geq 0\,°\mathrm{C}$; or (b) bed shear velocity is either below a critical value u^*_{cr1} or above a critical value u^*_{cr2}.

The rate of melting or growth at the bottom of the anchor ice can be calculated by:

$$\frac{dh_{sb}}{dt} = \frac{1}{(1 - e_{an})\rho_i L_i}\left[h_{iws}(T_w - T_m) + \phi_{pz}\right] \tag{9.82}$$

in which, h_{sb} = substrate flow depth at the bottom of the anchor ice; ϕ_{pz} = short wave radiation absorbed by the anchor ice; h_{iws} = heat exchange coefficient between anchor ice and the substrate flow.

Anchor ice release could occur during the daytime, when water is warmed up, due to the melting of the bond between the anchor ice and the bed material. This melting of the bond is a complicated process. When the water temperature is supercooled, any such melting will be prevented. The melting will only occur when the water temperature rises above 0 °C. Anchor ice release is assumed to occur when the thermal erosion from its bottom is significant, i.e. when the undersurface of the anchor ice raises to a level of 0.1 bed gravel diameter below the crown. Anchor ice can also be detached from the bed when the buoyant force of anchor ice overcomes the resistance due to the submerged weight of the bed particle and the inter-particle resistance, $W_b + F_b < (1 - e_{an})\forall(\rho - \rho_i)$, with W_b = submerged weight of bed materials per unit area; F_b = inter-particle resistance per unit area; $\forall$ = volume of anchor ice per unit area. When anchor ice is released, it will rise to the water surface, and contribute to the surface ice run.

5.4 Ice Cover Formation

The formation of the main ice cover is initiated either by bridging or jamming of the surface ice run or by artificial obstructions, such as a dam, or an ice boom. The jamming of surface ice occurs when the surface ice discharge exceeds the surface ice transport capacity at a cross section. The ice transport capacity at a river cross section depends on factors such as flow velocity, channel top width between riverbanks or border ice boundaries, floe size, and surface slope [25]. The jamming condition is governed by the dynamics of ice motion as discussed in Sect. 4.4.

The progression of an ice cover or ice jam due to the accumulation of surface ice is an important freeze up process. This process also occurs during the breakup when an ice run is produced by mechanical breakup. If ice dynamics is ignored, the thickness profile of the initial ice cover or jam can be approximated by the theory of static equilibrium of floating ice accumulation [18–22].

Pariset and Hausser [18] showed that the cover thickness formed from the accumulation of surface ice floes depends on the flow and ice conditions. The modes of accumulation include particle juxtaposition, hydraulic thickening (commonly known as narrow river jam), and mechanical thickening (commonly known as wide river jam).

5.4.1 Floe Juxtaposition

The limiting condition for the particle juxtaposition mode is the entrainment of the incoming ice floe at the leading edge of the cover, which may be expressed in terms of the Froude number of the flow as

$$F_{rc} = \left[\frac{U}{\sqrt{gH}}\right]_c \leq F_l\left(\frac{H-h_e}{H}\right)\sqrt{2\left(1-\frac{\rho_i}{\rho}\right)(1-e_p)\frac{h_e}{H}} \tag{9.83}$$

in which, F_{rc} = a critical Froude number beyond which juxtaposition cannot occur; U and H = velocity and flow depth immediately upstream of the leading edge of the cover; g = gravity; F_l = a form factor, which is a function of the floe dimensions h_e/l_i; l_i = length of ice floe; e_p = porosity of ice floe. This equation implies that incoming ice floes will accumulate into a single layer ice cover if the Froude number is less than F_{rc}. This equation serves as a guideline for estimating the critical Froude number. Ashton [29] showed that the critical condition for entrainment is better described by a Froude number based on the floe thickness.

5.4.2 Hydraulic Thickening: Narrow River Jam

When the Froude number at the leading edge exceeds F_{rc}, incoming ice floes will submerge at the leading edge and accumulate into a cover of multi-layer thickness. This mode of cover formation is commonly known as a narrow river jam. The thickness t_i can be estimated by [18]:

$$\frac{U}{\sqrt{gH}} = \left[2\frac{t_i}{H}(1-e_j)\left(1-\frac{\rho_i}{\rho}\right)\right]^{\frac{1}{2}}\left(1-\frac{t_i}{H}\right) \tag{9.84}$$

in which, $e_j = e_c + (1-e_c)e_p$, overall porosity of the accumulation; e_c = porosity in the accumulation between ice floes, about 0.4; and t_i = cover thickness.

Equation (9.84) gives a limiting Froude number for cover progression, which occurs when $t_i/H = \frac{1}{3}$, as

$$F_{rp} = \frac{U_c}{\sqrt{gH}} = 0.158\sqrt{1-e_j} \tag{9.85}$$

Field observations [33] indicated that F_{rp} varies between 0.05 and 0.1. Data from the Yellow River and the upper St. Lawrence showed the value is about 0.09 [30, 31].

5.4.3 Mechanical Thickening: Wide River Jam

In a wide river, the increase in streamwise thrust acting on the floating ice accumulation may exceed the bank resistance. In this case, the ice accumulation will not be able to resist the increasing stress as it extends to the upstream. The cover will collapse and consolidate until an equilibrium thickness profile is reached [18, 19]. This process of mechanical thickening results in a wide river jam. It is of interest to point out that most surface ice jams are wide river jams resulting from collapse and shoving of narrow river jams [22, 75].

The thickness profile of a wide jam can be obtained as [20, 25]:

$$t_i\frac{dt_i}{dx} = \frac{1}{\tan^2\left(\frac{\pi}{4}+\frac{\varphi}{2}\right)\left(1-\frac{\rho_i}{\rho}\right)\rho_i g}\left[-\frac{2}{B}\tau_c t_i - \frac{1}{B}\mu_i\left(1-\frac{\rho_i}{\rho}\right)\rho_i g t_i^2 + \tau_a + \tau_w + \tau_g\right] \tag{9.86}$$

in which, φ = internal friction angle of surface ice particles; τ_c = cohesion; μ_i = an ice-over-ice friction coefficient; τ_a = wind drag; τ_w = water drag; and τ_g = weight component of the ice cover along the surface slope. Michel [32] suggested the cohesion term τ_c does not exist and it should be replaced by the additional strength gained by the solid ice layer in the accumulation during the freeze-up. The jam thickness equation can be written as

$$\frac{dt_i}{dx} = a' t_i + \frac{b'}{t_i} + c' \tag{9.87}$$

where, $a' = \frac{k_0 k_1}{B}$; $b' = \frac{\rho R_i g S_f}{2k_x \gamma_e}$; $c' = \frac{\rho_i g S_w}{2k_x \gamma_e}$; $k_o = \tan\phi$; $k_x = \tan^2(45^\circ + \frac{\varphi}{2})^2$; $k_1 = \frac{1-\sin^2\varphi}{1+\sin^2\varphi}$, coefficient of lateral thrust; R_i = hydraulic radius associated with the ice cover, which can be expressed as $R_i = \left(\frac{n_i}{n_c}\right)^{1.5} R$; S_f = friction slope; S_w = water surface slope; $\gamma_e = 0.5\rho_i g\left(1-\frac{\rho_i}{\rho}\right)(1-e_j)$. Flato and Gerard [23] and Beltaos [56] developed methods for solving the ice jam thickness profile based on Eq. (9.87). The equilibrium ice jam thickness, t_{eq}, defined as the jam thickness in the uniform reach, can be obtained by setting $\frac{dt_i}{dx} = 0$. Beltaos [75, 76] showed that the underside roughness of a jam increases linearly with the jam thickness and developed an equation for calculating the composite flow resistance coefficient for breakup ice jams.

5.4.4 Cover Progression

The rate of progression of the ice cover depends on the ice supply and the cover thickness. Based on the mass conservation of surface ice at the leading edge [20]:

$$\left(Q_s^i - Q_u\right)\frac{\left(U_s + V_{cp}\right)}{U_s} = B_0 t_i\left(1 - e_j\right)V_{cp} \tag{9.88}$$

in which, V_{cp} = rate of progression of leading edge; Q_s^i = incoming surface ice discharge; Q_u = volumetric rate of ice entertainment under the cover at the leading edge; and U_s = average velocity of the incoming surface ice particles. The ice cover cannot progress beyond a cross-section where the maximum local Froude number of the cross section exceeds the critical Froude number F_{rp}.

5.4.5 Cover Stability

Michel [32] noted that it takes a little freezing to form a solid crust in the newly formed ice cover from ice floe accumulation to prevent the occurrence of shoving or consolidation of the cover. The failure of an ice cover can take place at any time during the winter when the internal strength of the cover is not capable of withstanding the external forces. The failure of ice cover due to the external forces and the streamwise component of the cover weight can be numerically integrated to determine the internal stress and compare with the compressive strength of the cover to determine the stability of the cover [13].

6 Undercover Transport and Frazil Jams

Frazil jams often form under an ice cover, downstream of a rapid reach where ice cover cannot form, in the form of thick undercover accumulation. Critical velocity or critical Froude number criterion have been used in the past to determine the thickness of the undercover accumulation [33–35]. These theories were found to be inadequate, and were improved by the transport capacity concept [33]. According to the ice transport capacity theory, for a given flow condition there is a corresponding ice transport capacity. Deposition or erosion of the undercover ice will occur depending on whether the ice supply is larger or smaller than the ice transport capacity of the flow. When the incoming ice discharge equals the ice transport capacity, the ice transport is at a dynamic equilibrium condition such that erosion balances with deposition. When the flow strength is below a threshold value, the transport capacity will be zero, and all the incoming ice will be deposited. The ice transport capacity can be estimated by [33]:

$$\Phi_i = 5.487(\Theta_i - \Theta_{ci})^{1.5} \tag{9.89}$$

in which, Φ_i = the dimensionless ice transport capacity; Θ_i = dimensionless flow strength; and Θ_{ci} = 0.041, the critical flow strength, below which there is no ice transport. The dimensionless transport capacity and flow strength are defined as:

$$\Phi_i = \frac{q_c}{d_n \omega_B} = \frac{q_c}{d_n F \sqrt{g d_n \Delta_i}} \tag{9.90}$$

$$\Theta_i = \frac{\tau_i}{F^2 \rho g d_n \Delta_i} = \frac{u_{*i}^2}{F^2 g d_n \Delta_i} \tag{9.91}$$

where, $\omega_B = F\sqrt{\Delta_i g d_n}$, buoyant velocity of ice particles, m/s; q_c = volumetric rate of undercover ice transport per unit width; d_n = nominal diameter of ice particles; τ_i, u_{*i} = shear stress and shear velocity on the undersurface of the frazil jam, respectively; $\Delta_i = (\rho - \rho_i)/\rho$; and F is a shape factor. For spherical particles, the value of F is approximately 1.0. The ice cover shear velocity could be estimated by $u_{*,i} = \sqrt{\rho g R_i S}$; with $S = \frac{U^2 n_c^2}{(H'/2)^{4/3}}$; $R_i = \frac{H'}{1 + \left(\frac{n_b}{n_i}\right)^{3/2}}$; U = flow velocity; and H' = under ice water depth.

The thickness of the frazil jam can be calculated from the continuity equation for the frazil granules:

$$(1 - e_u)\frac{\partial t_f}{\partial t} + \frac{\partial q_c}{\partial x} - q_f^i = 0 \tag{9.92}$$

in which, e_u = porosity of the frazil jam, approximately equals 0.4; t_f = thickness of frazil jam; q_c = unit width ice discharge of the cover load; and q_f^i = net contribution to the cover load from the suspended ice, which is negligible in ice covered reaches.

7 Thermal Growth and Decay of Ice Cover

The rate of ice thickness growth or decay is governed by heat exchanges at the top and bottom surfaces, and heat conduction in the ice cover [41]. The effect of water accumulation over melting ice on the dissipation rate is insignificant [77]. The equation governing the temperature distribution over the thickness of an ice cover is

$$\rho_i C_i \frac{\partial T}{\partial t} = -\frac{\partial}{\partial z}\left(k_i \frac{\partial T}{\partial z}\right) + \phi_v(z,t) \tag{9.93}$$

in which, t = time; z = vertical distance measured downward from the top surface; T = temperature in the cover; k_i = thermal conductivity of ice; and ϕ_v = rate of internal heating of the cover due to absorption of penetrated short wave radiation. For a solid cover of thickness h, boundary conditions at the top and bottom surfaces of the cover are:

$$\rho_i L_i \frac{dh}{dt} = \phi^* - k_i \frac{\partial T}{\partial z} \quad \text{at z} = 0 \tag{9.94}$$

$$\rho_i L_i \frac{dh}{dt} = -\phi_{wi} + k_i \frac{\partial T}{\partial z} \quad \text{at z} = h \tag{9.95}$$

Time dependent ice thickness growth and decay can be determined by assuming one-dimensional quasi-steady state calculations at each time step.

At steady state, Eq. (9.93) gives a linear temperature distribution if the internal heating $\phi_v(z, t)$ is neglected. This quasi-steady state assumption has been shown to be acceptable for river ice covers because of the relatively small thickness [78]. For an ice cover with a snow cover above, the rate of change of ice cover thickness can be calculated by [40]:

$$\rho_i L_i \frac{dh}{dt} = \left[(T_m - T_a)/\left(\frac{h}{k_i} + \frac{h_s}{k_s} + \frac{1}{h_{sa}}\right)\right] - h_{wi}(T_w - T_m) \tag{9.96}$$

in which, h_s = snow cover thickness; k_s = thermal conductivity of snow; h_{sa} = heat transfer coefficient at the snow–air interface defined by $\phi_s^* = h_{ia}(T_{sn} - T_a)$; ϕ_s^* = rate of heat loss at the snow surface, and T_{sn} = snow surface temperature. Equation (9.96) provides a theoretical basis for degree-day methods [39].

As spring approaches, the ice cover may deteriorate due to the absorption of short wave radiation after the snow cover has been melted. The cover will lose its structural integrity due to the melting of the grain boundary of columnar ice crystals. A detailed analysis of the deterioration process has been made by Ashton [45] based on [79]. The deterioration process occurs at the time when ice is melting rapidly.

8 Breakup

River ice breakup can generally be classified as a thermal breakup or a mechanical breakup. In a thermal breakup, the ice cover melts in place with no significant movement. The physics of thermal breakup is relatively well understood. Due to the varying turbulent heat transfer at the ice–water interface across the river associated with the transverse velocity distribution, partial opening of the ice cover often starts from melting along the thalweg. Mechanical breakup is due to the fragmentation of a floating cover by hydraulic and mechanical forces associated with changes in river discharge and water level, and is affected by the boundary constraints of the river plane geometry in allowing the movement of the large ice plates. The increase in river discharge could occur as the result of a runoff increase, the release of channel storage from an ice cover or jam failure, or the operation of a reservoir upstream. The driving forces that contribute to the mechanical breakup of an ice cover include friction forces due to water current and wind; streamwise component of the weight of the cover; longitudinal and vertical forces acting on the cover by the brash ice accumulation at the leading edge; and pressure variation on the underside of the cover produced by the river wave. Beltaos [48] developed an empirical criterion for the threshold between mechanical and thermal breakup.

The mechanical breakup begins with a detachment of the cover from the banks with an increase in river discharge which leads to the formation of hinge cracks along the banks. This is followed by the formation of far-spaced and close-spaced transverse cracks [47, 80–82]. An ice run will occur after the fragmentation of the cover if the water level raises enough to allow the movement of the ice plates.

8.1 *Initiation of Breakup*

Beltaos [47, 82, 83] defined the initiation of breakup as the time when sustained ice movement occurs at a particular site and formulated an empirical method for predicting its occurrence:

$$\Pi \equiv \frac{8(B - W_i)\tilde{\tau}_i m_c^2}{(m_c - 0.5)h_{i,o}} = (C_{bk}\sigma_o)\frac{\sigma_f h_i}{\sigma_{fo} h_{io}} = \Pi_o f(S_5) \tag{9.97}$$

in which, the composite quantity π has dimensions of stress, and the subscript *o* denotes values prior to the start of melt; B = water surface width; W_i = width of ice sheet, i.e. the distance between hinge cracks; $m_c = R_c/W_i$ with Rc = radius of curvature of the centreline of the ice sheet; C_{bk} = a dimensionless coefficient; h_i = solid portion of the ice cover thickness; $\tilde{\tau}_i$ = driving stress on the cover; σ_f = flexural strength of the ice cover; and f is a function of the thermal index S_5 = cumulative degree-days above −5 °C, such that $f(0) = 1$. The driving stress, $\tilde{\tau}_i$, includes the flow shear stress applied on the ice cover plus the downslope component of the weight of the cover per unit surface area:

$$\tilde{\tau}_i \approx \rho g R_i S_f + \rho g\left(\frac{\rho_i}{\rho}h_i + 0.96h_{i,p}\right)S_W \tag{9.98}$$

in which, R_i = hydraulic radius associated with the ice cover; S_W = water surface slope; and S_f = friction slope; $h_{i,p}$ = thickness of the porous portion of the ice cover.

Empirically, the function $f(S_5)$ can be approximated by $f(S_5) = \exp(-\alpha_s S_5)$, and the ranges of π_o and α_s are 70~110 kPa and 0.005~0.009, respectively.

8.2 Breakup Ice Run and Jam

Intermittent stoppage and movement of the ice rubble, accompanied by ice jam formation and release, can occur during the downstream progression of the ice run. The primary unknowns in this process are the mechanisms which cause the ice jam release and the rubble front advancement in the ice sheet. Both of these are related to the mechanical and hydrodynamic forces acting on the ice rubble and the ice sheet. The ice rubble moving along the river can incorporate additional ice fragments by plowing through the stationary ice cover [84]. Thermal erosion of the ice jam [85, 86] and decay of the ice sheet are important factors which can significantly reduce the strength and mass of the ice cover and jam in the river. Sudden release or failure of ice jams accompanied by the release of channel storage often generates surges of water with high water velocity and a rapid increase in stage [87]. Shen et al. [28] showed that the ice resistance has significant effects on both stage and discharge hydrographs during an ice jam release. A flood wave produced by jam release propagates slower than a flood wave without ice. The ice resistance also lowers the peak discharge significantly.

9 Sediment Transport

The presence of river ice causes major changes in the hydraulics of the flow, which will affect the sediment transport in the channel. This change in sediment transport properties can have important effects on channel morphology, aquatic ecology, and water quality. Field examples of ice effects on sediment transport have been reported [88–90]. Despite its importance, studies on sediment transport under the influence of ice have been very limited, with the exception of several flume studies with floating covers, each with a limited range of flow and sediment transport conditions. Knack and Shen [91] developed methods for estimating sediment transport capacity and bed resistance in fluvial channels based on available flume studies [79, 92–95] by considering the changes in flow condition due to the existence of an ice cover and the change in bed shear stress.

9.1 Bed Load

Extensive literature exists on bed load transport formulas for open water conditions. All of these formulas can be transformed into relationships between the

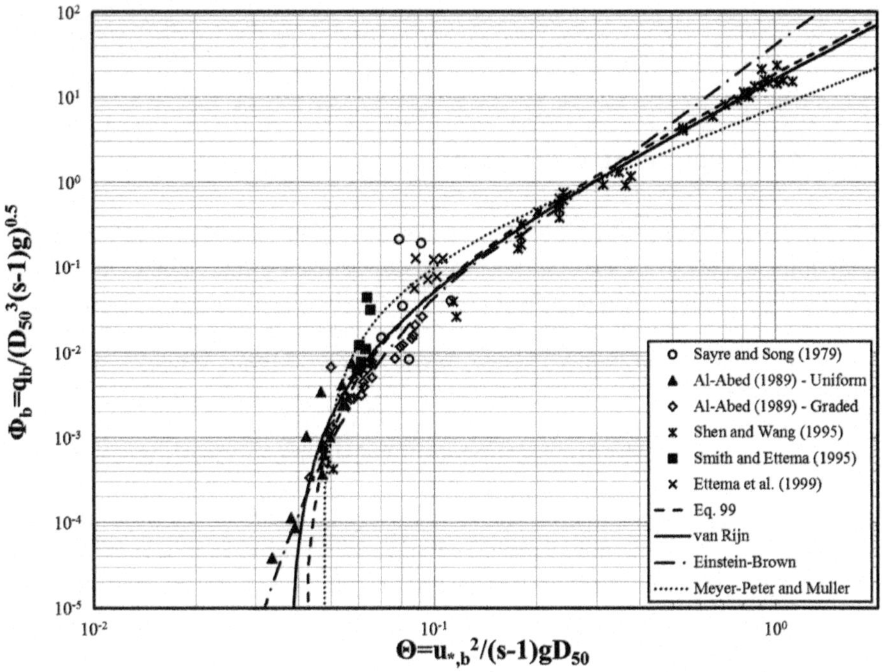

Fig. 9.2 Comparison of bed load data for ice-covered conditions

dimensionless bed load intensity $\Phi_b = \frac{q_b}{\sqrt{\Delta g D^3}}$ and the dimensionless flow strength $\Theta = \frac{\tau_b}{\Delta \rho g D}$; where q_b = volumetric rate of sediment transport per unit width; $\Delta = (\rho_s - \rho)/\rho = s - 1$; and D = particle diameter. Figure 9.2 shows the comparison of the ice-covered flume data with the formulations of Einstein and Brown [96], Meyer-Peter and Muller [97], and van Rijn [98]. This comparison shows that bed load transport in ice-covered channels can be described by conventional relationships for the equivalent free-surface flow if the flow strength is expressed in terms of the bed shear stress. An empirical equation fit to the data gives:

$$\Phi_b = 20(\Theta - \Theta_c)^{2.1} \tag{9.99}$$

where, Θ_c is the critical value of Θ for incipient motion. Figure 9.2 showed the van Rijn formulation compares well with the data through the entire range of flow strength. It is also preferred since it considers the effect of water temperature, which is important to sediment transport in winter when the water temperature is near the freezing point [99]. The Einstein–Brown formula fits well with the exception of the high flow strength region where the dimensionless bed load is over predicted. The Meyer-Peter and Muller formula does not match the data as well as the others. This is likely due to the equation being more applicable for coarse sediments with sizes larger than those used in the available flume experiments.

9.2 Suspended Load

The formation of surface ice changes the vertical distribution of the velocity [100]. The radical changes in the shear stress and diffusivity distributions over the depth have a significant effect on the suspended load. The method of Sayre and Song [101], which is an extension of the Rouse equation considering the ice cover effect, can be used to determine the suspended sediment discharge for ice covered conditions with some minor modifications. The sediment concentration can be calculated from:

$$C_s(z) = \begin{cases} C_o\left(\frac{z_o(Y_b-z)}{z(Y_b-a)}\right)^{z_b} & for \quad z_o \le z \le \frac{Y_b}{2} \\ C_{0,2}\left(\frac{\varepsilon_{si}+\frac{2}{H'}(\varepsilon_{si}-\varepsilon_{sb})\left(z-\left(Y_b+\frac{Y_i}{2}\right)\right)}{\varepsilon_{sb}}\right)^{-\frac{\omega_s H'}{2(\varepsilon_{si}-\varepsilon_{sb})}} & for \quad \frac{Y_b}{2} < z \le Y_b + \frac{Y_i}{2} \\ C_{0,3}\left(\frac{H'-z}{z-Y_b}\right)^{z_i\sqrt{\frac{d_b}{d_i}}} & for \quad Y_b + \frac{Y_i}{2} < z \le H' \end{cases} \tag{9.100}$$

where, $Z_b = \omega/\beta_b\kappa_b u_{*b}$ and $Z_i = \omega/\beta_i\kappa_i u_{*i}$; $\beta_b = 1 + 2[\omega/u_{*b}]^2$ for $0.1 < \omega/u_{*b} < 1$ and $\beta_i = 1 + 2[\omega/u_{*i}]^2$ for $0.1 < \omega/u_{*i} < 1$; $C_{0,2} = C(0.5Y_b) = C_o\left(\frac{z_o}{Y_b - z_o}\right)^b$;

$C_{0,3} = C(Y_b + 0.5Y_i) = C_{0,2}\left(\frac{\varepsilon_{si}}{\varepsilon_{sb}}\right)^{-\frac{\omega H'}{2(\varepsilon_{si}-\varepsilon_{sb})}}$; and C_o = a reference concentration at a near bed reference level z_o, which separates bed load and suspended load transport. Based on Van Rijn [102], C_o and z_o are calculated from $C_o = 0.015\,\frac{D_{50}}{z_o}\frac{T_t^{1.5}}{D_*^{0.3}}$, and $z_o = 0.5\Delta_d$ where Δ_d is the dune height if the bed form dimensions are known, otherwise $z_o = k_s$, and z_o must be at least $0.01H'$. The parameters T_t and D_* are defined as $D_* = D_{50}[(s-1)g/\nu^2]^{1/3}$, a dimensionless particle parameter; $T_t = \left[\left(u'_*\right)^2 - \left(u_{*,cr}\right)^2\right]/\left(u_{*,cr}\right)^2$, a transport stage parameter; ν = kinematic viscosity of water; $u'_* = \left(\tau'_b/\rho\right)^{1/2}$, shear velocity related to grains; $\tau'_b = \gamma\left(n'_b\right)^2\bar{u}^2/\left(\kappa_b H'\right)^{1/3}$, bed shear stress related to grains; $n'_b = D_{50}^{1/6}/20$, bed Manning's grain roughness; $u_{*,cr}$, critical bed shear velocity [98].

A comparison of the measured and calculated suspended sediment discharges are presented in Fig. 9.3. The calculated suspended discharge compares well with the measured values for both open water and ice covered flows.

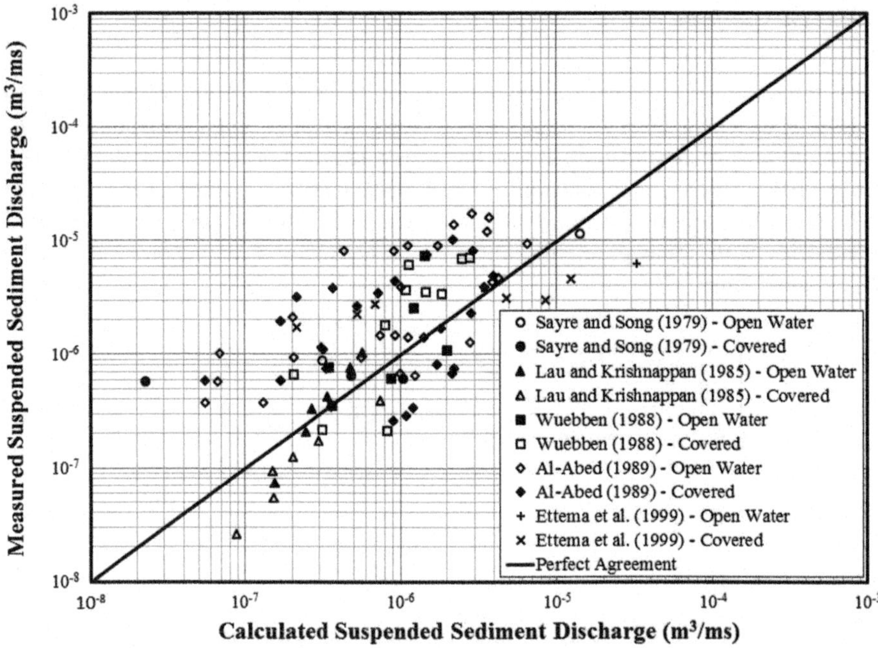

Fig. 9.3 Comparison of measured and calculated suspended sediment discharge rates

9.3 Bed Resistance

It is necessary to determine the bed resistance of an ice-covered alluvial channel for sediment transport analysis. Smith and Ettema [103] developed a regression relationship for the ratio between the bed shear stress and cover shear stress:

$$\frac{\tau_i}{\tau_b} = 0.84\left(\widetilde{\eta}\,\frac{D_{50}}{k_{ic}}\right)^{-0.2} \qquad (9.101)$$

where, $\widetilde{\eta} = \frac{\Theta}{\Theta_c}$; and k_{ic} = ice cover roughness height. Figure 9.4 shows the ratio of the Manning coefficients of the ice cover and the bed [91], which is an extension of Eq. (9.101).

10 Summary

River ice is an important aspect of wintertime water resources engineering in cold regions. The existence of river ice can have major impacts on hydropower operations, inland navigation, water transfer, and various other hydraulic engineering

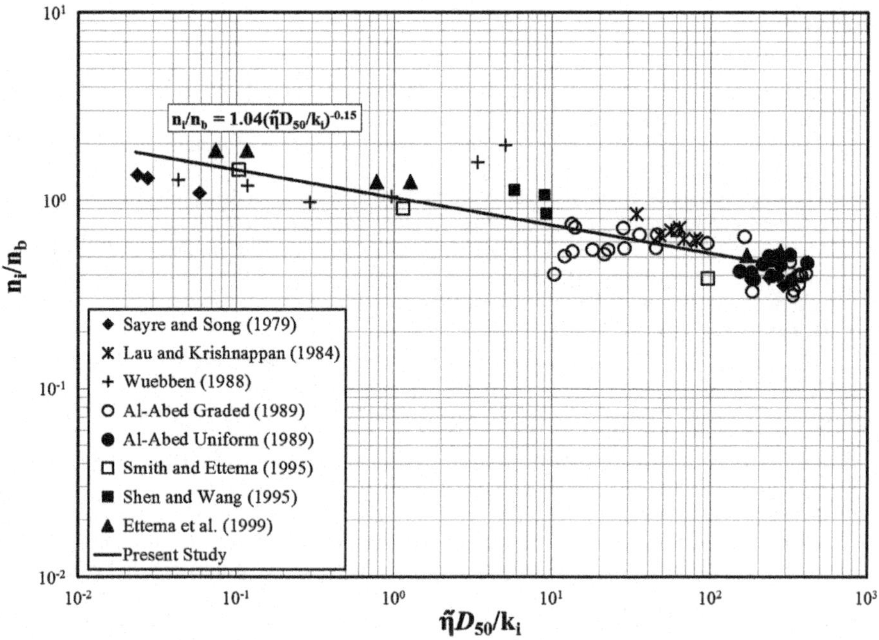

Fig. 9.4 Ratio of Manning's coefficients for ice-covered alluvial channels

projects, in addition to the concern on severe floods caused by ice jams. River ice also affects environmental and ecological conditions of rivers, as well as the channel morphology. Significant progress on river ice research has been made in the last few decades. This chapter discussed the current state of knowledge on river ice processes from freeze up to breakup. An overview of river ice processes, including the formation, evolution, transport, accumulation, growth, deterioration, and dissipation of various forms of ice, is presented first. It is followed by detailed discussions on the theory and formulation of various river ice and related hydraulic and sediment transport phenomena. These discussions include the river hydraulics with floating ice covers and ice jams, the energy budget related to river ice evolution, variations of water temperature and surface and suspended frazil ice concentrations along the river, the development of various types of ice cover during freeze up, the formation of anchor ice and undercover frazil jams, thermal growth and decay of ice cover, and ice cover breakup. Ice jams can develop during both freeze up and breakup. Static and dynamics ice jam theories are discussed. In alluvial channels, the presence of river ice can cause major changes on sediment transport in the channel. Recent development on sediment transport with ice conditions is presented.

References

1. Shen, H. T. (1996). River ice processes – State of research. In *Proceedings, 13th IAHR Symposium on Ice* (pp. 825–833). Beijing.
2. Carsterns, T. (1966). Experiments with supercooling and ice formation in flowing water. *Geofysiske Publikasjoner, 26*(9), 3–18.
3. Ye, S. Q., Deoring, J., & Shen, H. T. (2004). A laboratory study of frazil evolution in a counter-rotating flume. *Canadian Journal of Civil Engineering, 31*, 899–914.
4. Matousek, V. (1984). Types of ice run and conditions for their formation. In *Proceedings, IAHR Ice Symposium* (pp. 315–327).
5. Osterkemp, T. E. (1978). Frazil ice formation: A review. *Journal of Hydraulics Division, 104* (HY9), 1239–1255.
6. Daly, S. F. (1984). *Frazil ice dynamics*. CRREL Monograph 84-1. U.S. Army CRREL, Hanover.
7. Hammar, L., & Shen, H. T. (1995). Frazil evolution in channels. *Journal of Hydraulic Research, 33*(3), 291–306.
8. Kerr, D. J., Shen, H. T., & Daly, S. F. (2002). Evolution and hydraulic resistance of anchor ice on gravel bed. *Cold Regions Science and Technology, 35*(2), 101–114.
9. Prowse, T. D., & Gridley, N. C. (Eds.). (1993). *Environmental aspects of river ice* (NHRI Science Report No. 5, 155 pp). Saskatoon: National Hydrology Research Institute.
10. Tsang, G., & Lau, Y. (1995). Frazil and anchor ice laboratory investigations. In *Proceedings, 8th Workshop on Hydraulics of Ice Covered Rivers*. Kamloops
11. Yamazaki, M., Hirayama, K., Sakai, S., Sasamoto, M., Kiyohara, M., & Takiguchi, H. (1996). Formation of frazil and anchor ice. In *Proceedings, 13th IAHR International Symposium on Ice* (pp. 488–496). Beijing.
12. Doering, J. C., Berkeris, L. E., Morris, M. P., & Girling, W. C. (2001). Laboratory study of anchor ice growth. *Journal of Cold Regions Engineering, 15*(1), 60–66.
13. Lal, A. M. W., & Shen, H. T. (1991). A mathematical for river ice processes. *Journal of Hydraulic Engineering, 117*(7), 851–867.
14. Andres, D. (1995). Floe development during frazil generation in large rivers. In *Proceedings, 8th Workshop on the Hydraulics of Ice Covered Rivers*. Kamloops.
15. Michel, B., Marcotte, N., Fonseca, F., & Rivard, G. (1980). Formation of border ice in the St. Anne River. In *Proceedings of Workshop on Hydraulics of Ice-Covered Rivers* (pp. 38–61). Edmonton.
16. Shen, H. T., & Van DeValk, W. A. (1984). Field investigation of St. Lawrence River hanging ice dams. In *Proceedings, IAHR Ice Symposium* (pp. 241–249). Hamburg.
17. Matousek, V. (1990). *Thermal processes and ice formation in rivers* (Papers and Studies No. 180, 146 pp). Prague: Water Research Institute.
18. Pariset, R., & Hausser, H. (1961). Formation and evolution of ice covers on rivers. *Transactions, Engineering Institute of Canada, 5*(1), 41–49.
19. Pariset, R., Hausser, H., & Gagnon, A. (1966). Formation of ice cover and ice jams in rivers. *Journal of the Hydraulics Division, 92*(HY6), 1–24.
20. Uzuner, M. S., & Kennedy, J. F. (1976). Theoretical model of river ice jams. *Journal of the Hydraulics Division, 102*(HY9), 1365–1383.
21. Tatinclaux, J.-C. (1977). Equilibrium thickness of ice jams. *Journal of the Hydraulics Division, ASCE, 103*(HY9), 959–974.
22. Beltaos, S. (1983). River ice jams: Theory, case studies and applications. *Journal of Hydraulic Engineering, 109*(10), 1338–1359.
23. Flato, G. M., & Gerard, R. (1986). Calculation of ice jam profiles. In *Proceedings, 4th Workshop on River Ice*. Montreal, Paper C-3.
24. Beltaos, S., & Wong, J. (1986). Downstream transition of river ice jams. *Journal of Hydraulic Engineering, 112*(2), 91–110.

25. Shen, H. T., Shen, H., & Tsai, S. M. (1990). Dynamic transport of river ice. *Journal of Hydraulic Research, 28*(6), 659–671.
26. Shen, H. T., Su, J., & Liu, L. (2000). SPH simulation of river ice dynamics. *Journal of Computational Physics, 165*(2), 752–771.
27. Shen, H. T., & Liu, L. (2003). Shokotsu River ice jam formation. *Cold Regions Science and Technology, 37*(1), 35–49.
28. Shen, H. T., Gao, L., Kolerski, T., & Liu, L. (2008). Dynamics of ice jam formation and release. *Journal of Coastal Research, S52*(2008), 25–32.
29. Ashton, G. D. (1974). Froude criterion for ice block stability. *Journal of Glaciology, 13*(689), 307–313.
30. Shen, H. T., & Ho, C. F. (1986). Two-dimensional simulation of ice cover formation in a large river. In *Proceedings, IAHR Ice Symposium* (pp. 547–558). Iowa City.
31. Sun, Z. C., & Shen, H. T. (1988). A field investigation of frazil jam in Yellow River. In *Proceedings, 5th Workshop on Hydraulics of River Ice/Ice Jams* (pp. 157–175). Winnipeg.
32. Michel, B. (1984). Comparison of field data with theories on ice cover progression in large rivers. *Canadian Journal of Civil Engineering, 11*(4), 798–814.
33. Kivisild, H. R. (1959). Hydrodynamic analysis of ice floods. In *Proceedings, 8th IAHR Congress*. Montreal, Paper 23F.
34. Tesaker, E. (1975). Accumulation of frazil ice in an intake reservoir. In *Proceedings, IAHR Symposium on Ice Problems*. Hanover, NH.
35. Michel, B., & Drouin, M. (1975). *Equilibrium of a Hanging Dam at the La Grande River* (Report GCS-75-03-01). Quebec: Laval University.
36. Shen, H. T., & Wang, D. S. (1995). Under cover transport and accumulation of frazil granules. *Journal of Hydraulic Engineering, 120*(2), 184–194.
37. White, K. D., & Acone, S. E. (1995). Assessing the effects of dam operation on upstream ice conditions: Aroostook River at Fort Fairfield, Maine. In *Proceedings, 8th Workshop on Hydraulics of Ice Covered Rivers*. Kamloops.
38. Mao, Z.-Y., Zhao, X. F., Wang, A.-M., Xu, X., & Wu, J.-J. (2008). Two-dimensional numerical model for river-ice processes based upon boundary-fitted coordinate transformation method. In *Proceedings, 19th International Symposium on Ice* (pp. 141–152). Vancouver.
39. Shen, H. T., & Yapa, P. D. (1985). A unified degree-day method for river ice cover thickness simulation. *Canadian Journal of Civil Engineering, 12*, 54–62.
40. Ashton, G. D. (Ed.). (1986). *River and lake ice engineering*. Littleton, CO: Water Resources Publications.
41. Shen, H. T., & Chiang, L. A. (1984). Simulation of growth and decay of river ice cover. *Journal of Hydraulics Division, 110*(7), 958–971.
42. Calkins, D. J. (1979). *Accelerated ice growth in rivers* (CRREL Report 79-14, 4 pp). Hanover: US Army Cold Regions Research and Engineering Laboratory.
43. Shen, H. T., & Lal, A. M. W. (1986). Growth and decay of river ice covers. In *Proceedings, Cold Regions Hydrology Symposium* (pp. 583–591). Fairbanks.
44. Ashton, G. D., & Kennedy, J. F. (1972). Ripples on underside of river ice covers. *Journal of Hydraulic Engineering, 98*, 1603–1624.
45. Ashton, G. D. (1985). Deterioration of floating ice covers. *Journal of Energy Resources Technology, 107*, 177–182.
46. Prowse, T. D., Demuth, M. N., & Chew, H. A. M. (1990). The deterioration of freshwater ice under radiation decay. *Journal of Hydraulic Research, 28*(6), 685–691.
47. Beltaos, S. (1984). A conceptual model of river ice breakup. *Canadian Journal of Civil Engineering, 17*(2), 173–183.
48. Beltaos, S. (2003). Threshold between mechanical and thermal breakup of river ice cover. *Cold Regions Science and Technology, 37*, 1–13.
49. Foltyn, E. P., & Shen, H. T. (1986). St. Lawrence River freeze-up forecast. *Journal of Waterway, Port, Coastal and Ocean Engineering, 112*(4), 467–481.

50. Potok, A. J., & Quinn, F. H. (1979). A hydraulic transient model of the Upper St. Lawrence River for water resources research. *Water Resources Bulletin, 15*(6).
51. Hicks, B. B. (1972). Some evaluation of drag and bulk transfer coefficients over water. *Boundary Layer Meteorology, 3*, 211–213.
52. Wake, A., & Rumer, R. R., Jr. (1979). Modeling ice regime in Lake Erie. *Journal of Hydraulics Division, 105*(HY7), 827–844.
53. Yapa, P. D., & Shen, H. T. (1986). Unsteady flow simulation for an ice-covered river. *Journal of Hydraulic Engineering, 112*(11), 1036–1049.
54. Nezhikovskiy, R. A. (1964). Coefficients of roughness of bottom surface of slush ice cover. *Soviet Hydrology: Selected Papers, 2*, 127–148.
55. Beltaos, S. (1993). Flow through breakup jams. In *Proceeding, 11th Canadian Hydrotechnical Conference* (pp. 643–652). Fredericton.
56. Beltaos, S. (1993). Numerical computation of river ice jams. *Canadian Journal of Civil Engineering, 20*(1), 88–99.
57. Lu, S., Shen, H. T., & Crissman, R. D. (1999). Numerical study of ice jam dynamics in Upper Niagara River. *Journal of Cold Regions Engineering, 13*(2), 78–102.
58. Shen, H. T. (2010). Mathematical modeling of river ice processes. *Cold Regions Science and Technology, 62*(1), 3–13.
59. Wu, J. (1973). Prediction of near-surface drift currents from wind velocity. *Journal of Hydraulics Division, 99*(9), 1291–1301.
60. Paily, P. P., Macagon, E. O., & Kennedy, J. F. (1974). *Winter-regime surface heat loss from heated streams* (IIHR Report No. 155, 137 pp). Iowa City, IA: Iowa Institute of Hydraulic Research.
61. Iqbal, M. (1983). *An introduction to solar radiation*. Down Mills: Academic Press Canada, Inc.
62. Bolsenga, S. J. (1969). Total albedo of Great Lakes ice. *Water Resources Research, 5*(5), 1132–1133.
63. Pivovarov, A. A. (1973). *Thermal conditions in freezing lakes and rivers*. New York: Wiley.
64. Satterland, D. R. (1979). An improved equation for estimating long-wave radiation from the atmosphere. *Water Resources Research, 15*(6), 1649–1650.
65. Rimsha, V. A., & Donchenko, R. V. (1957). The investigation of heat loss from free water surfaces in wintertime (in Russian). *Trudy Leningrad Gosud. Gidrol. Inst., 65*, 58–83.
66. Dingman, S. L., & Assur, A. (1969). *The effects of thermal pollution on river-ice conditions, Part II* (Technical Report). Hanover, NH: US Army Cold Regions Research and Engineering Laboratory.
67. Haynes, F. D., & Ashton, G. D. (1979). *Turbulent heat transfer in large aspect channels* (CRREL Report 79-13, 5 pp). Hanover, NH: US Army Cold Regions Research and Engineering Laboratory.
68. Shen, H. T., Wang, D. S., & Lal, A. M. W. (1995). Numerical simulation of river ice processes. *Journal of Cold Regions Engineering, ASCE, 9*(3), 107–118.
69. Hibler, W. D. (1979). A dynamic thermodynamic sea ice model. *Journal of Physical Oceanography, 9*(4), 815–846.
70. Wake, A., & Rumer, R. R., Jr. (1983). Great Lakes ice dynamics simulation. *Journal of Waterway, Port, Coastal, and Ocean Engineering, 109*(1), 86–102.
71. Ji, S., Shen, H. T., Wang, Z., Shen, H. H., & Yue, Q. (2005). A viscoelastic-plastic constitutive model with Mohr-Coulomb yielding criterion for sea ice dynamics. *Acta Oceanologica Sinica, 24*(4), 54–65.
72. Arden, R. S., & Wigle, T. E. (1972). Dynamics of ice formation in the Upper Niagara River. In *Proceedings, The Role of Snow and Ice in Hydrology* (pp. 1296–1313). IAHS-UNESCO-WMO, Banff.
73. Marcotte, N., & Roberts, S. (1984). Elementary mathematical modeling of anchor ice. In *Proceedings, IAHR Ice Symposium* (pp. 151–161). Hamburg.

74. Girling, W. C., & Groeneveld, J. (1999). Anchor ice formation below Limestone Generating Station. In *Proceedings of the 10th River Ice Workshop, River Ice Management with a Changing Climate* (pp. 160–173). Winnipeg.
75. Beltaos, S. (2008). Progress in the study and management of river ice jams. *Cold Regions Science and Technology, 51*, 2–19.
76. Beltaos, S. (2001). Hydraulic roughness of breakup ice jams. *Journal of Hydraulic Engineering, 127*(8), 650–656.
77. Wake, A., & Rumer, R. R., Jr. (1979). Effect of surface meltwater accumulation on the dissipation of lake ice. *Water Resources Research, 15*(2).
78. Greene, G. M. (1981). *Simulation of ice-cover growth and decay in one-dimensional on the Upper St. Lawrence River* (NOAA Tech. Memo. ERL GLERL-36). Ann Arbor, MI.
79. Smith, B. T., & Ettema, R. (1995). *Ice-cover influence on flow and bedload transport in dune-bed channels* (IIHR Rep. No. 374). Iowa City, IA: Iowa Inst. of Hydraulic Research, University of Iowa.
80. Xia, X., & Shen, H. T. (2002). Nonlinear interaction of ice cover with shallow water wave in channels. *Journal of Fluid Mechanics, 467*, 259–267.
81. Billfalk, L. (1982). *Breakup of solid ice covers due to rapid water level variations* (CRREL Report 82-3). Hanover, NH: US Army Cold Regions Research and Engineering Laboratory.
82. Beltaos, S. (1990). Fracture and breakup of river ice cover. *Canadian Journal of Civil Engineering, 17*(2), 173–183.
83. Beltaos, S. (Ed.). (2008). *River ice breakup*. Highlands Ranch, CO: Water Resources Publications.
84. Shen, H. T., & Lu, S. (1996). Dynamics of river ice jam release. In *Proceedings, 8th International Conference on Cold Regions Engineering, ASCE* (pp. 594–605). Fairbanks.
85. Prowse, T. D. (1990). Heat and mass balance of an ablating ice jam. *Canadian Journal of Civil Engineering, 17*(4), 629–635.
86. Beltaos, S., & Burrell, B. C. (2006). Water temperature decay under breakup ice jams. *Cold Regions Science and Technology, 45*, 123–136.
87. Jasek, M. (2003). Ice jam release surges, ice runs, and breaking fronts: Field measurements, physical descriptions, and research needs. *Canadian Journal of Civil Engineering, 30*, 113–127.
88. Mercer, A. G., & Cooper, R. H. (1977). Bed scour related to the growth of a major ice jam. In *Proceedings of the 3rd National Hydrotechnical Conference* (pp. 291–208). Quebec, Canada.
89. Lawson, D. E., Chacho, E. F., Brockett, B. E., Wuebben, J. L., & Collins, C. M. (1986). *Morphology, hydraulics, and sediment transport of an ice covered river: Field techniques and initial data* (CRREL Report 86-11). Hanover, NH: Cold Regions Research and Engineering Laboratory.
90. Yang, X., Zhang, B., & Shen, H. T. (1993). Simulation of wintertime fluvial processes in the lower Yellow River. *Journal of Sedimentary Research, 2*, 36–43.
91. Knack, I. M., & Shen, H. T. (2013). A numerical model for sediment transport and bed change in rivers with ice. In *Proceedings, 12th International Symposium on River Sedimentation*. Kyoto.
92. Lau, Y. L., & Krishnappan, B. G. (1985). Sediment transport under ice cover. *Journal of Hydraulic Engineering, 111*(6), 934–950.
93. Al-Abed, N. M. A. (1989). *Sediment transport under ice covered channels*. Thesis Submitted in Partial Fulfillment of the Requirements for the Degree of Doctor of Philosophy, Civil and Environmental Engineering, University of Kashmir.
94. Ettema, R., Braileanu, F., & Muste, M. (1999). *Laboratory study of suspended sediment transport in ice-covered flow* (IIHR Rep. No. 404). Iowa City, IA: Iowa Institute of Hydraulic Research, University of Iowa.
95. Ettema, R., Braileanu, F., & Muste, M. (2000). Method for estimating sediment transport in ice-covered channels. *Journal of Cold Regions Engineering, 14*(3), 130–144.

96. Brown, C. B. (1950). Sediment transport. In H. Rouse (Ed.), *Engineering hydraulics*. New York: John Wiley & Sons.
97. Meyer-Peter, E., & Muller, R. (1948). Formulas for bed-load transport, Paper No. 2. In *Proceedings of the Second Meeting* (pp. 39–64). IAHR.
98. van Rijn, L. C. (1984). Sediment transport, Part I: Bed load transport. *Journal of Hydraulic Engineering, 110*(10), 1431–1456.
99. Hong, R. J., Karim, M. F., & Kennedy, J. F. (1984). Low-temperature effects on flow in sand-bed streams. *Journal of Hydraulic Engineering, 110*(2), 109–125.
100. Larsen, P. A. (1973). Hydraulic roughness of ice covers. *Journal of the Hydraulics Division, 99*(HY1), 111–119.
101. Sayre, W. W., & Song, G. B. (1979). *Effects of ice covers on alluvial channel flow and sediment transport processes* (IIHR Rep. No. 218). Iowa City, IA: Iowa Inst. of Hydraulic Research, University of Iowa.
102. van Rijn, L. C. (1984). Sediment transport, Part II: Suspended load transport. *Journal of Hydraulic Engineering, 110*(11), 1613–1641.
103. Smith, B. T., & Ettema, R. (1997). Flow resistance in ice-covered channels. *Journal of Hydraulic Engineering, 123*(7), 592–599.

Chapter 10
Mathematical Modeling of Water Properties

Mu-Hao S. Wang, Lawrence K. Wang, Ching-Gung Wen, and David Terranova Jr.

Contents

M.-H.S. Wang, Ph.D., P.E. (✉) • L.K. Wang, Ph.D., P.E.
Department of Environmental Engineering, College of Engineering,
National Cheng Kung University, Tainan, Taiwan

Lenox Institute of Water Technology Newtonville, NY, USA
e-mail: lenox.institute@gmail.com

C.-G. Wen, Ph.D., P.E.
College of Engineering, National Cheng Kung University, Tainan, Taiwan

D. Terranova Jr., M.E.
Department of Mechanical Engineering, Stevens Institute of Technology,
Hoboken, NJ 07030, USA

L.K. Wang, C.T. Yang, and M.-H.S. Wang (eds.), *Advances in Water Resources Management*, Handbook of Environmental Engineering, Volume 16,
DOI 10.1007/978-3-319-22924-9_10

Abstract Mathematical models describing physical water property data are presented. These models define each of the following water properties as a function of the water temperature. The water property models developed include the specific weight, density, modulus of elasticity, dynamic viscosity, kinematic viscosity, surface tension and vapor pressure. Along with each model, the actual data and calculated values are presented for comparison. The method of least squares fit, which was used to model each set of data, is reviewed. The source of FORTRAN computer program used to generate the equations can be obtained from the Lenox Institute of Water Technology.

Keywords Environmental engineering • Water quality control • Mathematical models • Water property • Specific weight • Density • Modulus of elasticity • Dynamic viscosity • Kinematic viscosity • Surface tension • Vapor pressure • Computer analysis

Nomenclature

A	A coefficient
B	A coefficient
C	A coefficient
D	Density, slug/ft^3 or kg/m^3
D.V.	Dynamic viscosity, lb-s/ft^2 or N-s/m^2
dP	Differential change in pressure, lb/in^2, or kN/m^2
dv	Velocity change, ft/s, or m/s
dV	Differential change in volume, ft^3, or m^3
dy	Length change, ft, or m
E	A constant = 2.718
F_s	Shearing force, lb/ft, or N/m
F_{sf}	Stretching force, lb, or N
g	Acceleration of gravity = 32.174 ft/s^2 = 9.81 m/s^2
K.V.	Kinematic viscosity, ft^2/sec or m^2/sec
L	Unit length, ft, or m
M.E.	Modulus of elasticity, lb/in^2 or kN/m^2
S.T.	Surface tension, N/m or lb/ft
S.W.	Specific weight, lb/ft^3 or kN/m^3
T	Temperature, °F or °C or K
V	Volume, ft^3, or m^3
V_g	Specific volume, ft^3/slug, or m^3/kg
V.P.	Vapor pressure, lb/in^2 or kN/m^2

1 Introduction

This chapter introduces the mathematical models for calculating the following physical properties of water:

		Metric units	British unit
Specific weight	Y	kN/m^3	lb/ft^3
Density	ρ	kg/m^3	$slug/ft^3$
Modulus of elasticity	E	kN/m^2	lb/in^2
Dynamic viscosity	μ	$N\text{-}sec/m^2$	$lb\text{-}sec/ft^2$
Kinematic viscosity	ν	m^2/sec	ft^2/sec
Surface tension	σ	N/m	lb_f/ft
Vapor pressure	P_V	kN/m^2	lb/in^2

Each developed model is illustrated by means of practical examples.

The experimental data found in Section 7 Appendices were adapted from the literature [1–3] and used as the actual values from which the authors' models were generated. The water property equations were generated with the specific water property as a function of the water temperature.

$$\text{Water Property} = \phi(\text{Temperature in C}^\circ \text{ and F}^\circ)$$

In all but one instance, only one equation was needed to describe the water property data. The water quality that could not be defined by one equation was Surface Tension = ϕ (Temperature in °F), which was broken down into three individual curves and studied. It was determined that three equations, each have upper and lower boundary conditions, could define the true experimental data. The technique of mathematical modeling is presented in Section 2 of this chapter. All derived mathematical equations are presented in Section 3, and summarized in Section 4. A FORTRAN computer program used for mathematical modeling can be obtained from the Lenox Institute of Water Technology.

2 Mathematical Modeling

A FORTRAN computer program written by the authors was used to fit the data with the method of least squares. A polynomial function in Eq. (10.1) was selected to represent the data.

$$C = a_0 + a_1T + a_2T^2 + a_3T^3 + \ldots + a_uT^u \quad (10.1)$$

The shape of the resulting mathematical curve depends both upon the value of the coefficients a; and the degree of the polynomial u. Equations (10.2) and (10.3) are for the calculations of two coefficients in the linear fit equation, u = 1.

$$\text{Linear fit:} \qquad C = a_0 + a_1T$$

$$a_0N + a_1\Sigma T = \Sigma C \tag{10.2}$$

$$a_0\Sigma T + a_1\Sigma T^2 = \Sigma CT \tag{10.3}$$

Equations (10.4)–(10.6) and (10.7)–(10.10) can be used for the quadratic fit ($u = 2$) and the cubic fit ($u = 3$), respectively.

$$\text{Quadratic fit:} \qquad C = a_0 + a_1T + a_2T^2$$

$$a_0N + a_1\Sigma T + a_2\Sigma T^2 = \Sigma C \tag{10.4}$$

$$a_0\Sigma T + a_1\Sigma T^2 + a_2\Sigma T^3 = \Sigma CT \tag{10.5}$$

$$a_0\Sigma T^2 + a_1\Sigma T^3 + a_2\Sigma T^4 = \Sigma CT^2 \tag{10.6}$$

$$\text{Cubic fit:} \qquad C = a_0 + a_1T + a_2T^2 + a_3T^3$$

$$a_0N + a_1\Sigma T + a_2\Sigma T^2 + a_3\Sigma T^3 = \Sigma C \tag{10.7}$$

$$a_0\Sigma T + a_1\Sigma T^2 + a_2\Sigma T^3 + a_3\Sigma T^4 = \Sigma CT \tag{10.8}$$

$$a_0\Sigma T^2 + a_1\Sigma T^3 + a_2\Sigma T^4 + a_3\Sigma T^5 = \Sigma CT^2 \tag{10.9}$$

$$a_0\Sigma T^3 + a_1\Sigma T^4 + a_2\Sigma T^5 + a_3\Sigma T^6 = \Sigma CT^3 \tag{10.10}$$

For the data presented in Tables 10.1 and 10.2, T = water temperature and C = water property of interest. The values of ΣT, ΣT^2, ΣT^3, ΣT^4, ΣT^5, ΣT^6, ΣC, ΣCT, ΣCT^2 and ΣCT^3 can be calculated mathematically. The coefficients of linear fit (a_0 and a_1) can be calculated by solving simultaneous Eqs. (10.2) and (10.3); the coefficients of quadratic fit (a_0, a_1 and a_2) can be calculated by solving simultaneous Eqs. (10.4)–(10.6); and finally the coefficients of cubic fit (a_0, a_1, a_2 and a_3) can be determined by solving simultaneous Eqs. (10.7)–(10.10).

Using and expanding further the above equations, an nth degree equation can be computed.

All simultaneous linear equations were solved by the Gaussian elimination method with a computer.

To determine, to which power the best curve fit would be, two other numerical techniques were derived and used. They are the standard error, and the coefficient of correlation. A brief derivation of these two general equations are as follows:

$$C = a_0 + a_1T + a_2T^2 + \ldots\ldots + a_uT^u$$

Table 10.1 Temperature T (°F) versus specific weight S.W. (lb/ft^3)

The degree of fit for this run is 3			
	Coefficients		
	x (0) =	0.62294E + 02	
	x (1) =	0.82009E − 02	
	x (2) =	−0.12783E − 03	
	x (3) =	0.16262E − 06	
Variable	Actual value	Calculated value	Difference
32.000	62.420	62.431	−0.011
40.000	62.430	62.428	0.002
50.000	62.410	62.405	0.005
60.000	62.370	62.361	0.009
70.000	62.300	62.297	0.003
80.000	62.220	62.215	0.005
90.000	62.110	62.115	−0.005
100.000	62.000	61.998	0.002
110.000	61.860	61.866	−0.006
120.000	61.710	61.718	−0.008
130.000	61.550	61.557	−0.007
140.000	61.380	61.383	−0.003
150.000	61.200	61.197	0.003
160.000	61.000	61.000	0.000
170.000	60.800	60.793	0.007
180.000	60.580	60.577	0.003
190.000	60.360	60.353	0.007
200.000	60.120	60.122	−0.002
212.000	59.830	59.837	−0.007
Standard error is 61.50789400		Correlation coefficient is .99996	

By defining:

$$\Sigma C^2 = C^2 - C_{mean}\Sigma C \tag{10.11}$$

$$C_{mean} = \Sigma C/N \tag{10.12}$$

then,

$$\Sigma C_{\times c}{}^2 = -C_{mean}\Sigma C + a_0\Sigma C + a_1\Sigma TC + a_2\Sigma T^2C + \ldots\ldots + a_u\Sigma T^uC \tag{10.13}$$

$$\Sigma C_{sc}{}^2 = \Sigma C^2 - \Sigma C_{\times c}{}^2 \tag{10.14}$$

Table 10.2 Temperature T (°C) versus specific weight S.W. (kN/m^3)

The degree of fit for this run is 3			
	Coefficients		
	x (0) =	0.98068E + 01	
	x (1) =	0.18341E − 03	
	x (2) =	−0.58132E − 04	
	x (3) =	0.15575E − 06	
Variable	Actual value	Calculated value	Difference
0.000	9.805000	9.806765	−0.001765
5.000	9.807000	9.806243	0.000752
10.000	9.804000	9.802941	0.001059
15.000	9.798000	9.796962	0.001038
20.000	9.789000	9.788426	0.000574
25.000	9.777000	9.777451	−0.000451
30.000	9.764000	9.764153	−0.00153
40.000	9.730000	9.731057	−0.001057
50.000	9.689000	9.690073	−0.001073
60.000	9.642000	9.642135	−0.000135
70.000	9.589000	9.588178	0.000822
80.000	9.530000	9.529135	0.000865
90.000	9.466000	9.465943	0.000057
100.000	9.399000	9.399534	−0.000534
Standard error is 9.68492870		Correlation coefficient is 1.00000	

The standard error is derived as

$$S_c = \left[\left(\Sigma C_{sc}{}^2 \right) / N \right]^{0.5} \tag{10.15}$$

and the coefficient of correlation is derived as:

$$R_c = \left[\left(\Sigma C_{\times c}{}^2 \right) / \left(\Sigma C^2 \right) \right]^{0.5} \tag{10.16}$$

A FORTRAN listing of the program used to generate the mathematical equations presented, can be obtained from the Lenox Institute of Water Technology. This program was run on a DEC-SYSTEM 10 at Stevens Institute of Technology. An in-house plotting routine, RSPLOT (HOUSTON INSTRUMENTS DP7) was used to plot the data generated by the program, along with the actual input data. This allowed the authors to visually inspect each set of data for accuracy.

3 Derivation of Water Property Models

For each set of data, two equations were generated; one in °C the other °F. The rest of this section presents these equations, computer output of the modeling data, including the actual and calculated data.

Tables 10.1, 10.2, 10.3, 10.4, 10.5, 10.6, 10.7, 10.8, 10.9, 10.10, 10.11, 10.12, 10.13, 10.14, 10.15, 10.16, and 10.17 document the tabulated computer output data and the graphically illustrated computer output data.

4 Summary of Water Property Models

Unless otherwise noted the following derived equations are good in the following temperature ranges:

degrees F = 32 to 212 °F
degrees C = 0 to 100 °C

4.1 *Specific Weight*

$$\begin{aligned}\text{S.W. } \left(1\text{b}/\text{ft}^3\right) &= 0.62294 \times 10^2 + 0.82009 \times 10^{-2}\text{T} \\ &\quad - 0.12783 \times 10^{-3}\text{T}^2 + 0.16262 \times 10^{-6}\text{T}^3\end{aligned} \tag{10.17}$$

where T = temperature (°F)

$$\begin{aligned}\mathbf{S.W.}\ \left(\mathbf{kN/m^3}\right) &= 0.98068 \times 10 + 0.18341 \times 10^{-3}\mathbf{T} - 0.58132 \times 10^{-4}\mathbf{T^2} \\ &\quad + 0.15575 \times 10^{-6}\text{T}^3\end{aligned} \tag{10.18}$$

where T = temperature (°C)

4.2 *Density*

$$\begin{aligned}\mathbf{D}\ \left(\mathbf{slug/\ ft^3}\right) &= \mathbf{0.19354 \times 10 + 0.27304 \times 10^{-3}T - 0.40986 \times 10^{-5}T^2} \\ &\quad + 0.53217 \times 10^{-8}\text{T}^3\end{aligned} \tag{10.19}$$

Table 10.3 Temperature T (°F) versus density D (slug/ft^3)

The degree of fit for this run is 3			
	Coefficients		
	x (0) =	0.19354E+01	
	x (1) =	0.27304E − 03	
	x (2) =	−0.40986E − 05	
	x (3) =	0.53217E − 08	
Variable	Actual value	Calculated value	Difference
32.000	1.940	1.940	−0.000
40.000	1.940	1.940	−0.000
50.000	1.940	1.939	0.001
60.000	1.938	1.938	−0.000
70.000	1.936	1.936	−0.000
80.000	1.934	1.934	0.000
90.000	1.931	1.931	0.000
100.000	1.927	1.927	−0.000
110.000	1.923	1.923	0.000
120.000	1.918	1.918	−0.000
130.000	1.913	1.913	−0.000
140.000	1.908	1.908	0.000
150.000	1.902	1.902	−0.000
160.000	1.896	1.896	0.000
170.000	1.890	1.890	0.000
180.000	1.883	1.883	0.000
190.000	1.876	1.876	0.000
200.000	1.868	1.869	−0.001
212.000	1.860	1.860	0.000
Standard error is 1.91173690		Correlation coefficient is 0.999	

where T = temperature (°F)

$$\mathbf{D\left(kg/m^3\right) = 0.99997 \times 10^3 + 0.21908 \times 10^{-1}T - 0.59813 \times 10^{-2}T^2} + 0.16133 \times 10^{-4}T^3 \quad (10.20)$$

where T = temperature (°C)

4.3 Modulus of Elasticity

$$\text{M.E.}\left(10^{-3}\,1\text{b/in}^2\right) = 0.24333 \times 10^3 + 0.16532 \times 10\,T - 0.8989 \times 10^{-2}T^2 + 0.11640 \times 10^{-4}T^3 \quad (10.21)$$

Table 10.4 Temperature T (°C) versus density D (kg/m^3)

The degree of fit for this run is 3			
	Coefficients		
	x (0) =	0.9999 + 03	
	x (1) =	0.21908E − 01	
	x (2) =	−0.59813E − 02	
	x (3) =	0.16133E − 04	
Variable	Actual value	Calculated value	Difference
0.000	999.800000	999.969220	−0.169212
5.000	1000.000000	999.931230	0.068771
10.000	999.700000	999.606290	0.093704
15.000	999.100000	999.006480	0.093521
20.000	998.200000	998.143900	0.056099
25.000	997.000000	997.030650	−0.030655
30.000	995.700000	995.678830	0.021164
40.000	992.200000	992.307910	−0.107910
50.000	988.000000	988.127910	−0.127907
60.000	983.200000	983.235620	−0.035622
70.000	977.800000	977.727870	0.072136
80.000	971.800000	971.701450	0.098557
90.000	965.300000	965.253140	0.046860
100.000	958.400000	958.479760	−0.079758

where T = temperature (°F)

$$\text{M.E.}\left(10^{-6}\text{kN/m}^2\right) = 0.19825 \times 10 + 0.13210 \times 10^{-1}\text{T} - 0.1594 \times 10^{-3}\text{T}^2 + 0.35871 \times 10^{-6}\text{T}^3 \tag{10.22}$$

where T = temperature (°C)

4.4 Dynamic Viscosity

$$\textbf{D.V.}\left(\mathbf{10^5 lb{-}sec / ft^2}\right) = 0.72021 \times 10 - 0.15908 \times \text{T} + 0.20304 \times 10^{-2}\text{T}^2 - 0.15985 \times 10^{-4}\text{T}^3 + 0.75655 \times 10^{-7}\text{T}^4 - 0.19625 \times 10^{-9}\text{T}^5 + 0.21364 \times 10^{-12}\text{T}^6 \tag{10.23}$$

Table 10.5 Temperature T (°F) versus modulus of elasticity M.E. (10^{-3} lb/in^2)

The degree of fit for this run is 3			
	Coefficients		
	x (0) =	0.24333E + 03	
	x (1) =	0.16532E + 01	
	x (2) =	−0.89890E − 02	
	x (3) =	0.11640E − 04	
Variable	Actual value	Calculated value	Difference
32.000	287.000	287.412	−0.412
40.000	296.000	295.823	0.177
50.000	305.000	304.975	0.025
60.000	313.000	312.679	0.321
70.000	319.000	319.004	−0.004
80.000	324.000	324.020	−0.020
90.000	328.000	327.796	0.204
100.000	331.000	330.404	0.596
110.000	332.000	331.912	0.088
120.000	332.000	332.391	−0.391
130.000	331.000	331.910	−0.910
140.000	330.000	330.539	−0.539
150.000	328.000	328.348	−0.348
160.000	326.000	325.407	0.593
170.000	322.000	321.785	0.215
180.000	318.000	317.554	0.446
190.000	313.000	312.781	0.219
200.000	308.000	307.538	0.462
212.000	300.000	300.722	−0.722

where T = temperature (°F)

$$\textbf{D.V.}\ (\mathbf{10^3 N{-}sec/m^2}) = 0.17798 \times 10 - 0.57931 \times 10^{-1}T + 0.12343 \times 10^{-2}T^2 - 0.16454 \times 10^{-4}T^3 + 0.11946 \times 10^{-6}T^4 - 0.35408 \times 10^{-9}T^5 \quad (10.24)$$

where T = temperature (°C)

4.5 *Kinematic Viscosity*

$$\textbf{K.V.}\ (\mathbf{10^5 ft^2/sec}) = 0.37214 \times 10 - 0.82655 \times 10^{-1}T + 0.10631 \times 10^{-2}T^2 - 0.84268 \times 10^{-5}T^3 + 0.40144 \times 10^{-7}T^4 - 0.10478 \times 10^{-9}T^5 + 0.11471 \times 10^{-12}T^6 \quad (10.25)$$

Table 10.6 Temperature T (°C) versus modulus of elasticity M.E. (10^{-6} kN/m^2)

The degree of fit for this run is 3			
	Coefficients		
	x (0) =	0.19825E+01	
	x (1) =	0.13210E – 01	
	x (2) =	–0.15940E – 03	
	x (3) =	0.35871E – 06	
Variable	Actual value	Calculated value	Difference
0.000	1.980000	1.982473	–0.002473
5.000	2.050000	2.044583	0.005417
10.000	2.100000	2.098992	0.001008
15.000	2.150000	2.145968	0.004032
20.000	2.170000	2.185781	–0.015781
25.000	2.220000	2.218701	0.001299
30.000	2.250000	2.244995	0.005005
40.000	2.280000	2.278784	0.001216
50.000	2.290000	2.289302	0.000698
60.000	2.280000	2.278700	0.001300
70.000	2.250000	2.249131	0.000869
80.000	2.200000	2.202747	–0.002747
90.000	2.140000	2.141700	–0.001700
100.000	2.070000	2.068143	0.001857
Standard error is 2.17357710		Correlation coefficient is .99857	

where T = temperature (°F)

$$\mathbf{K.V.}\ (\mathbf{10^6 m^2/sec}) = 0.17835 \times 10 - 0.58676 \times 10^{-1}T + 0.12796 \times 10^{-2}T^2 - 0.17407 \times 10^{-4}T^3 + 0.12836 \times 10^{-6}T^4 - 0.38482 \times 10^{-9}T^5 \qquad (10.26)$$

where T = temperature (°C)

4.6 Vapor Pressure

$$\mathbf{V.P.}\ (\mathbf{lb/in^2}) = 0.26344 - 0.13404 \times 10^{-1}\mathbf{T} + 0.32548 \times 10^{-3}\mathbf{T^2} - 0.25608 \times 10^{-5}T^3 + 0.13386 \times 10^{-7}T^4 \qquad (10.27)$$

where T = temperature (°F)

Table 10.7 Temperature T (°F) versus dynamic viscosity D.V. (10^5 lb-sec/ft^2)

The degree of fit for this run is 6			
	Coefficients		
	x (0) =	0.72021E + 01	
	x (1) =	−0.15908E + 00	
	x (2) =	0.20304E − 02	
	x (3) =	−0.15985E − 04	
	x (4) =	0.75655E − 07	
	x (5) =	−0.19625E − 09	
	x (6) =	0.21364E − 12	
Variable	Actual value	Calculated value	Difference
32.000	3.746	3.740	0.006
40.000	3.229	3.239	−0.010
50.000	2.735	2.741	−0.006
60.000	2.359	2.351	0.008
70.000	2.050	2.044	0.006
80.000	1.799	1.797	0.002
90.000	1.595	1.596	−0.001
100.000	1.424	1.429	−0.005
110.000	1.284	1.288	−0.004
120.000	1.168	1.169	−0.001
130.000	1.069	1.067	0.002
140.000	0.981	0.979	0.002
150.000	0.905	0.903	0.002
160.000	0.838	0.837	0.001
170.000	0.780	0.779	0.001
180.000	0.726	0.727	−0.001
190.000	0.678	0.680	−0.002
200.000	0.637	0.637	−0.000
212.000	0.593	0.592	0.001
Standard error is 1.50505530		Correlation coefficient is 1.00000	

$$\mathbf{V.P.}\ (\mathbf{kN/m^2}) = 0.60991 + 0.44337 \times 10^{-1}\mathbf{T} + 0.14364 \times 10^{-2}\mathbf{T^2} + 0.2680 \times 10^{-4}T^3 + 0.27022 \times 10^{-6}T^4 + 0.28101 \times 10^{-8}T^5 \tag{10.28}$$

where T = temperature (°C)

Table 10.8 Temperature T (°C) versus dynamic viscosity D.V. (10^3 N-sec/m^2)

The degree of fit for this run is 5			
	Coefficients		
	x (0) =	0.17798E+01	
	x (1) =	−0.57931E − 01	
	x (2) =	0.12343E − 02	
	x (3) =	−0.16454E − 04	
	x (4) =	0.11946E − 06	
	x (5) =	−0.35408E − 09	
Variable	Actual value	Calculated value	Difference
0.000	1.781000	1.779821	0.001179
5.000	1.518000	1.519043	−0.001043
10.000	1.307000	1.308652	−0.001652
15.000	1.139000	1.138831	0.000169
20.000	1.002000	1.001286	0.000714
25.000	0.890000	0.889121	0.000879
30.000	0.798000	0.796695	0.001305
40.000	0.653000	0.654024	−0.001024
50.000	0.547000	0.548341	−0.001341
60.000	0.466000	0.466393	−0.000393
70.000	0.404000	0.402352	0.001648
80.000	0.354000	0.353568	0.000432
90.000	0.315000	0.316319	−0.001319
100.000	0.282000	0.281564	0.000436
Standard error is 0.81828707		Correlation coefficient is .99999	

4.7 Surface Tension

$$\mathbf{S.T.}\ (\mathbf{N/m}) = 0.76353 \times 10^{-1} - 0.26673 \times 10^{-3}\mathbf{T} + 0.68479 \times 10^{-5}\mathbf{T^2} - 0.16667 \times 10^{-6}\mathrm{T}^3 + 0.1697 \times 10^{-8}\mathrm{T}^4 - 0.62319 \times 10^{-11}\mathrm{T}^5 \tag{10.29}$$

where T = temperature (°C)

$$\mathbf{S.T.(1b/ft)} = 0.134 \times 10^{-2} + 0.12 \times 10^{-3}\mathbf{T} \tag{10.30}$$

where T = water temperature (°F) at the range of 32 °F < T < 40 °F

$$\mathbf{S.T.(1b/ft)} = 0.1034 \times 10^{-1} - 0.105 \times 10^{-3}\mathbf{T} \tag{10.31}$$

Table 10.9 Temperature T (°F) versus kinematic viscosity K.V. (10^5 ft^2/sec)

The degree of fit for this run is 6			
	Coefficients		
	x (0) =	0.37214E + 01	
	x (1) =	−0.82655E − 01	
	x (2) =	0.10631E − 02	
	x (3) =	−0.84268E − 05	
	x (4) =	0.40144E − 07	
	x (5) =	−0.10478E − 09	
	x (6) =	0.11471E − 12	
Variable	Actual value	Calculated value	Difference
32.000	1.931	1.928	0.003
40.000	1.664	1.669	−0.005
50.000	1.410	1.413	−0.003
60.000	1.217	1.213	0.004
70.000	1.059	1.056	0.003
80.000	0.930	0.929	0.001
90.000	0.826	0.826	−0.000
100.000	0.739	0.741	−0.002
110.000	0.667	0.670	−0.003
120.000	0.609	0.609	−0.000
130.000	0.558	0.557	0.001
140.000	0.514	0.513	0.001
150.000	0.476	0.475	0.001
160.000	0.442	0.442	0.000
170.000	0.413	0.412	0.001
180.000	0.385	0.386	−0.001
190.000	0.362	0.363	−0.001
200.000	0.341	0.341	−0.000
212.000	0.319	0.319	0.000
Standard error is 0.78221232		Correlation coefficient is .99999	

where T = water temperature (°F) at the range of 40 °F < T < 50 °F

$$\mathbf{S.T.\ (1b/ft)} = 0.53708 \times 10^{-2} - 0.52817 \times 10^{-5}\mathbf{T} - 0.46523 \times 10^{-8}\mathbf{T^2} \tag{10.32}$$

where T = water temperature (°F) at the range of 50 °F < T < 212 °F

Table 10.10 Temperature T (°C) versus kinematic viscosity K.V. (10^6 m^2/sec)

The degree of fit for this run is 5			
	Coefficients		
	x (0) =	0.17835E + 01	
	x (1) =	−0.58676E − 01	
	x (2) =	0.12796E − 02	
	x (3) =	−0.17407E − 04	
	x (4) =	0.12836E − 06	
	x (5) =	−0.38482E − 09	
Variable	Actual value	Calculated value	Difference
0.000	1.785000	1.783548	0.001452
5.000	1.519000	1.520060	−0.001060
10.000	1.306000	1.308585	−0.002585
15.000	1.139000	1.138776	0.000224
20.000	1.003000	1.001922	0.001078
25.000	0.893000	0.890806	0.002194
30.000	0.800000	0.799557	0.000443
40.000	0.658000	0.659059	−0.001059
50.000	0.553000	0.554950	−0.001950
60.000	0.474000	0.474090	−0.000090
70.000	0.413000	0.411054	0.001946
80.000	0.364000	0.363518	0.000482
90.000	0.326000	0.327641	−0.001641
100.000	0.294000	0.293444	0.000556
Standard error is 0.82335910		Correlation coefficient is .99999	

5 Discussion and Examples

Fluid mechanics is a study of fluids including liquids and gases. It involves various properties of the fluid, such as density, specific weight, modulus of elasticity, dynamic viscosity, kinematic viscosity, surface tension, vapor pressure, velocity, pressure, volume, temperature, etc. as functions of space and time. To environmental water resources engineers, the most important liquid and gas are water and air, respectively.

Density is defined as the mass of a substance, such as water, per unit volume. Specific weight is defined as weight per unit volume. Weight is a force (N/m^3 or lb/ft^3).

Modulus of elasticity is a material property characterizing the compressibility of a fluid, such as water. The modulus of elasticity can be mathematically modeled as M.E. = −dP/(dV/V) = dD/(dD/D) where M.E. = modulus of elasticity (psi, lb/in^2, Pa, kPa, or N/m^2); dP = differential change in pressure on the object (lb/in^2, or N/m^2); dV = differential change in volume of the object (ft^3, or m^3); V = initial

Table 10.11 Temperature T (°F) versus vapor pressure V.P. (lb/in^2)

The degree of fit for this run is 4			
	Coefficients		
	x(0) =	0.26349E+00	
	x(1) =	−0.13404E–01	
	x(2) =	0.32548E–03	
	x(3) =	−0.25608E–05	
	x(4) =	0.13386E–07	
Variable	Actual value	Calculated value	Difference
32.000	0.09000	0.09793	−0.00793
40.000	0.12000	0.11842	0.20158
50.000	0.18000	0.17053	0.00950
60.000	0.26000	0.25127	0.00873
70.000	0.36003	0.36305	−0.00305
80.000	0.51000	0.51134	−0.00134
90.000	0.70000	0.70488	−0.00488
100.000	0.95000	0.95561	−0.40561
110.000	1.27000	1.27868	−0.00868
120.000	1.69000	1.69247	−0.00247
130.000	2.22000	2.21854	0.00146
140.000	2.89000	2.88170	0.00830
150.000	3.72000	3.70995	0.01005
160.000	4.74000	4.73450	0.00550
170.000	5.99000	5.98978	0.00022
180.000	7.51000	7.51344	−0.00344
190.000	9.34000	9.34633	−0.00633
200.000	11.52000	11.53252	−0.01252
212.000	14.70000	14.68909	0.01091
Standard error is 3.61895380		Correlation coefficient is 1.00000	

volume of the object (ft^3, or m^3); dD = differential change in density of the object (slug/ft^3, or kg/m^3); and D = initial density of the object (slug/ft^3, or kg/m^3).

Dynamic viscosity is the property of a fluid whereby it tends to resist relative motion within itself. It is the shear stress, i.e., the tangential force on unit area, between two infinite horizontal planes at unit distance apart, one of which is fixed while the other moves with unit velocity. In other words, it is the shear stress divided by the velocity gradient, i.e., (N/m^2)/(m/sec/m) = N-sec/m^2.

Kinematic viscosity is the dynamic viscosity of a fluid divided by its density, i.e., (N-sec/m^2)/(kg/m^3) = m^2/sec.

Surface tension is a tensile force that attracts molecules to each other on a liquid's surface. Thus, a barrier is created between the air and the liquid.

Water vapor pressure is the pressure at which water vapor is in thermodynamic equilibrium with its condensed state. At higher vapor pressures water would condense. The water vapor pressure the partial pressure of water vapor in any gas

Table 10.12 Temperature T (°C) versus vapor pressure V.P. (kN/m^2)

The degree of fit for this run is 5			
	Coefficients		
	x (0) =	0.60991E + 00	
	x (1) =	0.44337E − 01	
	x (2) =	0.14364E − 02	
	x (3) =	0.26800E − 04	
	x (4) =	0.27022E − 06	
	x (5) =	0.28101E − 08	
Variable	Actual value	Calculated value	Difference
0.000	0.610000	0.609909	0.000091
5.000	0.870000	0.871036	−0.001036
10.000	1.230000	1.226712	0.003288
15.000	1.700000	1.704436	−0.004436
20.000	2.340000	2.337867	0.002133
25.000	3.170000	3.167875	0.002125
30.000	4.240000	4.243604	−0.003604
40.000	7.380000	7.376451	0.003549
50.000	12.330000	12.334949	−0.004949
60.000	19.920000	19.917375	0.002625
70.000	31.160000	31.155464	0.004536
80.000	47.340000	47.348126	−0.008126
90.000	70.100000	70.095174	0.004827
100.000	101.330000	101.331040	−0.001036
Standard error is 21.69428600		Correlation coefficient is 1.00000	

mixture in equilibrium with solid or liquid water. As for other substances, water vapor pressure is a function of temperature.

The readers are referred to the following literature sources for more technical information [1–17]. The examples in this section show how the basic water properties (density, specific weight, modulus of elasticity, dynamic viscosity, kinematic viscosity, surface tension and vapor pressure) can be calculated using various water property models. The examples for determination of dissolved oxygen concentration in water [10, 11], the reaeration coefficient of water [17] and others [16] can be found from the literature.

The following conversion factors are useful for water resources engineers.

1 g/cm^3 = 1 g/mL = 1000 kg/m^3 = 1 kg/L = 1 kg/dm^3 = 1 $tonne/m^3$ = 8.34 lb/gallon = 62.4 lb/ft^3

Table 10.13 Temperature T (°C) versus surface tension S.T. (N/m)

The degree of fit for this run is 5			
	Coefficients		
	x (0) =	0.76353E – 01	
	x (1) =	−0.26673E – 03	
	x (2) =	0.68479E – 05	
	x (3) =	−0.16667E – 06	
	x (4) =	0.16970E – 08	
	x (5) =	−0.62319E – 11	
Variable	Actual value	Calculated value	Difference
0.000	0.076500	0.076353	0.000147
5.000	0.074900	0.075171	−0.000271
10.000	0.074200	0.074220	−0.000020
15.000	0.073500	0.073412	0.000088
20.000	0.072800	0.072676	0.000124
25.000	0.072000	0.071963	0.000037
30.000	0.071200	0.071238	−0.000038
40.000	0.069600	0.069680	−0.000080
50.000	0.067900	0.067962	−0.000062
60.000	0.066200	0.066150	0.000050
70.000	0.064400	0.064342	0.000058
80.000	0.062600	0.062598	0.000002
90.000	0.060800	0.060859	−0.000059
100.000	0.058900	0.058877	0.000023
Standard error is 0.06896435		Correlation coefficient is .99984	

5.1 *Calculation of Density*

Density is defined as an object's mass per unit volume. The density can be determined using Eq. (10.33):

$$D = \mathbf{M/V=1/V_g} \tag{10.33}$$

where

D = density (kg/m^3, slugs/ft^3)
M = mass (kg, slugs)
V = volume (m^3, ft^3)
V_g = specific volume (m^3/kg, ft^3/slug)

The SI unit for density is **kg/m^3**. The US customary unit is slug/**ft^3**, in which slug is the correct measure of mass. One can multiply slugs by **32.2** for a rough value in pounds. Density is a physical property constant at a given temperature and density.

Table 10.14 Temperature T (°F) versus surface tension S.T. (lb/ft)

The degree of fit for this run is 5			
	Coefficients		
	x (0) =	0.47401E − 02	
	x (1) =	0.64525E − 04	
	x (2) =	−0.17569E − 05	
	x (3) =	0.17739E − 07	
	x (4) =	−0.79242E − 10	
	x (5) =	0.12998E − 12	
Variable	Actual value	Calculated value	Difference
32.000	0.005180	0.005508	−0.000328
40.000	0.006140	0.005456	0.000664
50.000	0.005090	0.005337	−0.000247
60.000	0.005040	0.005192	−0.000152
70.000	0.004980	0.005048	−0.000068
80.000	0.004920	0.004920	−0.000000
90.000	0.004860	0.004817	0.000043
100.000	0.004600	0.004738	0.000062
110.000	0.004730	0.004682	0.000048
120.000	0.004670	0.004640	0.000030
130.000	0.004600	0.004603	−0.000003
140.000	0.004540	0.004564	−0.000024
150.000	0.004470	0.004512	−0.000042
160.000	0.004410	0.004444	−0.000034
170.000	0.004340	0.004359	−0.000019
180.000	0.004270	0.004261	0.000009
190.000	0.004200	0.004164	0.000056
200.000	0.004130	0.004089	0.000041
212.000	0.004040	0.004075	−0.000035
Standard error is 0.00470960		Correlation coefficient is .91584	

5.1.1 Example 1. Determination of the Density of an Unknown Liquid

An unknown liquid substance at a unknown temperature has a mass of 999.7 kg (2203.94 lb) and occupies a volume of 1000 L (35.316 ft^3). What is the density of this unknown liquid?

Solution 1 (SI System):

M = 999.7 kg
V = 1000 L = 1 m^3
D = M/V = 999.7 kg/m^3

Table 10.15 Temperature T (°F) versus surface tension S.T. (lb/ft) when 32 °F < T < 40 °F

The degree of fit for this run is 1			
	Coefficients		
	x (0) =	0.13400E – 02	
	x (1) =	0.12000E – 03	
Variable	Actual value	Calculated value	Difference
32.00	0.00518	0.00518	−0.00000
40.00	0.00614	0.00614	0.00000

Solution 2 (US Customary System):

M = 2203.94 lb = 68.4985 slug
V = 35.316 ft^3
D = M/V = 68.4985 slug/35.316 ft^3 = 1.94 slug/ft^3
(Note: 1 lb force = 0.03108 slug mass; 1 slug mass = 32.174 lb force)

5.1.2 Example 2. Calculation of the Density of Water at 50 °F

Solution

Using Eq. (10.19), the density of water at 50 °F can be determined.

$$\begin{aligned} &D\ (\text{slug/ft}^3)\\ &= 0.19354 \times 10 + 0.27304 \times 10^{-3}T - 0.40986 \times 10^{-5}T^2 + 0.53217 \times 10^{-8}T^3\\ &= 0.19354 \times 10 + 0.27304 \times 10^{-3}(50) - 0.40986 \times 10^{-5}(50)^2 + 0.53217 \times 10^{-8}(50)^3\\ &= 1.939\ \text{slug/ft}^3 \end{aligned}$$

See Tables 10.3 and 10.19 for verification of the results. The advantage of using an equation (such as Eq. 10.19) is that the density of water at any temperature (such as 52.68 °F, etc.) can be quickly determined.

5.1.3 Example 3. Calculation of the Density of Water at 10 °C

Solution

Using Eq. (10.20), the density of water at 10 °C can be determined.

Table 10.16 Temperature T (°F) versus surface tension S.T. (lb/ft) when 40 °F < T < 50 °F

The degree of fit for this run is 1			
	Coefficients		
	x (0) =	0.10340E – 01	
	x (1) =	–0.10500E – 03	
Variable	Actual value	Calculated value	Difference
40.00	0.00614	0.00614	0.00000
50.00	0.00509	0.00509	–0.00000

Table 10.17 Temperature T (°F) versus surface tension S.T. (lb/ft) when 50 °F < T < 212°F

The degree of fit for this run is 2			
	Coefficients		
	x (0) =	0.53708E – 02	
	x (1) =	–0.52817E – 05	
	x (2) =	–0.46523E – 08	
Variable	Actual value	Calculated value	Difference
50.00	0.00509	0.00510	–0.00001
60.00	0.00504	0.00504	0.00000
70.00	0.00498	0.00498	0.00000
80.00	0.00492	0.00492	0.00000
90.00	0.00486	0.00486	0.00000
100.00	0.00480	0.00480	0.00000
110.00	0.00473	0.00473	–0.00000
120.00	0.00467	0.00467	0.00000
130.00	0.00460	0.00461	–0.00001
140.00	0.00454	0.00454	–0.00000
150.00	0.00447	0.00447	–0.00000
160.00	0.00441	0.00441	0.00000
170.00	0.00434	0.00434	0.00000
180.00	0.00427	0.00427	0.00000
190.00	0.00420	0.00420	0.00000
200.00	0.00413	0.00413	0.00000
212.00	0.00404	0.00404	–0.00000
Standard error is 0.00459353		Correlation coefficient is .99997	

$$
\begin{aligned}
&D\ (\mathrm{kg/m^3}) \\
&= 0.99997 \times 10^3 + 0.21908 \times 10^{-1}T - 0.59813 \times 10^{-2}T^2 + 0.16133 \times 10^{-4}T^3 \\
&= 0.99997 \times 10^3 + 0.21908 \times 10^{-1}(10) - 0.59813 \times 10^{-2}(10)^2 + 0.16133 \times 10^{-4}(10)^3 \\
&= 999.6\ \mathrm{kg/m^3}
\end{aligned}
$$

See Tables 10.4 and 10.18 for verification of the calculated results using an equation.

5.2 Calculation of Specific Weight

Specific weight is defined as weight per unit volume, in which weight is a force. Specific weight can be calculated using Eq. (10.34).

$$\text{S.W.} = \text{D g} \tag{10.34}$$

where

S.W. = specific weight (N/m^3, lb/ft^3)
D = density (kg/m^3, $slugs/ft^3$)
g = acceleration of gravity (9.81 m/s^2, 32.174 ft/s^2)

The SI-unit of specific weight is N/m^3. The US customary unit is lb/ft^3. The local acceleration g is under normal conditions 9.807 m/s^2 in SI-units and 32.174 ft/s^2 in US customary units. 1 N = 1 Newton = the force that, when applied to a body of 1 kg mass, gives it an acceleration of 1 m/s^2. 1 N = 1 kg-m/s^2. 1 kN = 1000 N.

5.2.1 Example 4. Determination of the Pure Water's Specific Weight at 4 °C (39.2 °F)

At 4 °C (39.2 °F), pure water has its highest density 1000 kg/m^3 (1.940 $slug/ft^3$). What is this pure water's specific weight at this temperature?

Solution 1 (SI System)

Using Eq. (10.34), the specific weight of pure water at 4 °C can be determined

$$\begin{aligned} \text{S.W.} &= \text{D g} \\ &= (1000\ \text{kg/m}^3)(9.81\ \text{m/s}^2) \\ &= (9810\ \text{kg-m/s}^2)/\text{m}^3 \\ &= 9810\ \text{N/m}^3 \\ &= 9.81\ \text{kN/m}^3 \end{aligned}$$

Solution 2 (US Customary System)

Using Eq. (10.34), the specific weight of pure water at 39.2 °F can be determined

$$\begin{aligned}\text{S.W.} &= \text{D g} \\ &= (1.940\ \text{slug/ft}^3)(32.174\ \text{ft/s}^2) \\ &= 62.4\ (\text{slug-ft/s}^2)/\text{ft}^3 \\ &= 62.4\ \text{lb/ft}^3\end{aligned}$$

where

1 lb force = the force that, when applied to a body of 1 slug mass, gives it an acceleration of 1 ft/s^2.

5.2.2 Example 5. Calculation of the Specific Weight of Water at 50 °F

Solution

Using Eq. (10.17), the specific weight of water at 50 °F can be determined.

$$\begin{aligned}&\text{S.W.}(\text{lb/ft}^3) \\ &= 0.62294 \times 10^2 + 0.82009 \times 10^{-2}\text{T} - 0.12783 \times 10^{-3}\text{T}^2 + 0.16262 \times 10^{-6}\text{T}^3 \\ &= 0.62294 \times 10^2 + 0.82009 \times 10^{-2}(50) - 0.12783 \times 10^{-3}(50)^2 + 0.16262 \times 10^{-6}(50)^3 \\ &= 62.405\ \text{lb/ft}^3\end{aligned}$$

See Tables 10.1 and 10.19 for verification of the calculated results.

5.2.3 Example 6. Calculation of the Specific Weight of Water at 10 °C

Solution

Using Eq. (10.18), the specific weight of water at 10 °C can be determined.

$$\begin{aligned}&\text{S.W.}\ (\text{kN/m}^3) \\ &= 0.98068 \times 10 + 0.18341 \times 10^{-3}\text{T} - 0.58132 \times 10^{-4}\text{T}^2 + 0.15575 \times 10^{-6}\text{T}^3 \\ &= 0.98068 \times 10 + 0.18341 \times 10^{-3}(10) - 0.58132 \times 10^{-4}(10)^2 + 0.15575 \times 10^{-6}(10)^3 \\ &= 9.802941\ \text{kN/m}^3\end{aligned}$$

See Tables 10.2 and 10.18 for verification of the calculated results.

5.3 Calculation of Modulus of Elasticity

The modulus of elasticity is a material property characterizing the compressibility of a fluid, such as water. The modulus of elasticity can be mathematically modeled as

$$\text{M.E.} = -\text{dP}/(\text{dV}/\text{V}) = \text{dP}/(\text{dD}/\text{D}) \quad (10.35)$$

where M.E. = modulus of elasticity (psi, lb/in^2, Pa, kPa, or N/m^2); dP = differential change in pressure on the object (psi, lb/in^2, Pa, kPa, or N/m^2); dV = differential change in volume of the object (ft^3 or m^3); V = initial volume of the object (ft^3 or m^3); dD = differential change in density of the object (slug/ft^3, or kg/m^3); and D = initial density of the object (slug/ft^3, or kg/m^3).

An increase in the pressure will decrease the volume. A decrease in the volume will increase the density. The SI unit of the modulus of elasticity is N/m^2 (Pa), while the US customary unit is lb/in^2 (psi). Here 1 lb/in^2 (psi) = 6.894×10^3 N/m^2 (Pa)

5.3.1 Example 7. Calculation of the Modulus of Elasticity of Water at 50 °F

Solution

Using Eq. (10.21), the modulus of elasticity of water at 50 °F can be determined

$$\begin{aligned}
&\text{ME.}\ (10^{-3}\,\text{lb/in}^2) \\
&= 0.24333 \times 10^3 + 0.16532 \times 10\text{T} - 0.8989 \times 10^{-2}\text{T}^2 + 0.11640 \times 10^{-4}\text{T}^3 \\
&= 0.24333 \times 10^3 + 0.16532 \times 10\,(50) - 0.8989 \times 10^{-2}(50)^2 + 0.11640 \times 10^{-4}(50)^3 \\
&= 304.975 \\
&\text{M.E.} = 304975\ \text{lb/in}^2
\end{aligned}$$

See Tables 10.5 and 10.19 for verification of the calculated result.

5.3.2 Example 8. Calculation of the Modulus of Elasticity of Water at 10 °C

Solution

Using Eq. (10.22), the modulus of elasticity of water at 10 °C can be determined

$$\begin{aligned}
&\text{M.E.}\ (10^{-6}\text{kN/m}^2) \\
&= 0.19825 \times 10 + 0.13210 \times 10^{-1}\text{T} - 0.1594 \times 10^{-3}\text{T}^2 + 0.35871 \times 10^{-6}\text{T}^3 \\
&= 0.19825 \times 10 + 0.13210 \times 10^{-1}(10) - 0.1594 \times 10^{-3}(10)^2 + 0.35871 \times 10^{-6}(10)^3 \\
&= \mathbf{2.098992} \\
&\textbf{M.E.} = \mathbf{2098992}\ \text{kN/m}^2 = \mathbf{2098992}\ \text{kPa}
\end{aligned}$$

See Tables 10.6 and 10.18 for verification of the calculated result.

5.4 *Calculation of Dynamic Viscosity*

Viscosity of a fluid is that property of a fluid which determines the amount of its resistance to a shearing force. Both the dynamic viscosity (absolute viscosity) and the kinematic viscosity are used to measure this resistance to flow. Equation (10.36), known as the Newtons Law of Friction, defines the dynamic (absolute) viscosity:

$$\text{D.V.} = F_s(dy/dv) \tag{10.36}$$

where

D.V. = dynamic viscosity (lb-s/ft^2; N-s/m^2)
Fs = shearing force (lb/ft^2; N/m^2)
dy = length change (ft; m)
dv = velocity change (ft/s; m/s)

In the SI system, the dynamic viscosity units are N-s/m^2, Pa-s or kg/m-s where 1 N-s/m^2 = 1 Pa-s (Pascal-second) = 1 kg/m-s = 10 g/cm-s = 10 dyne-s/cm^2 = 10 P (poise) = 10 × (100 cP) = 1000 cP (centipoises)

5.4.1 Example 9. Calculation of the Dynamic Viscosity of Water at 50 °F

Solution

Using Eq. (10.23), the dynamic viscosity of water at 50 °F can be determined D.V.

$$\begin{aligned}
&\text{D.V.}\left(10^5\,\text{1b-s/ft}^2\right)\\
&= 0.72021 \times 10 - 0.15908 \times T + 0.20304 \times 10^{-2}T^2 - 0.15985 \times \text{I}0^{-4}T^3\\
&\quad + 0.75655 \times 10^{-7}T^4 - 0.19625 \times 10^{-9}T^5 + 0.21364 \times 10^{-12}T^6\\
&= 0.72021 \times 10 - 0.15908 \times (50) + 0.20304 \times 10^{-2}(50)^2 - 0.15985 \times 10^{-4}(50)^3\\
&\quad + 0.75655 \times 10^{-7}(50)^4 - 0.19625 \times 10^{-9}(50)^5 + 0.21364 \times 10^{-12}(50)^6\\
&= 2.741\\
&\text{D.V.} - 2.741 \times 10^{-5}\,\text{1b-s/ft}^2
\end{aligned}$$

See Tables 10.7 and 10.19 for verification of the calculated result.

5.4.2 Example 10. Calculation of the Dynamic Viscosity of Water at 10 °C

Solution

Using Eq. (10.24), the dynamic viscosity of water at 10 °C can be determined

$$\begin{aligned}
&\text{D.V.}\left(10^{3}\text{N-s/m}^{2}\right)\\
&= 0.17798 \times 10 - 0.57931 \times 10^{-1}\text{T} + 0.12343 \times 10^{-2}\text{T}^{2} - 0.16454 \times 10^{-4}\text{T}^{3}\\
&\quad + 0.11946 \times 10^{-6}\text{T}^{4} - 0.35408 \times 10^{-9}\text{T}^{5}\\
&= 0.17798 \times 10^{1} - 0.57931 \times 10^{-1}(10) + 0.12343 \times 10^{-2}(10)^{2}\\
&\quad - 0.16454 \times 10^{-4}(10)^{3} + 0.11946 \times 10^{-6}(10)^{4} - 0.35408 \times 10^{-9}(10)^{5}\\
&= 1.308652\\
&\text{D.V.} = 1.308652 \times 10^{-3}\text{N-s/m}^{2}
\end{aligned}$$

See Tables 10.8 and 10.18 for verification of the calculated result.

5.5 Calculation of Kinematic Viscosity

Equation (10.37) defines the kinematic viscosity:

$$\text{K.V.} = \text{D.V./D} = \text{D.V./(S.W./g)} \tag{10.37}$$

where

K.V. = kinematic viscosity (ft^2/s; m^2/s)
D.V. = dynamic viscosity (lb-s/ft^2; N-s/m^2)
D = density (slug/ft^3; kg/m^3)
S.W. = specific weight (lb/ft^3; kN/m^3)
g = 32.2 ft/s^2 = 9.806 m/s^2.

In the SI system the unit of theoretical kinematic viscosity is m^2/s, or commonly used Stoke (St) where 1 St (Stokes) = 10^{-4} m^2/s = 1 cm^2/s = 100 cSt (centiStokes). 1 cSt (centistokes) = 10^{-6} m^2/s = 1 mm^2/s.

5.5.1 Example 11. Calculation of the Kinematic Viscosity of Water at 50 °F

Solution

Using Eq. (10.25), the kinematic viscosity of water at 50 °F can be determined

$$\begin{aligned}
&\text{K.V.}\left(10^5\text{ft}^2/\text{s}\right)\\
&= 0.37214 \times 10 - 0.82655 \times 10^{-1}\text{T} + 0.10631 \times 10^{-2}\text{T}^2 - 0.84268 \times 10^{-5}\text{T}^3\\
&\quad + 0.40144 \times 10^{-7}\text{T}^4 - 0.10478 \times 10^{-9}\text{T}^5 + 0.11471 \times 10^{-12}\text{T}^6\\
&= 0.37214 \times 10 - 0.82655 \times 10^{-1}(50) + 0.10631 \times 10^{-2}(50)^2 - 0.84268 \times 10^{-5}(50)^3\\
&\quad + 0.40144 \times 10^{-7}(50)^4 - 0.10478 \times 10^{-9}(50)^5 + 0.11471 \times 10^{-12}(50)^6\\
&= 1.413\\
&\text{K.V.} = 1.413 \times 10^{-5}\text{ft}^2/\text{s}
\end{aligned}$$

See Tables 10.9 and 10.19 for verification of the calculated result.

5.5.2 Example 12. Calculation of the Kinematic Viscosity of Water at 10 °C

Solution

Using Eq. (10.26), the kinematic viscosity of water at 10 °C can be determined.

$$\begin{aligned}
&\text{K.V.}\left(10^6\text{m}^2/\text{s}\right)\\
&= 0.17835 \times 10 - 0.58676 \times 10^{-1}\text{T} + 0.12797 \times 10^{-2}\text{T}^2 - 0.17407 \times 10^{-4}\text{T}^3\\
&\quad + 0.12836 \times 10^{-6}\text{T}^4 - 0.38482 \times 10^{-9}\text{T}^5\\
&= 0.17835 \times 10 - 0.58676 \times 10^{-1}(10) + 0.12797 \times 10^{-2}(10)^2\\
&\quad - 0.17407 \times 10^{-4}(10)^3 + 0.12836 \times 10^{-6}(10)^4 - 0.38482 \times 10^{-9}(10)^5\\
&= 1.308585\\
&\text{K.V.} = 1.308585 \times 10^{-6}\text{m}^2/\text{s}
\end{aligned}$$

See Tables 10.10 and 10.18 for verification of the calculated result.

5.6 Calculation of Vapor Pressure

The water vapor pressure is the pressure at which gaseous water vapor is in thermodynamic equilibrium with its condensed state. Gaseous water vapor would condense becoming liquid at higher pressures. The water vapor pressure is the partial pressure of water vapor in any gas mixture in equilibrium with solid or liquid water. The water vapor pressure is a function of temperature, and may be approximated by the following Eqs. (10.38)–(10.40) as well as Eqs. (10.27)–(10.28).

$$\text{V.P. (mm Hg)} = \exp\left(20.386 - 5132 \times \text{T}^{-1}\right) \tag{10.38}$$

where

V.P. = water vapor pressure, mm Hg
T = temperature, Kelvin.

$$\text{Log}_{10}(\text{V.P.}) = \text{A} - \text{B}/(\text{C} + \text{T}) \tag{10.39}$$

where

V.P. = water vapor pressure, mm Hg
T = temperature, °C
A = 8.07131; if T is in the range of 1–100 °C
B = 1730.63; if T is in the range of 1–100 °C
C = 233.426; if T is in the range of 1–100 °C
A = 8.14019; if T is in the range of 100–374 °C.
B = 1810.94; if T is in the range of 100–374 °C.
C = 244.485; if T is in the range of 100–374 °C.

Equation (10.39) is also knows as Antoine equation.

$$\text{V.P. (max)} = \exp\left(77.3450 + 0.0057\text{T} - 7235\text{T}^{-1}\right)/\text{T}^{8.2} \tag{10.40}$$

where

V.P.(max) = maximum saturation pressure of water vapor in moist air, Pa
T = temperature, K
e = constant 2.718

5.6.1 Example 13. Calculation of Maximum Water Vapor Pressure in Moist Air at Dry Bulb Temperature 25 °C

Solution

$$
\begin{aligned}
&\text{V.P. (max)} \\
&= \exp\left(77.3450 + 0.0057T - 7235\,T^{-1}\right)/T^{8.2} \\
&= \exp[77.3450 + 0.0057\,(273 + 25) - 7235/(273 + 25)]/(273 + 25)^{8.2} \\
&= e^{[77.3450 + 0.0057\,(298) - 7235\,/(298)]}/(298)^{8.2} \\
&= 3130\ \text{Pa}
\end{aligned}
$$

5.6.2 Example 14. Calculation of the Vapor Pressure of Water at 50 °C

Solution

Using Eq. (10.27), the water vapor pressure at 50 °F can be determined.

$$
\begin{aligned}
&\text{V.P. } \left(1\text{b/in}^2\right) \\
&= 0.26344 - 0.13404 \times 10^{-1}T + 0.32548 \times 10^{-3}T^2 \\
&\quad - 0.25608 \times 10^{-5}T^3 + 0.13386 \times 10^{-7}T^4 \\
&= 0.26344 - 0.13404 \times 10^{-1}(50) + 0.32548 \times 10^{-3}(50)^2 \\
&\quad - 0.25608 \times 10^{-5}(50)^3 + 0.13386 \times 10^{-7}(50)^4 \\
&= 0.17050 \\
&\text{V.P.} = 0.1705\ 1\text{b/in}^2
\end{aligned}
$$

See Tables 10.11 and 10.19 for verification of the calculated result.

5.6.3 Example 15. Calculation of the Vapor Pressure of Water at 10 °C

Solution

Using Eq. (10.28), the vapor pressure of water at 10 °F can be determined.

$$\begin{aligned}
&\text{V.P. } (\text{kN/m}^2)\\
&= 0.60991 + 0.44337 \times 10^{-1}T + 0.14364 \times 10^{-2}T^2 + 0.2680 \times 10^{-4}T^3\\
&\quad + 0.27022 \times 10^{-6}T^4 + 0.28101 \times 10^{-8}T^5\\
&= 0.60991 + 0.44337 \times 10^{-1}(10) + 0.14364 \times 10^{-2}(10)^2 + 0.2680 \times 10^{-4}(10)^3\\
&\quad + 0.27022 \times 10^{-6}(10)^4 + 0.28101 \times 10^{-8}(10)^5\\
&= 1.226712\\
&\text{V.P.} = 1.226712 \text{ kN/m}^2
\end{aligned}$$

See Tables 10.12 and 10.18 for verification of the calculated result.

5.7 *Calculation of Surface Tension*

Surface tension of a liquid is caused by the inward attraction of molecules at a boundary, such as liquid–solid. It is the energy required to stretch a unit change of a surface area. Since the sphere offers the smallest area for a definite volume, the surface tension will form a drop of liquid to a sphere. Mathematically surface tension is defined by Eq. (10.41).

$$\text{S.T.} = F_{sf}/L \qquad (10.41)$$

where

S.T. = surface tension, N/m or lb/ft
F_{sf} = stretching force, N or lb;
L = unit length, m or ft

Commonly surface tension is reported in dyne/cm, which is the force in dynes required to break a film of length 1 cm. The other common surface tension units are: 1 dyne/cm = 0.001 N/m = 0.0000685 lb/ft = 0.571×10^{-5} lb/in = 1 mN/m = 0.001 J/m^2 = 1 erg/cm^2 = 0.00010197 kg/m

5.7.1 Example 16. Calculation of the Surface Tension of Water at 10 °C

Solution

Using Eq. (10.29), the surface tension of water at 10 °C can be determined.

S.T. (N/m)

$= 0.76353 \times 10^{-1} - 0.26673 \times 10^{-3}T + 0.68479 \times 10^{-5}T^2 - 0.16667 \times 10^{-6}T^3$

$+ 0.1697 \times 10^{-8}T^4 - 0.62319 \times 10^{-11}T^5$

$= 0.76353 \times 10^{-1} - 0.26673 \times 10^{-3}(10) + 0.68479 \times 10^{-5}(10)^2 - 0.16667 \times 10^{-6}(10)^3$

$+ 0.1697 \times 10^{-8}(10)^4 - 0.62319 \times 10^{-11}(10)^5$

$= 0.074220$

V.P. $= 0.074220$ N/m

See Tables 10.13 and 10.18 for verification of the calculated result.

5.7.2 Example 17. Calculation of the Surface Tension of Water at 50 °F

Solution

Using Eq. (10.31), the vapor pressure of water at 50 °F can be determined.

S.T. (1b/ft)

$= 0.1034 \times 10^{-1} - 0.105 \times 10^{-3}T$

$= 0.1034 \times 10^{-1} - 0.105 \times 10^{-3}(50)$

$= 0.00509$

S.T. $= 0.00509$ 1b/ft

See Tables 10.16 and 10.19 for verification of the calculated result. Additional mathematical modeling data related to water temperature and surface tension can be found from Tables 10.14, 10.15 and 10.17.

6 Glossary of Water Properties [5–9, 18]

Bulk modulus and fluid elasticity See modulus of elasticity.

Density The mass of a substance, such as water, per unit volume.

Dynamic viscosity It is the property of a fluid whereby it tends to resist relative motion within itself. It is the shear stress, i.e., the tangential force on unit area, between two infinite horizontal planes at unit distance apart, one of which is fixed while the other moves with unit velocity. In other words, it is the shear stress divided by the velocity gradient, i.e., $(N/m^2)/(m/sec/m) = N\text{-}sec/m^2$.

Fluid mechanics A study of fluids including liquids and gases. It involves various properties of the fluid, such as density, specific weight, modulus of elasticity, dynamic viscosity, kinematic viscosity, surface tension, vapor pressure, velocity, pressure, volume, temperature, etc. as functions of space and time.

Kinematic viscosity It is the dynamic viscosity of a fluid divided by its density, i.e., $(\text{N-sec/m}^2)/(\text{kg/m}^3) = \text{m}^2/\text{sec}$.

Modulus of elasticity A material property characterizing the compressibility of a fluid, such as water. The modulus of elasticity can be mathematically modeled as M.E. $= -\, dP/(dV/V) = dD/(dD/D)$ where M.E. = modulus of elasticity (psi, lb/in^2, Pa, kPa, or N/m^2); dP = differential change in pressure on the object (lb/in^2, or N/m^2); dV = differential change in volume of the object (ft^3, or m^3); V = initial volume of the object (ft^3, or m^3); dD = differential change in density of the object (slug/ft^3, or kg/m^3); D = initial density of the object (slug/ft^3, or kg/m^3).

Newton (N) It is the force that, when applied to a body of mass of 1 kg, gives it an acceleration of 1 m/s^2.

Specific weight It is defined as weight per unit volume. Weight is a force (N/m^3 or lb/ft^3.)

Surface tension The attraction of molecules to each other on a liquid's surface. Thus, a barrier is created between the air and the liquid.

Vapor pressure The water vapor pressure is the pressure at which water vapor is in thermodynamic equilibrium with its condensed state. At higher vapor pressures water would condense. The water vapor pressure the partial pressure of water vapor in any gas mixture in equilibrium with solid or liquid water. As for other substances, water vapor pressure is a function of temperature.

7 Appendices

Table 10.18 Physical properties of water (SI units)[a]

Temperature, °C	Specific weight γ, kN/m^3	Density, ρ, kg/m^3	Modulus of elasticity[b], $E/10^6$, kN/m^2	Dynamic viscosity $\mu \times 10^3$ $\text{N}\cdot\text{s/m}^2$	Kinematic viscosity $\upsilon \times 10^6$, m^2/s	Surface tension[c], σ, N/m	Vapor pressure, p_v, kN/m^2
0	9.805	999.8	1.98	1.781	1.785	0.0765	0.61
5	9.807	1000.0	2.05	1.518	1.519	0.0749	0.87
10	9.804	999.7	2.10	1.307	1.306	0.0742	1.23
15	9.798	999.1	2.15	1.139	1.139	0.0735	1.70
20	9.789	998.2	2.17	1.002	1.003	0.0728	2.34
25	9.777	997.0	2.22	0.890	0.893	0.0720	3.17
30	9.764	995.7	2.25	0.798	0.800	0.0712	4.24
40	9.730	992.2	2.28	0.653	0.658	0.0696	7.38
50	9.689	988.0	2.29	0.547	0.553	0.0679	12.33
60	9.642	983.2	2.28	0.466	0.474	0.0662	19.92

(continued)

Table 10.18 (continued)

Temperature, °C	Specific weight γ, kN/m^3	Density, ρ, kg/m^3	Modulus of elasticity[b], $E/10^6$, kN/m^2	Dynamic viscosity $\mu \times 10^3$ $N \cdot s/m^2$	Kinematic viscosity $\upsilon \times 10^6$, m^2/s	Surface tension[c], σ, N/m	Vapor pressure, p_v, kN/m^2
70	9.589	977.8	2.25	0.404	0.413	0.0644	31.16
80	9.530	971.8	2.20	0.354	0.364	0.0626	47.34
90	9.466	965.3	2.14	0.315	0.326	0.0608	70.10
100	9.399	958.4	2.07	0.282	0.294	0.0589	101.33

[a]Adapted from [1, 2]
[b]At atmospheric pressure
[c]In contact with air

Table 10.19 Physical properties of water (U.S. customary units)[a]

Temperature, °F	Specific weight, γ, lb/ft^3	Density[b], ρ, $slug/ft^3$	Modulus of elasticity[b], $E/10^3$, lb_f/in^2	Dynamic viscosity $\mu \times 10^5$ $lb \cdot s/ft^2$	Kinematic viscosity, $\upsilon \times 10^5$, ft^2/s	Surface tension[c], σ, lb/ft	Vapor pressure, p_v, lb_f/in^2
32	62.42	1.940	287	3.746	1.931	0.00518	0.09
40	62.43	1.940	296	3.229	1.664	0.00614	0.12
50	62.41	1.940	305	2.735	1.410	0.00509	0.18
60	62.37	1.938	313	2.359	1.217	0.00504	0.26
70	62.30	1.936	319	2.050	1.059	0.00498	0.36
80	62.22	1.934	324	1.799	0.930	0.00492	0.51
90	62.11	1.931	328	1.595	0.826	0.00486	0.70
100	62.00	1.927	331	1.424	0.739	0.00480	0.95
110	61.86	1.923	332	1.284	0.667	0.00473	1.27
120	61.71	1.918	332	1.168	0.609	0.00467	1.69
130	61.55	1.913	331	1.069	0.558	0.00460	2.22
140	61.38	1.908	330	0.981	0.514	0.00454	2.89
150	61.20	1.902	328	0.905	0.476	0.00447	3.72
160	61.00	1.896	326	0.838	0.442	0.00441	4.74
170	60.80	1.890	322	0.780	0.413	0.00434	5.99
180	60.58	1.883	318	0.726	0.385	0.00427	7.51
190	60.36	1.876	313	0.678	0.362	0.00420	9.34
200	60.12	1.868	308	0.637	0.341	0.00413	11.52
212	59.83	1.860	300	0.593	0.319	0.00404	14.70

[a]Adapted from [1, 2]
[b]At atmospheric pressure
[c]In contact with air

References

1. Tchobanoglous, G., Burton, F. L., & Stensel, H. D. (2003). *Wastewater engineering*. NYC, NY: McGraw-Hill Book Company.
2. Brooks, N. H. (1960). Diffusion of sewage effluent in an ocean current. In *Proceedings First International Conference on Waste Disposal in the Marine Environment*. Berkeley/NYC, NY: University of California/Pergamon.
3. Wang, M. H. S., Wang, L. K., & Chan, W. Y. W. (1979, November). *Technical manual for engineers and scientists*. Stevens Institute of Technology, Hoboken, NJ, Technical Manual No. SIT-ME-/11-79/3, p. 79.
4. Wang, L. K., Wang, M. H. S., & Vielkind, D. (1978). Mathematical models of dissolved oxygen concentration in fresh water. *Ecological Modeling, 5*, 115–123.
5. Wang, M. H. S., & Wang, L. K. (2014). Glossary and conversion factors for water resources engineers. In: L. K. Wang & C. T. Yang (Eds.), *Modern water resources engineering* (pp. 759–851). New York: Humana Press.
6. Symons, J. M., Bradley, L. C. J., & Cleveland, T. C. (Eds.). (2000). *The drinking water dictionary* (p. 506). Denver, CO: American Water Works Association.
7. Wang, L. K. (1974). *Environmental engineering glossary* (p. 1–439). Buffalo, NY: Calspan Corporation.
8. Wang, M. H. S., & Wang, L. K. (2015). Environmental and water engineering glossary. In: C. T. Yang & L. K. Wang (Eds.), *Advances in water resources engineering*. (pp. 471–556) New York: Springer.
9. AWWA, APHA, & WEF. (2012). *Standard methods for the examination of water and wastewater* (22nd ed.). Washington, DC: American Publish Health Association. ISBN-10: 0875530133.
10. Wang, L. K., & Pereira, N. C. (1986). *Water resources and natural control processes* (Handbook of environmental engineering, Vol. 4, pp. 447–492). Totowa, NJ: Humana Press.
11. Wang, L. K., Shammas, N. K., Selke, W. A., & Aulenbach, D. B. (2010). Gas dissolution, release, and bubble formation in flotation systems. In: L. K. Wang, N. K. Shammas, W. A. Selke, & D. B. Aulenbach (Eds.), *Flotation technology* (pp. 49–83). Totowa, NJ: Humana Press.
12. Vega, C., & DeMiguel, E. (2007). Surface tension of the most popular models of water by using the test-area simulation method. *The Journal of Chemical Physics, 126*(15), 154707-1.
13. Editor. (2014). Surface tension of water in contact with air. The Engineering Tool Box. editor. engineeringtoolbox@gmail.com
14. Onn, F. (2006, November). *Modeling water vapor using GPS with application to mitigating in SAR atmospheric distortions* (PhD dissertation). Stanford University, Stanford, CA.
15. Frierson, D. M. W. (2013, September 30). *Fundamentals of climate change: Water vapor*. Seattle, WA: Department of Atmospheric Sciences, University of Washington.
16. US EPA. (2014). *Water models*. Washington, DC: US Environmental Protection Agency.
17. Wen, C. G., Kao, J. F., Liaw, C. C., Wang, M. H. S., & Wang, L. K. (2016). Determination of reaeration coefficient of saline receiving water for water quality management. In: L. K. Wang, C. T. Yang, & M. H. S. Wang (Eds.), *Advances in water resources management*. NYC, NY: Springer Science + Business Media.
18. Shammas, N. K., & Wang, L. K. (2016). *Water engineering: Hydraulic, distribution and treatment*. Hoboken, NJ: Wiley. 806 p.

Index

L.K. Wang, C.T. Yang, and M.-H.S. Wang (eds.), *Advances in Water Resources Management*, Handbook of Environmental Engineering, Volume 16,
DOI 10.1007/978-3-319-22924-9

Printed by Printforce, the Netherlands